University of California Publications in

MATHEMATICS

VOLUME I
1912-1924

M. W. HASKELL
D. N. LEHMER
J. H. McDONALD
EDITORS

UNIVERSITY OF CALIFORNIA PRESS
BERKELEY, CALIFORNIA

The University of California Press
Berkeley, California

Cambridge University Press
London, England

CONTENTS

UNIVERSITY OF CALIFORNIA PUBLICATIONS
IN
MATHEMATICS

Vol. 1, No. 1, pp. 1-26 June 13, 1912

ON NUMBERS WHICH CONTAIN NO FACTORS OF THE FORM $p(kp+1)$*

BY

HENRY W. STAGER

"If p^a is the highest power of a prime p which divides the order of a group G, the subgroups of G of order p^a form a single conjugate set and their number is congruent to unity, mod p."

The above well-known theorem, which was first established by Sylow,† suggests the segregation of numbers into two distinct classes: the one class to contain all numbers which have factors of the form $p(kp+1)$ where p may be any prime except unity and k may have any integral value greater than zero; the other class to contain all remaining numbers. Numbers of the second class will be found to have certain properties analogous to the properties of primes and will be denoted by P, numbers of the first class will be found to have certain properties analogous to the properties of composite numbers and will be denoted by C. If a C should be of the form $p(kp+1)$, and not merely a multiple of a number of that form, we will call it a "*fundamental*" C. Certain C's are fundamental for certain p's and not for others; *e.g.*,

$$12 = 2\{2(1\cdot 2+1)\} = 3(1\cdot 3+1),$$

and 12 is therefore fundamental for $p=3$, but not for $p=2$; whether a given C is to be considered a fundamental C or not will be made clear by the context. Two fundamental C's will be called "*different*" if the values of the p's and k's respectively are not both the same in the two C's. The present investigation is intended to develop some of the more fundamental properties of the P's and C's and to establish certain formulae for the number of P's within a given limit.

Corollaries from the Definitions

It follows directly from the definitions above, that:

1. *Any number which is a positive integral power, including the first power, of a prime p is a* P.

---

* Dissertation in partial fulfillment of the requirements for the degree of Doctor of Philosophy in the University of California.

† Sylow, *Théorèmes sur les groupes de substitutions.* Math. Ann. (1872). For the particular statement of the theorem used above, compare Burnside, *The Theory of Groups of Finite Order*, § 78 et seq.

2. *Every even number not a positive integral power of* 2 *is a* C. Hence the only even P's are the positive integral powers of 2.

3. *Every number whose last two digits are* 05 *or* 55 *is a* C.

Such numbers are evidently of the form $5(5n+1)$.

Theorem

Every number which is divisible by any power of 3 *and contains at least two other factors not multiples of* 3 *is a* C.

Let

$$\mathrm{N} = 3^n p^\alpha q^\beta \cdots\cdots, \quad n \gtreqless 1, \quad \alpha + \beta + \cdots\cdots \gtreqless 2.$$

Every odd prime is of the form, $3m+1$, or $3m+2$.

Hence, N is a C, if any prime is 2, or of the form $3m+1$.

Also, since

$$(3m_1 + 2)(3m_2 + 2) = 3m + 1,$$

the theorem follows in any case.

It is evident that in the expression

$$\alpha + \beta + \cdots\cdots \gtreqless 2$$

all the terms, except one, may be zero, provided that term is not less than 2.

Theorem

The product of two consecutive odd primes cannot be a C.

Suppose

$$p_1 p_2 = lp(kp+1),$$

where p_1 and p_2 are consecutive odd primes and $p_2 > p_1 > 2$.

Then, since we have only two factors, p_1 and p_2, and $p_2 > p_1$, $l = 1$, and $p = p_1$.

Hence

$$p_1 p_2 = p_1(kp_1 + 1),$$

and

$$p_2 = kp_1 + 1.$$

But, since p_1 and p_2 are both odd, k must be even.

Therefore

$$p_2 = 2np_1 + 1,$$

or

$$p_2 \gtreqless 2p_1 + 1.$$

Now, by Bertrand's Postulate,* between x and $2x - 2$, $(x > 7/2)$, there is at least one prime, whence p_2 is not consecutive to p_1, as was assumed. If $p_1 = 3$, $p_2 = 5$, we have $p_1 p_2 = 15$, which is not a C, and the theorem holds for all odd primes.

* This theorem was first conjectured by Bertrand (Journal de l'École Polyt., cah. 30) and afterwards proved by Tchébicheff in 1850 in his *Mémoire sur les nombres premiers* presented to the Academy of St. Petersburg and reprinted in *Liouville*, vol. XVII (1852).

Corollary

It follows directly from the proof of the theorem that the product of two odd primes, p_1 and p_2, so related that $p_2 < 2p_1$, cannot be a C.

For, by the theorem

$$p_2 \gtreqless 2p_1 + 1,$$

or

$$p_2 > 2p_1,$$

which is contrary to hypothesis.

Further, the *necessary* and *sufficient* condition that the product of two odd primes, p_1 and p_2, $p_2 > p_1$, should be a C is that $p_2 = 2np_1 + 1$.

Theorem

No rational integral algebraic polynomial can represent only numbers which are P's.

For, suppose

$$f(x) \equiv a_0x^n + a_1x^{n-1} + \cdot \cdot \cdot \cdot \cdot + a_{n-1}x + a_n,$$

should be such a polynomial, then for some value of x, say ζ, the formula would represent a P.

Whence

$$f(\zeta) \equiv a_0\zeta^n + a_1\zeta^{n-1} + \cdot \cdot \cdot \cdot \cdot \cdot \cdot \cdot \cdot \cdot + a_{n-1}\zeta + a_n = \mathrm{P}.$$

But

$$f(\zeta + \mathrm{P}^2) = a_0(\zeta + \mathrm{P}^2)^n + a_1(\zeta + \mathrm{P}^2)^{n-1} + \cdot \cdot \cdot \cdot \cdot + a_{n-1}(\zeta + \mathrm{P}^2) + a_n,$$
$$= M\mathrm{P}^2 + \mathrm{P} = \mathrm{P}(M\mathrm{P} + 1).$$

The right member is a C whether P be a prime or a composite number.

Theorem

If x and y are both odd or both even numbers, provided they are not both equal to 1 *or* 2^κ, *then $x^n + y^n$ is a* C, *for $n > 1$.*

Since any odd number is of the form $(2k+1)$, it is only necessary to show that the sum contains at least one odd and one even factor to prove the theorem.

1° Consider

$$S = x^n + y^n, \text{ where } x \text{ and } y \text{ are odd and } x \neq y = 1.$$

(a). Let

$$n = 2\nu + 1.$$

Then

$$S = x^{2\nu+1} + y^{2\nu+1},$$
$$= (x+y)(x^{2\nu} - x^{2\nu-1}y + \cdot \cdot \cdot \cdot \cdot \cdot \cdot + y^{2\nu}).$$

Now, the first factor is the sum of two odd numbers and therefore is even, and the second factor is odd since it has an odd number of odd terms.

Therefore the sum S is a C.

(*b*). Let

$$n = 2\nu.$$

Then

$$S = x^{2\nu} + y^{2\nu},$$
$$= (2x' + 1)^{2\nu} + (2y' + 1)^{2\nu}.$$

If we expand the parentheses, there will be $(2\nu + 1)$ terms in each expansion, of which the last term is unity and all the others contain the factor 2^2. Therefore, we may write

$$S = 2^2X + 1 + 2^2Y + 1,$$
$$= 2\{2(X + Y) + 1\}.$$

Hence S is a C.

It is necessary to exclude the case where x and y are both unity, for in that case

$$S = 1^n + 1^n = 2$$

and S is a P.

2° Next consider

$S = x^n + y^n$, where x and y are both even and $x \neq y = 2^\kappa$.

(*a*). Let

$$x = 2^\alpha x' \text{ and } y = 2^\beta y',$$

where x' and y' are both odd and $\beta > \alpha$.

Then

$$S = 2^{\alpha n}(x'^n + 2^{(\beta - \alpha)n}y'^n).$$

The first factor is even and the second is odd. Therefore S is a C.

(*b*). Let

$x = 2^\alpha x'$ and $y = 2^\alpha y'$, where x' and y' are both odd.

Then

$$S = 2^{\alpha n}(x'^n + y'^n).$$

But the second factor has already been proved a C by Case 1, therefore S is a C.

The case excluded by the theorem, $x = y = 2^\kappa$, gives

$$S = 2^{\kappa n} + 2^{\kappa n} = 2^{\kappa n + 1};$$

and S is a P in this case.

In connection with this theorem, it is to be noted that if x is odd and y is even, then $x^n + y^n$ may be either a P or a C. Two illustrations will serve to prove this statement.

If $x = 11$, $y = 4$, then $x^2 + y^2 = 121 + 16 = 137 \equiv$ P.
$x = 17$, $y = 20$, then $x^2 + y^2 = 289 + 400 = 689 = 13(4 \cdot 13 + 1) \equiv$ C.

Theorem

If p, q are odd primes, and α, β are positive integers, $p^\alpha q^\beta$ is a C, *provided $\alpha \gtreqless \frac{\phi(q)}{\delta}$, where δ is the greatest common factor of $\phi(q)$ and* ind *p with respect to q.*

Let

$$N=p^{\alpha}q^{\beta},$$

then must

$$p^{\alpha}q^{\beta}=p^{\alpha'}q^{\beta-1}\{q(kq+1)\}, \text{ where } \alpha'<\alpha,$$

whence

$$p^{\alpha-\alpha'}=kq+1,$$

and

$$p^{\alpha-\alpha'}\equiv 1 \ (mod\ q).$$

But, by Fermat's theorem,

$$p^{\phi(p)}\equiv 1 \ (mod\ q),$$
$$\therefore\ p^{\alpha-\alpha'}\equiv p^{\phi(q)} \ (mod\ q),$$

whence

$$(\alpha-\alpha') \text{ ind } p\equiv 0 \ (mod\ \overline{q-1}),$$

or

$$(\alpha-\alpha')\equiv 0 \ (mod\ \frac{q-1}{\delta}), \text{ where } (q-1, \text{ ind } p)=\delta.$$

Hence, it follows that the least value of $(\alpha-\alpha')$, and consequently of α, is $\frac{q-1}{\delta}$, a quantity independent of the particular primitive root selected.

Having discussed some of the fundamental properties of the P's and C's, we will now take up the problem of deducing certain formulae of enumeration. It is very evident from the nature of the C's that, if we attempt to carry out the analogy between P's and prime numbers, C's and composite numbers, a new interpretation must be placed upon many of the ordinary operations with numbers, in particular those operations which involve the processes of multiplication. This is due to the fact that the very important theorem, that a number may be resolved into prime factors in only one way, does not apply when we consider fundamental C's as our fundamental factors; thus

$$105=5(4\cdot 5+1)=7(2\cdot 7+1).$$

While these two divisors are entirely different C's in the sense of our definition, they are each equal to 105 and their product in the ordinary sense is equal to 105^2. It seems most convenient for many purposes to define the product of two or more different C's as their least common multiple and to designate the operation by the usual symbols of multiplication; the powers of a C will have the ordinary interpretation. In case of division, in which divisor or dividend or both are products of C's, all multiplications must be performed first; thus, if $C_1=2(1\cdot 2+1)$ and $C_2=3(1\cdot 3+1)$, $C_1C_2=12$, $\frac{C_1C_2}{C_1}=\frac{12}{6}=2$, and not, $\frac{C_1C_2}{C_1}=C_2=12$. The operation of cancellation of C's before multiplying will be indicated by the symbol $\overline{\overline{}}$; thus, $\frac{C_1C_2}{\overline{\overline{C_1}}}=C_2$.

The divisors of a product of C's will consist, unless otherwise stated, of unity, the number itself, and any product of the C's therein contained.

THEOREM

The number of P's *not greater than* n *is obtained by expanding the double product*

$$[n]\left\{\begin{array}{l}\left(1-\frac{1}{C_{1,1}}\right)\left(1-\frac{1}{C_{1,2}}\right)\cdots\cdots\cdots\left(1-\frac{1}{C_{1,K_1}}\right)\\ \left(1-\frac{1}{C_{2,1}}\right)\left(1-\frac{1}{C_{2,2}}\right)\cdots\cdots\cdots\left(1-\frac{1}{C_{2,K_2}}\right)\\ \vdots\\ \left(1-\frac{1}{C_{i,1}}\right)\left(1-\frac{1}{C_{i,2}}\right)\cdots\cdots\cdots\left(1-\frac{1}{C_{i,K_i}}\right)\\ \vdots\\ \left(1-\frac{1}{C_{\lambda,1}}\right)\left(1-\frac{1}{C_{\lambda,2}}\right)\cdots\cdots\cdots\left(1-\frac{1}{C_{\lambda,K_\lambda}}\right)\end{array}\right\}$$

in which $C_{i,k}\equiv p_i(kp_i+1)$, *and then forming the sum by taking the integral part of each term with its proper sign.*

It is evident, in the first place, that for a given p and k, the number of integers not greater than n which contain the factor $p_1(jp_1+1)$, where p_1 is a given prime, is given by

$$\left[\frac{n}{p_1(jp_1+1)}\right],$$

where $[x]$ has its usual interpretation, the greatest integer in x. If we define $\phi_P(n, j, p_1)$ as the number of integers not greater than n, which do not contain the factor $p_1(jp_1+1)$, where p_1 and j have given values, then

$$\phi_P(n,\ j,\ p_1)=n-\left[\frac{n}{p_1(jp_1+1)}\right].$$

Now, let p remain fixed, but let k have a sequence of integral values, 1, 2, 3 . . . Many numbers may contain two or more factors of the form $p(kp+1)$, as for instance, $48=4[3(1\cdot 3+1)]=3(5\cdot 3+1)$, and, moreover, the largest factor cannot exceed $[n]$. It will, therefore, be necessary to find the largest value of k for a given prime p and a given number n. It is evident that this maximum value of k will be obtained when, if possible, the factor is exactly equal to n, or

$$\begin{aligned} n &= p_1(kp_1+1),\\ &= kp_1^2+p_1,\end{aligned}$$

whence

$$k=\frac{n-p_1}{p_1^2}.$$

Therefore, the maximum value of k for a given p and n will be given by

$$k=\left[\frac{n-p}{p^2}\right].$$

$\phi_P(n, p_1)$ will now be defined as the number of integers not greater than n which contain no factors of the form $p_1(kp_1+1)$, where p_1 is a given prime and $k=1$ to $\left[\frac{n-p_1}{p_1^2}\right]$, and therefore $\phi_P(n, p_1)$ will necessarily consist of the sum

$$\sum_{k=1}^{\left[\frac{n-p_1}{p_1^2}\right]} \left[\frac{n}{p_1(kp_1+1)}\right]$$

plus some function which will provide for such cases as have two or more different factors of the given form.

Let

$$n_1 = l_1 p_1(k_1 p_1 + 1) = l_2 p_1(k_2 p_1 + 1), \text{ where } n_1 > n,\ k_1 \neq k_2.$$

Then

$$l_1(k_1 p_1 + 1) = l_2(k_2 p_1 + 1).$$

And since $k_1 \neq k_2$, the two sets l_1, k_1p_1+1 and l_2, k_2p_1+1 have factors common to each other. If δ is the least common multiple of (k_1p_1+1) and (k_2p_1+1), any number of the form $mp_1\delta$ will contain factors of the two forms $p_1(k_1p_1+1)$ and $p_1(k_2p_1+1)$. A similar result holds for any number of factors of the form $p(kp+1)$, and also for the case in which both the p's and k's differ.

We have already defined the product of two or more different C's as their least common multiple. If then we use the notation

$$C_{1,k} \equiv p_1(kp_1+1),$$

the sequence

$$C_{1,1},\ C_{1,2},\ \cdots\cdots\ C_{1,K}$$

will correspond exactly with the sequence

$$p_1(p_1+1),\ p_1(2p_1+1), \ldots\ldots p_1(Kp_1+1)$$

where k and consequently the second subscript k of the C's has the values

$$1, 2, \ldots\ldots \left[\frac{n-p_1}{p_1^2}\right].$$

It follows then from analogy with the manner of forming the function $\phi(n; p, q, r)$ of ordinary primes that

$$\phi_P(n, p_1) = n - \left\{\left[\frac{n}{C_{1,1}}\right] + \left[\frac{n}{C_{1,2}}\right] \cdots\cdots\cdots + \left[\frac{n}{C_{1,K}}\right]\right\}$$
$$+ \left\{\left[\frac{n}{C_{1,1}C_{1,2}}\right] + \left[\frac{n}{\text{Product of different C's two at a time}}\right]\right\}$$
$$- \left\{\left[\frac{n}{C_{1,1}C_{1,2}C_{1,3}}\right] + \left[\frac{n}{\text{Product of different C's three at a time}}\right]\right\}$$
$$+ \text{ etc.}$$

We may write the formula in the neater form

$$\phi_P(n,\ p_1)=[n]\left(1-\frac{1}{C_{1,1}}\right)\left(1-\frac{1}{C_{1,2}}\right)\cdot\cdot\cdot\cdot\cdot\cdot\left(1-\frac{1}{C_{1,K}}\right),$$

if we interpret this form to mean that we shall first expand the product and then take the sum of the integral parts of all the terms with their proper signs. Thus,

$$\phi_P(50,\ 2)=50-27+9-2=30,$$

which is easily verified as the number of integers not greater than 50 which contain no factors of the form $2(2k+1)$.

We will next define $\phi_P(n,\lambda)$ as the number of integers not greater than n which contain no factors of the form $p_i(kp_i+1)$, where p_i is any one of the first first λ primes, $p_1=2$, $p_2=3, \ldots\ldots p_i \ldots\ldots p_\lambda$, with a limit depending on n, and k runs independently for each prime, being limited as already shown. To find the limit of p for any n we assume $k=1$, since this will give us the maximum value of p, such that $p(kp+1)\leqq[n]$.

Then

$$p\ (p+1)\leqq n,$$
$$p^2+p\leqq n,$$
$$(p+\tfrac{1}{2})^2\leqq n+\tfrac{1}{4},$$
$$p\leqq\frac{\sqrt{4n+1}\ \ -1}{2},$$

whence the maximum value of p will be that of the largest prime not greater than $\left[\frac{\sqrt{4n+1}\ -1}{2}\right]$. If, then, we consider the double sequence of C's,

$$C_{i,\,k}\equiv p_i(kp_i+1),$$

where p_i is any one of the first λ primes, $p_1=2$, $p_2=3$, ... p_i . . . . p_λ; $p_\lambda\leqq\frac{\sqrt{4n+1}\ -1}{2}$, and where k runs independently for each p from 1 to $\left[\frac{n-p}{p^2}\right]$, we can write down the desired formula at once, in that it must have the same general form as that for $\phi_P(n,\ p_1)$. Therefore,

$$\phi_P(n,\ \lambda)=n-\sum\left\{\left[\frac{n}{C_{i,K}}\right]\right\}$$
$$+\sum\left\{\left[\frac{n}{\text{Product of different C's two at a time}}\right]\right\}$$
$$-\sum\left\{\left[\frac{n}{\text{Product of different C's three at a time}}\right]\right\}$$
$$+\text{ etc.}$$

Or, as before, we may write

$$\phi_P(n,\ \lambda) = [n]\left\{\begin{matrix} \left(1-\frac{1}{C_{1,1}}\right)\left(1-\frac{1}{C_{1,2}}\right) \cdots\cdots \left(1-\frac{1}{C_{1,K_1}}\right) \\ \left(1-\frac{1}{C_{2,1}}\right)\left(1-\frac{1}{C_{2,2}}\right) \cdots\cdots \left(1-\frac{1}{C_{2,K_2}}\right) \\ \vdots \\ \left(1-\frac{1}{C_{i,1}}\right)\left(1-\frac{1}{C_{i,2}}\right) \cdots\cdots \left(1-\frac{1}{C_{i,K_i}}\right) \\ \vdots \\ \left(1-\frac{1}{C_{\lambda,1}}\right)\left(1-\frac{1}{C_{\lambda,2}}\right) \cdots\cdots \left(1-\frac{1}{C_{\lambda,K_\lambda}}\right) \end{matrix}\right\},$$

if we interpret this form to mean that we are to expand the double product and then form the sum by taking the integral part of each term with its proper sign. If p_λ has the maximum value of the limit already found, the formula will give the number of P's not greater than n.

We will now define an important function $\mu_C(N)$, analogous to Merten's function.* Let $N \equiv C_1\, C_2 \ldots\ldots C_n$ be the product of any number of fundamental C's, not necessarily different, then we will define $\mu_C(N)$ as follows:

$\mu_C(N) = 0$, if two or more of the C's are equal.

$\mu_C(N) = +1$, if all the C's are different and their number is even, also if $N = 1$.

$\mu_C(N) = -1$, if all the C's are different and their number is odd.

Lemma

$\sum \mu_C(D) = 0$, *if the sum is extended over all divisors* D (1 *and* N *included*) *of a number,* N, *which is a product of fundamental* C's.

Let us assume $N' = C_1\, C_2 \ldots\ldots C_n$, a product of n fundamental C's, then the total number of their products with an even number of factors (among this number we include the case of zero factors) is equal to the total number of their products with an odd number of factors. For the number of products with an even number of factors is

$$1 + \frac{n\,(n-1)}{1\cdot 2} + \frac{n\,(n-1)\,(n-2)\,(n-3)}{1\cdot 2\cdot 3\cdot 4} + \cdots\cdots,$$

and with an odd number of factors is

$$\frac{n}{1} + \frac{n\,(n-1)\,(n-2)}{1\cdot 2\cdot 3} + \cdots\cdots$$

Their difference is zero, being equal to $(1-1)^n$. It follows, then, that in the expanded product

$$(1-C_1)\,(1-C_2)\ldots\ldots(1-C_n)$$

the number of positive terms is equal to the number of negative terms.

* See Crelle, vol. 77, p. 289.

Whence

$$\sum \mu_c(\mathrm{D}) = 0.$$

when this sum is extended over all the divisors D, a product of C's, of N′. Finally, let

$$\mathrm{N} = \mathrm{C}_1^{\alpha_1}\mathrm{C}_2^{\alpha_2} \;.\;.\;.\;.\;.\;.\; \mathrm{C}_n^{\alpha_n}.$$

Then all the divisors of N′ are divisors of N, while all the remaining divisors of N contain at least the second power of a C, for which $\mu_{\mathrm{C}}(\mathrm{D}) = 0$. It follows, therefore, that

$$\sum \mu_c(\mathrm{D}) = 0,$$

if the sum is extended over all divisors D (1 and N included) of N.

Theorem

The number of P's not greater than x is given by the formula

$$\sum_{\mathrm{D}} \mu_c(\mathrm{D})\left[\frac{x}{\mathrm{D}}\right],$$

where D *is a divisor of* Π, *the product of all possible fundamental* C's *not greater than x.*

Let $f(n)$ and $F(n)$ be two "*zahlentheoretische Functionen*"* which are only different from zero, when n is a divisor of a given number Π, the product of any number of different fundamental C's. Thus, assume

$$f(\mathrm{D}) = \sum F(\mathrm{KD})$$

for every divisor D of Π, the sum being formed for all divisors K of $\frac{\Pi}{\mathrm{D}}$, where all numbers are subject to our definitions of products, divisors, etc., of C's.

Then we may form the sum

$$\sum_{\mathrm{D}} \mu_c(\mathrm{D}) f(\mathrm{D}) = \sum_{\mathrm{K}\cdot\mathrm{D}} \mu_c(\mathrm{D}) F(\mathrm{KD}).$$

The left member is to be summed for all divisors D of Π.

The right member is a double sum which is to be extended over all divisors D of Π and also all divisors K of $\frac{\Pi}{\mathrm{D}}$.

Now KD will become equal to every divisor of Π and if the terms of the double sum are collected in which KD is equal to the same divisor Δ of Π, then the entire double sum becomes $\sum_{\Delta} \mathrm{G}_{\Delta} F(\Delta)$, where $\mathrm{G}_{\Delta} = \sum \mu_c(\mathrm{D})$, and is summed for all divisors of Δ. But, by the lemma just proved, $\mathrm{G}_{\Delta} = 0$ for all values of Δ, except $\Delta = 1$, when $\mathrm{G}_{\Delta} = 1$. It follows, then, that

$$\sum_{\mathrm{D}} \mu_c(\mathrm{D}) f(\mathrm{D}) = \sum_{\Delta} \mathrm{G}_{\Delta} F(\Delta) = F(1).$$

* Compare Bachmann, *Zahlentheorie*, Part II, p. 308, for the definition of this term.

Now, let x be any positive quantity and let $\Pi \equiv C_1C_2 \ldots\ldots C_n$ be the product of n different fundamental C's. We may arrange the integers 1, 2, $3 \cdots\cdots [x]$ in groups such that each group contains all the integers which have the same greatest common factor* with Π, and no integer will appear in more than one group.

Defining $\psi(x, \Pi, K)$ as the number of integers not greater than x which have with Π the greatest common factor, K, as already defined, we shall have

$$[x] = \sum_K \psi(x, \Pi, K),$$

the summation being taken over all divisors K of Π. Let D be a given divisor of Π and K a divisor of $\frac{\Pi}{D}$, then KD is also a divisor of Π. But all the numbers not greater than x which have the greatest common factor KD† with Π will be found among the following numbers:

$$1 \cdot D,\ 2 \cdot D,\ 3 \cdot D \cdots\cdots \left[\frac{x}{D}\right] \cdot D;$$

and there will be just as many such numbers as there are integers not greater than $\frac{x}{D}$ which have the greatest common factor, $\frac{KD}{D}$‡, with $\frac{\Pi}{D}$. It follows that

$$\psi(x, \Pi, KD) = \psi\left(\frac{x}{D}, \frac{\Pi}{D}, \frac{KD}{D}\right).$$

But

$$\left[\frac{x}{D}\right] = \sum_K \psi\left(\frac{x}{D}, \frac{\Pi}{D}, \frac{KD}{D}\right),$$

where K is a divisor of $\frac{\Pi}{D}$.

Whence

$$\left[\frac{x}{D}\right] = \sum_K \psi(x, \Pi, KD),$$

where K is a divisor of $\frac{\Pi}{D}$.

---

* In determining the greatest common factor of two numbers, either or both of which are C's or products of C's, the greatest common factor may be any rational function of the C's, provided that in numerical calculations it is integral and a divisor in the ordinary sense of the quantities involved. In the particular case of any number n and Π, the greatest common factor may be only a divisor of Π as already defined, and, moreover, since two or more divisors of Π, while different as products of C's, may be numerically equal, the greatest common factor may be only some one of the numerically distinct divisors of Π. To illustrate this paragraph, let $\Pi \equiv C_1C_2C_3C_4 = 6 \cdot 10 \cdot 12 \cdot 18$. Then the greatest common factor of any number n and Π may be only some one of the numbers 1, 6, 10, 12, 18, 30, 36, 60, 90, 180; thus, the greatest common factor of 50 and Π is 10. Again of the numbers 1, 2, 3, 6, only 3 has $\frac{C_1C_3C_4}{C_1C_3} = \frac{6 \cdot 12 \cdot 18}{6 \cdot 12} = \frac{36}{12} = 3$ as greatest common factor with $C_2C_4 = 10 \cdot 18 = 90$.

† We only consider the distinct divisors KD.

‡ $\frac{KD}{D}$ does not necessarily equal K.

We have here an equation between two "*zahlentheoretische Functionen*" of the type already defined and so we can apply the results obtained.

Hence it follows that

$$\psi(x,\ \Pi,\ 1) = \sum_{\mathrm{D}} \mu_{\mathrm{C}}(\mathrm{D})\left[\frac{x}{\mathrm{D}}\right],$$

where the sum is extended over all divisors D of Π, a product of C's, *i.e.*, the number of integers not greater than x which contain no one of the C's which form the product Π is given by

$$\sum_{\mathrm{D}} \mu_{\mathrm{C}}(\mathrm{D})\left[\frac{x}{\mathrm{D}}\right].$$

If, now, we take Π as the product of all C's,

$$\prod_{i,k} \mathrm{C}_{i,k} \equiv \prod_{p_i,k} \{p_i(kp_i + 1)\},$$

where p_i is any one of of the first λ primes, $p_1 = 2,\ p_2 = 3 \ldots . p_i \ldots .. p_\lambda$, and $k = 1$ to $\left[\frac{x - p_i}{p_i^2}\right]$, running independently for each prime, then

$$\phi_{\mathrm{P}}(x,\lambda) = \sum_{\mathrm{D}} \mu_{\mathrm{C}}(\mathrm{D})\left[\frac{x}{\mathrm{D}}\right],$$

where D is a divisor of Π.

Finally, if p_λ is the largest prime not greater than $\left[\frac{\sqrt{4x+1} - 1}{2}\right]$ and the k's are determined as before, then $\phi_{\mathrm{P}}(x, \lambda)$ will denote the number of P's not greater than x. Therefore, the number of P's not greater than x is given by the formula

$$\sum_{\mathrm{D}} \mu_{\mathrm{C}}(\mathrm{D})\left[\frac{x}{\mathrm{D}}\right],$$

where D is a divisor of Π, the product of all C's not greater than x.

As an illustration of the method, let us find

$$\psi(x, \Pi, 1),$$

where $x = 50$ and $\Pi = \{2(2+1)\}\{2(2\cdot 2+1)\}\{3(3+1)\}\{2(4\cdot 2+1)\}$
$= 6\cdot 10\cdot 12\cdot 18.$

$$\psi(50,\ \Pi,\ 1) = \sum_{\mathrm{D}} \mu_{\mathrm{C}}(\mathrm{D})\left[\frac{50}{\mathrm{D}}\right], \text{ where D is a divisor of } \Pi = 6\cdot 10\cdot 12\cdot 18,$$

$$= \left[\frac{50}{1}\right] - \left[\frac{50}{6}\right] - \left[\frac{50}{10}\right] + \left[\frac{50}{30}\right] \text{ (All the other terms destroyed each other.)}$$

$$= 50 - 8 - 5 + 1 = 38.$$

Whence 38 of the first 50 integers are not divisible by 6, 10, 12, or 18.

Computation Formulae for $\phi_P(n, \kappa, p_1)$ and $\phi_P(n, \lambda)$

While the formulae already obtained furnish the solution of the problem in question, the actual computations necessary to obtain numerical results when n increases soon becomes too extended and complicated for practical use. Therefore in the succeeding paragraphs we will obtain formulae which lend themselves more readily to numerical computation.

In accordance with our previous definitions, we will define $\phi_P(n, \kappa, p_1)$ as the number of integers not greater than n which contain no factor of the form $p_1(kp_1+1)$ where p_1 is a given prime and $k = 1, 2, 3 \cdots\cdots \kappa$. If we form the sequence of natural numbers in order from 1 to $[n]$ and then exclude all those numbers which are divisible by factors of the form $p_1(kp_1+1)$ where

$$k = 1, 2, 3 \cdots\cdots (\kappa - 1),$$

the number of integers remaining will be equal to $\phi_P(n, \overline{\kappa - 1}, p_1)$. It is still necessary to exclude the numbers which are divisible by $p_1(\kappa p_1 + 1)$, which are the following:

$$p_1(\kappa p_1+1),\ 2p_1(\kappa p_1+1) \cdots\cdots \left[\frac{n}{p_1(\kappa p_1+1)}\right]\{p_1(\kappa p_1+1)\}.$$

However, we have already excluded all the numbers of this sequence which are divisible by other factors of the given form, so that for our purpose it is only necessary to consider those numbers of this sequence which are divisible by no other factors of the given form than $p_1(\kappa p_1+1)$. We may obtain this number by determining how many integers of the sequence

$$(\kappa p_1+1),\ 2(\kappa p_1+1),\ 3(\kappa p_1+1) \cdots\cdots \left[\frac{n}{p_1(\kappa p_1+1)}\right](\kappa p_1+1)$$

are divisible by any factor of the form (kp_1+1), where $k = 1, 2 \ldots\ldots (\kappa - 1)$, a result which is readily obtainable by a sieve process. For convenience we will define

$$\phi_P\{\overset{\nu}{\underset{1}{A}}(\kappa p_1+1),\ (\kappa-1),\ p_1\}$$

as the number of terms of the arithmetical sequence, whose common difference is (kp_1+1) and the number of whose terms is $\nu = \left[\dfrac{n}{p_1(\kappa p_1+1)}\right]$, which have no factors of the form (κp_1+1), where $k = 1, 2, 3 \cdots\cdots\cdots (\kappa - 1)$.

Then we shall have

$$\phi_P(n, \kappa, p_1) = \phi_P(n, \overline{\kappa-1}, p_1) - \phi_P\{\overset{\nu}{\underset{1}{A}}(\kappa p_1+1),\ (\kappa-1),\ p_1\}.$$

If we put

$$\kappa = \left[\frac{n-p_1}{p_1^2}\right],$$

then

$$\phi_P(n, \kappa, p_1) = \phi_P(n, p_1) = \phi_P(n, \overline{\kappa-1}, p_1) - \phi_P\{\overset{\nu}{\underset{1}{A}}(\kappa p_1+1),\ (\kappa-1),\ p_1\}.$$

As an illustration, we will apply the method to determine

$$\phi_P(100, 3) = 81,$$

Now

$$\kappa = \left[\frac{100-3}{9}\right] = 10.$$

$$\therefore\ \phi_P(100, 3) = \phi_P(100, 10, 3).$$

$$\phi_P(100, 10, 3) = \phi_P(100, 9, 3) - \phi_P\{\overset{1}{\underset{1}{A}}(31), 9, 3\} = 82 - 1 = 81.$$

$$\phi_P(100, 9, 3) = \phi_P(100, 8, 3) - \phi_P\{\overset{1}{\underset{1}{A}}(28), 8, 3\} = 82 - 0 = 82.$$

$$\phi_P(100, 8, 3) = \phi_P(100, 7, 3) - \phi_P\{\overset{1}{\underset{1}{A}}(25), 7, 3\} = 83 - 1 = 82.$$

$$\phi_P(100, 7, 3) = \phi_P(100, 6, 3) - \phi_P\{\overset{1}{\underset{1}{A}}(22), 6, 3\} = 84 - 1 = 83.$$

$$\phi_P(100, 6, 3) = \phi_P(100, 5, 3) - \phi_P\{\overset{1}{\underset{1}{A}}(19), 5, 3\} = 85 - 1 = 84.$$

$$\phi_P(100, 5, 3) = \phi_P(100, 4, 3) - \phi_P\{\overset{6}{\underset{1}{A}}(16), 4, 3\} = 85 - 0 = 85.$$

$$\phi_P(100, 4, 3) = \phi_P(100, 3, 3) - \phi_P\{\overset{2}{\underset{1}{A}}(13), 3, 3\} = 87 - 2 = 85.$$

$$\phi_P(100, 3, 3) = \phi_P(100, 2, 3) - \phi_P\{\overset{3}{\underset{1}{A}}(10), 2, 3\} = 89 - 2 = 87.$$

$$\phi_P(100, 2, 3) = \phi_P(100, 1, 3) - \phi_P\{\overset{4}{\underset{1}{A}}\ (7), 1, 3\} = 92 - 3 = 89.$$

$$\phi_P(100, 1, 3) = 92.$$

The numerical calculations must be made in reverse order, beginning with the last line instead of with the first.

Interpreting $\phi_P(n, \lambda)$ as already defined, we will proceed to find a formula for it in exactly the same manner as employed in the previous case. If we exclude from the sequence of integers, $1, 2 \cdots\cdots [n]$ all those integers which are divisible by factors of the form $p(kp+1)$, where p is any one of the first $(\lambda - 1)$ primes and k is determined in the usual way, there will be left just $\phi_P(n, \lambda - 1)$ numbers. The integers divisible by $p_\lambda(kp_\lambda + 1)$ are the following:

$$p_\lambda(p_\lambda+1),\ 2p_\lambda(p_\lambda+1) \cdots\cdots \left[\frac{n}{p_\lambda(p_\lambda+1)}\right]\left\{p_\lambda(p_\lambda+1)\right\},$$

$$\vdots$$

$$p_\lambda(\kappa p_\lambda+1),\ 2p_\lambda(\kappa p_\lambda+1) \cdots\cdots \left[\frac{n}{p_\lambda(\kappa p_\lambda+1)}\right]\left\{p_\lambda(\kappa p_\lambda+1)\right\}.$$

From this double sequence of numbers, by a sieve process, we can exclude all those numbers which contain factors of the form $p(kp+1)$, p being any one of the first $(\lambda-1)$ primes. We will then define

$$\phi_P\left\{\underset{\nu=1\quad k=1}{\overset{\nu=\nu_k\quad k=K}{A}} p_\lambda(kp_\lambda+1),\ \lambda-1\right\},$$

where p_λ is the λth prime, $k=1, 2\cdot\cdot\cdot\left[\frac{n-p_\lambda}{p_\lambda^2}\right]$, and $\nu=1, 2\cdot\cdot\cdot\left[\frac{n}{p_\lambda(kp_\lambda+1)}\right]$, determined independently for each value of k, as the number of integers of the double sequence which contain no factors of the given form, where p is any one of the first $(\lambda-1)$ primes. We may then write our formula, which is analogous to Meissel's computation formula for primes,*

$$\phi_P(n,\lambda)=\phi_P(n,\lambda-1)-\phi_P\left\{\underset{\nu=1\quad k=1}{\overset{\nu=\nu_k\quad k=K}{A}} p_\lambda(kp_\lambda+1),\ \lambda-1\right\}.$$

If p_λ is the largest prime not greater than $\left[\frac{\sqrt{4n+1}-1}{2}\right]$, then $\phi_P(n,\lambda)$ will give the number of P's not greater than n.

As an illustration, we will calculate the number of P's not greater than 100. In this case

$$p_\lambda \overline{\overline{<}} \left[\frac{\sqrt{400+1}-1}{2}\right] \overline{\overline{<}} 9, \textit{ i.e.}, p_\lambda=7 \text{ and } \lambda=4.$$

$$\therefore\ \phi_P(100,4)=\phi_P(100,3)-\phi_P\left\{\underset{\nu=1\quad k=1}{\overset{\nu=\nu_1\quad k=1}{A}} 7(7+1),\ 3\right\}=49-0=49.$$

$$\phi_P(100,3)=\phi_P(100,2)-\phi_P\left\{\underset{\nu=1\quad k=1}{\overset{\nu=\nu_3\quad k=3}{A}} 5(k_j5+1),\ 2\right\}=50-1=49.$$

$$\phi_P(100,2)=\phi_P(100,1)-\phi_P\left\{\underset{\nu=1\quad k=1}{\overset{\nu=\nu_{10}\quad k=10}{A}} 3(k_j3+1),\ 1\right\}=56-6=50.$$

$$\phi_P(100,1)=56.$$

The numerical calculations were made in reverse order, beginning with the last line instead of with the first. It is to be noted that

$$\phi_P(n,1)=\left[\frac{n+1}{2}\right]+\left[\frac{\log n}{\log 2}\right],$$

since $\phi_P(n,1)$ is the number of integers less than n which contain no factors of the form $2(2n+1)$ and, therefore, consists of all the odd integers not greater than n plus the powers of 2 which are not greater than n.

Therefore, the number of P's not greater than 100 is 49, a result which is easily verified.

While the enumeration of the P's within a given limit, either by computation or by an asymptotic formula, is a matter of much theoretical interest, in many

* Meissel: *Math. Annalen* II and III.

cases, especially in applications to the theory of groups, to know whether a given number is a P or a C is most important. This information will necessarily be best supplied by a table showing for each integer all its factors of the form $p(kp+1)$. The tabulation of the P's could be made mechanically by a process similar to that of Eratosthenes's sieve, which consists in writing down the integers in their natural order and then cutting out the successive primes, 2, 3, 5 ......; but, as in the construction of factor-tables, various devices are available which permit of simpler and more ready computation.* Such a table may be constructed in the following manner. Arrange the numbers by tens, as in a table of common logarithms of numbers, on a sheet of paper sufficiently long to contain the entire list and ruled in rows and columns. In the proper rectangle write the prime factors of each number, the p's of the C's. It then remains only to write in the various values of k for its corresponding p and the table is complete. If the paper has been accurately ruled, the k's may be written in very readily by measurement, for a given C repeats itself every 5C or 10C lines in the same column, according as the C is an even or odd number. Then, in order, take each C of the double sequence of C's, $C_{i,k} \equiv p_i(kp_i+1)$ where the p's and k's are limited as in the preceding paragraphs. Compute its first appearance in each column, and finally enter it by measurement in the remainder of the column, writing the value of k after its corresponding prime. Many devices for checking the results will suggest themselves. On page 17 we give a sample of such a table.†

* For an account of these devices, see the introduction to Professor Lehmer's *Factor Table for the First Ten Millions.*

† For further details, the reader is referred to the writer's *A Sylow Factor Table of the First Twelve Thousand Numbers* which will be published by the Carnegie Institution of Washington and is now in press.

TABLE OF NUMBERS FROM 1 TO 99, SHOWING FACTORS OF THE FORM $p(kp+1)$

0	1	2	3	4	5	6	7	8	9
				2^2		2 1 3		2^3	3^2
2 2 5		2^2 3 1		2 3 7	3 5	2^4		2 4 3^2	
2^2 5	3 2 7	2 5 11		2^3 3 1	5^2	2 6 13	3^3	2^2 7	
2 7 3 3 5 1		2^5	3 11	2 8 17	5 7	2^2 3^2 1		2 9 19	3 4 13
2^3 5		2 10 3 2 7		2^2 11	3^2 5	2 11 23		2^4 3 5 1	7^2
2 12 5^2	3 17	2^2 13		2 13 3^3	5 2 11	2^3 7 1	3 6 19	2 14 29	
2^2 3 1, 3 5 1		2 15 31	3^2 2 7	2^6	5 13	2 3 7 11		2^2 17	3 23
2 17 5 7		2^3 3^2 1		2 18 37	3 8 5^2	2^2 19	7 11	2 19 3 4 13	
2^4 5 3	3^4	2 20 41		2^2 3 9 1, 2 7	5 17	2 21 43	3 29	2^3 11	
2 22 3^2 3 5 1	7 13	2^2 23	3 10 31	2 23 47	5 19	2^5 3 1, 5		2 24 7^2	3^2 11

For each number in the above table the following data are given:

1. In the first column of each rectangle, the representation of the number as a product of powers of primes, thus $90 = 2 \cdot 3^2 \cdot 5$.

2. In the second column the value of k greater than zero for each prime greater than unity such that $N = p(kp+1)$, thus $90 = 2(22 \cdot 2 + 1)$, but there are no values of k greater than zero, for which $90 = 3(3k+1)$ or

$$90 = 5(5\text{k} + 1).$$

3. In the remaining columns, the values of k which give divisors of N of the form $p(pk+1)$, $k > 0$, $p > 2$; thus 90 is divisible by $3(3 \cdot 3 + 1)$, also by

$$5(1 \cdot 5 + 1).$$

Since there is a close analogy between ordinary primes and P's, a comparison between the number of primes and the number of P's within a given limit naturally suggests itself. In the following table (page 19), the number of primes and the number of P's for each century and for each thousand are listed in adjacent columns. The arrangement of the table is so simple that no explanation is necessary. In the appendix a list of all the P's less than 12,229 is given, so arranged that the number of P's between any two limits less than 12,229 may be easily obtained. The cut, facing the table, shows the number of P's and the number of primes by centuries plotted side by side. The abscissas give the century and the ordinates give the number of primes, or the number of P's, for that century. These two curves are approximate curves of frequency for the P's and primes.

NUMBER OF PRIMES AND P'S FOR EACH CENTURY AND FOR EACH THOUSAND

0-3999

Century	Primes*	P's†	Century	Primes	P's	Century	Primes	P's	Century	Primes	P's
1‡	26	49	11	16	32	21	14	33	31	12	34
2	21	40	12	12	35	22	10	31	32	10	33
3	16	37	13	15	35	23	15	34	33	11	31
4	16	34	14	11	36	24	15	34	34	15	33
5	17	37	15	17	32	25	10	34	35	11	30
6	14	38	16	12	35	26	11	35	36	14	34
7	16	34	17	15	35	27	15	30	37	13	33
8	14	36	18	12	33	28	14	33	38	12	33
9	15	36	19	12	33	29	12	37	39	11	32
10	14	33	20	13	36	30	11	31	40	11	33
	169	374		135	342		127	332		120	326

4000-7999

Century	Primes	P's	Century	Primes	P's	Century	Primes	P's	Century	Primes	P's
41	15	33	51	12	33	61	12	29	71	9	33
42	9	34	52	11	31	62	11	36	72	10	34
43	16	35	53	10	28	63	13	33	73	11	34
44	9	31	54	10	35	64	15	30	74	9	33
45	11	33	55	13	30	65	8	35	75	11	32
46	12	33	56	13	32	66	11	30	76	15	24
47	12	32	57	12	32	67	10	31	77	12	32
48	12	31	58	10	31	68	12	31	78	10	33
49	8	33	59	16	33	69	12	33	79	10	31
50	15	33	60	7	33	70	13	31	80	10	34
	119	328		114	318		117	319		107	320

8000-11999

Century	Primes	P's	Century	Primes	P's	Century	Primes	P's	Century	Primes	P's
81	11	35	91	11	31	101	11	31	111	10	36
82	10	31	92	12	33	102	12	28	112	11	27
83	14	30	93	11	32	103	10	28	113	10	31
84	9	34	94	11	30	104	12	33	114	10	35
85	8	32	95	15	31	105	10	31	115	11	33
86	12	35	96	7	31	106	8	31	116	9	32
87	13	31	97	13	29	107	12	35	117	8	31
88	11	32	98	11	32	108	11	31	118	9	32
89	13	32	99	12	32	109	10	28	119	12	30
90	9	32	100	9	35	110	10	31	120	13	30
	110	324		112	316		106	307		103	317

* The number of primes was obtained from similar tables in the introduction of Glaisher's *Factor Table of the Sixth Million.*

† The number of P's was obtained from the list of P's given in the appendix.

‡ The first century, 0-99, includes 1 as a prime and a P.

Addenda

A careful study of the curves of approximate frequency and the tables from which they were obtained, together with a study, as n increases, of the increase of the following four different types of P's; namely, (1) primes; (2) powers of primes; (3) products of two consecutive odd primes; (4) all other P's; brings out very clearly what appears to be an important relation between the number of P's and the number of primes within a given limit. Various methods of empirically noting the increase of the last three types of P's within the limits of the list of P's given in the appendix, especially of the logarithms of the prime factors involved, have brought out very interesting data, but so far it has been impossible to correlate these data into definite theorems. However, the data obtained and a study of the curve suggest the following theorem, no proof of which has been obtained:

The number of primes and the number of P's within a given limit differ asymptotically by a constant.

APPENDIX

List of P's*

In the following pages is given a list of P's in their order from 1 to 12,229 and so arranged for the ready computation of the number of P's between any two limits less than 12,229. The columns are numbered consecutively from 1 to the last column of the list in the first line at the top of the pages, and the lines are numbered from 1 to 50 in the first column on the left-hand side of each page. The last three digits of the P's are given in the columns, the hundredth's digit only being given for the first P in the hundred, the remaining digits are to be found at the top of the column; *i.e.*, in the second line at the top of the page. In case the entry for any P is in heavy type, preceded by an asterisk, the other digits for the rest of the P's in that column are to be found at the top of the next column to the right. Thus the P at the intersection of column 35 and row 43 is 5127.

* The P's for this list were obtained from the writer's *A Sylow Factor Table of the First Twelve Thousand Numbers*, already referred to (p. 16). I am indebted to Professor D. N. Lehmer for the method of listing the P's, which is identical with the arrangement to be used by him in a *List of Prime Numbers from One to Ten Millions*, of which he kindly showed me the manuscript.

List of P's, 1–2257

	1	2	3	4	5	6	7	8	9	10	11	12	13	14	15	16
									1	1	1	1	1	1	1	2
1	1	103	235	371	511	643	787	929	079	229	367	517	663	813	957	103
2	2	07	39	73	12	47	89	31	87	31	69	19	67	17	59	11
3	3	09	41	77	15	49	93	33	91	33	73	23	69	19	61	13
4	4	13	43	79	17	53	97	37	93	35	77	27	71	23	63	17
5	5	15	45	83	19	59	99	41	97	37	81	29	75	25	69	19
6	7	19	47	89	21	61	801	43	99	41	83	31	79	29	73	23
7	8	21	49	91	23	65	03	47	103	43	85	35	81	31	75	25
8	9	23	51	93	27	67	07	49	07	47	87	37	85	35	77	29
9	11	25	56	95	29	71	09	51	09	49	91	41	87	37	79	31
10	13	27	57	97	31	73	11	53	11	53	93	43	89	41	81	37
11	15	28	59	401	33	77	15	59	15	57	97	47	91	43	85	41
12	16	31	61	03	35	79	17	61	17	59	99	49	93	47	87	43
13	17	33	63	07	37	81	21	63	21	61	401	53	97	49	91	47
14	19	35	65	09	39	83	23	65	23	67	03	57	99	51	93	49
15	23	37	67	11	41	85	27	67	27	69	09	59	707	53	97	51
16	25	39	69	13	45	91	29	71	29	71	11	61	09	59	99	53
17	27	41	71	15	47	95	33	73	33	73	15	63	15	61	***003**	57
18	29	43	77	19	51	97	35	77	35	77	17	65	17	63	09	59
19	31	45	81	21	53	99	39	83	39	79	23	67	19	65	11	61
20	32	49	83	23	57	701	41	85	41	83	27	71	21	67	17	65
21	33	51	87	25	59	03	43	89	45	85	29	73	23	71	21	67
22	35	53	89	27	63	07	45	91	47	89	31	77	27	73	23	71
23	37	57	93	31	65	09	47	95	49	91	33	79	29	77	27	73
24	41	59	95	33	69	13	51	97	51	93	37	83	33	79	29	77
25	43	61	97	37	71	17	53	***001**	53	95	39	85	35	83	31	79
26	45	63	99	39	73	19	57	03	57	97	41	89	39	85	33	83
27	47	67	303	43	75	21	59	07	59	301	45	91	41	89	39	87
28	49	69	07	45	77	25	63	09	63	03	47	93	45	91	43	91
29	51	73	11	47	81	27	65	13	65	07	51	97	47	95	45	95
30	53	75	13	49	83	29	69	17	67	09	53	601	53	97	47	97
31	59	77	17	51	87	31	71	19	69	13	57	03	57	901	48	201
32	61	79	19	57	89	33	75	21	71	15	59	07	59	03	49	03
33	64	81	21	59	91	39	77	24	75	19	65	09	61	07	51	07
34	65	85	23	61	93	43	79	31	77	21	69	11	63	09	53	09
35	67	87	25	63	99	45	81	33	79	25	71	13	65	13	57	13
36	69	91	29	67	601	47	83	37	81	27	73	15	69	15	59	15
37	71	93	31	69	07	49	87	39	87	29	75	19	73	17	63	19
38	73	97	35	73	11	51	91	41	89	31	81	21	77	19	69	21
39	77	99	37	75	13	53	93	43	93	33	83	25	79	21	71	25
40	79	207	39	77	17	57	95	49	95	37	87	27	81	23	75	27
41	81	09	41	79	19	61	99	51	99	39	89	31	83	27	77	29
42	83	11	43	81	21	63	901	57	201	41	93	33	87	31	81	31
43	85	13	47	85	23	67	07	59	03	43	95	37	89	33	83	37
44	87	15	49	87	25	69	09	61	07	45	99	39	93	37	87	39
45	89	17	53	91	29	71	11	63	11	47	501	43	95	39	89	41
46	91	21	59	93	31	73	13	67	13	49	03	45	97	41	93	43
47	95	23	61	99	35	79	17	69	17	51	07	49	99	43	95	45
48	97	27	65	501	37	81	19	73	19	57	09	51	801	45	97	49
49	99	29	67	03	39	83	23	75	23	61	11	57	07	49	99	51
50	101	33	69	09	41	85	25	77	25	63	13	61	11	51	101	57

List of P's, 2259–4687

	17	18	19	20	21	22	23	24	25	26	27	28	29	30	31	32
	2	2	2	2	2	3	3	3	3	3	3	3	4	4	4	4
1	259	407	551	713	857	011	151	317	473	623	777	929	085	227	381	531
2	63	11	57	19	59	13	61	19	75	25	79	31	87	29	85	33
3	67	13	61	23	61	17	63	21	81	31	81	35	91	31	87	35
4	69	17	63	25	63	19	67	23	87	35	85	37	93	37	91	37
5	73	19	67	27	67	23	69	25	89	37	87	41	96	41	93	41
6	79	21	69	29	69	29	73	27	91	43	91	43	97	43	97	43
7	81	23	71	31	73	31	75	29	93	47	93	47	99	47	99	47
8	83	25	73	33	75	35	77	31	97	49	97	49	101	49	403	49
9	85	27	75	35	79	37	79	35	99	51	99	53	03	53	09	53
10	87	29	79	37	81	39	81	37	501	53	803	57	09	59	11	59
11	91	35	81	41	85	41	83	41	03	59	07	59	11	61	15	61
12	93	37	87	43	87	43	87	43	09	61	09	61	15	65	19	67
13	97	41	89	47	89	47	91	47	11	65	11	67	17	67	21	69
14	99	43	91	49	91	49	93	49	15	67	15	71	19	71	23	71
15	303	47	93	53	93	51	99	53	17	69	17	73	21	73	27	73
16	09	49	97	59	97	53	203	59	21	71	21	77	27	77	29	77
17	11	53	99	61	99	57	09	61	23	73	23	79	29	79	33	79
18	13	59	603	65	903	59	11	65	27	77	27	83	31	81	35	81
19	15	61	09	67	09	61	15	67	29	79	31	85	33	83	39	83
20	17	63	11	71	11	65	17	71	33	83	33	87	35	85	41	89
21	19	65	15	73	13	67	21	73	37	87	35	89	39	89	43	91
22	21	67	17	77	17	71	23	77	39	89	39	91	41	91	47	95
23	23	71	21	79	21	73	27	79	41	91	41	95	45	93	51	97
24	27	73	23	85	23	77	29	83	43	95	45	97	47	95	53	601
25	33	77	27	87	27	79	31	85	45	97	47	*001	49	97	57	03
26	35	79	29	89	29	83	33	89	47	99	49	03	51	99	59	07
27	39	81	33	91	31	85	35	91	51	701	51	07	53	303	61	19
28	41	83	37	95	33	89	39	95	53	03	53	09	57	07	63	21
29	45	89	41	97	35	91	47	97	57	07	59	13	59	09	69	25
30	47	91	43	99	39	93	51	401	59	09	63	19	63	11	71	27
31	49	95	45	801	41	95	53	07	61	13	65	21	69	13	75	31
32	51	97	47	03	49	97	57	09	63	15	67	23	71	15	77	33
33	53	501	51	07	51	101	59	13	69	19	69	27	75	17	79	37
34	57	03	57	09	53	03	63	15	71	21	71	31	77	21	81	39
35	63	07	59	13	57	07	65	19	77	25	77	33	81	25	83	43
36	67	09	61	15	59	09	69	21	79	27	79	37	83	27	87	45
37	69	13	63	19	63	13	71	25	81	33	81	39	87	31	89	49
38	71	15	69	21	65	15	73	27	83	37	87	41	89	33	93	51
39	75	17	71	23	69	19	77	31	87	39	89	43	93	37	97	57
40	77	19	75	25	71	21	81	33	89	43	93	45	95	39	99	59
41	81	21	77	27	77	23	83	39	93	45	99	49	99	43	501	61
42	83	27	81	31	81	25	87	43	95	49	901	51	201	49	07	63
43	87	29	83	33	83	27	91	47	99	51	03	57	03	53	09	67
44	89	31	87	37	87	31	93	49	601	57	07	61	07	57	11	71
45	91	33	89	39	89	33	95	53	07	61	11	67	11	61	13	73
46	93	37	93	41	93	37	99	57	09	63	17	73	13	63	17	77
47	95	39	99	43	95	39	301	61	11	67	19	75	17	69	19	79
48	99	43	701	45	99	45	07	63	13	69	21	79	19	73	23	81
49	401	45	07	51	*001	47	09	67	17	71	23	81	23	75	27	85
50	03	49	11	53	07	49	13	69	19	73	25	83	25	79	29	87

List of P's, 4689–7181

	33	**34**	**35**	**36**	**37**	**38**	**39**	**40**	**41**	**42**	**43**	**44**	**45**	**46**	**47**	**48**
	4	4	4	5	5	5	5	5	5	6	6	6	6	6	6	7
1	689	841	997	153	321	471	629	783	939	103	251	407	561	715	877	039
2	91	43	99	57	23	73	31	85	41	07	53	09	63	19	81	43
3	93	47	***001**	61	27	77	33	91	47	09	57	13	69	21	83	45
4	97	49	03	63	29	79	39	93	51	13	59	15	71	23	87	47
5	99	53	09	65	33	83	41	801	53	15	61	19	75	25	89	49
6	703	57	11	67	37	85	45	03	57	17	63	21	77	27	91	51
7	09	59	13	71	39	89	47	07	59	19	65	23	81	29	93	53
8	13	61	15	73	41	91	51	09	63	21	69	25	83	31	99	57
9	17	65	17	77	45	97	53	13	65	25	71	27	87	33	901	61
10	21	67	21	79	47	501	57	15	69	27	77	31	89	37	07	67
11	23	71	23	83	51	03	59	21	71	29	81	33	93	39	11	69
12	27	73	27	89	53	07	63	23	75	31	83	37	95	43	13	71
13	29	77	29	91	57	09	67	25	77	33	87	39	99	49	17	73
14	33	79	33	95	59	15	69	27	81	37	89	43	601	51	19	75
15	35	83	35	97	61	19	75	31	83	39	91	45	07	61	25	79
16	39	85	39	99	63	21	77	33	87	43	95	49	11	63	27	81
17	41	89	41	203	65	27	81	37	89	45	97	51	13	67	29	85
18	47	91	45	07	69	31	83	39	91	47	99	53	17	77	31	87
19	49	97	47	09	71	33	87	43	93	49	301	57	19	79	35	91
20	51	901	51	13	75	39	89	45	99	51	07	59	21	81	37	93
21	53	03	53	15	77	41	93	47	***001**	57	09	63	23	85	39	97
22	57	09	57	21	81	43	95	49	07	61	11	67	25	91	41	99
23	59	11	59	27	83	45	99	51	09	63	13	69	29	93	43	101
24	63	13	63	31	87	49	701	57	11	67	17	71	31	97	47	03
25	65	15	65	33	89	51	03	61	13	69	19	73	35	99	49	09
26	69	19	67	37	91	53	07	63	17	73	23	79	37	801	57	15
27	71	21	69	39	93	57	11	67	19	75	29	81	39	03	59	17
28	77	25	71	43	95	61	13	69	23	79	31	85	41	09	61	21
29	79	27	77	45	99	63	17	73	29	85	33	87	43	11	67	23
30	81	31	81	49	401	67	21	75	31	87	37	91	47	15	71	27
31	83	33	87	51	07	69	23	77	37	89	41	93	49	17	73	29
32	87	37	89	57	11	73	25	79	41	91	43	97	53	19	77	33
33	89	39	91	61	13	75	29	81	43	93	47	99	59	21	79	35
34	93	43	95	63	17	79	31	91	47	97	49	503	61	23	83	39
35	95	45	99	67	19	81	37	93	49	99	53	09	67	27	89	41
36	99	49	101	69	23	85	39	97	53	203	59	11	73	29	91	43
37	801	51	07	73	29	87	41	99	59	07	61	15	77	33	95	45
38	03	57	11	79	31	89	43	903	65	09	65	21	79	35	97	47
39	11	61	13	81	33	91	47	09	67	11	67	23	83	39	99	51
40	13	63	19	83	35	97	49	11	71	17	73	27	87	41	***001**	53
41	17	67	21	87	37	99	51	17	73	21	79	29	89	45	03	57
42	19	69	23	91	41	603	53	19	77	27	83	33	91	47	07	59
43	21	73	27	93	43	09	61	21	79	29	85	35	95	49	09	63
44	25	75	29	97	47	11	65	23	81	33	87	39	97	51	13	65
45	29	79	31	303	49	13	67	27	85	35	89	41	701	57	17	69
46	31	81	37	09	53	15	69	29	89	39	91	47	03	59	19	71
47	33	85	41	11	59	17	71	31	91	41	95	51	07	63	27	73
48	35	87	43	15	61	21	73	33	93	43	97	53	09	65	31	77
49	37	91	47	17	65	23	77	35	95	45	401	57	11	69	33	79
50	39	93	49	19	69	27	79	37	101	47	03	59	13	71	37	81

List of P's, 7183–9697

	49	50	51	52	53	54	55	56	57	58	59	60	61	62	63	64
	7	7	7	7	7	7	8	8	8	8	8	8	9	9	9	9
1	183	331	487	669	823	979	129	287	437	583	743	899	061	215	373	533
2	87	33	89	73	25	81	31	91	41	87	47	903	65	17	77	35
3	93	39	93	75	27	83	33	93	43	91	49	07	67	21	79	39
4	95	41	95	79	29	85	35	97	47	93	53	09	69	23	83	47
5	97	43	99	81	31	87	37	99	53	97	59	11	71	27	85	49
6	99	45	501	85	35	89	43	303	57	99	61	15	73	29	89	51
7	201	49	07	87	37	91	47	09	59	603	65	17	77	33	91	53
8	07	51	17	91	41	93	49	11	61	09	67	21	79	35	95	57
9	09	61	19	93	49	99	53	15	65	11	71	23	83	39	97	59
10	11	63	23	97	53	*003	59	17	67	15	73	27	85	41	401	63
11	13	67	29	99	59	09	61	21	69	17	77	29	89	45	03	65
12	17	69	37	703	61	11	67	27	71	21	79	33	91	47	07	69
13	19	73	41	09	63	15	71	29	73	23	81	35	95	49	09	71
14	23	75	43	11	67	17	73	31	77	27	83	39	97	51	11	73
15	25	77	47	13	71	21	77	33	79	29	89	41	101	53	13	75
16	29	79	49	15	73	23	79	35	83	33	91	45	03	57	19	77
17	31	81	51	17	77	27	81	39	85	37	93	47	07	59	21	83
18	33	87	59	21	79	29	83	41	89	39	95	51	09	63	25	87
19	35	89	61	23	83	33	85	45	91	41	97	53	13	67	27	89
20	37	91	65	27	85	35	87	47	95	47	803	59	15	69	31	93
21	41	93	71	29	89	39	89	51	97	51	07	63	17	71	33	99
22	43	97	73	33	91	41	91	53	501	53	09	69	19	77	37	601
23	47	99	77	37	95	45	92	57	03	57	13	71	23	79	39	07
24	49	403	83	39	97	47	97	59	07	59	17	75	25	81	41	09
25	51	09	87	41	99	51	99	61	09	63	19	77	27	83	45	13
26	53	11	89	45	901	53	201	63	11	65	21	81	31	87	51	17
27	61	15	91	47	03	57	03	67	13	67	25	83	33	89	57	19
28	63	17	93	51	07	59	07	69	19	69	31	87	37	93	61	23
29	65	21	97	53	09	61	09	71	21	71	33	89	43	99	63	27
30	67	23	601	57	11	63	13	75	23	75	37	93	49	301	67	29
31	69	27	03	59	13	65	19	77	27	77	39	95	51	07	69	31
32	73	31	07	63	15	69	21	81	29	81	43	97	53	11	73	37
33	77	33	09	65	19	71	23	83	31	89	47	99	57	13	75	43
34	79	35	13	67	21	75	25	87	37	91	49	*001	61	19	79	47
35	81	41	15	69	25	77	27	89	39	93	51	07	63	23	81	49
36	83	43	19	71	27	79	31	93	43	95	57	11	67	25	87	59
37	89	45	21	73	29	81	33	95	45	99	61	13	69	29	91	61
38	91	47	27	77	31	83	37	97	49	707	63	17	71	35	93	63
39	95	51	29	81	33	87	43	99	51	09	67	19	73	37	97	65
40	97	53	31	83	37	89	45	401	57	11	69	29	75	41	99	67
41	301	57	33	89	39	93	49	03	59	13	71	31	79	43	501	71
42	03	59	37	93	43	95	57	07	61	17	73	33	81	47	03	73
43	07	63	39	95	49	97	59	11	63	19	79	35	87	49	09	77
44	09	65	43	99	51	99	61	13	67	25	81	37	91	53	11	79
45	13	69	47	801	57	101	63	17	69	27	85	41	93	57	17	83
46	19	71	49	07	63	11	67	19	73	31	87	43	97	59	21	85
47	21	75	57	11	67	13	69	23	75	35	89	47	99	65	23	89
48	23	77	61	13	69	17	73	25	77	37	91	49	203	67	27	93
49	25	81	63	17	71	19	79	29	79	39	93	53	09	69	29	95
50	27	83	67	19	73	23	85	31	81	41	97	59	11	71	31	97

LIST OF P'S, 9701–12229

	65	**66**	**67**	**68**	**69**	**70**	**71**	**72**	**73**	**74**	**75**	**76**	**77**	**78**	**79**	**80**
	9	9	10	10	10	10	10	10	10	11	11	11	11	11	11	12
1	701	865	007	171	345	501	657	813	981	125	301	447	601	761	929	079
2	03	69	09	77	47	03	61	17	85	29	03	49	03	63	33	83
3	07	71	13	81	49	07	63	19	87	31	09	53	09	69	39	85
4	13	73	15	83	51	11	67	23	91	41	11	59	11	71	41	89
5	15	75	19	87	57	13	69	25	93	47	13	61	17	73	45	91
6	19	77	21	89	59	17	71	27	97	49	17	63	21	77	47	97
7	21	81	27	93	61	19	73	31	*003	53	21	65	23	79	51	101
8	25	83	31	95	63	23	79	37	07	57	23	67	29	83	53	03
9	27	87	33	201	67	29	81	41	09	59	27	71	33	85	57	07
10	31	89	37	07	69	31	83	43	11	61	29	77	35	87	59	09
11	33	93	39	11	75	37	85	47	13	71	31	79	37	89	61	13
12	37	95	41	13	79	41	87	51	15	73	33	83	39	91	65	15
13	39	97	43	17	81	43	91	53	17	77	35	85	41	97	67	19
14	43	99	49	19	83	47	93	59	21	83	37	89	43	801	69	21
15	45	901	51	21	87	49	97	61	23	85	39	91	47	07	71	23
16	49	07	57	23	91	53	99	67	27	89	41	93	51	09	75	25
17	51	11	61	25	93	59	703	69	29	91	45	97	53	13	81	27
18	53	13	63	29	97	61	09	71	31	95	47	99	57	15	83	31
19	57	17	67	31	99	65	11	73	33	97	51	501	59	19	87	33
20	61	19	69	35	401	67	15	77	35	201	53	03	63	21	89	37
21	63	23	73	37	03	71	17	83	39	03	57	07	65	27	93	39
22	67	25	77	39	09	73	21	89	41	07	59	09	71	31	95	43
23	69	27	79	43	11	77	23	91	45	09	63	13	75	33	99	45
24	71	29	81	47	15	79	27	95	47	13	65	15	77	37	*001	47
25	73	31	85	49	23	81	29	97	51	15	69	19	81	39	03	49
26	81	35	87	53	27	83	33	903	53	19	71	21	83	41	07	51
27	85	37	91	59	29	85	35	07	57	21	77	27	87	45	11	57
28	87	41	93	61	33	89	37	09	59	25	81	31	89	49	13	61
29	91	43	97	65	35	95	39	13	61	27	83	33	93	51	17	63
30	93	49	99	67	39	97	41	15	63	31	87	37	95	57	19	67
31	97	53	103	71	41	99	43	19	65	33	89	39	99	61	21	69
32	99	59	09	73	45	601	45	21	69	37	91	43	701	63	23	73
33	803	61	11	77	47	07	47	27	71	39	93	45	07	67	29	75
34	09	63	13	79	51	09	51	31	75	43	95	47	09	71	31	79
35	11	65	17	89	53	11	53	33	77	45	99	49	17	73	35	85
36	17	67	23	91	57	13	57	37	83	49	401	51	19	75	37	87
37	19	69	29	301	59	17	63	39	87	51	07	53	23	79	39	91
38	21	71	33	03	63	19	65	43	89	57	09	61	25	81	41	93
39	23	73	35	07	67	21	71	49	93	61	11	67	29	85	43	97
40	27	77	39	09	69	25	77	51	95	63	13	69	31	87	47	99
41	29	79	41	13	71	27	79	53	97	67	17	73	33	93	49	203
42	33	81	45	15	73	29	81	57	99	69	19	75	35	97	53	09
43	39	83	47	19	75	31	83	61	101	73	21	79	37	99	57	11
44	41	85	51	21	77	33	89	63	03	79	23	81	41	903	59	15
45	47	87	57	27	81	37	93	67	07	81	25	87	43	09	65	17
46	51	89	59	31	83	39	95	69	11	83	29	89	47	11	67	19
47	53	91	63	33	87	43	99	73	13	87	35	91	49	15	69	21
48	57	95	65	37	89	45	801	75	17	91	37	93	51	17	71	23
49	59	97	67	41	95	49	07	77	19	93	41	97	53	23	73	27
50	63	*001	69	43	99	51	11	79	23	99	43	99	59	27	77	29

VITA

I, Henry Walter Stager, was born in the year 1879 at Nutley, New Jersey, the son of John Willis and Bertha Stager. My elementary education was obtained in the public schools of Nutley and the neighboring town of Montclair. I was prepared for college at the Montclair High School and graduated from the Leland Stanford Junior University in 1902 with the degree of A.B. In 1905, I returned to Stanford for graduate study, pursuing work under Professors Allardice and G. A. Miller, and received the degree of A.M. in 1906. In September, 1907, I entered the University of California and continued graduate work under the direction of Professors Haskell, Lehmer, MacDonald, and Schilling.

I wish to express my appreciation for the kindly interest of all my instructors. My thanks are especially due to the members of my Committee, Professors Lehmer, Haskell, and Schilling; and more especially to Professor Lehmer, whose encouragement and sympathetic advice have been a constant source of help in preparing my dissertation.

UNIVERSITY OF CALIFORNIA PUBLICATIONS
IN
MATHEMATICS
Vol. 1, No. 2, pp. 27-54, 31 text-figures September 24, 1912

CONSTRUCTIVE THEORY OF THE UNICURSAL PLANE QUARTIC BY SYNTHETIC METHODS

BY
ANNIE DALE BIDDLE

INTRODUCTION

In the following discussion the unicursal quartic is regarded from two points of view. Chapter I treats of the curve in its correspondence to a conic section through a quadratic reciprocal transformation. This leads to an interesting classification of unicursal quartics and affords a convenient and ready method for determining the form of the curve. Incidentally, it brings to light a geometrical application of the well known "Group of Four." In Chapter II the curve is defined as the locus of intersection of corresponding rays of two projective pencils of the second order. This develops properties of the curve not readily obtained in the other treatment. The discussion shows that the two definitions are not independent, but that each is supplementary to the other.

CHAPTER I

THE UNICURSAL QUARTIC IN CORRESPONDENCE TO A CONIC SECTION

1. *A Quadratic Reciprocal Transformation.*—The following well known geometrical construction of the general quadratic reciprocal transformation is of fundamental importance for the study of the unicursal quartic. Consider in the plane any two conic sections α and α' and a point P. The polars p and p' of P with respect to α and α', respectively, will intersect in a point P'. The polars of P' must likewise intersect in P. To P corresponds P', and vice versa. We say that P and P' are conjugate points of the transformation. By correlating all such pairs of conjugate points an involutory transformation is established between the points of the plane.

2. *Theorem.—As a point P describes a straight line, its conjugate point P' generates a conic section as the locus of the intersection of corresponding rays of two projective pencils of the first order.*

As P describes a straight line l, the polars p and p' of P with respect to α and α' describe projective pencils of rays L and L' of the first order.

3. *Theorem.—As a point P describes a conic its conjugate point P' generates a unicursal quartic, as the locus of the intersection of corresponding rays of two projective pencils of the second order.*

As P describes a conic γ, its polars p and p' with respect to a and a', respectively, describe projective pencils of rays κ and κ' of the second order. To determine the degree of the locus of P', the conjugate point of P, that is, the locus of the point of intersection of corresponding rays of the two projective pencils κ and κ' of the second order, cut it with a straight line l. To l corresponds a conic λ. The points in which l intersects the locus of P' correspond to the points in which λ intersects γ, the locus of P. But λ and γ can intersect in at most four points. The locus of P' is, therefore, a quartic. In establishing a one-to-one correspondence between the points of the quartic and a conic section, we have shown that the quartic is unicursal.

4. *Theorem.—Any curve of degree n is transformed into a curve of degree $2n$ as the locus of the intersection of correspondings rays of two projective pencils each of order n.*

As a point P describes a curve of degree n, its polars p and p' with respect to a and a', respectively, generate projective pencils of rays each of order n. To determine the degree of the locus of the point of intersection of corresponding rays, that is, of the point P', conjugate to P, we proceed as before in § 3, cutting the locus of P' with a straight line l. To l corresponds a conic λ. The intersections of l and the locus of P' correspond to the intersections of λ and the locus of P. But of these there can be at most $2n$.

5. *Theorem.—Any point on a side of the self-polar triangle of a and a' corresponds to the opposite vertex and vice versa.*

For if P lies on a side s_1 of the triangle $S_1\ S_2\ S_3$, self-polar with respect to a and a', its polars p and p' must meet in S_1, the vertex opposite to s_1.

6. *Theorem.—A curve of degree n corresponds by the above transformation to a curve of degree $2n$ having the vertices of the self-polar triangle as n-fold points.*

For from § 5 it follows that to each intersection of the curve of degree n described by P with a side of the self-polar triangle corresponds the opposite vertex as a point of the curve described by P'.

In special cases the curve of P may pass through a vertex of the self-polar triangle. The curve of P' then degenerates, the opposite side of the triangle appearing as a part of the curve. In general, a curve of degree n which passes in all k times through the vertices of the self-polar triangle goes into a degenerate curve of degree $2n$, which contains the sides of the triangle counted k times and a curve of degree $2n-k$. In particular, a line l passing through a vertex of the triangle corresponds to a line l' passing through the same vertex.

7. *The Quartics Q_a and Q_a'.*—To the conic a (or a') itself corresponds a quartic Q_a (or Q_a') which passes through the four intersections of a and a', and also through the four points of contact on a' (or a) of the tangents common to a and a'. These curves can be traced at once for any given conics a and a'. They are, therefore, of great assistance in the construction of any curve C', for to the intersections of C (the corresponding curve) and Q_a (or Q_a') correspond the intersections of C' and a (or a'). These points of C' on a (or a') are the points of contact on a (or a') of one of the tangents drawn to a (or a') from the intersections of C and Q_a (or Q_a').

8. *Theorem.—Conjugate points P and P′ of the transformation project to any vertex of the self-polar triangle in an involution of rays.*

This follows from § 6, where it was shown that to a line l passing through a vertex of the triangle corresponds a line l' passing through the same vertex, and vice versa. From this we see that it is possible to regard P' as the point determined by the two rays which correspond to the rays in which a point P projects to two arbitrary involution centers, S_1 and S_2. The line $S_1\ S_2$ will correspond to some line s_2 at S_1 and to a line s_1 at S_2, $s_1\ s_2$ determining the point S_3. $S_1\ S_2\ S_3$ is then a singular triangle of which each vertex corresponds to the opposite side and vice versa. The quadratic involutory transformation can be completely worked out from this point of view.*

9. *Theorem.—The double or focal rays of the involutions at the vertices of the self-polar triangle are in each case the two common chords of α and α' which intersect at that vertex.*

If this is true as P moves along c, a chord common to α and α', P' must also describe c. The four intersections of α and α' are the self-corresponding, or invariant, points of the transformation. c is then a line passing through a vertex of the self-polar triangle and through two invariant points, I_1 and I_2. It must then correspond to a line passing through the same vertex and through I_1 and I_2, that is, to itself.

If α and α' intersect in four real points, the double or focal rays are all real, that is, the involutions at the three vertices of the self-polar triangle are all hyperbolic; if they intersect in four imaginary points, at two of the three real vertices the involution is elliptic, at the third it is hyperbolic; if they intersect in two real and two imaginary points, the involution at the one real vertex is hyperbolic.

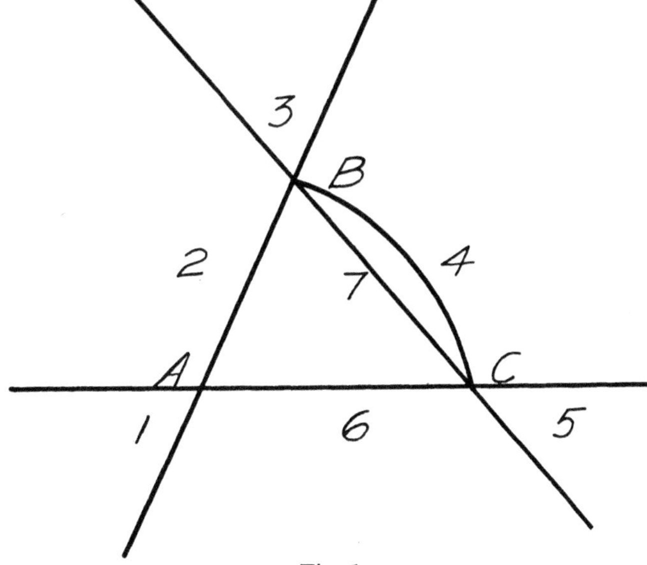

Fig. 1

* See D. N. Lehmer, "On the Combination of Involutions," *Am. Math. Mo.*, vol. 18, no. 3 (March, 1911).

10. *The Form of the Curve.*—If we regard the plane as divided into seven regions by the self-polar or singular triangle, we can, knowing the character of the involutions at the vertices of the triangle, determine in what region a point P' must lie that is conjugate to any given point P. Number the regions 1, 2, 3, 4, 5, 6, 7, and denote the vertices A, B, C, as in figure 1. If the involutions are all hyperbolic, a ray projected from a point P in 1 corresponds at A to a ray passing through the regions 1, 7, and 4; at B to a ray passing through the regions 2, 4, and 1 or 5; at C to a ray passing through the regions 6, 4, and 1 or 3. Any two of these three rays are sufficient to determine P' as a point in 4 or 1. In the same manner we can locate P' for any other point P in the plane, whether the involutions are all hyperbolic, or two are elliptic. If the involution is hyperbolic at A, B, and C, we find that:

P in 1 or 4 projects to a ray
at A passing through 4, 7 and 1
at B passing through 4, 2 and 1 or 5 } P' is, therefore, in 4 or 1
at C passing through 4, 6 and 1 or 3

P in 2 or 5
A — 2, 6, 3 or 5
B — 2, 4, 1 or 5 } P' in 2 or 5
C — 2, 7, and 5

P in 3 or 6
A — 6, 2, 5 or 3
B — 6, 7, and 3 } P' in 6 or 3
C — 6, 4, 1 or 3

P in 7
A — 7, 1, 4
B — 7, 2, 5 } P' in 7
C — 7, 3, 6

In the same manner we locate P' for P in any region when the involution is hyperbolic at C while elliptic at A and B, or hyperbolic at A while elliptic at B and C, or hyperbolic at B while elliptic at C and A. The complete results may be tabulated as follows:

Involution is hyperbolic at			$A\ B\ C$	C	A	B
P lies in	1 or 4	P' lies in	4 or 1	6 or 3	7	2 or 5
	2 or 7		2 or 5	7	6 or 3	4 or 1
	3 or 6		6 or 3	4 or 1	2 or 5	7
	7		7	2 or 5	4 or 1	6 or 3

This investigation shows that the regions 1 and 4 are interchangeable; likewise, 2 and 5, 3 and 6. Indeed, we see that one can pass from 1 to 4 without crossing a side of the triangle; similarly, from 2 to 5, and from 3 to 6. We may, therefore, regard 1 and 4 as one region, R_1; 2 and 5 as one region, R_2; and 3 and 6 as one region, R_3. 7 stands by itself as R_4. The above table may then be written:

Involution is hyperbolic at			$A\ B\ C$	C	A	B
P lies in	R_1	P' lies in	R_1	R_3	R_4	R_2
	R_2		R_2	R_4	R_3	R_1
	R_3		R_3	R_1	R_2	R_4
	R_4		R_4	R_2	R_1	R_3

The transformation of P indicated by the first column leaves R_1, R_2, R_3, R_4 all unaltered; that is, it is effected by the substitution (1) (2) (3) (4). The transformation of the second column is effected by the substitution (1, 3) (2, 4); that of the third column by the substitution (1, 4) (2, 3); and that of the fourth by the substitution (1, 2) (3, 4). These four substitutions

(1) (2) (3) (4)
(1, 3) (2, 4)
(1, 4) (2, 3)
(1, 2) (3, 4)

constitute the well known "Group of Four."

Any given curve C intersects the sides of the triangle in a definite order, thus establishing the order in which the corresponding curve C' must pass through the vertices. This enables us to determine for a given branch of the curve C how the corresponding branch of the curve C' must pass through the region in which it lies. We can, therefore, for a given curve C determine at once the form of the corresponding curve C'. Consider, for example, the quartic that will correspond to the conic C in figure 2, where the involution is hyperbolic at every vertex. Reading from right to left, C intersects the sides of the triangle in the order $a\ b\ c\ a\ b\ c$. Denote the portions of C which lie in 1, 2, 3, 4, 5, 6, by c_1, c_2, c_3, c_4, c_5, and c_6, respectively, and the corresponding portions of C', c'_1, c'_2, c'_3, etc.,

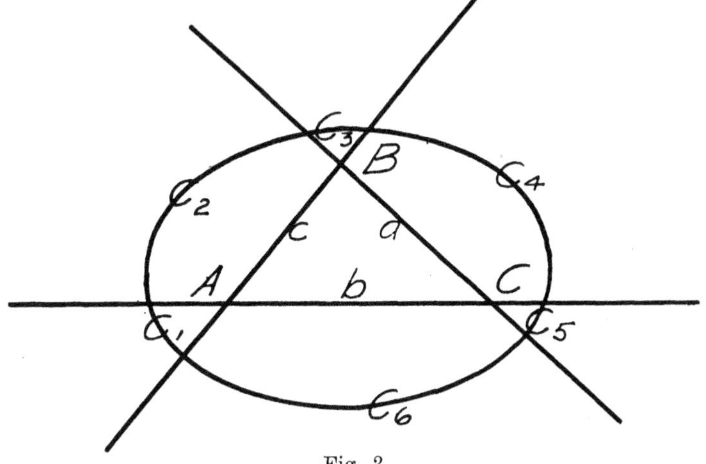

Fig. 2

respectively. c_1 lies in 1 and between b and c. c'_1, then, must lie in 4 or 1 and between B and C. Moving continuously along C' we may pass from B to C directly so that c'_1 lies entirely in 4, as in figure 1, or indirectly by way of infinity so that c'_1 lies partly in 4 and partly in 1, as in figure 3. In the same way we determine c'_2, c'_3, c'_4, c'_5, and c'_6, and, excluding the possibility of infinite branches of the kind shown in figure 3, we find that C' must for the conic C in figure 2 have the form given in figure 6.

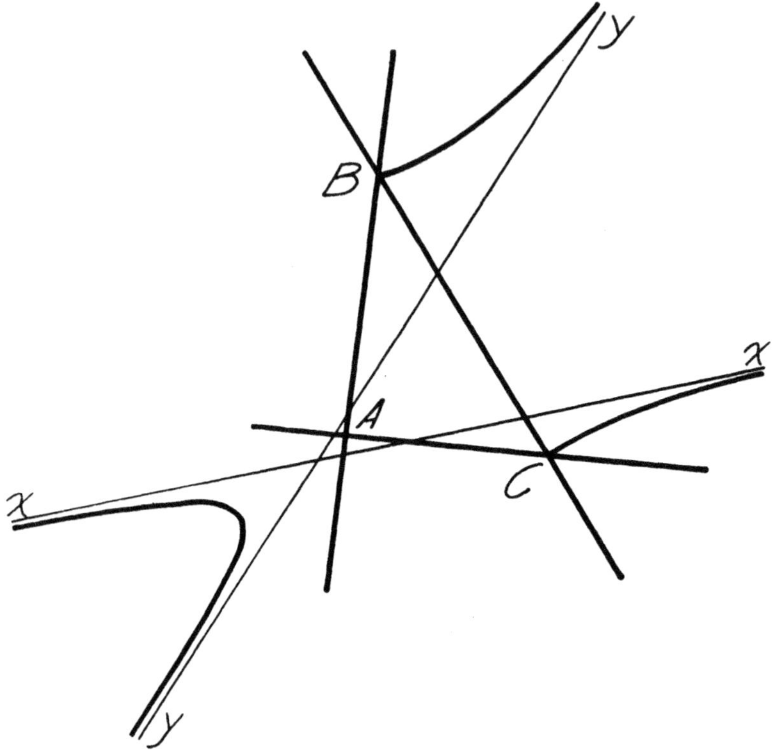

Fig. 3

11. *Classification of Unicursal Trinodal Quartics.*—If the given curve C is a conic or a curve of degree greater than the second, the position of the curve C with reference to the triangle affords a basis for classification of the curves C', since the order in which the curve C intersects the sides of the triangle is the order in which C' must pass through the vertices. The unicursal quartic is of particular interest here.

A conic section according to its relative position may intersect the sides of the triangle in five essentially different orders. These are:

(1) $a\ a\ b\ b\ c\ c$

(2) $a\ b\ c\ a\ b\ c$

(3) $a\ b\ a\ b\ c\ c$

(4) $a\ a\ c\ b\ b\ c$

(5) $a\ c\ a\ b\ c\ b$

We may describe the position of the conic by noting the successive regions through which it passes. Thus the position of the conic in figure 2 is given by noting that it passes successively through the regions 1, 2, 3, 4, 5, 6.

The same order may be given by more than one position of the conic. Order (1) is given by four positions, namely,

4 7 6 7 2 7,
3 2 1 2 7 2,
7 4 5 4 3 4,
5 6 7 6 1 6;

order (2) by one position,

6 5 4 3 2 1;

order (3) by three positions,

4 7 6 5 4 3,
6 5 4 7 6 1,
7 4 5 6 7 2;

order (4) by four,

3 2 3 4 5 4,
6 5 6 1 2 1,
2 3 2 7 6 7,
7 4 7 2 1 2;

and order (5) by one position,

2 3 4 7 6 1.

Any particular order represents more than one class of unicursal quartics, since each of the four possible forms of the involution may give rise to a different kind of curve for a given position of the conic. An analysis of all cases shows that a change of the conic from one position to another representing the same order is equivalent to a possible change in the character of the involution. It follows that all the classes represented by one order may be obtained from any one position of the conic which produces that order by changing the character of the involution.

It is not difficult by the method indicated above to ascertain for the conic in any given position and under any form of the involution the number of classes represented by each order and the form of the curve (excluding the possibility of infinite branches such as those shown in figure 3). The complete results may be conveniently arranged in the following table:

Order	No. of Classes	Position	Involution Hyperbolic at	Form of Quartic
a a b b c c	2	476727	*A*, *B* and *C*	Fig. 4
			C, or *A*, or *B*	Fig. 5
		321272	*C*	Fig. 4
			A, or *B*, or *A*, *B* and *C*	Fig. 5
		745434	*A*	Fig. 4
			B, or *C*, or *A*, *B* and *C*	Fig. 5
		567616	*B*	Fig. 4
			C, or *A*, or *A*, *B* and *C*	Fig. 5
a b c a b c	2	654321	*A*, *B* and *C*	Fig. 6
			A, or *B*, or *C*	Fig. 7
a b a b c c	3	745672	*A*, *B* and *C*	Fig. 8
			C	Fig. 9
			A, or *B*	Fig. 10
		476543	*A*	Fig. 8
			B	Fig. 9
			C, or *A*, *B* and *C*	Fig. 10
		654761	*B*	Fig. 8
			A	Fig. 9
			C, or *A*, *B* and *C*	Fig. 10
a a c b b c	3	323454	*A*, *B* and *C*	Fig. 11
			C	Fig. 12
			A, or *B*	Fig. 13
		656121	*A*, *B* and *C*	Fig. 11
			C	Fig. 12
			A, or *B*	Fig. 13
		232767	*A*	Fig. 11
			B	Fig. 12
			C, or *A*, *B* and *C*	Fig. 13
		747212	*B*	Fig. 11
			A	Fig. 12
			C. or *A*, *B* and *C*	Fig. 13
a c a b c b	2	234761	*A*, *B* and *C*, or *C*	Fig. 14
			A, or *B*	Fig. 15

FIG. 4
aabbcc
FIG 5
aabbcc
FIG 6
abcabc
FIG. 9
ababcc
FIG 8
ababcc
FIG. 7
abcabc
FIG 10
ababcc
FIG 11
aacbbc
FIG 12
aacbbc
FIG 15
acabcb
FIG. 14
acabcb
FIG 13
aacbbc

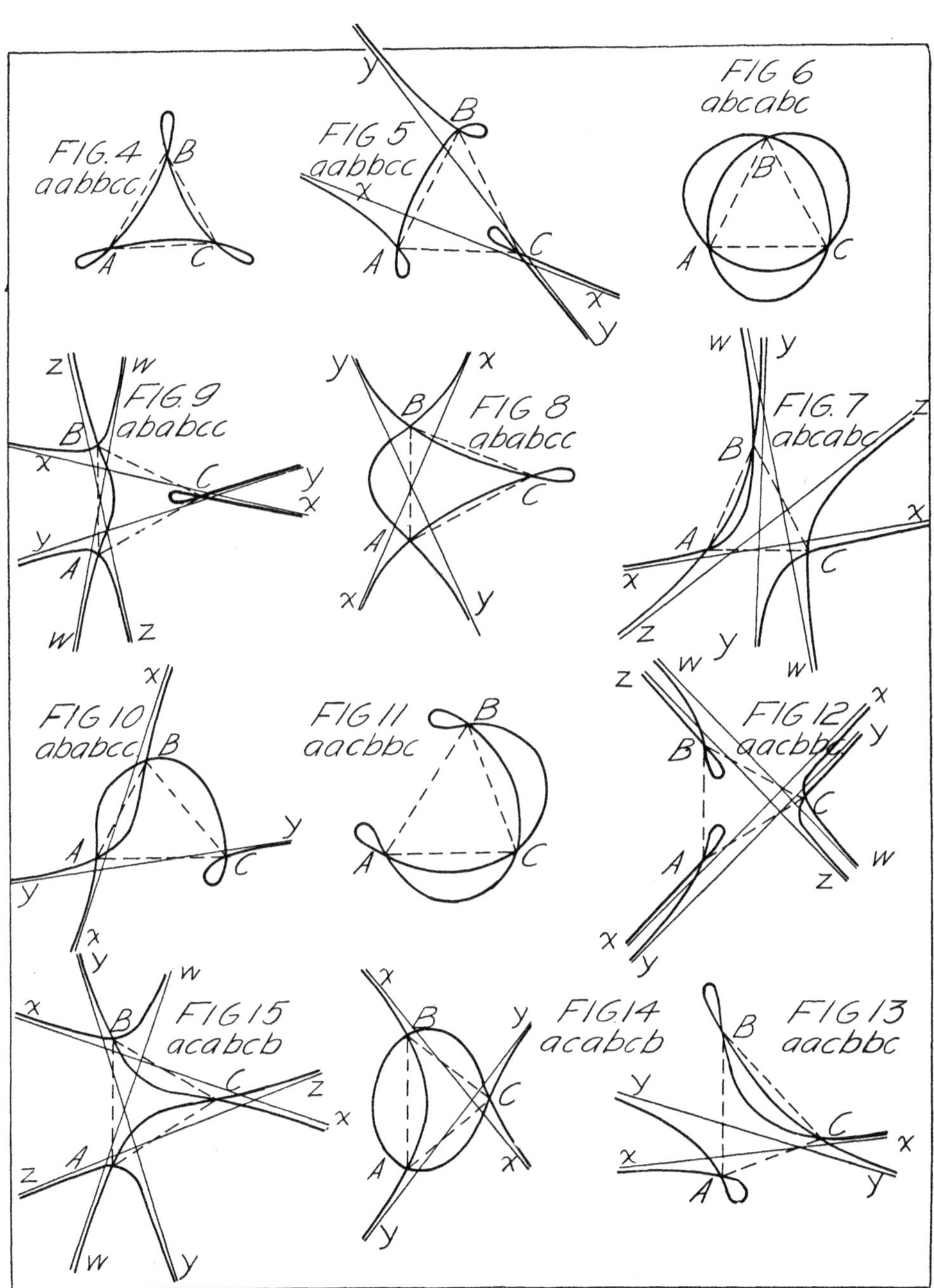

Figs. 4–15

In all, twelve classes of unicursal trinodal quartics are obtained in this way. Further distinction may be made, according as the intersections of the conic with the sides of the triangle are real and distinct, real and coincident, or conjugate-imaginary.

Special forms of the self-polar triangle give rise, of course, to special kinds of curves.

CHAPTER II

THE UNICURSAL QUARTIC AS THE LOCUS OF THE INTERSECTION OF CORRESPONDING RAYS OF TWO PROJECTIVE PENCILS OF THE SECOND ORDER

1. The curve is defined as the locus of the points of intersection of corresponding rays of two projective pencils of the second order.

2. *Theorem.—The locus described is a unicursal curve having, in general, three nodes, and intersecting any line in the plane in at most four points.*

See Chapter I, §§ 3 and 6.

3. *Notations.*—One pencil of the second order (and also the conic enveloped by it) will be denoted by κ; the other by κ'. The rays of each pencil will be denoted by α, β, γ, ... and α', β', γ', . . . respectively. The tangents to κ and κ' from the points of the quartic and other than α, β, γ, ... and α', β', γ', ... will be denoted by a, b, c, ... and a', b', c', ... respectively. The quartic itself will be denoted by Q.

4. *Theorem.—No point of Q lies within κ or κ'.*

This is obvious from the definition of the curve given in § 1.*

5. *Theorem.—The quartic Q touches each conic κ and κ' in at most four points.*

Take a point S on κ for the center of a pencil of the first order perspective to κ. This generates with κ' a cubic with a double point or node at S.† The cubic can intersect κ in at most four other points besides S. These are easily seen to be points of the locus Q.

6. *Problem.*—Given κ and κ' with three pairs of corresponding rays, to construct Q. (See figure 16.)

Let the three pairs of rays be α, α'; β, β'; and γ, γ'. The intersection of α and α' will be a point A on Q. Likewise β and β' determine B on Q and γ and γ', C on Q. Draw through A to κ the other tangent a; also to κ' the other tangent a'. Take a perspective to κ and a' perspective to κ'. a and a' are two point-rows in perspective position since they have a self-corresponding point, A. Their center of perspectivity may be found by joining any two pairs of corresponding points on a and a', such as (a, β) with (a', β') and (a, γ) with (a', γ'). Denote this center of perspectivity by Σ_A. Then to find the point X of Q determined by any ray ξ of κ, we first join the point (ξ, a) with Σ_A. This line will intersect a' in the point (ξ, a'). The intersection of ξ and ξ' is the desired point X.

---

* Unless otherwise stated, the sections referred to are those of Chapter II.

† See Drasch, "Beitrag zur synthetischen Theorie der ebenen Curven dritter Ordnung mit Doppelpunkt," *Wiener Berichte*, vol. 85 (1882), p. 534. Also see D. N. Lehmer, "Constructive Theory of the Unicursal Cubic by Synthetic Methods," *Trans. Am. Math. Soc.*, vol. 3, p. 372.

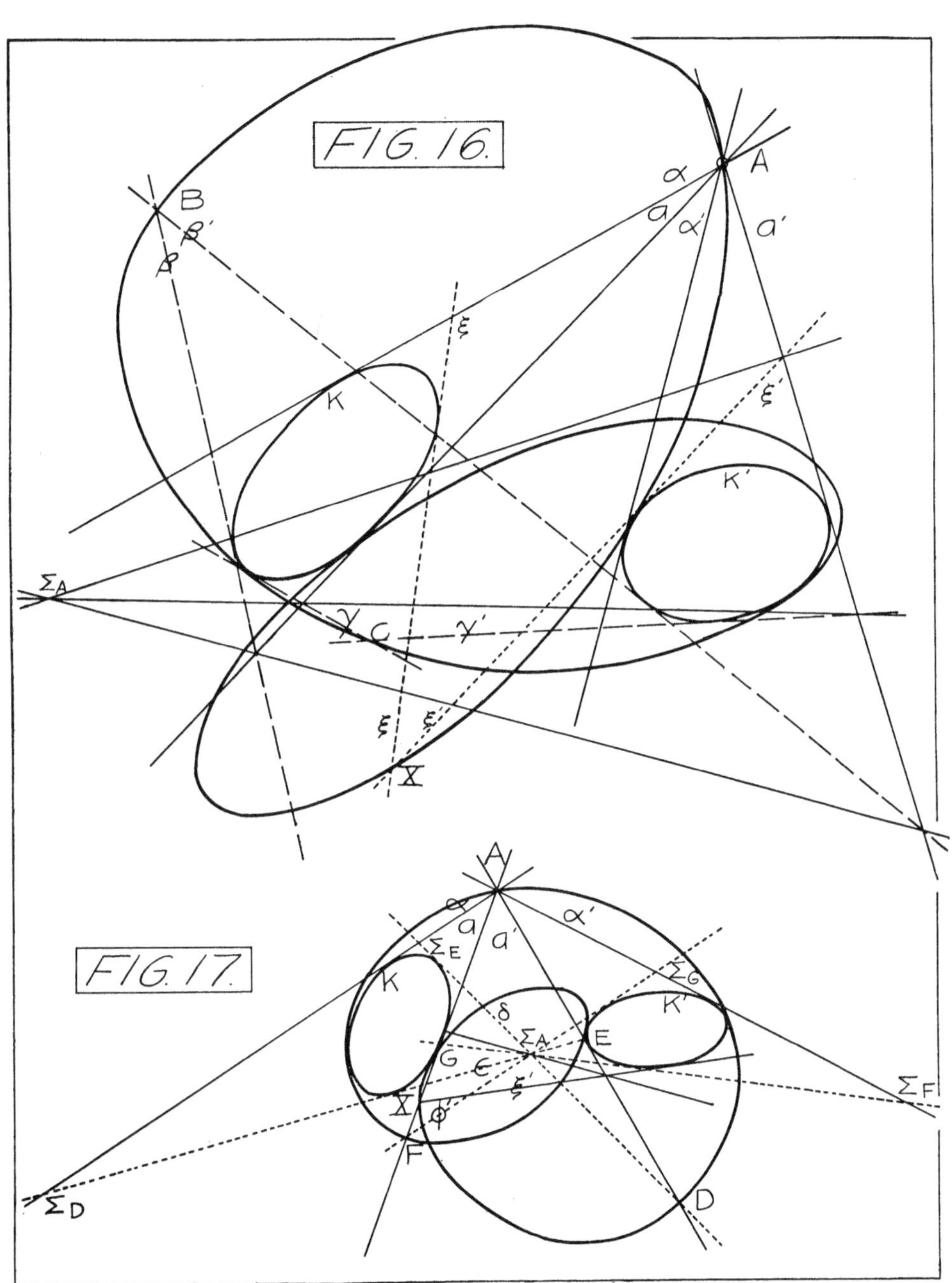

Figs. 16 and 17

7. *Theorem.—Every point P of the quartic Q has its corresponding point Σ_P.* The locus of all such points will be denoted by Σ.

8. *Problem.*—Given κ and κ' with one pair of corresponding rays and the corresponding point of Σ, to construct Q.

Let the given pair of corresponding rays be α, α'. The intersection of α and α' will be the point A of Q. As in § 6, draw a and a', taking them perspective to κ and κ', respectively. Using Σ_A, which is given, as the center of perspectivity, the ray of κ' corresponding to any ray of κ, or vice versa, may at once be found, as before.

9. *Theorem.—The tangents from Σ_A to κ meet a' in points of Q. Likewise, the tangents from Σ_A to κ' meet a in points of Q.* (See figure 17.)

This is seen to be true by drawing the rays of κ' which correspond to the tangents from Σ_A to κ; similarly, those of κ which correspond to the tangents from Σ_A to κ'.

10. *Problem.*—To find the fourth point of Q upon a or a'. (It is the point X in figure 17.)

To do this we consider a as a ray ξ of κ. ξ meets a in the point of contact of a and κ. Join this point with Σ_A. The line so obtained intersects a' in the point (ξ', a'). The intersection of ξ' with ξ (or a) is the desired point.

Instead of the point-rows a and a' and their center of perspectivity, Σ_A, we may make use of any other such set, as b, b' and Σ_B. In this case a, considered as a ray ξ of κ, appears as an ordinary ray and ξ' is found in the usual manner indicated in § 6.

The fourth point of Q upon a' is found in precisely the same manner as the fourth point of Q upon a.

11. We are now in a position to state the following:

Theorem.—The tangents from Σ_A to κ meet a' in points of Q and α in points of Σ. Likewise, the tangents from Σ_A to κ' meet a in points of Q and α' in points of Σ. The point of Σ on α (or α') determined by one tangent from Σ_A to κ (or κ') corresponds to the point of Q on a' (or a) in which a' (or a) is met by the other tangent from Σ_A to κ (or κ'). (See figure 17.)

The first part of the theorem was proved in § 9.

Denote one tangent to κ from Σ_A by δ and the other one by ϵ. Then it follows from § 9 that (δ, a') is D and (ϵ, a') is E. d' and e' coincide with a'. In finding Σ_D, draw the line $(\alpha, d - a'\ d')$. This is α itself. Therefore, Σ_D lies on α. Similarly, we show that Σ_E lies on α.

Instead of Σ_A start with Σ_D, drawing the tangents from Σ_D to κ. One of these is α. The other must meet d', which is a', in a point of Q. Obviously, it cannot meet d' in A or D. Nor can it meet d' in the point X found in § 10, since for X, a' is ξ' and not x'. It must then meet d' in E and, therefore, the second tangent from Σ_D to κ is ϵ. But ϵ passes through Σ_A. That is, Σ_A and Σ_D lie on ϵ. In like manner, starting with Σ_E, it may be shown that Σ_E and Σ_A lie on δ. Thus one tangent δ from Σ_A to κ meets a' in a point D of Q and α in the point Σ_E, while the other tangent ϵ from Σ_A to κ meets a' in the point E of Q and α in Σ_D.

In precisely the same manner the theorem is proved for the tangents from Σ_A to κ' instead of κ.

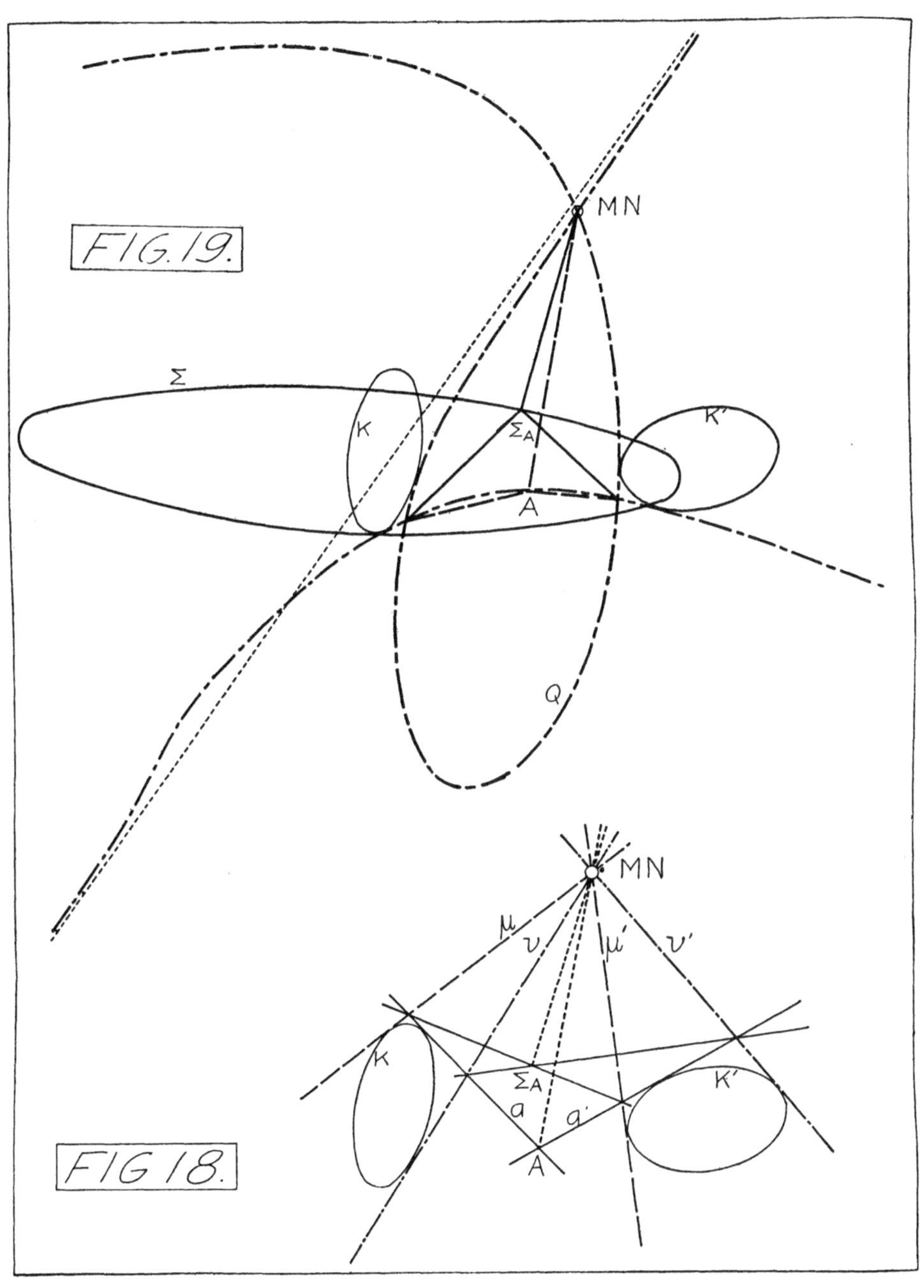

Figs. 18 and 19

12. *Problem.*—Given κ and κ' with one pair of corresponding rays, and the corresponding point of Σ, to find the point of Q corresponding to any particular point of Σ.

Let the given pair of rays be β β', and the corresponding point of Σ, Σ_B. Since β and β' are known, b and b' are known. Let the particular point of Σ be Σ_A. To find A, draw the tangents from Σ_A to κ and κ', calling them δ, ϵ, θ', and ι'. Using Σ_B as a center of perspectivity of the point rows b and b', construct δ', ϵ', θ, and ι, thus obtaining D, E, H, and I. D and E determine a' and H and I determine a. (a, a') is A.

13. *Theorem.*—*The locus of the points Σ, the centers of perspectivity corresponding to the points of Q, is a conic section from which Q is generated by means of a quadratic reciprocal transformation.* (See figures 18 and 19.)

A quartic generated from a conic by means of the quadratic reciprocal transformation discussed in Chapter I had its nodes at the vertices of the singular triangle, the involution centers.

Consider the four tangents from a node of Q, two to κ and two to κ'. (See figure 18.) They are two pairs of corresponding rays of the pencils κ and κ'. Call them μ, μ', and ν, ν'. One pair determines the node as M, considered as lying on one branch of the curve, the other determines it as N, considered as lying on the other branch. Using μ, μ', and ν, ν', find Σ_A for A. This is done by marking the intersection of the line $(a, \mu - a', \mu')$ with the line $(a, \nu - a', \nu')$. The points

$$A, \Sigma_A; (a, \mu), (a', \nu'); (a, \nu), (a', \mu')\cdot$$

are three pairs of opposite vertices of a complete quadrilateral and therefore project to any point of the plane in an involution of rays. They project to the node MN in three pairs of rays, two of which are μ, ν', and μ', ν. But these lines are fixed and do not depend upon A or Σ_A.

Now consider the involutions at any two of the three nodes of Q. The lines from any given point Σ_A of Σ to the nodes correspond in involution to rays which intersect in the point A of Q corresponding to Σ_A. As Σ_A moves on Σ, A describes the quartic Q. Thus Σ is exhibited as that curve from which the quartic is generated by the quadratic reciprocal transformation and is, therefore, a conic section. (Chapter I, §§ 3 and 8.)

14. *Theorem.*—*The points of intersection of Q and Σ lie on the tangents common to κ and κ'.* (See figure 20.)

Let a common tangent of κ and κ' be a ray β in the pencil κ. It corresponds to a ray β' of κ', in general distinct from β. b' then coincides with β. The tangents from Σ_B to κ meet β in points of Σ and b' in points of Q. But β and b' coincide.

15. *Theorem.*—*κ (or κ') and Σ are two conics such that a triangle circumscribed about κ (or κ') is inscribed in Σ.*

This follows from § 11, the tangents there denoted a, δ, and ϵ forming such a triangle, of which the vertices are Σ_E, Σ_A, and Σ_D. (See figures 17 and 21, triangles Σ_A Σ_D Σ_E and Σ_A Σ_F Σ_G.)

For the invariant relation connecting two conics related as in this theorem see Salmon, *Conic Sections*, § 376.

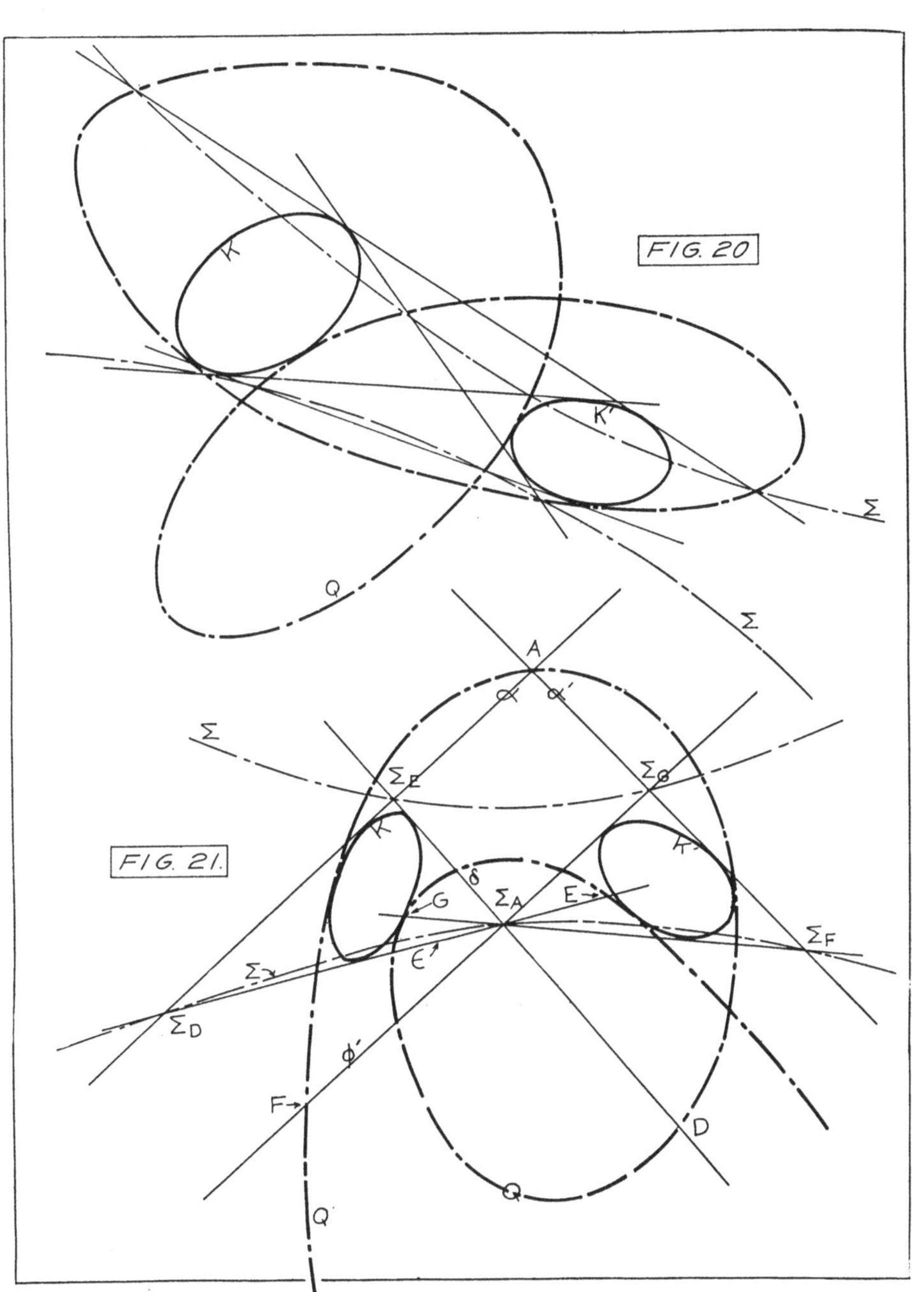

Figs. 20 and 21

16. *Theorem.—When Σ_A lies on κ, a' is tangent to Q and α is tangent to Σ; similarly, when Σ_A lies on κ', a is tangent to Q and α' is tangent to Σ.* (See figure 22.)

In this case the two tangents from Σ_A to κ, δ and ϵ, coincide. Consequently, a' meets Q in two coincident points, D and E, and α meets Σ in two coincident points Σ_E and Σ_D. Similarly, for Σ_A on κ'.

Differently stated, the latter part of this theorem tells us:

The tangents common to κ (or κ') and Σ are obtained by drawing the tangents to κ (or κ') from those points of Σ where the tangents to κ (or κ') at the intersections of κ (or κ') and Σ again intersect Σ.

Thus κ (or κ') and Σ have as many common tangents as intersections.

17. *Theorem.—If κ (or κ') and Σ do not intersect κ (or κ') lies wholly within Σ.* (See figures 23 and 24.)

If κ (or κ') and Σ do not intersect, either Σ lies wholly within κ (or κ') or κ (or κ') lies wholly within Σ. Otherwise common tangents could be drawn. But Σ cannot lie wholly within κ (or κ') since a triangle circumscribed about κ (or κ') is inscribed in Σ. Therefore, if κ (or κ') and Σ do not intersect, κ (or κ') lies wholly within Σ.

18.—*Theorem.—a and α', a' and α determine an involution of rays at A in which $A\,\Sigma_A$ corresponds to the tangent to Q at A.* (See figures 24 and 25.)

Let D and E be the points of Q on a determined by the tangents from Σ_A to κ. κ' and F and G the points of Q on a' determined by the tangents from Σ_A to κ. Consider, then, the three pairs of points A and Σ_A, E and Σ_E, F and Σ_F. They are obviously not the three pairs of opposite vertices of a complete quadrilateral, since A, Σ_A, E, and Σ_E determine a quadrilateral of which D and Σ_D are the third pair of opposite vertices; and A, Σ_A, F and Σ_F determine a quadrilateral of which G and Σ_G are the third pair of opposite vertices. The locus of points from which the three given pairs of points are seen in involution is a general plane cubic C, passing through all the six given points and through D, Σ_D, G and Σ_G.* From § 13 we know that the nodes of Q also satisfy the condition for points of C.

The cubic C is tangent to Q at A and to Σ at Σ_A.

In the quadratic reciprocal transformation a point A of Q goes into Σ_A of Σ. The tangent line to Q at A goes into a conic through the three involution centers, the nodes of Q (see Chapter I, § 6), and tangent to Σ at Σ_A. A point of the cubic C passing through the three involution centers goes into the conjugate point of C.† A and Σ_A are conjugate points of the cubic C. The tangent line to C at A corresponds to a conic through the three involution centers and tangent to C at Σ_A. If the tangent lines to C and to Q at A coincide, the corresponding conics must coincide; that is, if C is tangent to Q at A it must be tangent to Σ at Σ_A.

The cubic C is tangent to Σ at Σ_A.

Construct the tangent to Σ at Σ_A by means of Pascal's theorem, using the points Σ_A, Σ_D, Σ_E, Σ_F, Σ_G, and numbering them 16, 2, 3, 4, 5, respectively. (See figure 25.)

* See Schroeter, *Ebene Curven Dritter Ordnung*, Leipzig, 1888; § 1, 6; § 2, 1–4.

† See D. N. Lehmer, "On the Combination of Involutions," *Am. Math. Mo.*, vol. 18, no. 3 (March, 1911).

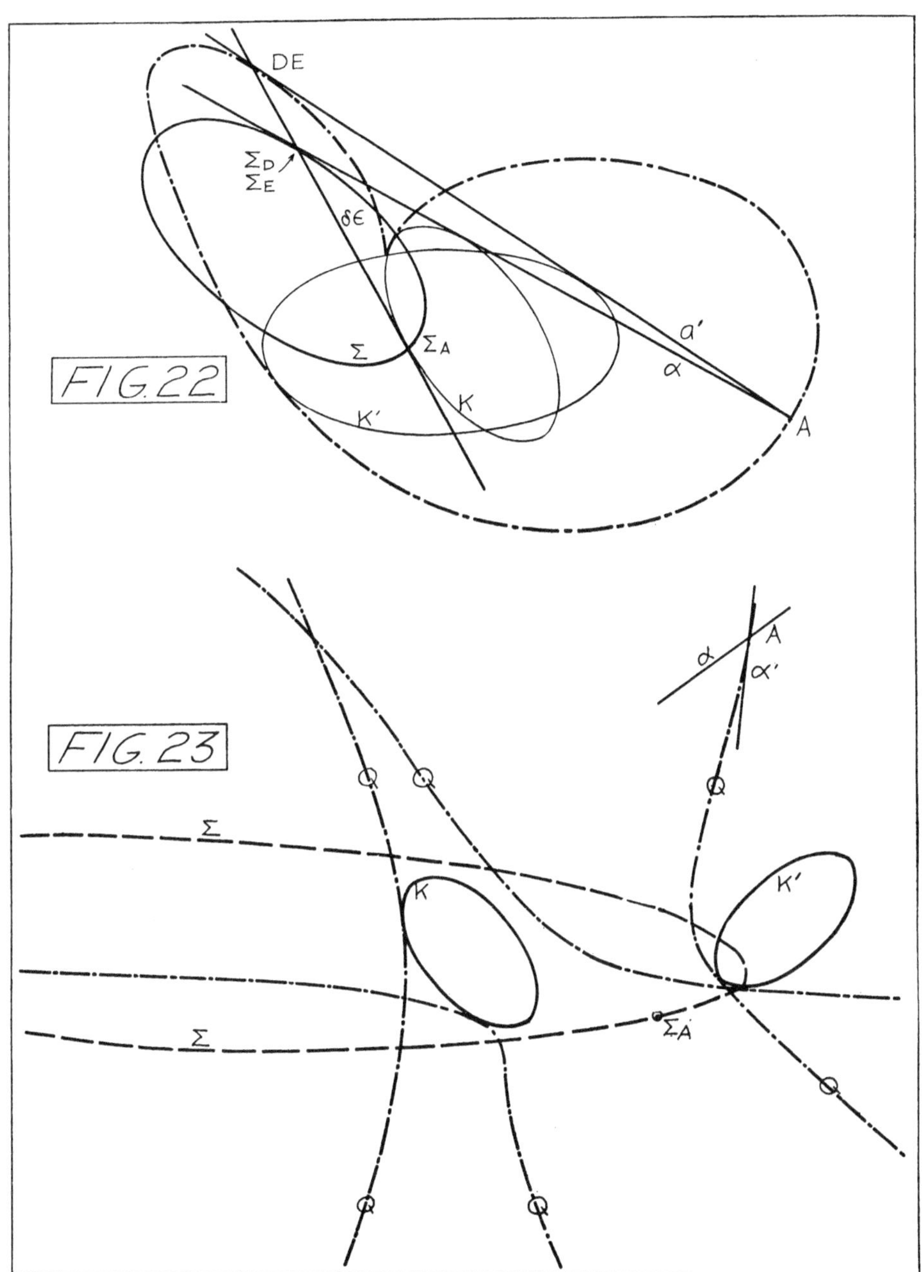

Figs. 22 and 23

The tangent line to the cubic C at Σ_A is that ray which corresponds to the ray $\Sigma_A A$ in the involution determined at Σ_A by two pairs of conjugate rays, as:

$$(\Sigma_A E) \text{ and } (\Sigma_A \Sigma_E);\ (\Sigma_A F) \text{ and } (\Sigma_A \Sigma_F).^*$$

To obtain this tangent line to C at Σ_A, construct a complete quadrilateral, two of whose pairs of opposite vertices will project to Σ_A in $(\Sigma_A E)$ and $(\Sigma_A \Sigma_E)$; $(\Sigma_A F)$ and $(\Sigma_A \Sigma_F)$; and of the third pair, one vertex in $(\Sigma_A A)$. (See figure 25.) The other vertex must lie on the ray $(\Sigma_A \Sigma_A)$, the tangent to C at Σ_A. Such a complete quadrilateral is determined by the four lines: a, a', $(\Sigma_E \Sigma_F)$, and the Pascal line found in constructing the tangent to Σ at Σ_A. But the vertex which projects to Σ_A giving the tangent to C at Σ_A is the intersection of the Pascal line and the line (3, 4) in the construction of the tangent to Σ at Σ_A; that is to say, the point which with Σ_A gave (6, 1), the tangent to Σ at Σ_A. The tangent to Σ at Σ_A and the tangent to C at Σ_A are, therefore, one and the same line; that is, the cubic C is tangent to Σ at Σ_A. It is, therefore, tangent to Q at A.

Hence, to construct the tangent to Q at A, we need only to construct the tangent to C at A. This is done, as in the case of the tangent to C at Σ_A, by constructing the ray which corresponds to the ray $A\,\Sigma_A$ in the involution determined at A by two pairs of conjugate rays, as:

$$(A\,E) = a, \text{ and } (A\,\Sigma_E) = a';$$
$$(A\,F) = a' \text{ and } (A\,\Sigma_F) = a.$$

This may be done in the following manner. (See figures 24 and 25.) On the line $A\,\Sigma_A$ select any point other than A. Through it draw two lines l and l'. Mark the intersections of l with a and a' and those of l' with a and a'. The intersection of the lines $(la - l'a)$ and $(la' - l'a')$ must lie on the tangent to Q at A.

19. *The Cubic C.*—For every point of Q there is a cubic C such as that described in § 18. It may be denoted by C_A, C_B, etc., the subscript indicating the point of Q at which the cubic is tangent.

We observe that we have at once all the intersections of C_A, both with the quartic Q and the conic Σ. The three nodes of Q account for six intersections of Q and C_A, the tangency at A for two more, and the remaining four are at the points denoted by D, E, F, and G in § 18. C_A is tangent to Σ at Σ_A and intersects it in the four remaining points Σ_D, Σ_E, Σ_F, and Σ_G.

Incidentally, the existence of the cubics C gives a method of determining the nodes of Q since through them all the cubics C must pass.

20. *Theorem.—The class of Q is 6.*

A tangent line to Σ from a node of Q, that is, from an involution center S_1, corresponds to a tangent to Q from S_1. Since from any point in the plane only two tangents may be drawn to a conic section, only two tangents may be drawn from S_1 to Σ and, hence, only two from S_1 to Q. There are two tangents to Q at S_1, each of which counts as two tangents to Q from a point of the plane not on Q. In all, then, at most four tangents may be drawn to Q from a node S_1, or at most six from an ordinary point of the plane.

Consistent with this is the number of tangents common to κ (or κ') and Q.

* See Schroeter, *op. cit.*; § 2, 6.

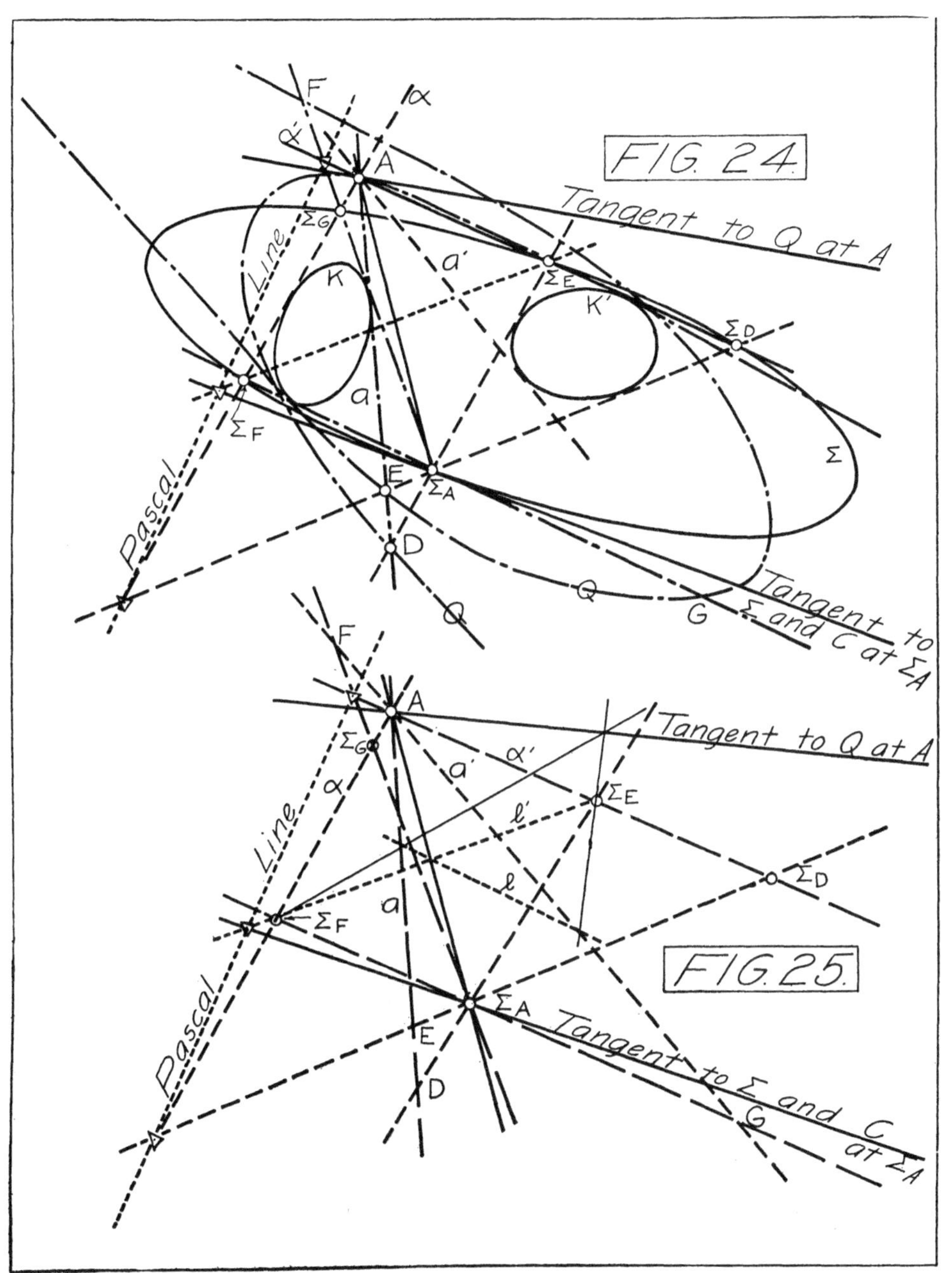

Figs. 24 and 25

From § 16 we know that every intersection of κ' (or κ) and Σ yields a tangent common to κ (or κ') and Q. Since Σ and κ' (or κ) can intersect in at most four points, this gives rise to only four tangents common to κ (or κ') and Q. The remaining eight lie at the four points where Q may touch κ (or κ'). (See § 5.)

21. *Theorem.—The six tangents drawn from the three nodes to the quartic are tangent to one and the same conic.* (See figure 26, conic I.)

The lines in question correspond by involution to the six tangents from the three nodes to the Σ conic. Denote the nodes by S_1, S_2, and S_3; the tangents from them to Σ by 1, 2; 3, 4; and 5, 6, respectively; and the lines corresponding to 1, 2, 3, 4, 5, 6, by 1′, 2′, 3′, 4′, 5′, 6′, respectively. Since the lines 1, 2, 3, 4, 5, 6 are tangent to a conic, the lines

$$(1, 2 — 4, 5),\ (2, 3 — 5, 6),\ (3, 4 — 6, 1)$$

are concurrent. (Brianchon.) But these lines are

$$(S_1 — 4, 5),\ (S_2 — 5, 6),\ (S_3 — 6, 1).$$

To them correspond

$$(S_1 — 4', 5'),\ (S_2 — 5', 6'),\ (S_3 — 6', 1'),$$

respectively. Since the first three are concurrent the second three must be. But the second three are

$$(1', 2' — 4', 5'),\ (2', 3' — 5', 6'),\ (3', 4' — 6', 1'),$$

and therefore 1′, 2′, 3′, 4′, 5′, 6′ are tangent to one and the same conic section. (Brianchon.)

22. *Theorem.—The tangents to Q at a node correspond in involution to the lines joining the node with those points of Σ which lie on the line joining the other two nodes.*

For from Chapter I, § 5, it follows that Σ_M and Σ_N, corresponding to the node $M\ N$, lie on the side of the singular triangle opposite to $M\ N$, that is, on the line joining the other two nodes. $M\ \Sigma_M$ corresponds to $M\ M$, the tangent to Q at M, and $N\ \Sigma_N$ corresponds to $N\ N$, the tangent to Q at N.

23. *Theorem.—The six tangent lines to Q at the three nodes are tangent to a conic section.* (See figure 26, conic II.)

As before in § 21, denote the nodes of Q by S_1, S_2, and S_3. From Brianchon's theorem we know that if the six lines in question circumscribe a conic section the three lines joining the opposite vertices of the hexagon they from all pass through one and the same point. Number the two tangents to Q at S_1, 1 and 2; those at S_2, 3 and 4; and those at S_3, 5 and 6. The lines

$$(1, 2 — 4, 5),\ (2, 3 — 5, 6),\ (3, 4 — 6,1)$$

join opposite vertices of the hexagon.

In the involutions at S_1, S_2, and S_3 there correspond to 1, 2, 3, 4, 5, 6, lines which we shall call 1′, 2′, 3′, 4′, 5′, and 6′, respectively.

If the lines 1, 2, 3, 4, 5, 6 are tangent to a conic section, the lines 1′, 2′, 3′, 4′, 5′, 6′ are also tangent to a conic section, and vice versa. For the point (4, 5) goes into the point (4′, 5′) and the line

FIG 26
S2
S3
S1
I
II
Σ
I
FIG 28
S2
N
S1
L
M
S3
S1
C3
C'3
1'
2'
5
1
3
C'2
C1
C2
C'1
FIG. 27
2
3'
6'
6
4'
5'
S2
S3
4

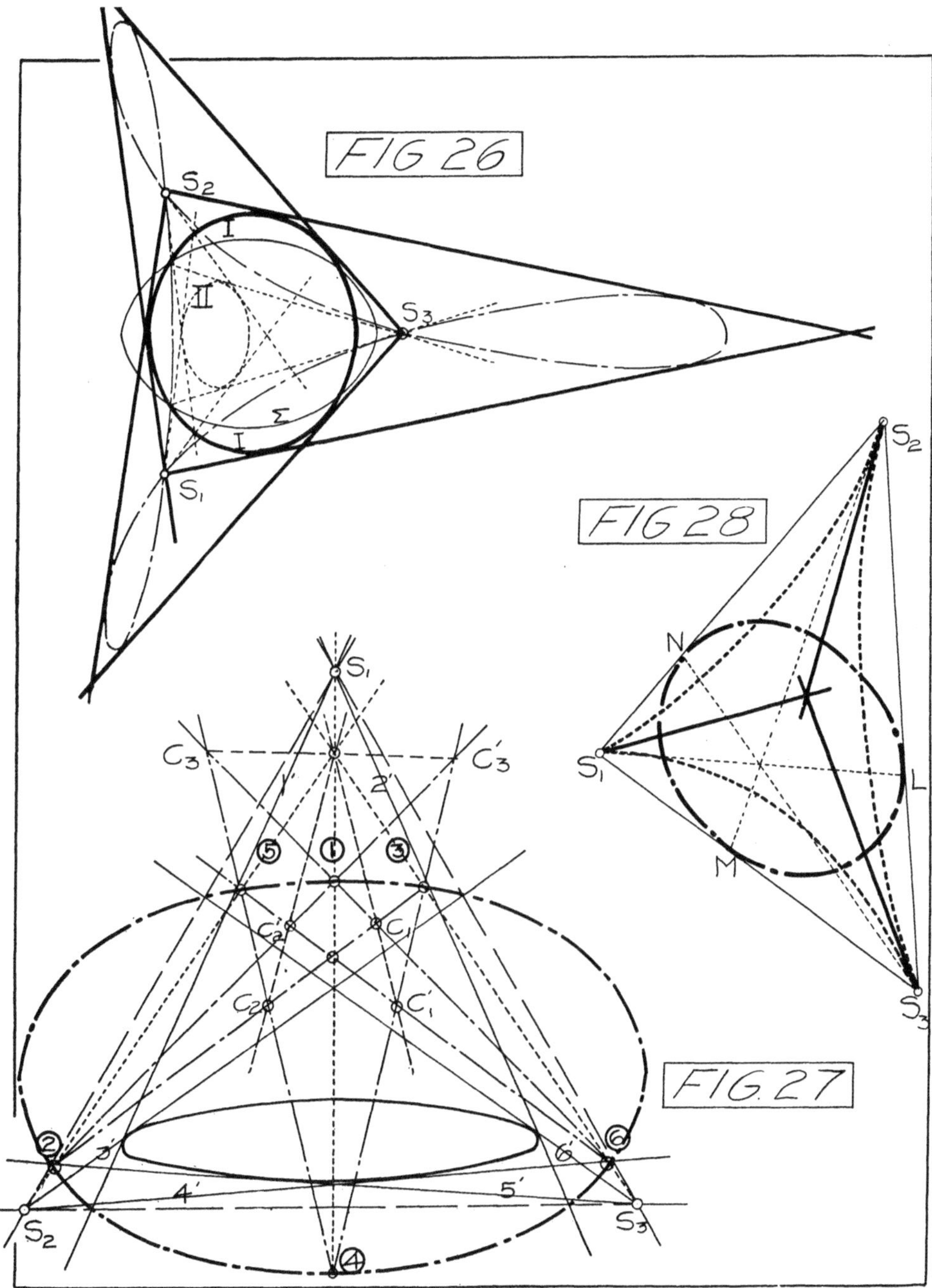

Figs. 26–28

$$(S_1 — 4, 5) \quad \text{corresponds to} \quad (S_1 — 4', 5').$$

Similarly,

$$(S_2 — 6, 1) \quad \text{corresponds to} \quad (S_2 — 6', 1')$$

and

$$(S_3 — 2, 3) \quad \text{to} \quad (S_3 — 2', 3').$$

The intersection of any two of the three,

$$(S_1 — 4, 5),\ (S_2 — 6, 1),\ (S_3 — 2, 3),$$
$$\text{as } (S_1 — 4, 5) \text{ and } (S_2 — 6, 1),$$

corresponds to the intersection of the corresponding two, as

$$(S_1 — 4', 5') \text{ and } (S_2 — 6', 1').$$

It follows that if the three lines

$$(S_1 — 4', 5'),\ (S_2 — 6', 1'),\ (S_3 — 2', 3'),$$

that is, the lines

$$(1', 2' — 4', 5'),\ (3', 4' — 6', 1'),\ (5', 6' — 2', 3'),$$

are concurrent, the three corresponding lines

$$(S_1 — 4, 5),\ (S_2 — 6, 1),\ (S_3 — 2, 3),$$

that is, the lines

$$(1, 2 — 4, 5),\ (3, 4 — 6, 1),\ (5, 6 — 2, 3)$$

are also concurrent, and vice versa. If the lines

$$(1', 2' — 4', 5'),\ (3', 4' — 6', 1'),\ (5', 6' — 2', 3')$$

are concurrent, the lines 1′, 2′, 3′, 4′, 5′, 6′ are tangent to a conic section. In that case, the lines 1, 2, 3, 4, 5, 6 are also tangent to a conic section.

1′, 2′, 3′, 4′, 5′, 6′ are lines joining the vertices of a triangle, each with the points in which the opposite side intersects a given conic section. (See § 22.) *Any six such lines circumbscribe a conic section.* (See figure 27.)

Mark the six intersections of the three lines

$$(1', 2', — 4', 5'),\ (2', 3' — 5', 6'),\ (3', 4' — 6', 1')$$

with the given conic, numbering them in any order (1), (2), (3), (4), (5), (6); say, so that (1) and (4) lie on the line (1′, 2′ — 4′, 5′),. (2) and (5) on the line (3′, 4′ — 6′, 1′), and (3) and (6) on the line (2′, 3′ — 5′, 6′). These six points form an inscribed hexagon whose opposite sides must, therefore, intersect in three collinear points. (Pascal.) The three collinear points are:

$$\{(1)\ (2) — (4)\ (5)\}, \{(2)\ (3) — (5)\ (6)\}, \{(3)\ (4) — (6)\ (1)\}.$$

Now make use of the following notation:*

$$\begin{array}{lll}
(1) = A_1 & (2) = B_1 & \{(2)(3) — (6)(1)\} = C_1 \\
(4) = A'_1 & (5) = B'_1 & \{(5)(6) — (3)(4)\} = C'_1 \\
(3) = A_2 & (4) = B_2 & \{(2)(3) — (4)(5)\} = C_2 \\
(6) = A'_2 & (1) = B'_2 & \{(5)(6) — (1)(2)\} = C'_2 \\
(5) = A_3 & (6) = B_3 & \{(4)(5) — (6)(1)\} = C_3 \\
(2) = A'_3 & (3) = B'_3 & \{(1)(2) — (3)(4)\} = C'_3
\end{array}$$

* Care must be taken to have corresponding vertices lie on the lines (1′, 2′ — 4′, 5′), (2′, 3′ — 5′, 6′), and (3′, 4′ — 6′, 1′).

$A_1\,B_1\,C_1$ and $A'_1\,B'_1\,C'_1$ are two triangles so situated that their corresponding sides intersect in three collinear points. Therefore, the lines joining corresponding vertices, that is, the lines

$$A_1\,A'_1 = (1', 2' - 4', 5'),\ B_1\,B'_1 = (3', 4' - 6', 1') \text{ and } C_1\,C'_1$$

are concurrent. (Desargues.) For a similar reason the lines

$$A_2\,A'_2 = (2', 3' - 5', 6'),\ B_2\,B'_2 = (1', 2' - 4', 5'), \text{ and } C_2\,C'_2$$

are concurrent; likewise the lines

$$A_3\,A'_3 = (3', 4' - 6', 1'),\ B_3\,B'_3 = (2', 3' - 5', 6'), \text{ and } C_3\,C'_3.$$

But the triangles $C_1\,C_2\,C_3$ and $C'_1\,C'_2\,C'_3$ also fulfill the condition for Desargues' theorem, and therefore the lines

$$C_1\,C'_1,\ C_2\,C'_2, \text{ and } C_3\,C'_3$$

are concurrent. Accordingly, the lines

$$(1', 2' - 4', 5'),\ (2', 3' - 5', 6'), \text{ and } (3', 4' - 6', 1')$$

are concurrent and the lines 1′, 2′, 3′, 4′, 5′, 6′ circumscribe a conic section. But this is the condition that the lines 1, 2, 3, 4, 5, 6 should also circumscribe a conic section.

24. *Theorem.—If Q is a tricuspidal quartic the three cuspidal tangents meet in one and the same point.* (See figure 28.)

To a tricuspidal quartic corresponds a conic tangent to each of the three sides of the triangle $S_1\,S_2\,S_3$. If the three points of contact on the sides s_1, s_2, s_3 be denoted by L, M, and N, respectively, then to the lines $S_1\,L$, $S_2\,M$, and $S_3\,N$ correspond the three cuspidal tangents. But $S_1\,L$, $S_2\,M$, and $S_3\,N$ are three concurrent lines (Brianchon), and therefore the three corresponding lines, the three cuspidal tangents, are concurrent.

25.—*Theorem.—If Q has three bitangents, their points of intersection may be joined with the three nodes, one with each node, in such a way that the three joining lines pass through one and the same point.* (See figure 29.) *If Q has four bitangents their points of intersection in sets of three may be joined with the nodes in such a way that the joining lines form a complete quadrangle of which each pair of opposite sides intersects in a node.* (See figure 30.)

Each bitangent of Q corresponds by the quadratic involutory transformation to a conic passing through the three nodes, S_1, S_2, S_3, and having double contact with the Σ conic. Denote the four bitangents to Q by 1, 2, 3, 4; their points of contact on Q by A, B; C, D; E, F; and G, H, respectively; and their corresponding conics through S_1, S_2, S_3, by I, II, III, and IV, respectively. Denote the fourth intersection of I and II by U; of II and III by V; of III and I by W; of I and IV by X; of II and IV by Y; and of III and IV by Z.

Consider first the case of three bitangents, 1, 2, 3. Then the lines

$$\Sigma_A\,\Sigma_B,\quad \Sigma_C\,\Sigma_D,$$

and one side of the triangle $S_1\,S_2\,S_3$, say $S_2\,S_3$, and $S_1\,U$ all pass through one and the same point and form a harmonic pencil.* Similarly, for the lines

* See Salmon, *Conic Sections*, § 263.

1
A
B
F
C
S_1
3
2
S_2
S_3
E
D
G
H
4
FIG. 30.
FIG. 31.
FIG. 29.
a
a'
ε
σ
Σ_A
K
K'
A
α α'
S_2
S_3
S_1

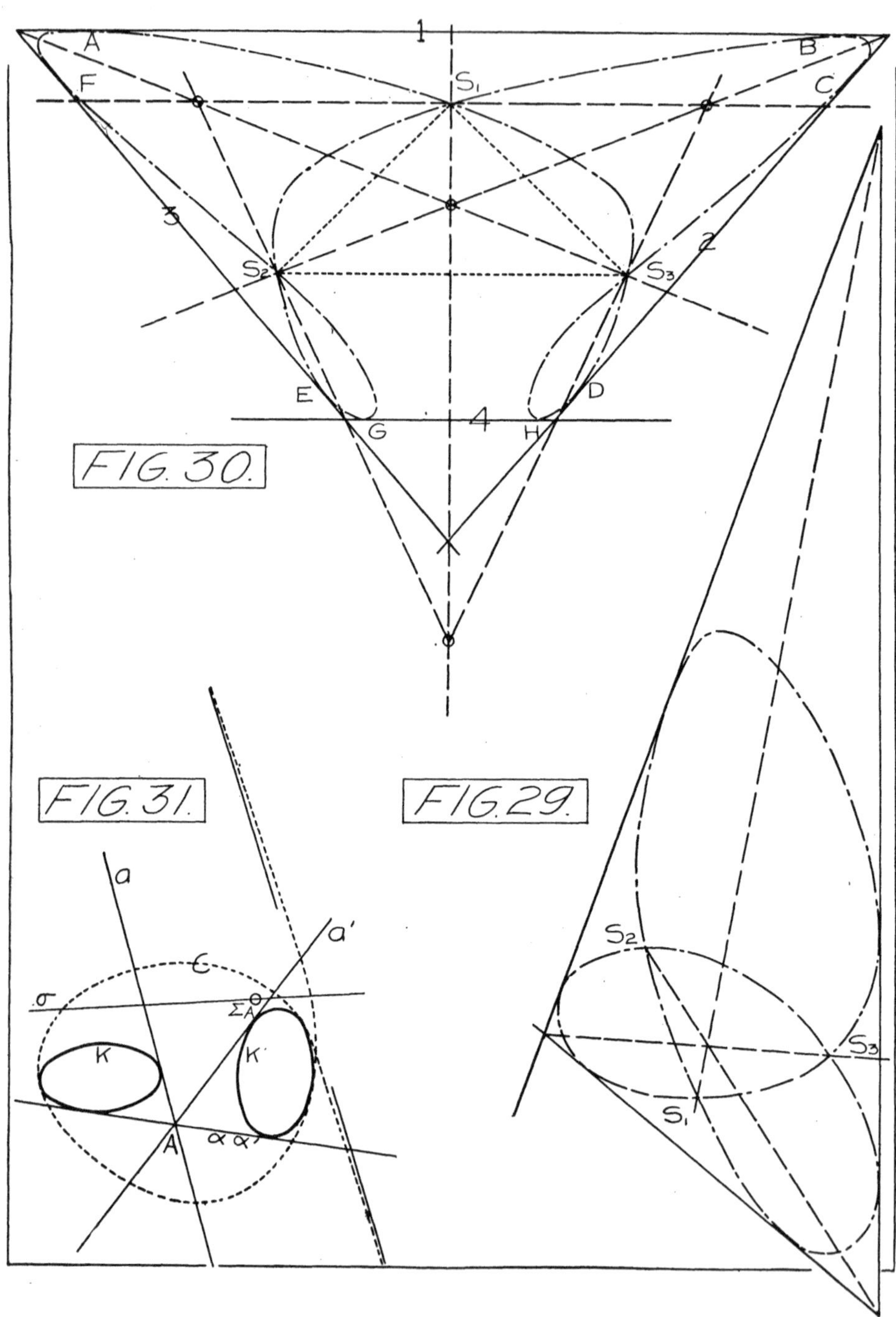

Figs. 29–31

$$\Sigma_C\,\Sigma_D,\ \Sigma_E\,\Sigma_F,\ S_3\,S_1, \text{ and } S_2\,V;$$

likewise, for the lines

$$\Sigma_E\,\Sigma_F,\ \Sigma_A\,\Sigma_B,\ S_1\,S_2, \text{ and } S_3\,W.$$

From this it follows that the lines

$$S_1\,U,\ S_2\,V, \text{ and } S_3\,W$$

are concurrent. For on $S_3\,W$ the lines

$$\Sigma_A\,\Sigma_B,\ \Sigma_C\,\Sigma_D,\ S_2\,S_3, \text{ and } S_1\,U$$

determine four harmonic points. Likewise

$$\Sigma_E\,\Sigma_F,\ \Sigma_C\,\Sigma_D,\ S_1\,S_3, \text{ and } S_2\,V$$

determine four harmonic points on $S_3\,W$. But on $S_3\,W$, $\Sigma_C\,\Sigma_D$ in each case determines the same point; $S_2\,S_3$ and $S_1\,S_3$ both determine S_3; $\Sigma_A\,\Sigma_B$ and $\Sigma_E\,\Sigma_F$ determine the same point since the lines

$$\Sigma_E\,\Sigma_F,\ \Sigma_A\,\Sigma_B,\ S_1\,S_2, \text{ and } S_3\,W$$

are concurrent. Therefore, $S_1\,U$ and $S_2\,V$ determine the same point on $S_3\,W$; that is, $S_1\,U$, $S_2\,V$, and $S_3\,W$ are concurrent.

But if these three lines are concurrent, the lines to which they correspond in involution must be concurrent. That is, the lines

$$(S_1 - 1, 2),\ (S_2 - 2, 3),\ (S_3 - 3, 1)$$

pass through one and the same point.

Consider now the case of four bitangents to Q, 1, 2, 3, 4. Proceeding as before, we find the following sets of harmonic lines:

For I, II, and III—

I and II	II and III	III and I
$S_2\,S_3$	$S_3\,S_1$	$S_1\,S_2$
$S_1\,U$	$S_2\,V$	$S_3\,W$
$\Sigma_A\,\Sigma_B$	$\Sigma_C\,\Sigma_D$	$\Sigma_E\,\Sigma_F$
$\Sigma_C\,\Sigma_D$	$\Sigma_E\,\Sigma_F$	$\Sigma_A\,\Sigma_B$

For I, II, and IV—

I and II	II and IV	IV and I
$S_2\,S_3$	$S_1\,S_2$	$S_1\,S_3$
$S_1\,U$	$S_3\,Y$	$S_2\,X$
$\Sigma_A\,\Sigma_B$	$\Sigma_C\,\Sigma_D$	$\Sigma_G\,\Sigma_H$
$\Sigma_C\,\Sigma_D$	$\Sigma_G\,\Sigma_H$	$\Sigma_A\,\Sigma_B$

For I, III, and IV—

I and III	III and IV	I and IV
$S_1\,S_2$	$S_2\,S_3$	$S_1\,S_3$
$S_3\,W$	$S_1\,Z$	$S_2\,X$
$\Sigma_E\,\Sigma_F$	$\Sigma_E\,\Sigma_F$	$\Sigma_G\,\Sigma_H$
$\Sigma_A\,\Sigma_B$	$\Sigma_G\,\Sigma_H$	$\Sigma_A\,\Sigma_B$

For II, III, and IV—

II and III	III and IV	IV and II
$S_3\,S_1$	$S_2\,S_3$	$S_1\,S_2$
$S_2\,V$	$S_1\,Z$	$S_3\,Y$
$\Sigma_C\,\Sigma_D$	$\Sigma_E\,\Sigma_F$	$\Sigma_C\,\Sigma_D$
$\Sigma_E\,\Sigma_F$	$\Sigma_G\,\Sigma_H$	$\Sigma_G\,\Sigma_H$

From this we obtain the following four sets of three concurrent lines:

$$\left.\begin{matrix} S_1 U \\ S_2 V \\ S_3 W \end{matrix}\right\} \quad \left.\begin{matrix} S_1 U \\ S_2 X \\ S_3 Y \end{matrix}\right\} \quad \left.\begin{matrix} S_1 Z \\ S_2 X \\ S_3 W \end{matrix}\right\} \quad \left.\begin{matrix} S_1 Z \\ S_2 V \\ S_3 Y \end{matrix}\right\}$$

The configuration determined by these six lines is a complete quadrangle of which each pair of opposite sides intersects in a node of Q. Each set of three concurrent lines, one of which passes through each involution center, corresponds in involution to another set of three concurrent lines, one of which passes through each involution center. Therefore, the figure corresponding in involution to a complete quadrangle of which each pair of opposite sides intersects in an involution center is another complete quadrangle of the same character; and the lines $S_1 U$, $S_2 V$, $S_3 W$, $S_1 Z$, $S_2 X$, $S_3 Y$ correspond in involution to lines joining the points of intersection of the bitangents of Q to the nodes.

26. *Degenerate Cases.*—In general, the pencils κ and κ' have no self-corresponding rays. That is to say, a tangent common to κ and κ', considered as a ray of κ, corresponds in general to some other ray of κ', and considered as a ray of κ', to some other ray of κ. But it may happen that of the three pairs of corresponding rays of κ and κ' given to construct Q (§ 6), one, two, or all three may be self-corresponding. Also, in general, the conics κ and κ' are distinct. But the two pencils of the second order may envelope the same base conic; and in that case, too, there may or may not be self-corresponding rays. In all these special cases both Q and Σ assume degenerate forms.

A. If κ and κ' have one self-corresponding ray, $\alpha\,\alpha'$, Q consists of a cubic C and the ray $\alpha\,\alpha'$; Σ consists of a line σ and the ray $\alpha\,\alpha'$; points of Q on $\alpha\,\alpha'$ correspond to points of Σ on σ and points of Q on C to points of Σ on $\alpha\,\alpha'$. (See figure 31.)

Since α and α' coincide, the four tangents from Σ_A to κ and κ' meet $\alpha\,\alpha'$ in four points of Σ, showing at once that all the points of $\alpha\,\alpha'$ are points of Σ, since more than two of the points of $\alpha\,\alpha'$ are points of Σ. That is, Σ degenerates into two straight lines, one of which is $\alpha\,\alpha'$.

If from points of Σ as Σ_D, Σ_E, Σ_F, etc., on $\alpha\,\alpha'$ we draw tangents to κ and κ', such tangents must meet the lines d', d, e', e, f', f, etc., in points of Q. But since always $\alpha\,\alpha'$ itself will be a tangent from such points both to κ and κ', it follows that half of the points so obtained will lie on $\alpha\,\alpha'$. $\alpha\,\alpha'$ is then a part of Q. The remaining part is, of course, a cubic C.

The construction shows at once the correspondence between the different parts of Q and Σ.

B. If κ and κ' have two self-corresponding rays, $\alpha\,\alpha'$ and $\beta\,\beta'$, Q consists of the two lines $\alpha\,\alpha'$ and $\beta\,\beta'$ and a conic λ; Σ consists of the two lines $\alpha\,\alpha'$ and $\beta\,\beta'$; points of Q on $\alpha\,\alpha'$ correspond to points of Σ on $\beta\,\beta'$ and points of Q on $\beta\,\beta'$ to points of Σ on $\alpha\,\alpha'$, while the points of Q on the conic λ all have their corresponding points of Σ at the intersection of $\alpha\,\alpha'$ and $\beta\,\beta'$.

C. If κ and κ' have three self-corresponding rays, $\alpha\,\alpha'$, $\beta\,\beta'$, and $\gamma\,\gamma'$, Q consists of the four tangents common to κ and κ'; Σ consists of the three vertices of the triangle formed by $\alpha\,\alpha'$, $\beta\,\beta'$, and $\gamma\,\gamma'$. Points of Q on one of the three lines

$\alpha\ \alpha'$, $\beta\ \beta'$, or $\gamma\ \gamma'$ correspond to the opposite vertex of the triangle formed by those lines. Points of Σ corresponding to points of Q on the fourth tangent common to κ and κ' are indeterminate.

To prove that the fourth tangent common to κ and κ' is a part of Q, find the point P on a for which a is a ray π of κ. p and p' must coincide. Suppose they do not. p, considered as a ray ρ of κ, determines a point R of Q, and p', considered as a ray ξ' of κ', determines a point X of Q. None of the points P, R, and X can lie on any of the rays $\alpha\ \alpha'$, $\beta\ \beta'$, $\gamma\ \gamma'$, since every point of Q on a self-corresponding ray is determined by that ray. Since Q is of the fourth degree, P, R, and X must lie in one and the same straight line. This can happen only if p and p' coincide.

D. The two projective pencils of the second order, κ' and κ'', envelope the same base conic κ.

(1) If there are no self-corresponding rays, Σ degenerates into two straight lines σ_1 and σ_2 tangent to κ, and Q into these same two lines and a conic section λ. The construction shows that points of Q on σ_1 correspond to points of Σ on σ_2, and points of Q on σ_2 to points of Σ on σ_1, while points of Q on λ all have their corresponding points of Σ at the intersection of σ_1 and σ_2. The conic λ is tangent to κ at the points of tangency of σ_1 and σ_2.

Here a coincides with α', a' with α, b with β', b' with β, etc. It follows that the line $(a, \beta - a', \beta')$ coincides with $(b', \alpha' - b, \alpha)$; $(a, \gamma - a', \gamma')$ with $(c', \alpha' - c, \alpha)$; and $(b, \gamma - b', \gamma')$ with $(c', \beta' - c, \beta)$. These are three lines joining the opposite vertices of a hexagon circumscribed to a conic section and are therefore concurrent. (Brianchon.) Therefore, Σ_A, Σ_B, Σ_C all coincide. The tangents from Σ_A, Σ_B, Σ_C, etc., to κ determine points of Σ on α, α', β, β', γ, γ', etc., and points of Q on a, a', b, b', c, c', etc. Because of the coincidences mentioned all these points lie on two lines, σ_1 and σ_2, the tangents from $\Sigma_{ABC, etc.}$, to κ.

(2) If κ' and κ'' have one self-corresponding ray $\alpha\ \alpha'$, $\alpha\ \alpha'$ coincides with σ_1 or σ_2, and Σ_A is indeterminate. In other respects Q and Σ are as described in (1).

(3) If κ' and κ'' have two self-corresponding rays $\alpha\ \alpha'$ and $\beta\ \beta'$, $\alpha\ \alpha'$ and $\beta\ \beta'$ are the lines σ_1 and σ_2, and Σ_A and Σ_B are indeterminate. In other respects Q and Σ are as described in (1).

(4) If κ' and κ'' have three self-corresponding rays, $\alpha\ \alpha'$, $\beta\ \beta'$, and $\gamma\ \gamma'$, everything is indeterminate.

27. *The Unicursal Curve of the Fourth Class.*—It is of interest to apply the principle of duality to the results of this chapter, considering then the unicursal curve Q' defined as the envelope of lines joining corresponding points in two projective point rows of the second order, κ and κ'. It has, in general, three bitangents and four nodes and is of the sixth order and fourth class. (Cf. §§ 1, 2, 20, 25.) Given three pairs of corresponding points in two projective point rows, we can construct the curve (cf. § 6) and establish properties entirely analogous to all those of the curve of the fourth order. The more interesting and important ones may be noted.

Every ray p of the rays enveloping Q' has its corresponding ray of perspectivity Σ'_p. The ensemble of rays Σ'_p envelope a curve Σ' of the second class. (Cf. §§ 7 and 13.)

Any ray a joining corresponding points A_1 and A'_1 has a second point A_2 in common with κ and a second point A'_2 in common with κ'. The lines joining the points of intersection of the ray Σ'_a and κ with A'_2 are rays enveloping Q'; with A_1, are rays enveloping Σ'. Likewise, the lines joining the points of intersection of the ray Σ'_a and κ' with A_2 are rays enveloping Q'; with A'_1, are rays enveloping Σ'. The ray tangent to Σ' which joins A_1 (or A'_1) with one point of intersection of Σ'_a and κ (or κ') corresponds to the ray tangent to Q' which joins A'_2 (or A_2) with the other point of intersection of Σ'_a and κ (or κ'). (Cf. § 11.)

The common rays of the sets enveloping Q' and Σ' pass through the points of intersection of κ and κ', two through each point. (Cf. § 14.)

The points A_2 and A'_1, A'_2 and A_1 determine an involution of points on a in which the point (a, Σ'_a) corresponds to the point of tangency of a on Q'. (Cf. § 19.)

The six points of Q' other than the points of tangency which lie on the three bitangents to Q' all lie on one and the same conic section. (Cf. § 21.)

The six points of tangency on the three bitangents to Q' all lie on one and the same conic section. (Cf. § 23.)

If the two points of tangency coincide on each bitangent the three points of tangency all lie on one and the same straight line. (Cf. § 24.)

If Q' has three nodes the lines joining them intersect the three bitangents, one each bitagent, in such a way that the three points of intersection lie on one and the same straight line. If Q' has four nodes the lines joining them in sets of three intersect the bitangents in such a way that the points of intersection form a complete quadrilateral of which each pair of opposite vertices lies on a bitangent. (Cf. § 25.)

Transmitted October 3, 1911.

UNIVERSITY OF CALIFORNIA PUBLICATIONS
IN
MATHEMATICS

Vol. 1, No. 3, pp. 55-85 February 28, 1913

A DISCUSSION BY SYNTHETIC METHODS OF TWO PROJECTIVE PENCILS OF CONICS*

BY

BALDWIN MUNGER WOODS

CONTENTS

* A dissertation submitted in partial satisfaction of the requirements for the degree of Doctor of Philosophy in the College of Natural Sciences of the University of California.

INTRODUCTORY OUTLINE

The Problems Treated in the Discussion

The following study of pencils of conics falls naturally into four parts, concerned each with the discussion of a particular problem.

The first part—with the exception of its introduction, which deals with the geometry of a one-to-one correspondence between pencils of conics—is concerned with the locus of intersections of corresponding elements of a pencil of rays of the first class and a pencil of conics. This locus is shown to be a cubic curve; and, from the construction, a discussion of the one-to-two involutory correspondence of points on a line is obtained. In this problem, Pascal's Theorem is found to be a useful tool; and, indeed, in the succeeding discussions, the method of attack is often by means of this theorem.

The second part is concerned with the locus of intersections of corresponding elements of two projective pencils of conics. This locus is shown to be the general quartic curve, and conditions are obtained for the double points. The solution is found to depend on the locus of intersections of two involutions of rays where there is a one-to-one correspondence between pairs of rays of the two.

This locus is studied in the third part and is shown to be a quartic curve with two double points. When the involutions are in half-perspective position, the locus is the single-branched cubic.

From the discussion of the problem of part three, a locus problem is suggested which is solvable with the aid of Pascal's Theorem, and which gives rise to a certain quadratic transformation of the plane which is studied by analytic methods. This is the essence of the fourth part.

I

The Cubic as a Locus of Intersections of a Pencil of Rays and a Pencil of Conics Projective to it

Introduction

In the following discussion, the term "pencil of conics" will be used to designate the totality of conics that can be constructed passing through four arbitrary fixed points in a plane. When two such pencils of conics are so related that there is a one-to-one correspondence between the conics of one pencil and those of the other, the two pencils will be said to be projective to each other. The geometrical construction used to determine this one-to-one correspondence will be set up as follows:

If an arbitrary line be drawn through one of the fixed points of a pencil of conics, it is obvious that every conic of the pencil will meet this line in one point besides the fixed point, which is common to all. Conversely, every point of the line will determine a single conic of the pencil passing through it. In particular, the fixed point of the pencil of conics considered as a point of the arbitrary line will be assumed to determine the conic of the pencil which is tangent to the line at the fixed point of the pencil. It will be shown later that two such point-rows are projective in the ordinary sense.

If this line, considered as a point-row, be projectively related to a pencil of rays of the first class, there will be a one-to-one correspondence between the rays of the pencil of rays and the conics of the pencil of conics. In this case the pencil of rays and the pencil of conics are said to be projective to each other. If, in a second pencil of conics, an arbitrary line be similarly drawn through one of the fixed points, the conics of the two pencils of conics can be put in one-to-one correspondence by merely considering the two arbitrary lines as projective point-rows. This construction, as outlined, is employed throughout the following discussion.

1. *Locus of Intersections of Pencil of Rays and Pencil of Conics*

The first problem to be discussed is the following: Required, the locus of intersections of corresponding elements of a pencil of rays of the first class projectively related to a pencil of conics.

Consider a pencil of conics through the points A_1, A_2, A_3, and A_4 (see fig. 1), and a pencil of rays through S, given projectively related to it. The ray by which the projectivity is set up is denoted by A_1Q. Hence, the pencil of rays at S is projectively related to the ray A_1Q, considered as a point-row. Consider an arbitrary cutting ray l in the plane. It may be taken as a point-row perspective to the pencil of rays S, and will therefore be projective to the point-row A_1Q. But points on l where a ray of S meets its corresponding conic are points of the required locus. Now, if a ray revolves about S, its corresponding point P moves along l, cutting out a point-row projective to the point-row Q cut out on A_1Q by the corresponding conic. If P falls on the conic of the pencil determined by Q, P is a point of the locus. In other words, whenever the six points P, Q, A_1, A_2, A_3, and A_4 lie on a conic, P is a point of the locus, and the six points named will satisfy Pascal's Theorem regarding a hexagon inscribed in a conic.

Number the points as indicated in the figure. Call the point of intersection of 12 and 45, L; of 23 and 56, M; of 34 and 61, N.

As P moves along l, 34 revolves about the fixed point 3, cutting out a point-row N on the fixed ray 61, perspective to the point-row P. Similarly L cuts out on a_1 a point-row perspective to P. Likewise, since the point-row Q is projective to P, M cuts out a point-row on the fixed ray 56 perspective to Q, and hence projective to P.

Hence, the point-rows L, M, and N are projectively related to one another, and there will, consequently, be at most three rays which pass through corresponding points of the three point-rows. For these three cases, P lies on the conic of the pencil determined by Q, and is a point of the locus. There are at most three such points on an arbitrary ray l. Hence, the

Theorem: The locus of intersections of corresponding elements of a pencil of rays of the first class and a pencil of conics projectively related to it, is a point-row of the third order.

That this locus is the general plane cubic is easily demonstrated both analytically* and synthetically.†

It should be noted that the four fixed points of the pencil of conics are on the locus; since one of the intersections of the line SA_1, for example, with its corresponding conic must be at A_1. Similarly, S is on the locus, since the conic of the pencil which passes through S must meet its corresponding ray there.

By the conditions of the problem, we observe that to any point A of l (see fig. 2) considered as a point of intersection with a ray of S, there correspond two points B_1 and B_2, considered as the points of intersection with l of the conic corresponding to that ray; and that, conversely, to this same pair of points B_1 and B_2 corresponds back again the starting-point A. Hence the following

Theorem: In an involutory one-to-two correspondence of points on a line, there are at most three points where corresponding points are coincident.

II

The Quartic Q as the Locus of Intersections of Two Projective Pencils of Conics

2. *Reference to Analytical Discussion*

The next problem to be studied is that of the locus of intersections of corresponding elements of two projective pencils of conics.

This locus is easily shown analytically to be the general quartic curve.* Let us denote it by Q. That it is a point-row of the fourth order will presently be demonstrated synthetically.

3. *Point-rows through a fixed point of a Pencil of Conics*

Before proceeding to this, however, let us examine a few properties of the figure, and enunciate a theorem that is of use later on.

Theorem: If, in a pencil of conics through four fixed points, rays are passed through the several fixed points, the point-rows described by the other intersections of the conics with the rays are projective to one another.

In fig. 3, represent the fixed points of the pencil of conics by A_1, A_2, A_3, and A_4, and two of the fixed rays by a_1 and a_3. Consider any conic of the pencil, and call the points in which it intersects a_1 and a_3, 2 and 5 respectively. The six points 2, 5, A_1, A_2, A_3, and A_4 must satisfy Pascal's Theorem. Hence, numbering the points as indicated in the figure, we have 12 and 45 intersecting at L, 23 and 56 at M, and 34 and 61 at N. Both L and N are fixed points, therefore this Pascal line of the pencil of conics is fixed. As 2 moves along the ray a_1

* See Emch, *Introduction to Projective Geometry and its Application*, p. 182.

† Schröter, *Theorie der Ebenen Curven dritter Ordnung*, p. 58.

* See Emch, *loc. cit.*, p. 181.

determining the various conics of the pencil, it projects to A_4 in a pencil of rays, giving a point-row M on the Pascal line perspective to the point-row 2. The point-row M projects to A_2 in a pencil of rays giving a point-row 5 on a_3 perspective to the point-row M and, hence, projective to the point-row 2. Therefore, the pencil of conics cuts out on a_1 and a_3 point-rows that are projectively related to each other. This may be extended to include rays through the other fixed points, or several rays through the same fixed point.

4. *Double Points of Quartic*

Consider further the two pencils of conics determined by the points A_1, A_2, A_3, and A_4 and B_1, B_2, B_3, B_4 (fig. 4), projectively related to each other by means of point-rows a_1 and b_1 through A_1 and B_1 respectively.

Theorem: The eight fixed points of the two projective pencils of conics are on the locus Q of intersections of corresponding conics.

This is evident since the conic of the first set passing through B_1, for example, must meet its corresponding conic of the second set there. Similarly, for the others.

Theorem: If the two projective pencils of conics have a fixed point in common, this is a double point of the locus Q.

Call the common point (A_1B_1) (see fig. 5). The projectivity of the two pencils of conics may now be referred to two point-rows through (A_1B_1), say a_1 and b_1, which are projectively related to each other—since these are projective to any other point-rows through any of the fixed points of either pencil.

Since the projectivity between the pencils of conics is arbitrary, the point-rows a_1 and b_1 are not, in general, perspective to each other, and the point (A_1B_1) is not, in general, self-corresponding. Hence (A_1B_1) considered as a point of a_1 corresponds to another point of b_1, say R, giving (A_1B_1) as a point on the locus Q, determined by the conic of the first pencil tangent to a_1, and the conic of the second pencil through R. Considered as a point of b_1, (A_1B_1) corresponds to another point of a_1, say P, and hence gives (A_1B_1) as a point on the locus Q determined by an entirely different pair of conics, the conic of the first pencil through P, and the conic of the second pencil tangent to b_1. Hence, (A_1B_1) occurs twice on the locus or, is a double point. There may be in this way as many as three double points. When there are three, the curve is evidently unicursal, since there is but one movable point of intersections of corresponding conics, and the correspondence is continuous. If there are four common fixed points, the locus degenerates, in general, either into these fixed points, or into two conics through them.

5. *Synthetic Discussion of Order of Q*

With this introduction, we proceed to the synthetic discussion of the order of the locus.

In figure 6, denote by $A_1A_2A_3A_4$ and $B_1B_2B_3B_4$ the fixed points of the two pencils of conics, by a_2 and b_2 the point-rows determining the projectivity of the two pencils of conics, and by l an arbitrary cutting ray of the plane. Let a pair of corresponding conics, determined by P of a_2 and R of b_2, cut l in A' and A'', and B' and B'' respectively. As P moves along a_2, A' and A'' will describe an involution of points on l; R will move along b_2, and B' and B'' will describe another involution of points on l. There is a one-to-one correspondence between pairs of points on l set up in this way, such that to any pair of points on l determined by a conic of one pencil corresponds a pair of points determined by the corresponding conic of the other pencil. Since the pencils of conics are projectively related, the correspondence of point-pairs on l is involutory.

Hence, our problem is reduced to that of discovering how many coincidences of corresponding points of an involutory two-to-two correspondence of points on a line can occur. For, if a point of l be at once a point of both involutions, it is an intersection point of corresponding conics and, therefore, a point of the locus. Join the points A' and A'' to an arbitrary point, say A_1, and B' and B'' to B_1. Now, as P moves along a_2 and R moves along b_2, projective to P, we shall have an involution of rays at A_1 and also one at B_1, with a one-to-one correspondence between pairs of rays of the two involutions. The locus of intersections of corresponding pairs of rays of these two involutions—call it L—will meet l in points where corresponding conics intersect; in other words, in points of the original locus Q of intersections of corresponding conics. Hence, the order of L is the same as that of the original locus, and we shall confine our attention for the moment to L.

6. *Proof of Order of L*

Theorem: The locus of intersections of corresponding pairs of rays of two involutions of rays, where there is a one-to-one correspondence between the pairs of rays of the two involutions, is a point-row of the fourth order with two double-points; or, a quartic curve of deficiency one.

In figure 7, denote by I_1 and I_2 the centers of the involutions of rays and let λ be an arbitrary cutting ray of the plane. Construct through I_1 and I_2 any conic—call it Γ—which is tangent to λ. There will be a double infinity of such conics. Now, the rays of I_1, for example, will cut out an involution of points on λ. Draw from each of these points the remaining tangent to Γ. Now, to a given pair of rays of I_1, there will be a pair of points on λ, say G_1 and G_2, and hence a pair of tangents to Γ, say g_1 and g_2. If four points G_1 of λ be taken, the four points of tangency of the four tangents g_1 will project to any point of Γ in four rays with the same anharmonic ratio as the points G_1, since the points G_1 are chosen on a tangent to Γ. Hence, the points of tangency of the various pairs of tangents of g_1 and g_2 will constitute an involution of points on Γ. Now, the rays joining corresponding points of an involution of points on a conic pass through a point,* say S_1. Hence, the pencil of rays S_1 will determine our involution of points on Γ, and, consequently, our involution of points on λ, determined by the rays of I_1. Of course, S_1 varies with the different conics Γ that may be taken. Similarly, there will be a pencil of rays S_2, determined by

* See Reye, *Geometrie der Lage*, p. 147.

tangents drawn to Γ from the involution of points H_1 and H_2, on λ, determined by the involution of rays I_2. Since there is a one-to-one correspondence between the line-pairs of the two involutions, there will be a one-to-one correspondence between the rays of S_1 and S_2. The point of intersection of corresponding rays of S_1 and S_2 will therefore describe a conic, say Σ, which will intersect Γ in, at most, four points. These points are coincident points of the two involutions of points on Γ and, consequently, since tangents g and h coincide here, the tangents to Γ at these points will meet λ in points where G and H coincide; that is, in points of L. There can be at most four such points on an arbitrary λ. Hence, the first part of the theorem above is established, and the two following theorems may be added as direct consequences of the discussion,—

Theorem: In a two-to-two involutory correspondence of points on a line there are at most four points where corresponding points coincide.

Theorem: The locus of intersections of corresponding conics of two projective pencils of conics is a point-row of the fourth order.

The conditions for double points have been established above. Hence, the locus is the general quartic curve.

III

Discussion of the Quartic L with two Double Points

7. *Deficiency of L*

In article 6 of the preceding part, we have shown that L is a point-row of the fourth order. Let us proceed to a discussion of this quartic beginning with the determination of its deficiency, which may be established as follows. Consider the ray I_1I_2 as a ray of the involution I_1. Considered as a ray I_1G_1, it meets its corresponding ray I_2H_1 at I_2; likewise, considered as a ray I_1G_2, it meets its corresponding ray I_2H_2 at I_2. I_2H_1 and I_2H_2 are in general different rays. Hence I_2 occurs twice on L or, is a double point.

Similarly with I_1. A third double point does not in general exist; but conditions for its existence are easily determined.

A pair of rays of I_1 has its corresponding pair of rays of I_2. The four intersection points—say P, Q, R, S—are points of L. If the two rays of I_2 should fall together as a double ray of the involution at I_2, the points P and Q, and R and S would fall together in pairs as indicated, and the rays of I_1 corresponding to a double ray of I_2 would be tangents to L. There are not more than two double rays of I_2. Hence,

Theorem: From either double point of L at most four tangents may be drawn to L. These are the rays of the involution at one double point corresponding to the double rays of the involution at the other double point.

If a double ray of I_1 corresponds to a double ray of I_2, they are each tangent to L at their point of intersection, and their point of intersection is, consequently, a double point of L. Hence,

Theorem: If a double ray of I_1 corresponds to a double ray of I_2, L has three double points, viz: I_1, I_2, and the point of intersection of the corresponding double rays.

8. *Bearing of Quadratic Transformation Theory*

Before discussing other cases of L, let us consider the bearing of the quadratic transformation theory of the plane on this discussion. Ordinarily, if two involutions of rays, I_1 and I_2, be given, and the point of intersection P of a ray of I_1 and one of I_2 move on a point-row of order n, the intersection point R of the corresponding rays of I_1 and I_2 will move on a point-row of order $2n$. I_1 and I_2 will be vertices of a fundamental triangle of the transformation, of which the third vertex, say I_3, will be the center of an involution of rays that may be used in place of either I_1 or I_2 in the construction. Now, each time the point P of the point-row of order n passes through one of the points I_1, I_2, or I_3, the point R of the point-row of order $2n$ describes a straight line as part of its locus and the order of the remainder is reduced by one. In the locus L, under discussion, the points P and R traverse the same point-row, a point-row of order four with two double points. Comparing this with the theory mentioned above, we find it to be consistent; for, if P move along L, Q will move on a point-row of order eight. However, P passes through I_1 twice and I_2 twice. Hence, the locus of Q degenerates into a point-row of the fourth order and doubly into the lines in the quadratic transformation which correspond to the points I_1 and I_2 of the fundamental triangle.

9. *Degeneracy of L with Double Rays Corresponding*

Certain degenerate cases of L are interesting and present themselves naturally. Suppose, first, that each double ray of I_1 corresponds to a double ray of I_2. Let the first double ray of I_1 meet its corresponding double ray of I_2 at A, and the second double ray of I_1 its corresponding double ray at B (see fig. 8).

By previous reasoning I_1, I_2, A and B are double points of L, hence—with four double points—degeneracy is to be expected. The ray AB has the same involution of points described on it by the rays of I_1 and I_2, since the double points A and B of the involutions are the same. Hence, the corresponding rays of I_1 and I_2 will meet on AB and it is a part of the locus. Let the points of intersections of corresponding pairs of rays of I_1 and I_2 be P, Q, R, and S. Two of these, say P and R, are obviously on AB. Now, as P moves along AB, it is always an intersection point of corresponding rays of I_1 and I_2. As it crosses I_1I_2, the point of intersection of the corresponding rays of I_1 and I_2 is indeterminate, and I_1I_2 is thus a part of the locus. The remainder is a conic through I_1, I_2, A, and B, and there are not only four but five double points to the locus in this case.

10. *Single-branched Cubic*

Another and more important case of degeneracy is that in which the ray I_1I_2 is self-corresponding, but is not a double ray of either involution. In this case, the locus L degenerates into the ray I_1I_2 and a point-row of the third order. This last is a single-branched cubic and has been discussed by Schröter from this point of view. This position of two involutions of rays is termed by him "half-perspective position."*

In figure 9, a case of this type is shown. From T_1, a point on a single-branched cubic, tangents t_1 and t_2 are drawn to the curve. (There are always points T_1 from which this is possible.) Calling the points of tangency A_1 and A_2, we know that the ray A_1A_2 meets the cubic in another point, say I_2, which is conjugate to T_1. If, now, any point P of the cubic be joined to A_1 and A_2, the lines so drawn will each meet the cubic in one other point. Call these B_1 and B_2, and join them to T_1 and T_2. As P moves along the curve, the rays T_1B_1 and T_1B_2 will give an involution of rays at T_1, and T_2B_1 and T_2B_2, an involution of rays at T_2. Since T_1T_2 is self-corresponding, it is part of the locus. The ray A_1A_2 is a double ray of T_2; hence, the corresponding rays of T_1 should be tangents to the curve as, indeed, they are.

11. *Conditions of Tangency of a Line to L*

Let us turn again to a discussion of the properties of L as revealed in figure 7. If the conic Σ, described by the intersection point of corresponding rays of S_1 and S_2 is tangent to the conic Γ, two of the points of intersection of the conics will be coincident, and two of the points of intersection of λ with L will be coincident. Hence λ will either be a tangent to L or a secant through a double point. That this last state of affairs might occur is readily shown.

In figure 10, let the double rays of I_1 meet λ in A and B, and the double rays rays of I_2 in C and D. Then, since the tangents to Γ from these points determine the double points of the two involutions of points on Γ, it is obvious, for instance, that the tangent to Γ from A joining, as it does, two corresponding points on Γ, contains S_1. Similarly, the tangent from B contains S_1. S_1 is therefore determined as the intersection point of the tangents to Γ from the points where the double rays of I_1 intersect λ. S_2 may be determined in like manner. Now, if one double ray of I_1 correspond to a double ray of I_2, their point of intersection A is a double point of L. Consider any cutting ray λ through this point, and construct the conic Γ as before. A tangent to Γ from this double point contains both S_1 and S_2. Now, since the point A is a self-corresponding double point of the involutions of points on λ, the common ray S_1S_2 of the pencils of rays S_1 and S_2 is self-corresponding, and the conic Σ degenerates into two lines, one of which is S_1S_2, the tangent to Γ from A. This is obviously true for any ray λ through A and, hence, Σ and Γ are tangent if λ be tangent to L, or pass through one of its double points.

* See Schröter, *Theorie der Ebenen Curven dritter Ordnung*, p. 148.

The same remarks apply to the general quartic determined by two projective pencils of conics, since the intersections of λ with the general quartic are the same as its intersections with the L determined for that particular cutting line. If the cutting line should pass through one of the double points of L, a single involution only of points would be determined upon it, and our reasoning regarding the coincidences of intersection points of Q and L with λ breaks down. Since the selection of double points for L is arbitrary, this difficulty is obviated by moving them.

As a consequence of the identity of the intersections of L with λ and of the general quartic with λ, we may revolve λ about a fixed point, and enunciate the following

Theorem: The general quartic curve may be described as the locus of intersections of a pencil of rays and of a pencil of quartics of deficiency one with arbitrary, fixed double points.

12. *One-to-Two Involutory Correspondence of Points on a Line from this Viewpoint*

Before leaving the discussion of L, let us apply this machinery to the discussion of the one-to-two involutory correspondence of points on a line, previously discussed with the aid of Pascal's Theorem.

In figure 11, let I_1 be the center of a pencil of rays of the first class, and I_2 the center of an involution of rays. Select any cutting ray λ and construct a conic Γ, containing I_1 and I_2, and tangent to λ. The points of tangency of tangents to Γ from the points H_1 and H_2, where pairs of rays of I_2 cut λ, will give an involution of points on Γ. The rays joining corresponding points of this involution will pass through a point, say S_2. The points of tangency of the tangents from the points G of λ where rays of I_1 meet λ will project to I_1 in a new pencil of rays projective to the original one. By the conditions of the problem, the new pencil of rays at I_1 and the pencil of rays at S_2 are projective to each other. They determine a conic Σ which cuts Γ in at most four points, one of which is I_1. Tangents to Γ at the other three points of intersection of Σ and Γ meet λ in points where corresponding points of the one-to-two involutory correspondence are coincident. The tangent at I_1 does not in general cut λ in such a point, for the tangent (from a point of λ) whose point of tangency is I_1, does not uniquely determine a ray of the new pencil at I_1, since any ray through I_1 answers the requirements. Hence the following

Theorem: The locus of intersections of corresponding elements of a pencil of rays of the first class and an involution of rays, where there is a one-to-one correspondence between the rays of the first and the pairs of rays of the second, is a point-row of the third order with one double-point; or, a unicursal cubic curve.

I_2 is the double point, since the ray I_1I_2, considered as a ray of I_1, meets two rays of I_2 at I_2. Hence, I_2 occurs twice on the locus, or, is a double point. I_1 is obviously also on the locus, since the ray I_2I_1 of I_2 meets its corresponding ray of I_1 at I_1. The last theorem may be stated inversely in the form in which it is given when approached from the other side.

*Theorem: The points of intersection with the curve of rays through any point of a unicursal cubic project to the double point in an involution of rays, the double rays of which are determined by the tangents from the arbitrary point to the curve. The rays of the involution corresponding to the ray joining the arbitrary point to the double point are the tangents of the curve at the double point.**

13. *Four-to-Four Transformation of the Plane*

The constructions of figures 7 and 10 by which L was discussed, contain and suggest several problems and loci. The points S_1 and S_2 are determined by constructing a conic through two points and tangent to a given line. Let us inquire whether it is possible to choose S_1 arbitrarily and, if so, how many points S_2 will there be corresponding to a given S_1. We note that with a given S_1, Γ is determined as a conic which shall be tangent to S_1A, S_1B, and λ, and shall, in addition, contain the points I_1 and I_2. There are, in general, four such conics.† For a given conic, it is obvious that S_2 is uniquely determined. Hence, to a given S_1, arbitrarily chosen, there are four S_2's, and conversely. This gives a four-to-four transformation of the plane, which may be called "one-quarter involutory," since any one of the points S_2 determined by an arbitrary S_1, gives back this same S_1 and three others.

14. *Two-to-Two Semi-involutory Correspondence of Two Pencils of Rays*

Suppose the point of tangency of Γ to λ be fixed and be denoted by K. Then the figure furnishes an example of what may be termed a "semi-involutory" two-to-two correspondence of rays, as follows,—suppose an arbitrary ray of A be chosen. On this ray there are in general two points S_1, since two conics of the pencil are in general tangent to the ray of A chosen. The conics Γ now constitute a pencil, since they all pass through I_1 and I_2 and are tangent to λ at K. Also, each conic determines a ray of B, tangent to it, and hence an S_1. To either of the rays of B so determined, two conics of the pencil are tangent, one of which is the conic tangent to the original ray of A. Hence, to each ray of A, there are two rays of B; and, to each of these rays of B, there are two rays of A, one of which is the ray of A with which we started.

An easy geometrical example of this is obtained by using either two points and two rays which do not contain them, or two points and a non-degenerate conic which does not contain them (see figures 12 and 13). To the ray a_1 of A, there are two rays b_1 and b_2 of B, and to the ray b_1 of B there are two rays a_1 and a_2 of A—one of which is a ray of A which determined the ray b_1 of B.

In figure 12, the rays joining A and B to the point of intersection R of the rays p and q of the construction are corresponding double rays. In figure 13, the tangents from A to the conic have double rays of B corresponding to them,

* Proved by D. N. Lehmer, "Constructive Theory of Unicursal Cubic by Synthetic Methods," *Transactions of American Mathematical Society*, 1902.

† See Salmon, *Conic Sections* (ed. 10), p. 389.

and, similarly, the tangents from B to the conic have double rays of A corresponding to them. This correspondence can be studied further from this point of view.

IV

Locus Problem Suggested by the Discussion of the Quartic L.

15. *Locus Problem Synthetically*

By altering slightly the construction and interpretation of S_1, we obtain a problem whose solution is interesting as furnishing another example of the method of using Pascal's Theorem in problems involving pencils of conics. The power of this method is obvious from the several uses of it already made in this discussion. Suppose that S_1 is a point determined as the point of intersection of tangents to Γ at the points where it meets the double rays of I_1. If the point of tangency K of Γ be fixed and Γ be determined as a conic through I_1 and I_2 and tangent to λ at K, what is the locus of S_1 as the various conics Γ of the pencil are taken? This problem is not directly connected with the study of L previously undertaken, but comes in naturally as a problem connected with this particular pencil of conics.

In figure 14, let Γ cut AI_1 in R, and BI_1 in R'. Call the tangents at R and R', α and β respectively. Their points of intersection is S_1. The points I_1, I_2, K and R must satisfy Pascal's Theorem. Number them, and call the intersection of 12 and 45, L; of 23 and 56, M; of 34 and 61, N, as indicated in the figure. The points L, M, N are on the Pascal line, and N is a fixed point of this line for the given pencil of conics. Similarly, construct the Pascal line $L'M'N'$, replacing R in the construction with R'. As R moves along AI_1, it cuts out a point-row determining the conics of the pencil. This point-row projects to I_2 in a pencil of rays which determines on KI_1 (a fixed line) a point-row L perspective to the point-row R. The point-row L of KI_1 projects to N in a pencil of rays which describes a point-row M on AB, perspective to the point-row L of KI_1, and hence projective to the point-row R on AI_1. Therefore, the ray RM envelopes a conic, as R moves on AI_1, determining the various conics of the pencil. Similarly, $R'M'$ envelopes another conic. Since there is a one-to-one correspondence between the tangents RM and $R'M'$, S_1 is determined as the locus of intersections of two projective pencils of rays of the second class. This curve is the unicursal quartic, and has been discussed from this point of view.* If a common ray of the two pencils is self-corresponding, it is, obviously, part of the locus, so that, if all four common rays be self-corresponding, they constitute the locus. Indeed, if three are self-corresponding, the locus consists of them and of one additional line. We shall show in this problem that the locus is degenerate and consists of the lines AB, KI_2—counted twice—and another line.

* See Annie Dale Biddle, "Constructive Theory of the Unicursal Plane Quartic by Synthetic Methods," *Univ. Calif. Publ. Math.*, vol. 1, no. 2, 1912.

Denote by Σ the conic enveloped by RM, or α, and by Σ', the conic enveloped by $R'M'$, or β. AB and AI_1 are tangents to Σ, and AB and BI_1 are tangents to Σ'. Hence, AB is a common tangent of Σ and Σ'. Now, in the conic Σ, the points corresponding to A, considered first as a point of AB and then as a point of AI_1, are the points of tangency to Σ of AI_1 and AB, respectively. If R moves to A, RI_2 becomes AI_2 and cuts KI_1 in a point denoted by Q. NQ determines M_1 as the corresponding position of M. This construction will be referred to later. Hence, M_1 is the point of tangency of AB to Σ. Moreover, if R moves to A, the conic Γ degenerates into AB and I_1I_2, and R' moves to B. Hence, the common tangent AB of Σ and Σ' is self-corresponding and is therefore part of the locus. When R' moves to B, $R'I_2$ becomes BI_2, and intersects KI_1 in Q'. $N'Q'$ determines M_1' as the point of tangency of AB to Σ'.

If R moves to N, L moves to K, M moves to K, and Γ degenerates into KI_1 and KI_2. Since R is at N and M at K, KN is a tangent of Σ, being a ray RM. Since, when R is at N, Γ degenerates into KI_1 and KI_2, R' is either at I_1 or N', as it is always on the conic Γ. If R' is at I_1, M' is at B, and $R'M'$ is not a tangent to Γ. As this is contrary to hypothesis, R' is at N'. Consequently, L' is at K, and M' is at K. Hence KN is a tangent of Σ', that is, KN is a self-corresponding common tangent of Σ and Σ', and is therefore a part of the locus of S_1. We shall apply Brianchon's Theorem in the following way to determine its points of tangency to Σ and Σ'.

Let AB, BC, and CA of figure 15 be three tangents to a conic and let S and T be the points of tangency of CA and AB, respectively. Then R, the point of tangency of BC, is found by drawing through A, a ray passing through the point of intersection of BS and CT.

In figure 14, the point of tangency of AB to Σ has been found to be M_1. The point of tangency of AI_1 is found by moving M to A. If this is done, L moves to I_1 and R moves to I_1. Hence AI_1 is tangent to Σ at I_1. To discover the point of tangency of KN to Σ, apply Brianchon's Theorem as follows, remembering that R is at N. Join M_1 to N and K to I_1, these lines meeting in Q. AQ cuts KN in the required point of tangency, which is seen to be I_2, from the way in which Q was previously determined.

To discover the point of tangency of KN to Σ', it is necessary to find the point of BI_1 to Σ'. If M' moves to B, L' moves to I_1, and R' moves to I_1. Hence, BI_1 is tangent to Σ' at I_1. The point of tangency of AB to Σ' has already been discovered to be M_1'.

Now, to discover the point of tangency of KN to Σ', apply Brianchon's Theorem, remembering that R' is at N'. Join M_1' to N' and K to I_1, giving Q'. $Q'B$ cuts KN in the required point, which is seen to be I_2. Hence, Σ and Σ' are tangent to each other at I_2, and KI_2 is a double common tangent which is self-corresponding.

Hence, finally, the locus of S_1 is composed of the lines AB, KI_2 (counted twice) and another. That is to say, S_1 moves in general on a straight line.

Similarly, if an S_2 be determined by tangents to Γ at the points where the double rays of I_2 cut Γ, S_2 will move on a line. As different points K of λ are taken as points of tangency for Γ, S_1 and S_2 will move on lines which will correspond in pairs. Of this, more will be said later.

16. *Locus Problem Analytically*

The analytic discussion of this problem reveals a further property of these rays. This is inserted temporarily, as a complete synthetic discussion has not yet been obtained.

In figure 16, denote by I_1 and I_2 the fixed points, by AI_1 and BI_1 the fixed lines through I_1 and by K the fixed point of tangency of conics through I_1 and I_2 to an arbitrary line 1. The tangents a and β at M and N will determine S_1. Choose I_2, I_1, and K as the points $(1:0:0)$, $(0:1:0)$, and $(0:0:1)$ respectively. The equation of Γ may be put in the form

$$yz + zx + \lambda xy = 0$$

AB is the tangent at $(0:0:1)$. Hence, its equation is

$$AB \qquad x + y = 0.$$

Choose for coördinates of A and B, $(1:-1:\mu)$ and $(1:-1:\nu)$ respectively.

The equations of AI_1 and BI_1 are:

$$AI_1 \qquad z - \mu x = 0$$
$$BI_1 \qquad z - \nu x = 0.$$

Hence, the coördinates of M are $(\mu + \lambda):-\mu:\mu(\mu+\lambda)$. The coördinates of N, similarly, are $(\nu+\lambda):-\nu:\nu(\nu+\lambda)$.

This gives as the equation of a,—

$$[\mu(\mu+\lambda) - \mu\lambda]x + (\mu+\lambda)^2 y + \lambda z = 0$$

or

$$\mu^2 x + (\mu+\lambda)^2 y + \lambda z = 0.$$

Similarly, the equation of β is

$$\nu^2 x + (\nu+\lambda)^2 y + \lambda z = 0.$$

The coördinates of S_1, the intersection point of a and β, reduce to—

$$\mu + \nu + 2\lambda : -(\mu+\lambda) : \lambda\mu + \lambda\nu + 2\mu\nu.$$

Let the absolute coördinates of S_1 be (x_1, y_1, z_1); then

$$\zeta x_1 = \mu + \nu + 2\lambda \qquad (1)$$
$$\zeta y_1 = -(\mu+\lambda) \qquad (2)$$
$$\zeta z_1 = \lambda(\mu+\nu) + 2\mu\nu \qquad (3)$$

whence

$$(\mu+\nu)^2 x_1 + (\mu-\nu)^2 y_1 - 2(\mu+\nu) z_1 = 0$$

results as the equation of the locus of S_1. This represents a line which cuts

$$x + y = 0$$

in the point $(1:-1:\dfrac{2\mu\nu}{\mu+\nu})$.

Now the line I_1I_2 cuts

$$x + y = 0$$

in the point $(1:-1:0)$.

The harmonic conjugate of $(1:-1:0)$ with respect to

$$A(1:-1:\mu)$$

and

$$B(1:-1:\nu)$$

is

$$(1:-1:\frac{2\mu\nu}{\mu+\nu}).$$

Hence the following

Theorem: As the point of tangency K of Γ moves along AB, the line which is the locus of S_1 revolves about the harmonic conjugate, with respect to A and B, of the intersection of I_1I_2 and AB. If a similar construction for S_2 be made, the fixed points on the tangent AB being C and D, the line which is the locus of S_2 will revolve about the harmonic conjugate, with respect to C and D, of the intersection of I_1I_2 and AB.

17. *Quadratic Transformation of Plane Resulting therefrom*

If, now, the point of tangency of Γ to AB be allowed to move and S_1 be chosen at random, we inquire as to the number of points S_2 to a given S_1, and, also, regarding the locus of S_2 when S_1 moves, for instance, on an arbitrary line.

In figure 17, let I_1, I_2, and A be the vertices of the fundamental triangle; the points $(1:0:0)$, $(0:1:0)$, and $(0:0:1)$ respectively. Let the fixed rays of I_1 meet, in A and B, the fixed line AB which is tangent to Γ; and the fixed rays of I_2 meet this line in C and D. Call the tangents at the several points of intersection, α, β, γ, δ, as indicated in the figure: α and β determine S_1 and γ and δ determine S_2.

The equation of any conic through I_1 and I_2 may be written

$$az^2 + byz + czx + dxy = 0.$$

The equation of the fixed line 1 through A is

$$y - \lambda x = 0$$

where λ is a constant.

That this line may be tangent to the conic Γ is represented by

$$(b\lambda + c)^2 = 4ad\lambda \qquad (1),$$

since the equation obtained by solving

$$az^2 + byz + czx + dxy = 0$$

and

$$y - \lambda x = 0 \quad \text{simultaneously,}$$

viz:

$$az^2 + b\lambda zx + czx + d\lambda x^2 = 0$$

must be a perfect square.

The equation of a tangent to Γ at a point $(x':y':z')$ on the curve is

$$(dy' + cz')x + (bz' + dx')y + (cx' + by' + 2az')z = 0.$$

The ray AI_1, whose equation is $y=0$, meets Γ at $(a:o:—c)$. Hence, the equation of a, the tangent at this point, is

$$c^2x+(bc-ad)y+acz=0.$$

Denote B by $(1:\lambda:\mu)$, since it is an arbitrary point on 1.

The line BI_1 has for its equation

$$\mu y-\lambda z=0.$$

Its point of intersection with Γ (besides I_1) is

$$[-\mu(a\mu+b\lambda):\lambda(c\mu+d\lambda):\mu(c\mu+d\lambda)]$$

Hence, the equation of β, the tangent to Γ at this point, is

$$(c\mu+d\lambda)^2x+(bc-ad)\mu^2y+(ac\mu^2+bd\lambda^2+2ad\lambda\mu)z=0.$$

The intersection point of a and β is S_1. Denote its coördinates by $(x_1:y_1:z_1)$. By absorbing whatever multiplier there may be (say ζ) into the constants of Γ, we may write

$$x_1=-(b\lambda+2a\mu) \qquad (2)$$
$$y_1=c\lambda \qquad (3)$$
$$z_1=2c\mu+d\lambda \qquad (4)$$

and the first condition:

$$(b\lambda+c)^2=4ad\lambda \qquad (1)$$

The equations (1), (2), (3), and (4) determine the parameters a, b, c, d of the conic Γ. Since only one of these is quadratic in a, b, c, and d, the others being linear, there are two conics to a given S_1, and hence two points S_2 to a given point S_1, as there is but one S_2 for a given conic.

Conversely, to a given S_2, there are two points S_1. This construction gives, then, a semi-involutory two-to-two transformation of the plane.

Suppose the point S_1 to move on an arbitrary line of the plane, let us discover the nature of the locus of S_2. Let the points where the fixed rays of I_2 meet the line 1 be $C(1:\lambda:\phi)$ and $D(1:\lambda:\rho)$, where ϕ and ρ are constants.

The equations of CI_2 and DI_2 follow:

$$CI_2 \qquad z-\phi x=0$$
$$DI_2 \qquad z-\rho x=0.$$

CI_2 meets Γ in the two points

$$\begin{cases} 0:1:0 \\ b\phi+d:-\phi(a\phi+c):\phi(b\phi+d) \end{cases}$$

DI_2 meets Γ in the two points

$$\begin{cases} 0:1:0 \\ b\rho+d:-\rho(a\rho+c):\rho(b\rho+d) \end{cases}$$

The equation of γ is

$$\phi^2(bc-ad)x+(b\phi+d)^2y+(ab\phi^2+2ad\phi+cd)z=0.$$

The equation of δ is

$$\rho^2(bc - ad)x + (b\rho + d)^2 y + (ab\rho^2 + 2ad\rho + cd)z = 0.$$

Denote S_2 by $(x_2 : y_2 : z_2)$, whence, by reduction

$$x_2 : y_2 : z_2 = 2d + b(\rho + \phi) : 2a\rho\phi + c(\rho + \phi) : 2b\rho\phi + b(\rho + \phi)$$

Let S_1 move on the line

$$Lx_1 + My_1 + Nz_1 = 0.$$

Its coördinates will satisfy this equation, and the several equations determining the locus of S_2 are

$$\begin{aligned}
&-L(b\lambda + 2a\mu) + Mc\lambda + N(2c\mu + d) = 0 && (1)\\
&(b\lambda + c)^2 = 4ad\lambda && (2)\\
&\tau x_2 = 2d + b(\rho + \phi) && (3)\\
&\tau y_2 = 2a\rho\phi + c(\rho + \phi) && (4)\\
&\tau z_2 = 2b\rho\phi + d(\rho + \phi) && (5)
\end{aligned}$$

The parameters to be eliminated are a, b, c, d, and τ. As all the equations but one are linear in these, and the exceptional one is quadratic, the locus of S_2 is a conic. Hence, we may write the

Theorem: The locus of S_2, as S_1 moves on a line, is a conic; or, in general, the locus of S_2, *as S_1 moves on a point-row of order n, is a point-row of order $2n$.*

From equation (2), we note that $d = 0$ gives the equation of a tangent to the S_2 conic. If $d = 0$, we have, by elimination from (3) and (5)

$$\frac{X}{Z} = \frac{\rho + \phi}{2\rho\phi} \text{ or } 2\rho\phi x - (\rho + \phi)z = 0.$$

This particular line is obtained regardless of which line S_1 moves on and is therefore tangent to all the S_2 conics. It is the line through I_2, which passes through the harmonic conjugate, with respect to C and D, of the intersection of I_1I_2 and AB.

Similarly, there is a line through I_1, which is tangent to all the S_1 conics which correspond to lines described by S_2. Hence, the following

Theorem: By means of the conics tangent to an arbitrary line and passing through two arbitrary points, through each of which two arbitrary rays are chosen, a quadratic transformation of the plane may be established. To every point S_1, there are two points S_2, and, conversely, the correspondence being semi-involutory. If S_1 move on a point-row of order n, S_2 moves on a point-row of order $2n$. All the S_2 loci are tangent to a certain invariant ray of the second of the two fixed points. Similarly, all the S_1 loci corresponding to arbitrary paths of S_2 are tangent to a certain invariant ray of the first of the two fixed points. In particular, however, if S_1 describe a ray passing through a certain point of the fixed tangent line, S_2 describes a ray passing through another certain point of the tangent line. Thus, in this quadratic transformation, there is a particular pencil of rays which goes into a pencil of rays.

VITA

I, BALDWIN MUNGER WOODS, was born in Lampasas, Texas, on September 22, 1887. I studied in the public schools of Fort Worth, Texas, until 1904, when I entered the University of Texas, from which I received the degree of Electrical Engineer in 1908. During the year 1907–08 I held the position of Assistant in Applied Mathematics in that institution.

In 1909, I was appointed John W. Mackay, Junior, Fellow in Electrical Engineering at the University of California. I held this position until January, 1910, when I was appointed Assistant in Mathematics. From July, 1910, to the present I have been Instructor in Mathematics in the University of California. In 1909, I received the degree of M.S. in Electrical Engineering from the University of California.

In the University of Texas, I studied under Professors Porter and Benedict and Mr. Rice in mathematics, and Professors Scott and Taylor in engineering. In the University of California, I have studied under Professors Stringham, Haskell, Lehmer, and Putnam in mathematics; under Professors Cory and LeConte in engineering; and under Professor Raymond in physics. To all these I wish to express my thanks,—especially to Professors Lehmer and Haskell, who, in their supervision of the present work, have been a constant source of inspiration.

On April 29, 1912, I passed the public final examination for the degree of Ph.D.

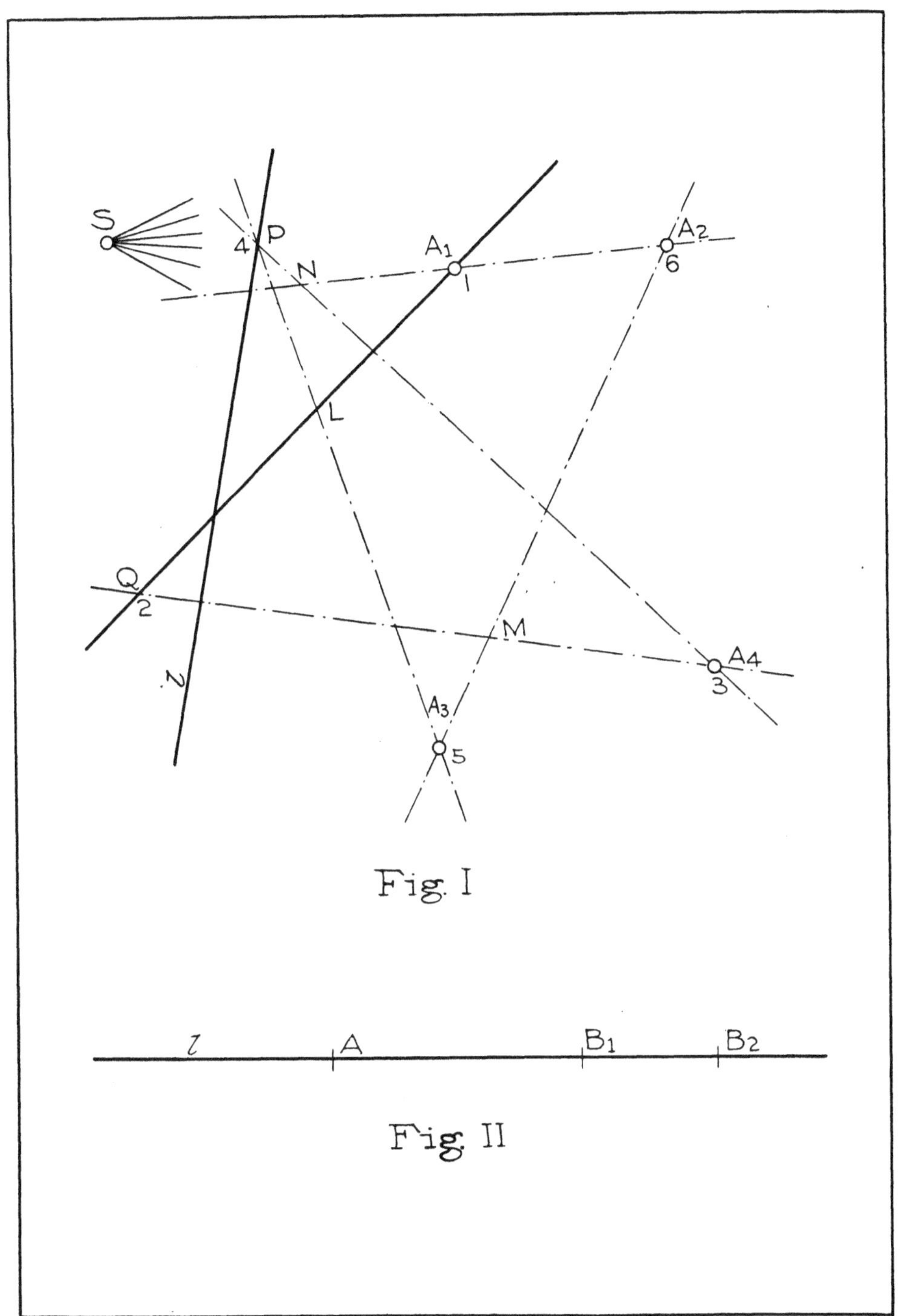

Figs. 1 and 2

A1
A2
N
6
1
2
M
a1
4
A3
5
a3
L
3
A4
Fig. III
P
R
A2
A1
B1
B2
b1
a1
B3
B4
A4
A3
Fig. IV
A1 B1
B2
R
A2
b1
P
a1
A3
A4
B4
B3
Fig. V

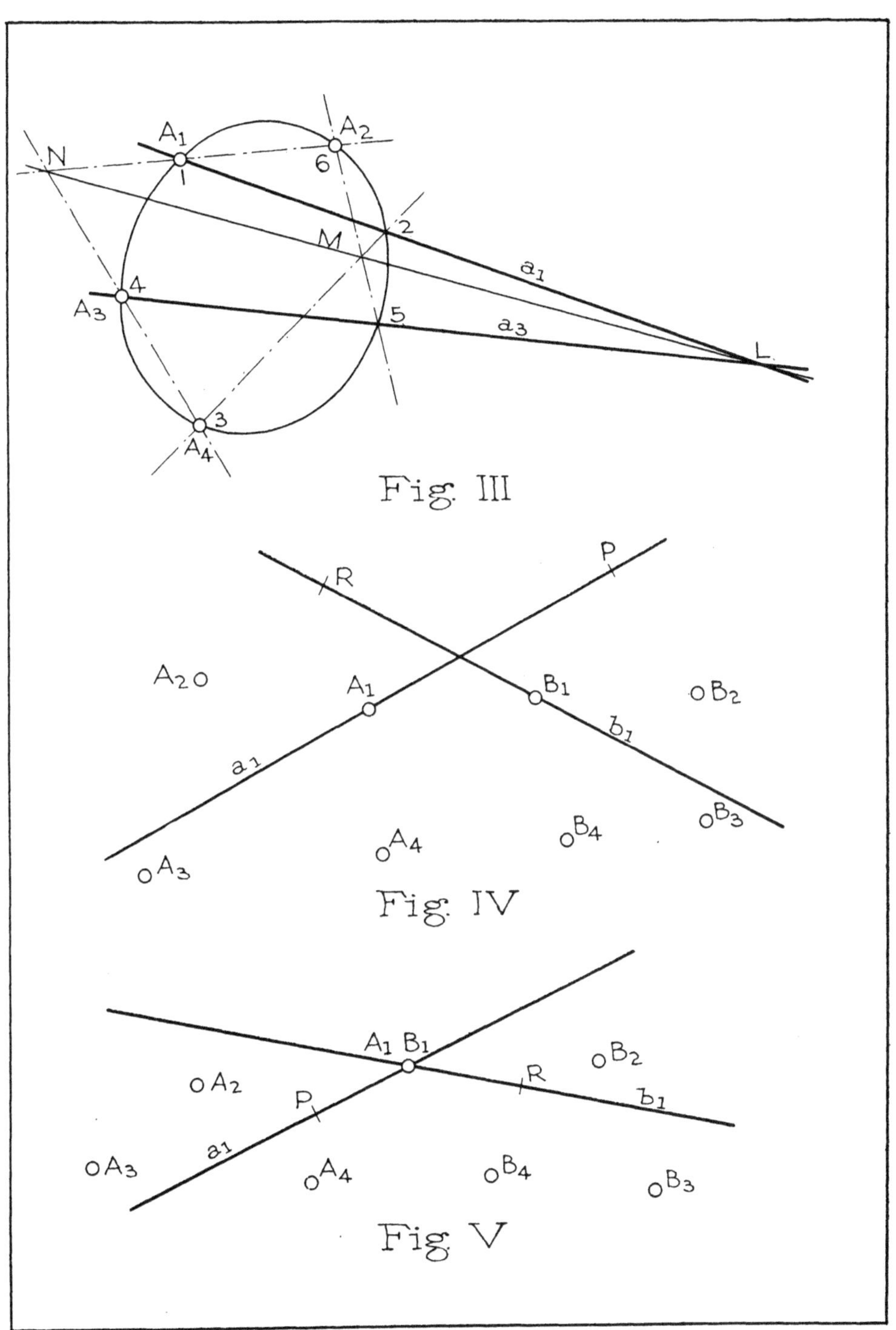

Figs. 3–5

A''
A'
B'
B''
l
A1
B1
A4
B4
A3
B3
B2
b2
a2
A2
Fig. VI
G1
G2
H1
H2
λ
g2
g1
h1
h2
Γ
I2
I1
S1
S2
Fig. VII
P
A
S
Q
R
I1
I2
B
Fig. VIII

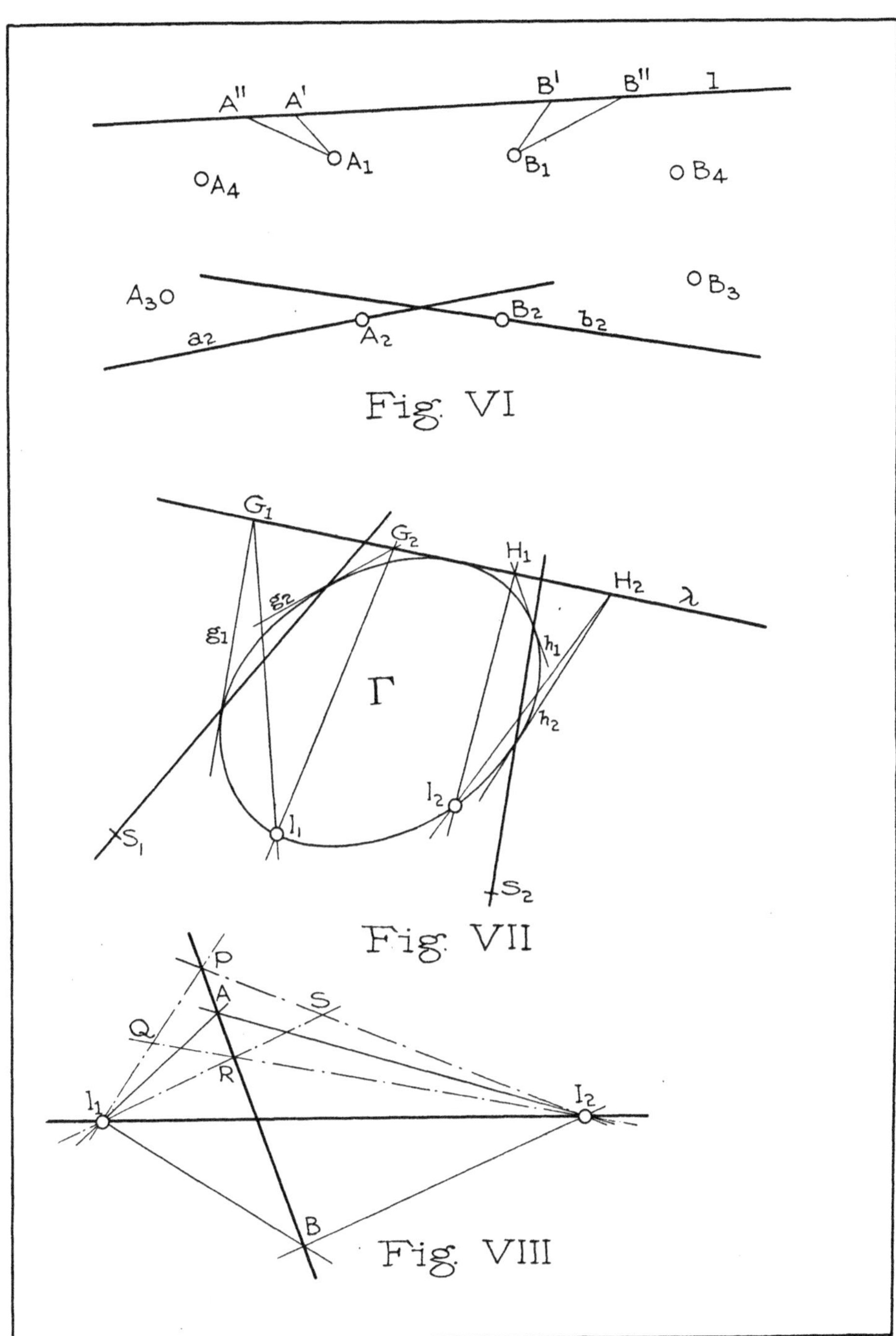

Figs. 6–8

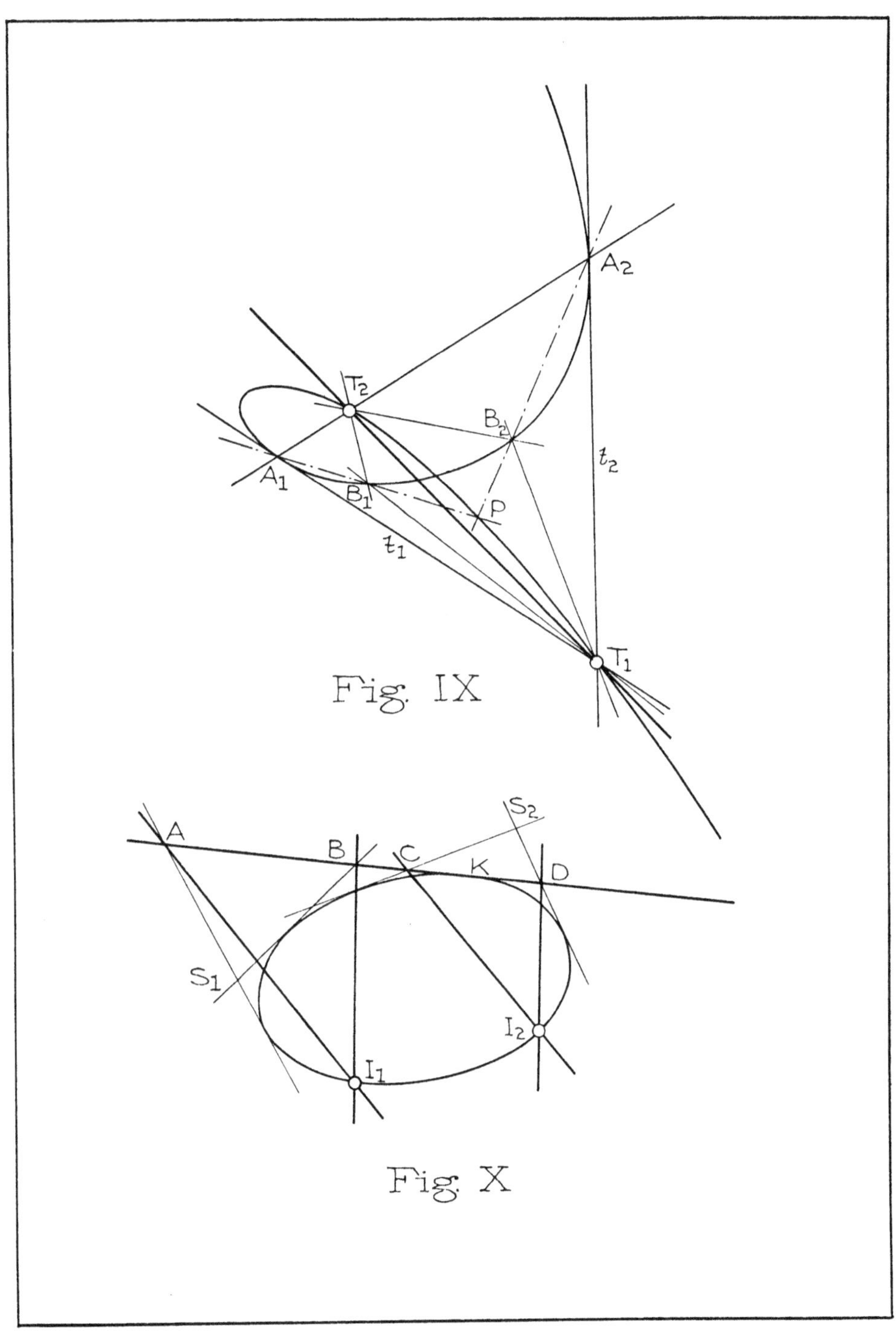

Figs. 9 and 10

G
H1
h1
h2
H2
λ
S2
g
Γ
I2
I1
Fig. XI
A
B
b2
a2
b1
p
a1
R
q
Fig. XII
A
B
b2
a2
a1
b1
Fig. XIII

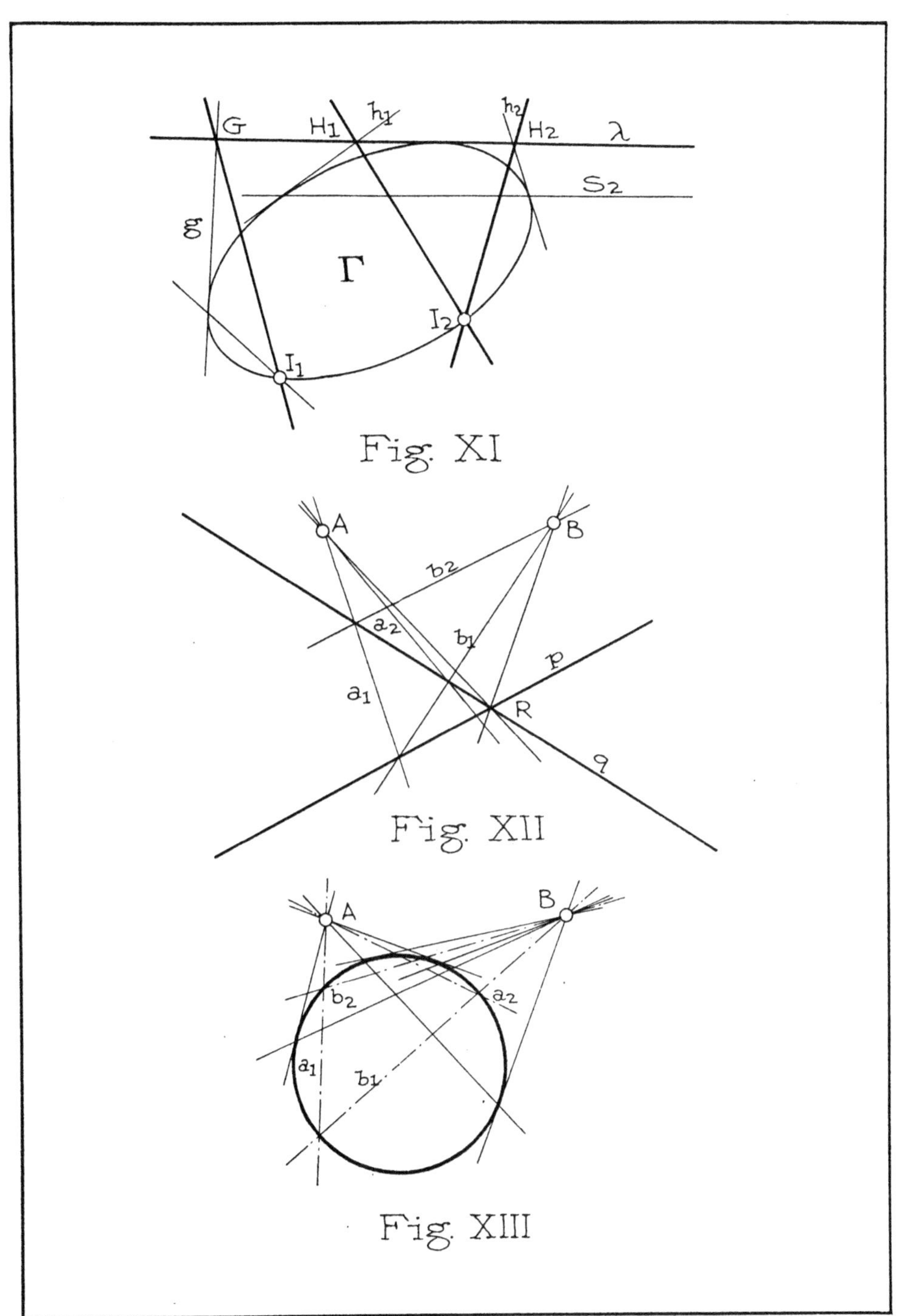

Figs. 11–13

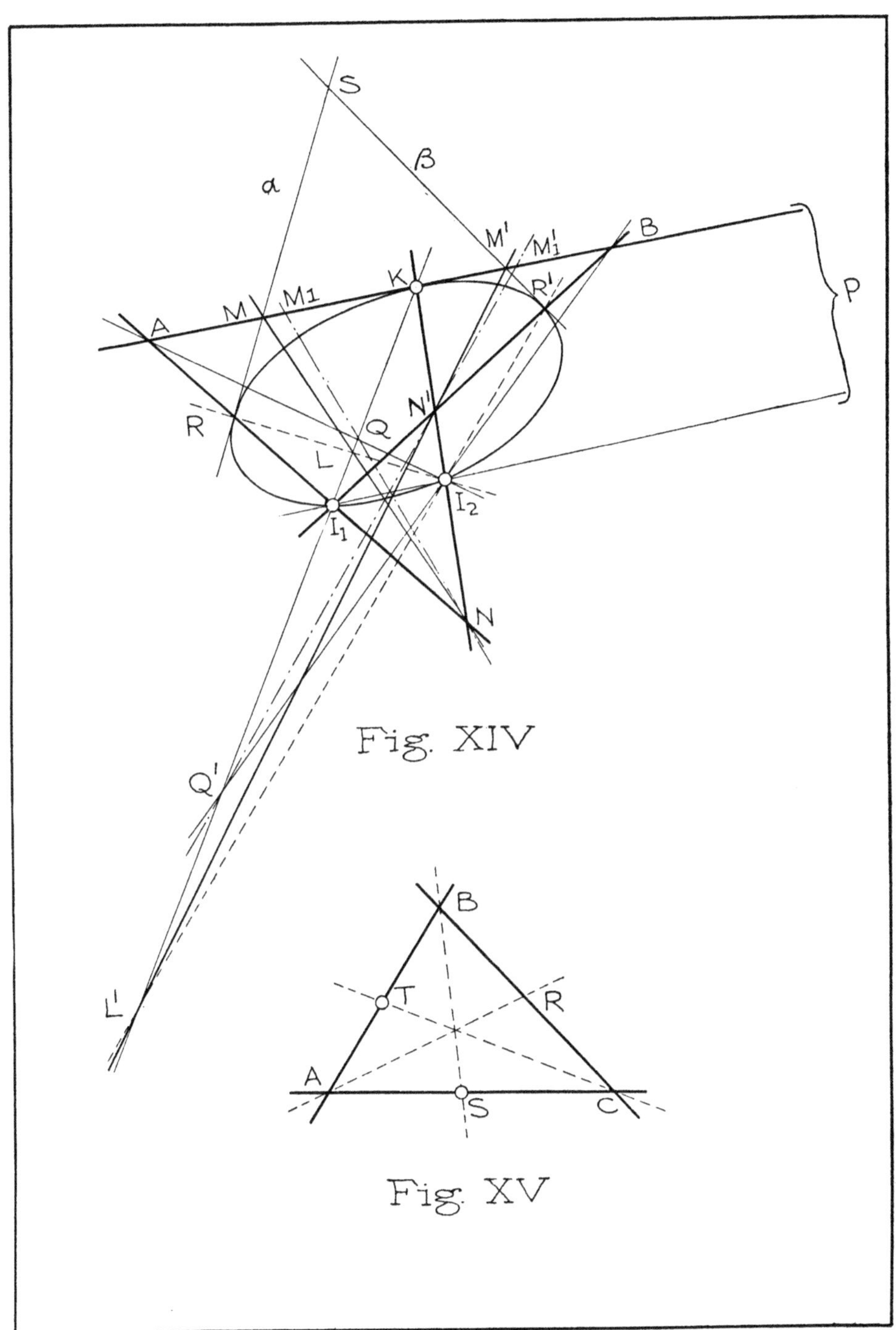

Figs. 14 and 15

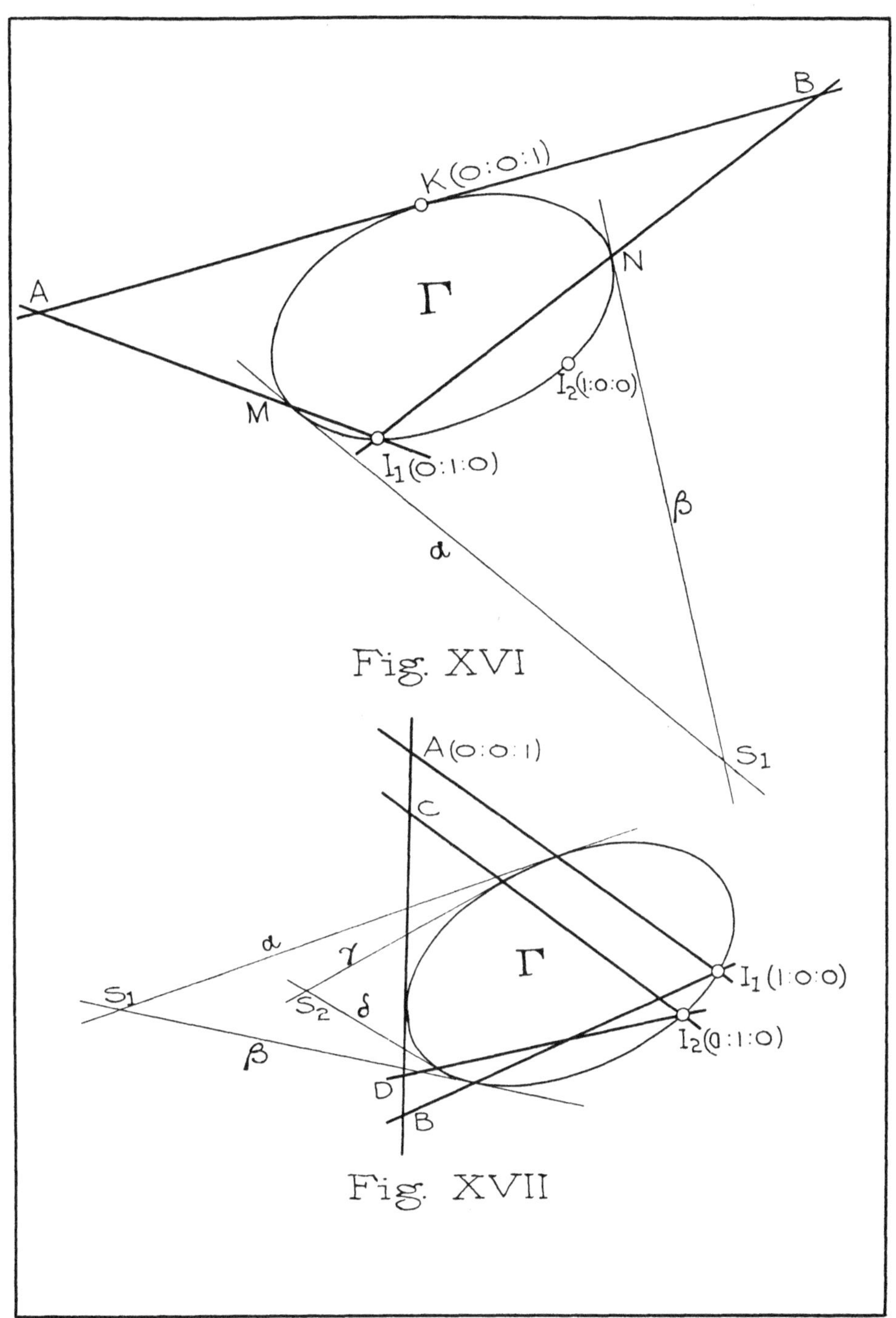

Figs. 16 and 17

UNIVERSITY OF CALIFORNIA PUBLICATIONS
IN
MATHEMATICS

Vol. 1, No. 4, pp. 87-96 May 15, 1914

A COMPLETE SET OF POSTULATES FOR THE LOGIC OF CLASSES EXPRESSED IN TERMS OF THE OPERATION "EXCEPTION", AND A PROOF OF THE INDEPENDENCE OF A SET OF POSTULATES DUE TO DEL RÉ*

BY
B. A. BERNSTEIN

INTRODUCTION

The first complete set of *independent* postulates for the logic of classes—the logic invented by Boole and perfected by Schröder—is due to E. V. Huntington. In his "Sets of independent postulates for the algebra of logic" (*Transactions of the American Mathematical Society,* July, 1904, p. 288) he presents, and proves the independence of, three sets of postulates for the logic of Boole: one is a modification of a set due to Whitehead, in which two operations, $\oplus$ and $\odot$, are assumed; another is founded on Schröder's axioms, postulates, and definitions, which are all expressed in terms of the relation $\ominus$; and a third is Huntington's own set, in which only the operation $\oplus$ of logical addition is assumed. Quite recently another set of independent postulates for Boolean logic was proposed. In a paper, "A set of independent postulates for the Boolean algebra,"[1] H. M. Sheffer presented a set of postulates all expressed in terms of the operation "per", denoted by the vertical line $|$, $a \mid b$ being ultimately identified with $\bar{a}\bar{b}$—"neither a nor b."

Of course, by virtue of Peirce's duality principle, the substitution of $\odot$ for $\oplus$, together with the interchange of V and Λ (the "whole" and "zero" respectively), will transform Huntington's third set of postulates into a set in which the single operation assumed is that of logical multiplication. Likewise, the

* Read before the American Mathematical Society (San Francisco section) October 25, 1913.

[1] Read before the American Mathematical Society December 31, 1912. An abstract of the paper is given in the *Bulletin of the American Mathematical Society,* March, 1913. Dr. Sheffer was kind enough to let me see his postulates. The complete paper appeared in the October (1913) number of the *Transactions of the American Mathematical Society.*

substitution of the operation $|_+$ (say) for $|$, $a \,|_+\, b$ reading "either *not-a* or *not-b*," will change Sheffer's set into one in which the single operation involved is the operation which forms from the logical elements a, b the element usually denoted by $\bar{a} + \bar{b}$. It is thus seen that of the six essentially different operations which form a new element from two given logical elements by means of logical addition alone or of logical multiplication alone, that is, of the operations which form from a, b the elements

$$a + b,\ a + \bar{b},\ \bar{a} + \bar{b},\ ab,\ a\bar{b},\ \bar{a}\bar{b},$$

Professor Huntington has chosen one and Dr. Sheffer another as basis for a set of postulates. Four of the six operations have thus been utilized for postulate building, the operations $\oplus$, $|$, carrying with them their respective duals $\odot$, $|_+$. It therefore suggests itself to one to complete this group of postulate sets by constructing the pair based on the two remaining operations. Accordingly, in part I of this paper I present a set of postulates involving the operation "exception", —, $a - b$ reading "a except b," or "a which is *not-b*"; which, also, carries with it the dual set involving the operation $\div$, which I call (having in mind Euler's diagrams) "adjunction", $a \div b$ reading "either a or *not-b*."[2]

In my set, as in Dr. Sheffer's, the existence of "zero" and of the "negative" is proved; the "whole", however, is postulated. (Professor Huntington postulates the existence of all three of these elements.)

In proving the independence of my postulates I employ the usual method—that of exhibiting for each postulate a "system" which contradicts this postulate but which satisfies all the rest.[3]

To show that my set is "sufficient" for the algebra of classes I deduce from it the set of postulates[4] from which Professor Del Ré develops his *Algebra della Logica* (Naples, 1907).

In part II of this paper I substantiate Professor Del Ré's (mere) assertion[5] of the independence of his postulates by exhibiting the necessary proof systems, hoping thereby to add to the completeness of his admirable *Algebra*.

---

[2] We have the identities $a - b = a\bar{b}$, $a \div b = a + \bar{b}$. So that the operations exception and adjunction are not logical subtraction and logical division, the inverses of logical addition and of logical multiplication respectively. $a - b$ and $a \div b$, as results of the operations exception and adjunction, are identical with the logical difference and the logical quotient only when, respectively, $b < a$ and $a < b$.

[3] Among the writers who used this method are to be mentioned especially (only the earliest papers known to me being cited):

PEANO, "Sul concetto di numero," *R. d. M.* (*Rivista di Matematica*), vol. 1 (1891), p. 93;

BURALI-FORTI, "Sulla teoria delle grandezze," *R. d. M.*, vol. 3 (1893), p. 78;

PADOA, *R. D. M.*, vol. 5 (1895), p. 185, note (in paper of Vailati, "Sulle proprietà caratteristiche delle varietà a una dimensione");

PIERI, "I principii della geometria di posizione," *Mem. Accad. Sci. Torino*, vol. 48 (1898), p. 60;

HILBERT, *Grundlagen der Geometrie*, Leipzig, 1899;

HUNTINGTON, "A complete set of postulates for the theory of absolute continuous magnitude," *Trans. Am. Math. Soc.*, vol. 3 (1902), p. 278;

VEBLEN, "A system of axioms for geometry," *Trans. Am. Math. Soc.*, vol. 5 (1904), p. 347.

[4] See below. The postulates are Huntington's first set in which the distributive law $x + yz = (x + y)\,(x + z)$ is replaced by the two laws: $x + x = x$, $x + (y + z) = (x + y) + z$.

[5] *Algebra della Logica*, p. 8.

I

A Set of Postulates in Terms of the Operation Exception

Let us take as undefined ideas a *class* K of *elements* $a, b, c, \ldots$ and an *operation* "—", $a - b$ reading "a except b." Let us understand by $a = b$ that a, b can be interchanged at pleasure; by $a \neq b$ that this interchange cannot take place. Then the logic of classes may be defined as a system $(K, -)$ which satisfies the following six postulates:

Postulates.

1. $a - b$ is an element of K whenever a, b are elements of K.

2. If postulate 1 hold, there exists an element $\mathbf{1}$ in K such that
$$a - (b - \mathbf{1}) = a,$$
whenever a, b are elements of K.

3. If postulate 1 hold,
$$(a - a) - c = (b - b) - d,$$
whenever a, b, c, d are elements of K.

4. From $a - b = b - a$ follows $a = b$,
whenever $a, b, a - b, b - a$ are elements of K.

Def. 1. $a' = \mathbf{1} - a$.[6]

5. If postulates 1, 2 hold, and if the element $\mathbf{1}$ is unique, then
$$(a - b) - (c - d) = \{[(a - b) - d']' - [(c' - a') - b]\}',$$
whenever a, b, c, d are elements of K.

6. If postulate 2 hold, there is an element a in K such that $a \neq \mathbf{1}$.

Consistency of the Postulates.

As a system $(K, -)$ satisfying all the postulates 1–6, thus proving their consistency, we have:

$K =$ the class of two elements e_0, e_1, with $e_i - e_j$ defined by the "exception" table:

—	e_0	e_1
e_0	e_0	e_0
e_1	e_1	e_0

The element $\mathbf{1}$ required by postulate 2 is e_1, which is unique. Postulate 5 is seen to hold by noting that the substitution of e_0 or of e_1 for a, b, c, or d will reduce 5 to an identity.

Theorems.

The following propositions, deducible from postulates 1–6, will throw into our hands Professor Del' Ré's eleven postulates as theorems, and will thus prove the sufficiency of our postulates for the algebra of classes.

6 In Peano's *Formulaire de Mathematiques*, $-a$ is used for a', the "negative" of a. Def. 1 makes clear the distinction between the *sign* "—" considered as *belonging to* an element and the *operation* "—" involving two elements. $-a$ is merely an abbreviation for $1 - a$.

7. $a - a = b - b$.
For,

$$(a - a) - (a - \mathbf{1}) = a - a, \text{ and } (b - b) - (a - \mathbf{1}) = b - b$$
(by 2).

Also,

$$(a - a) - (a - \mathbf{1}) = (b - b) - (a - \mathbf{1})$$
(by 3).

Therefore,

$$a - a = b - b.$$

Hence, we may define $\mathbf{0}$, "zero":

8. *Def. 2.* $a - a = \mathbf{0}$.

9. $a - \mathbf{1} = \mathbf{0}$.
For,

$$a - \mathbf{1} = (a - \mathbf{1}) - (a - \mathbf{1}) = \mathbf{0}$$
(by 2, 8).

10. $a - \mathbf{0} = a$.
For,

$$a - \mathbf{0} = a - (a - \mathbf{1}) = a$$
(by 9, 2).

11. $\mathbf{0} - a = \mathbf{0}$.
For,

$$\mathbf{0} - a = (\mathbf{0} - \mathbf{0}) - a = (\mathbf{0} - \mathbf{0}) - \mathbf{0} = \mathbf{0} - \mathbf{0} = \mathbf{0}$$
(by 8, 3).

12. The element $\mathbf{1}$ of postulate 2 is unique.
For, suppose two such elements: $\mathbf{1}_1$ and $\mathbf{1}_2$. Then $\mathbf{1}_1 - \mathbf{1}_2 = \mathbf{1}_2 - \mathbf{1}_1 = \mathbf{0}$
(by 9).

Therefore,

$$\mathbf{1}_1 = \mathbf{1}_2$$
(by 4).

Cor. 1. $a' = \mathbf{1} - a$ is unique.
Cor. 2. From $a = b$ follows $a' = b'$.

13. $\mathbf{1}' = \mathbf{0}$; $\mathbf{0}' = \mathbf{1}$.
For,

$$\mathbf{1}' = \mathbf{1} - \mathbf{1} = \mathbf{0}$$
(by Def. 1, 8);

and

$$\mathbf{0}' = \mathbf{1} - \mathbf{0} = \mathbf{1}$$
(by Def. 1, 10).

14. $(a')' = a$.
For, in 5, let $b = \mathbf{0}$, $c = \mathbf{1}$, $d = \mathbf{1}$; then the left member will reduce to a and the right to $(a')'$
(by 2, 10, 13, 11).

Def. 3. $a'' = (a')'$.

15. $(a - b) - c = (c' - a') - b$.

Proved by letting $d = \mathbf{0}$ in 5
(by 10, 13, 9, Def. 1, 14).

16. $a - b = b' - a'$.

Proved by letting $b = \mathbf{0}$, $c = b$ in 15
(by 10).

17. $a - (b - c) = [(a - b)' - (a - c')]'$.

Proved by writing a, $\mathbf{0}$, b, c respectively for a, b, c, d in 5
(by 10, 16, 14).

18. $a - a' = a$.

Proved by writing $b = \mathbf{0}$, $c = a$, $d = a$ in 5
(by 10, 8, 14).

19. $(a - b) - c = (a - c) - b$
(by 15, 16, 14).

20. $\mathbf{0} \neq \mathbf{1}$.

For if $\mathbf{0} = \mathbf{1}$, then, whatever a is,

$$a = a - \mathbf{0} = a - \mathbf{1} = \mathbf{0} = \mathbf{1}$$

(by 10, 9);

contrary to 6.

21. $a \neq a'$.

For if $a = a'$, then $a = a - a' = a - a = \mathbf{0}$
(by 18, 8);

hence

$$a' = \mathbf{0}' = \mathbf{1}$$

(by 12 Cor. 2, 13).

Therefore,

$$\mathbf{0} = a = a' = \mathbf{1};$$ contrary to 20.

22. *Def. 4.* $ab = a - b'$.

Def. 5. $a + b = (a' - b)'$.

Hence (by 22, 14),

23. $(a + b)' = a'b'$; $(ab)' = a' + b'$;

the well-known *De Morgan's formulae.*

Del Ré's Postulates.

The propositions I–VIII following, taken by Professor Del Ré as basis for his *Algebra della Logica,* may now be derived as theorems.

I. xy is an element of K, if x and y are elements of K
(by 22, 1).

I′. $x + y$ is an element of K, if x and y are elements of K
(by 22, 1).

II. There is an element T in K such that

$$xT = x$$

for every element x.

(The element T is **1**, by 22, 13, 10.)

II′. There is an element N in K such that

$$x + N = x$$

for every element x.

(The element N is **0**, by 22, 10, 14.)

III. $xy = yx$,

if x, y, xy, yx are elements of K

(by 22, 16, 14).

III′. $x + y = y + x$,

if x, y, $x + y$, $y + x$ are elements of K

(by 22, 16, 14).

IV. $x + (y + z) = (x + y) + z$,

if x, y, z, and their combinations are elements of K.

(For, $x + (y + z) = [x' - (y' - z)']' = [(y' - z) - x]' =$
$[(y' - x) - z]' = [(x' - y) - z]' = [(x' - y)'' - z]' = (x + y) + z$,
by 22, 16, 14, 19.)

V. $x\,(y + z) = xy + xz$,

if x, y, z, and their combinations are elements of K.

(For, $x\,(y + z) = x - (y' - z)'' = x - (y' - z) = [(x - y')' - (x - z')]'$
$= xy + xz$, by 22, 14, 17.)

VI. $x + x = x$,

if x, $x + x$ are elements of K

(by 22, 18, 14.)

VII. If T, N of postulates II, II′ exist and are unique, then for every element x in K there is an element $\bar{x}$ such that

$$x\bar{x} = N, \text{ and } x + \bar{x} = T.$$

(The elements **1**, **0**, x' will serve for T, N, $\bar{x}$ respectively, by 22, 8, 13.)

VIII. K has two distinct elements

(by 6).

The whole algebra of classes, having been derived from I–VIII, may thus also be derived from postulates 1–6.

It can easily be verified that propositions 1–6 can be derived from I–VIII, so that *the two sets of postulates are*, in fact, *equivalent*.

Independence of Postulates 1–6.

The systems $(K, -)$ following prove the independence of postulates 1–6, since there is for each postulate a system that is false for that postulate and not false for the rest. Denoting the elements of K by e_i, the following tables define these systems ($e_i \neq e_j$ if $i \neq j$).

1.

—	e_0	e_1
e_0	e_0	e_0
e_1	e_1	a

where a is not in K.

All the postulates, except 1, are satisfied by this system.

2.

—	e_1	e_2	e_3
e_1	e_1	e_1	e_1
e_2	e_3	e_1	e_2
e_3	e_3	e_3	e_1

Postulate 2 does not hold, since no element e_i can serve as the element **1**. In fact, $e_2 - (e_i - e_i) = e_3$, whatever e_i is.

Postulate 5 is not contradicted, since no element **1** exists.

3.

—	e_1	e_2
e_1	e_1	e_1
e_2	e_2	e_2

Postulate 3 is not satisfied; for, $(e_1 - e_1) - e_1 \neq (e_2 - e_2) - e_1$.

Postulate 5 is not contradicted, as the element **1** is not unique.

4

—	e_0	e_1	e_2	e_3
e_0	e_0	e_0	e_0	e_0
e_1	e_1	e_0	e_1	e_0
e_2	e_2	e_0	e_0	e_0
e_3	e_3	e_0	e_0	e_0

Here $e_2 - e_3 = e_3 - e_2$, but $e_2 \neq e_3$; so that 4 is false.

Postulate 2 holds, either e_1 or e_3 serving as the element **1**.

5.

—	e_0	e_1	e_2	e_3
e_0	e_0	e_0	e_0	e_0
e_1	e_1	e_0	e_1	e_0
e_2	e_2	e_0	e_0	e_1
e_3	e_3	e_0	e_0	e_0

Here $\mathbf{1} = e_1$, unique. Postulate 5 is false, since $(e_2 - e_0) - (e_0 - e_1) \neq \{[(e_2 - e_0) - e'_1]' - [(e'_0 - e_2)' - e_0]\}'$.

6. The class of one element e_0, with $e_0 - e_0 = e_0$.

Adjunction. The Duals of Postulates 1–6.

The operation which forms the element $a + b'$ from the elements a, b will be called *adjunction* and will be denoted by $\div$. We have:

Def. 6. $a \div b = a + b'$.

Since $a + b' = (a' - b')'$, and since $a - b = ab' = (a' + b)' = (a' \div b')'$, we get

$$(a \div b)' = a' - b'; \; (a - b)' = a' \div b';$$

formulae that play exactly the same role in connecting the operations exception and adjunction as De Morgan's formulae do in the case of logical addition and logical multiplication; indeed, they are De Morgan's formulae expressed in different language. From this follows (with the help of 12 Cor. 2):

The Law of Duality. For any proposition in the logic of classes expressed in terms of $-$ *and* $\div$ *there is a "dual" proposition which is obtained from the former by interchanging the signs* $-$, $\div$, *and the elements* **1**, **0**.

We thus have the following:

Duals of 1–6.

1′. $a \div b$ is an element of K whenever a, b are elements of K.

2′. If 1′ hold, there exists an element in **0** in K such that

$$a \div (b \div \mathbf{0}) = a,$$

whenever a, b are elements of K.

3′. If 1′ hold,

$$(a \div a) \div c = (b \div b) \div d,$$

whenever a, b, c, d are elements of K.

4′. From $a \div b = b \div a$ follows $a = b$,

whenever a, b, $a \div b$, $b \div a$ are elements of K.

Def. 1′. $a' = \mathbf{0} \div a$.

5′. If 1′, 2′ hold, and if the element **0** is unique, then

$$(a \div b) \div (c \div d) = \{[(a \div b) \div d']' \div [(c' \div a') \div b]\}',$$

whenever a, b, c, d are elements of K.

6′. If 2′ hold, there is an element a in K such that $a \neq \mathbf{0}$.

It is evident that propositions 1′–6′ might have been taken as postulates and propositions 1–6 derived from them.

II

PROOF OF THE INDEPENDENCE OF DEL RÉ'S POSTULATES

The following systems (K, $\odot$, $\oplus$) prove the independence of Del Ré's postulates I–VIII above. (The signs $\times$, $+$ without circles around them will denote arithmetical multiplication and addition.) I have aimed to make all the systems concrete. These "near algebras of logic" may be of interest in themselves. The systems are:

I. $K =$ the class of the two numbers 1, 2;

$x \odot y = xy$; $x \oplus y =$ G. C. D. of x, y.

Here $N = 2$, $T = 1$. Postulate I is false, because 2×2 is not an element of K.

I′. The same as for I, with the interpretations of $\odot$ and $\oplus$ interchanged.

II. $K =$ the class of concentric circles enclosed by a circle W of center C, including the point-circle C but excluding circle W;

$x \odot y =$ the largest circle contained in x, y;

$x \oplus y =$ the smallest circle containing x, y.

Postulate II does not hold, since no element T exists. II′ holds, the element N being the point-circle C. Postulate VII is not false, for T does not exist.

II′. K = the class of positive rationals;

$$\frac{a}{b} \odot \frac{c}{d} = \frac{a}{b} \cdot \frac{c}{d}; \qquad \frac{a}{b} \oplus \frac{c}{d} = \frac{a+c}{b+d}, \text{ the } \textit{mediant} \text{ of } \frac{a}{b}, \frac{c}{d}$$

Postulate II′ is false, since no element N exists.

The element T is the rational $\frac{1}{1}$.

III. K = the class of point-groups in a plane area A, including the "null" group of no points;

$x \odot y$ = group x; $x \oplus y$ = the smallest group containing the points in x and in y.

Postulate III does not hold.

The element T is not unique, and so VII is not contradicted.

III′. K = the same as for III;

$x \odot y$ = the largest group of points contained in x and in y; $x \oplus y$ = group x.

Here N is not unique.

IV. K = the class of letters e_i, the subscripts i being 0, 1, 2, 3, 4 and the combinations of 1, 2, 3, 4 taken two, three, and four at a time;

$e_i \odot e_j = e_p$, where p is composed of the integers common to i and j;

$e_i \oplus e_j = e_s$, where s is composed of the different integers found in i and j *with the exception that* $e_{12} \oplus e_3 = e_3 \oplus e_{12} = e_3$.

Postulate IV is not satisfied. For, $(e_1 \oplus e_2) \oplus e_3 = e_3$, while

$$e_1 \oplus (e_2 \oplus e_3) = e_{123}.$$

$T = e_{1234}$; $N = e_0$.

V. K = the class of linear spaces[7] passing through a point P in a space of n dimensions, say S_3 for concreteness, including the space S_3 and the space of zero dimensions P;

$x \odot y$ = the space of greatest dimensions contained in x and y;

$x \oplus y$ = the space of least dimensions containing x and y.

Postulate V is not satisfied. For, let x, y, z be any three *distinct co-planar* straight lines; then $x \odot (y \oplus z) = x$, while $(x \odot y) \oplus (x \odot z) = P$.

$T = S_3$; $N = P$; $\bar{P} = S_3$; $\bar{S}_3 = P$; for $x \neq P$, S_3, the element $\bar{x}$ is any space not containing x which determines with x the space S_3.

VI. K = the class of point-groups in a plane area A, including A and the "null" group N;

$x \odot y$ = the largest group common to x, y;

$x \oplus y$ = the smallest group of points in x and in y, excluding those points common to x and y.

Postulate VI does not hold, for $x \oplus x = N$ for all groups x.

---

7 A linear space is one which contains the straight line joining any two of its points.

VII. $(K, \odot, \oplus) =$ the system of circles given for II, with the exception that W is not excluded.

No element $\bar{x}$ exists for every x, such as postulate VII demands.

$N = C;\ T = W.$

VIII. $K =$ the class containing the single number 0;

$$x \odot y = xy;\quad x \oplus y = x + y.$$

Transmitted December 20, 1913.

UNIVERSITY OF CALIFORNIA PUBLICATIONS
IN
MATHEMATICS

Vol. 1, No. 5, pp. 97–114 July 6, 1914

ON A TABULATION OF REDUCED BINARY QUADRATIC FORMS OF A NEGATIVE DETERMINANT.

BY

HARRY N. WRIGHT.

The purpose of this paper is to study certain relationships among positive reduced binary quadratic forms of a negative determinant and to derive enumeration formulæ for the determination of the class number of such forms.

The discussion is arranged in four parts. Part I contains a description of a useful tabulation of the forms studied and a determination of the number of solutions of the congruence $x^2 \equiv \mathrm{D} \pmod{a}$, in which the usual restriction that $[\mathrm{D}, a] = 1$ is removed.

In Part II is derived, from the results of Part I, an expression for the number of reduced forms for given values of a and Δ. Then this is summed for all values of $a \leqq [\sqrt{\Delta}]$ and a method of computing the number of reduced forms for $a > \sqrt{\Delta}$ is described.

Part III is a development of formulæ giving the number of single forms on any value of Δ.

In Part IV theorems are developed showing the distribution of reduced forms of the first and second sorts, and the relationship between the number of single forms of the first sort and the number of those of the second sort is determined for values of Δ containing no square factor.

Part I.

The well-known conditions necessary for $ax^2 + 2bxy + cy^2$ to be reduced when $b^2 - ac < 0$ are that: $|2b| \leqq a$, $a \leqq c$ and if $|2b| = a$ or $a = c$ then $b \geqq 0$. Such forms may be arranged in a table with a and Δ as arguments, as shown in the accompanying tabulations, and in which it is necessary to enter only the values of b and c. The forms in a given column, i. e. having a given value of a, arrange themselves in periods, each period covering a values of Δ and containing a forms. This is seen from the following consideration: if the form (a_1, b_1, c_1) occurs on Δ_1 then $(a_1, b_1, c_1 + 1)$ occurs on $\Delta_2 = \Delta_1 + a_1$; for from the equation $ac - b^2 = \Delta$ we see that a change of one unit in c causes a corresponding change of a units in Δ. Therefore the forms of a given column with a given value of b occur at intervals of a places. In a column there are just a possible values of b, viz:

$$0, \quad \pm 1, \quad \pm 2, \quad \pm 3, \quad \cdots\cdots, \quad +\frac{a}{2}, \quad \text{or} \quad \pm\frac{a-1}{2},$$

the last named value depending upon whether a is even or odd. Now a form with each of these values of b will occur once, and only once, within a section of the column covering a consecutive values of Δ, where $\Delta \geqq a^2$. This follows from the fact just noted that the forms of the column having a given b occur at intervals of a places. Therefore this section contains a forms which are reproduced periodically as Δ varies. For the purposes of our discussion we define a period as beginning with the place occupied by the form $(a, 0, c)$ and ending with the place preceding that occupied by $(a, 0, c + 1)$.

The foregoing remarks apply to what we will call the complete periods of the table, viz: all periods for $\Delta \geqq a^2$. But in those periods where $\Delta < a^2$ the forms are missing, first in which $c < a$ and second in which $c = a$ and $b < 0$. From the equation $ac - b^2 = \Delta$ it is clear that these conditions may occur on any value of $\Delta < a^2$ but not on $\Delta \geqq a^2$. Therefore we call those periods incomplete which extend from the least value of Δ for the column to $\Delta = a^2$.

To determine this minimum value of Δ for a column we put the minimum value of c and the maximum value of b in $ac - b^2 = \Delta$ and get, for a even:

$$a^2 - \frac{a^2}{4} = \frac{3a^2}{4} = \Delta.$$

and for a odd:

$$a^2 - \left(\frac{a-1}{2}\right)^2 = \frac{3a^2 + 2a - 1}{4} = \Delta.$$

From these we may write the inequalities for a even:

$$\Delta \geqq \frac{3a^2}{4} \quad \text{and} \quad a \leqq \sqrt{\frac{4\Delta}{3}}$$

for a odd:

$$\Delta \geqq \frac{3a^2 + 2a - 1}{4} \quad \text{and} \quad a \leqq \frac{-1 + 2\sqrt{1 + 3\Delta}}{3}$$

The inequalities for a even are the ones usually given for a both odd and even, but they do not furnish exact limits for a odd as do the ones given here.

It is interesting to note that in the tabulation all forms, $(a, 0, a)$, occur on the parabola $\Delta = a^2$, which therefore is the boundary between the region of the complete periods and that of the incomplete periods. And, since in each complete period there are a forms, the number of forms in the region of such periods and for all values of $\Delta < \Delta_1$ is approximately the area included by the curve, its axis and the ordinate Δ_1.

One may devise various methods for rapid construction of the table. One of these is to construct a complete period for each value of a within the limits of the table, then repeat these as often as necessary. In writing the forms in the incomplete periods one needs to remember the inequalities defining reduced forms. Another method, which is of service especially in building a table covering a comparatively short range of large values in Δ, is to begin for each a on a Δ which is a multiple of a with the form $(a_1, 0, c_1)$ and from this locate the other forms $(a_1, 0, c)$ at intervals of a places. Then the form (a_1, b, c_1) is b^2 places above $(a_1, 0, c_1)$ and from it all forms (a_1, b, c) may be located. Or one may find it advantageous to write all forms (a_1, b, c_1) first, which may be done rapidly by using the fact that beginning at zero successive squares may be located by the addition of successive odd numbers. Combinations of these and other properties of the table will suggest themselves as being useful to one in tabulating the forms.

From the equation $b^2 - ac = -\Delta = D$ we may write the congruence $x^2 \equiv D \pmod{a}$. Assuming that the solutions of this congruence are expressed in such form that the absolute value of each is less than or equal to $a/2$, and that we refer only to forms in the complete periods, it follows that the solutions are the values of b for the forms of the given a and D and therefore that the number of forms with this a and D is the number of solutions of the congruence. Thus the problem of finding the number of reduced forms for given values of a and Δ resolves itself into the determination of the number of solutions of the quadratic congruence, with the understanding that the usual restriction that $[D, a] = 1$ does not apply. We now proceed to the solution of this problem.

Lemma.—If the congruence $x^2 \equiv D \pmod{a}$ has any solutions and if $a = a' \cdot p^n$ $D = D' \cdot p^\kappa$, where a' and D' are each relatively prime to p, then κ cannot be odd and less than n.

For let x_1 be a solution of the congruence and assume κ to be odd and $< n$, then x_1^2 must contain p as a factor at least $\kappa + 1$ times. Then from the equation $x_1^2 - a \cdot c = D$ it follows that p divides D at least $\kappa + 1$ times which contradicts the hypothesis.

Theorem I.—If $a = p^n$ (where p is an odd prime) and $[a, D] = p^\kappa$, then the number of solutions, if any, of $x^2 \equiv D \pmod{a}$

1) *is $2p^\mu$ when $\kappa = 2\mu < n$;*

2) *is p^μ when $\kappa = 2\mu$ or $2\mu + 1 = n$.*

We note by the above lemma that κ cannot be odd and less than n.

1) When $\kappa = 2\mu < n$. Put $D = D' \cdot p^\kappa$. To enumerate the solutions of (1) $x^2 \equiv D \pmod{p^n}$ consider those of (2) $x^2 \equiv D' \pmod{p^{n-2\mu}}$. Let $x_1 + l \cdot p^{n-2\mu}$ be a solution of (2): Then it follows that $p^\mu(x_1 + l \cdot p^{n-2\mu})$ is a solution of (1). Let l take various values, and as many of the resulting expressions are independent solutions of (1) as are incongruent modulus p^n.

Let r and s be two values of l, such that

$$p^\mu(x_1 + rp^{n-2\mu}) \equiv p^\mu(x_1 + s \cdot p^{n-2\mu}) \pmod{p^n};$$

then

$$(r - s)p^{n-\mu} \equiv 0 \pmod{p^n}$$

and

$$r - s \equiv 0 \pmod{p^\mu}.$$

Therefore, if $s = 0$, $r = p^\mu$ is the smallest value of l yielding a congruent solution modulus p^n. i. e., for $l = 0, 1, 2, \cdots p^\mu - 1$ we get solutions of (1) incongruent modulus p^n.

Therefore each solution of (2) yields p^μ solutions of (1). Moreover, all solutions of (1) are obtained in this way, for any one of them must be divisible by p^μ; and when divided by p^μ the quotient is a solution of (2). But by well-known results (2) has two solutions, if any.

$\therefore$ (1) has $2 \cdot p^\mu$ solutions, if any.

2) When $\kappa = 2\mu$ or $2\mu + 1 = n$.

(i) When $\kappa = 2\mu$. Here the argument is identical with that in case 1), excepting that $p^{n-2\mu} = p^0 = 1$, with the result that congruence (2) has but one solution, and therefore (1) has p^μ solutions.

(ii) When $\kappa = 2\mu + 1$. Here the argument is the same as that of case 1) modified for subcase (i) with the additional change: Instead of multiplying the solution of (2) by p^μ we must multiply it by $p^{(\kappa+1)/2} = p^{(n+1)/2}$. Then we get $r - s \equiv 0 \pmod{p^{(n-1)/2}}$. But $p^{(n-1)/2} = p^\mu$.

Therefore (1) has p^μ solutions.

Note that by the above theorem when $\kappa = n$ the number of solutions is one half of the number when $\kappa < n$.

Theorem II.—If $a = 2^n$ *and* $[a, \mathrm{D}] = 2^\kappa$, *then the number of solutions, if any, of* $x^2 \equiv \mathrm{D} \pmod a$ *is* $2^{\sigma+\mu}$, *where* $\mu = [\kappa/2]$ *and* $\sigma = 0$, 1 *or* 2, *according as* $n - \kappa <, = $ *or* > 2.

Recall that by the lemma κ cannot be odd and less than n. The argument of Theorem I, case 1) applies here with the change: Congruence (2) becomes $x^2 \equiv \mathrm{D}' \pmod{2^{n-\kappa}}$, which we know has 2^σ solutions where $\sigma = 0$, 1 or 2, according as $n - \kappa <, =$ or > 2. Therefore the number of solutions of $x^2 \equiv \mathrm{D} \pmod{2^n}$ is $2^\sigma \cdot 2^\mu = 2^{\sigma+\mu}$.

Theorem III.—If

$$a = \prod_{i=0}^{n} p_i^{\alpha_i} \qquad \text{(where } p_0 = 2\text{)},$$

the number of solutions, if any, of $x^2 \equiv \mathrm{D} \pmod a$ *is* $2^{n-\nu+\sigma} \prod_{i=0}^{n} p_i^{\mu_i}$ *where*

$$[a, \mathrm{D}] = \prod_{0}^{n} p_i^{\kappa_i}; \quad \mu_i = [\kappa_i/2];$$

$\nu =$ *the number of times* $\kappa_i = \alpha_i$ *for p an odd prime;* $\sigma = 0$, 1 *or* 2, *according as* $\alpha_0 - 2\mu_0 <, =$ *or* > 2.

We get a solution of (1) $x^2 \equiv \mathrm{D} \pmod a$ for each set of solutions of

$$(2) \qquad \begin{cases} x^2 \equiv \mathrm{D} \pmod{2^{\alpha_0}} \\ x^2 \equiv \mathrm{D} \pmod{p_1^{\alpha_1}} \\ \vdots \quad \vdots \quad \vdots \\ x^2 \equiv \mathrm{D} \pmod{p_n^{\alpha_n}} \end{cases}$$

By Theorems I and II the first member of (2) has $2^{\sigma+\mu_0}$ solutions, and each succeeding congruence has $2p_i^{\mu_i}$ solutions, except in the ν cases where $\kappa_i = \alpha_i$, when this number becomes $p_i^{\mu_i}$, then there are as many sets of solutions of (2) as the product of these results, which is

$$2^{\sigma+\mu_0} \cdot 2p_1^{\mu_1} \cdot 2p_2^{\mu_2} \cdots 2p_n^{\mu_n} \cdot 2^{-\nu}$$

which may be written in the form

$$2^{n-\nu+\sigma} \prod_{0}^{n} p_i^{\mu_i}.$$

Note that in case

$$[a, \mathrm{D}] = 1, \quad \nu = \mu_0 = \mu_1 = \cdots = \mu_n = 0$$

and $2^{n-\nu+\sigma}\prod_{0}^{n} p_i^{\mu_i}$ becomes $2,^{n+\sigma}$ which is the well-known formula for the number of solutions in this case.

PART II.

We have shown that the number of reduced forms, if any, for given values of Δ and a in the complete periods is $2^{n-\nu+\sigma}\prod_{0}^{n} p_i^{\mu_i}$. Until stated otherwise, we will assume $D = -\Delta$ to contain no square factor. Then

$$\mu_0 = \mu_1 = \cdots = \mu_n = 0,$$

and the possible number of forms is $2^{n-\nu+\sigma}$. This may be written in the form

$$\prod_{i=1}^{n}\left[1+\left(\frac{D}{p_i^{\alpha_i}}\right)\right]\cdot 2^{\sigma}$$

where $(D/p_i^{\alpha_i})$ is an extension of the Jacobian symbol for the quadratic character of D with respect to $p_i^{\alpha_i}$, and is equal to 1, 0, or -1 according as D is a residue, a multiple or a non-residue of $p_i^{\alpha_i}$. The difference between the Jacobian symbol and the one used here arises when $D \equiv 0 \pmod{p}$. In this case the Jacobian symbol, $(D/p_i^{\alpha_i})$, becomes 0, but as we shall use it, $(D/p_i^{\alpha_i}) = 0$ only if $D \equiv 0$ $(\text{mod } p_i^{\alpha_i})$ and is equal to 1 or -1 otherwise, depending upon whether D is a residue or non-residue in the extended sense (i. e. D not necessarily prime to $p_i^{\alpha_i}$) of $p_i^{\alpha_i}$. In particular, since we have assumed that D contains no square factor, $(D/p_i^{\alpha_i}) = -1$ when $D \equiv 0 \pmod{p}$ and $D \not\equiv 0 \pmod{p^{\alpha}}$; for if we could solve $x^2 \equiv D \pmod{p^{\alpha}}$ when $D \equiv 0 \pmod{p}$ and $\alpha > 1$, D would contain p^2 as a factor. Also we will later give meanings to $(D/p_0^{\alpha_0})$, i. e., for p^{α} = a power of 2.

The sum $1 + (D/p_i^{\alpha_i})$ equals 2 for each time that D is a residue of $p_i^{\alpha_i}$ and is not divisible by it, which happens $n - \nu$ times, unless $x^2 \equiv D \pmod{p_i^{\alpha_i}}$ has no solutions in which case $(D/p_i^{\alpha_i}) = -1$ and $1 + (D/p_i^{\alpha_i}) = 0$.

The factor 2^{σ}, which is the number of solutions, if any, of the partial congruence in which 2^{α_0} is the modulus, may be expressed as a binomial similar to the other terms of the formula by giving $(D/p_0^{\alpha_0})$ values as follows:

for $\alpha_0 < 2$ let $(D/p_0^{\alpha_0}) = 0$,
for $\alpha_0 = 2$ let $(D/p_0^{\alpha_0}) = 1$ for $D = 4n + 1$, and -1 for $D \not\equiv 4n + 1$.
for $\alpha_0 > 2$ let $(D/p_0^{\alpha_0}) = 3$ for $D = 8n + 1$, and -1 for $D \not\equiv 8n + 1$.

Then $1 + (D/p_0^{\alpha_0})$ is the number of solutions of $x^2 \equiv D \pmod{p_0^{\alpha_0}}$; for, in order that there should be any solutions, it is both necessary and sufficient that $D=4n+1$ or $8n + 1$, according as $\alpha_0 = 2$ or > 2; and if $\alpha_0 = 1$ there is always one and only one solution. By substituting the various values of $(D/p_0^{\alpha_0})$, it is seen that if there are solutions, $1 + (D/p_0^{\alpha_0}) = 2^{\sigma}$ and, if there are no solutions, $1+(D/p_0^{\alpha_0})=0$.

Therefore we may write $2^{n-\nu+\sigma}$ in the form $\prod_{0}^{n}[1 + (D/p_0^{\alpha_0})]$, which not only gives the number of reduced forms for given values of a and Δ when there are any, but is equal to zero when there are none.

To sum $\prod_0^n [1 + (D/p_0^{\alpha_0})]$ for the various values of a, it is of advantage to have it in terms of the quadratic symbols (D/p_i) rather than $(D/p_i^{\alpha_i})$, where $\alpha_i > 1$, for all values of p except $p = 2$. We will now make this change. Comparing values of (D/p_i) and $(D/p_i^{\alpha_i})$, we see that if $(D/p_i) = 0$, $(D/p_i^{\alpha_i}) = -1$, for since $\alpha_i > 1$, and D does not contain a square factor $(D/p_i^{\alpha_i}) \neq 0$, and since $D \equiv 0 \pmod{p}$ $(D/p_i^{\alpha_i}) \neq 1$. If $(D/p_i) = \pm 1$, $(D/p_i^{\alpha_i}) = \pm 1$. And conversely, if $(D/p_i^{\alpha_i}) = 1$, $(D/p_i) = 1$; and if $(D/p_i^{\alpha_i}) = -1$, $(D/p_i) = 0$ or -1 according as D is or is not divisible by p. If $(D/p_i^{\alpha_i}) = 0$, $\alpha_i = 1$. Now let

$$\Delta = \prod_0^r k_i \quad \text{and} \quad a = \prod_0^n p_i^{\alpha_i} = \prod_0^u h_i^{\beta_i} \prod_1^v k_i'^{\delta_i},$$

where h and k are primes, $h_0 = p_0 = 2$, $\delta_i > 1$ and the k''s are those odd k's which occur among the p's with exponents greater than one. Let

$$s_a = \prod_0^n \left[1 + \left(\frac{D}{p_i^{\alpha_i}}\right)\right].$$

Then

$$s_a = \left[1 + \left(\frac{D}{h_0^{\beta_0}}\right)\right] \prod_1^u \left[1 + \left(\frac{D}{h_i^{\beta_i}}\right)\right] \prod_1^v \left[1 + \left(\frac{D}{k_i'^{\delta_i}}\right)\right].$$

By the above comparison of values of $(D/h_i^{\beta_i})$ and (D/h_i) we get

$$s_a = \left[1 + \left(\frac{D}{h_0^{\beta_0}}\right)\right] \cdot \prod_1^u \left[1 + \left(\frac{D}{h_i}\right)\right] \prod_1^v \left[1 + \left(\frac{D}{k_i'^{\delta_i}}\right)\right] \tag{1}$$

and if there are no values of k' this becomes

$$s_a = \left[1 + \left(\frac{D}{h_0^{\beta_0}}\right)\right] \cdot \prod_1^u \left[1 + \left(\frac{D}{h_i}\right)\right]. \tag{2}$$

If there are one or more values of k', $s_a = 0$ for $(D/k_i'^{\delta_i}) = -1$; while if we replace $(D/k_i'^{\delta_i})$ by (D/k'_i) we get

$$s_a = \left[1 + \left(\frac{D}{h_0^{\beta_0}}\right)\right] \prod_1^u \left[1 + \left(\frac{D}{h_i}\right)\right] \quad \text{for} \quad \left(\frac{D}{k'_i}\right) = 0.$$

Therefore, making this substitution in (1), we must subtract the latter result as a corrective term, getting

$$s_a = \left[1 + \left(\frac{D}{h_0^{\beta_0}}\right)\right] \prod_1^u \left[1 + \left(\frac{D}{h_i}\right)\right] \prod_1^v \left[1 + \left(\frac{D}{k_i'}\right)\right] - \left[1 + \left(\frac{D}{h_0^{\beta_0}}\right)\right] \prod_1^u \left[1 + \left(\frac{D}{h_i}\right)\right]. \tag{3}$$

For greater simplicity in formulæ we now define (D/h_0) to mean $(D/h_0^{\beta_0})$, and (2) and (3) become respectively

$$s_a = \prod_0^u \left[1 + \left(\frac{D}{h_i}\right)\right] = \prod_0^n \left[1 + \left(\frac{D}{p_i}\right)\right] \tag{4}$$

and

$$(5) \qquad s_a = \prod_0^u \left[1 + \left(\frac{\mathrm{D}}{h_i}\right)\right] \prod_1^v \left[1 + \left(\frac{\mathrm{D}}{k_i'}\right)\right] - \prod_0^u \left[1 + \left(\frac{\mathrm{D}}{h_i}\right)\right]$$

$$= \prod_0^n \left[1 + \left(\frac{\mathrm{D}}{p_i}\right)\right] - \prod_0^u \left[1 + \left(\frac{\mathrm{D}}{h_i}\right)\right]$$

$$= 0 \quad \text{for } \left(\frac{\mathrm{D}}{k'}\right) = 0.$$

If there are no values of k', the expression for s_a in (4) is the number of reduced forms for a given a and Δ in the complete periods. It may be put in better form for computing as follows:

$$\prod_0^n \left[1 + \left(\frac{\mathrm{D}}{p_i}\right)\right] = 1 + \sum_0^n \left(\frac{\mathrm{D}}{p_i}\right) + \sum_{i \neq j} \left(\frac{\mathrm{D}}{p_i}\right)\left(\frac{\mathrm{D}}{p_j}\right) + \cdots + \left(\frac{\mathrm{D}}{p_0}\right)\left(\frac{\mathrm{D}}{p_1}\right) \cdots \left(\frac{\mathrm{D}}{p_n}\right).$$

Let

$$p_i \cdot p_j \cdots p_k = \mathrm{P}.$$

Then

$$\left(\frac{\mathrm{D}}{p_i}\right)\left(\frac{\mathrm{D}}{p_j}\right) \cdots \left(\frac{\mathrm{D}}{p_k}\right) = \left(\frac{\mathrm{D}}{p_i \cdot p_j \cdots p_k}\right) = \left(\frac{\mathrm{D}}{\mathrm{P}}\right).$$

If P is odd, $(\mathrm{D}/\mathrm{P}) = 1$ or -1, when $[\mathrm{D}, \mathrm{P}] = 1$, and equals zero when $[\mathrm{D}, \mathrm{P}] \neq 1$. If P is even and equals $p_0^{a_0} \cdot \mathrm{P}'$,

$$\left(\frac{\mathrm{D}}{\mathrm{P}}\right) = \left(\frac{\mathrm{D}}{p_0^{a_0}}\right) \cdot \left(\frac{\mathrm{D}}{\mathrm{P}'}\right),$$

in which (D/P') is evaluated as is (D/P) for P odd, and $(\mathrm{D}/p_0^{a_0})$ has the values given in its definition. Then, if we define $1 = (\mathrm{D}/1)$, we may write

$$\prod_0^n \left[1 + \left(\frac{\mathrm{D}}{p_i}\right)\right] = \sum_{\mathrm{P}} \left(\frac{\mathrm{D}}{\mathrm{P}}\right)$$

where P runs through unity, the odd factors of a containing no square and such of its even factors as are divisible by $p_0^{a_0}$ and contain no odd square. To put this result in general form so that it will apply whether or not there are values of k', we need only to multiply it by the symbol (D/k'), which is zero when we have a value of k' and does not exist when there are no values of k'. Therefore, in either case,

$$(6) \qquad s_a = \left(\frac{\mathrm{D}}{k'}\right) \cdot \sum_{\mathrm{P}} \left(\frac{\mathrm{D}}{\mathrm{P}}\right).$$

represents the number of reduced forms on a and Δ in the complete periods.

We proceed as follows to get a formula which sums this result for all values of a belonging to a given value of Δ and lying within the complete periods. For the present disregard the factor (D/k') in (6), and write

$$s_a' = \sum_{\mathrm{P}} \left(\frac{\mathrm{D}}{\mathrm{P}}\right).$$

Later correction will be made for the error thus introduced. Put $\mathrm{A} = [\sqrt{\Delta}]$, the largest value of a within the complete periods, and let

$$S' = \sum_{a=1}^{A} s_a'.$$

Then, recalling the range of values of P, we see that (D/1) occurs once in each s_a' or A times in S′; (D/3) occurs [A/3] times and in general (D/P), when P is odd, occurs [A/P] times in S′. When P is even (D/P) occurs [A/P] − [A/2P] times in S′, for, although P is a factor of the values of a which are multiples of 2P, it is not divisible by the highest power of 2 contained in such a's. Then, denoting the even values of P by P_e and the odd ones by P_o we have

$$S' = \sum_{a=1}^{A} s_a' = \sum_{P_o} \left(\frac{A}{P_o}\right)\left(\frac{D}{P_o}\right) + \sum_{P_e} \left[\left(\frac{A}{P_e}\right) - \left(\frac{A}{2P_e}\right)\right]\left(\frac{D}{P_e}\right). \tag{7}$$

To simplify this we now prove that

$$\left[\left(\frac{A}{x}\right) - \left(\frac{A}{2x}\right)\right] = \left(\frac{A+x}{2x}\right).$$

(i) When $[A/x]$ is even:

$$\frac{A}{x} = 2q + \frac{r}{x}; \quad \text{then} \quad \frac{A}{2x} = q + \frac{r}{2x}, \quad \text{and} \quad \left(\frac{A}{x}\right) - \left(\frac{A}{2x}\right) = q.$$

$$\frac{A+x}{2x} = \frac{A}{2x} + \frac{1}{2} = q + \frac{r}{2x} + \frac{1}{2}, \quad \text{but} \quad \frac{r}{2x} < \frac{1}{2}; \quad \text{therefore} \quad \left(\frac{A+x}{2x}\right) = q.$$

(ii) When $[A/x]$ is odd:

$$\frac{A}{x} = q + \frac{r}{x}, \quad \text{where } q \text{ is odd.} \quad \frac{A}{2x} = \frac{q-1}{2} + \frac{r+x}{2x}, \quad \text{and} \quad \left(\frac{A}{x}\right) - \left(\frac{A}{2x}\right) = \frac{q+1}{2}.$$

$$\frac{A+x}{2x} = \frac{A}{2x} + \frac{1}{2} = \frac{q-1}{2} + \frac{r+x}{2x} + \frac{1}{2} = \frac{q+1}{2} + \frac{r}{2x}.$$

Therefore

$$\left(\frac{A+x}{2x}\right) = \frac{q+1}{2}.$$

Using this in (7), we get

$$S' = \sum_{P_o} \left(\frac{A}{P_o}\right)\left(\frac{D}{P_o}\right) + \sum_{P_e} \left(\frac{A+P_e}{2P_e}\right)\left(\frac{D}{P_e}\right). \tag{8}$$

In making this summation getting S′ we assumed that none of the a's contain factors k'^2, thus omitting the symbol (D/k') in (6). With this symbol present $s_a = 0$ and without it

$$s_a = \sum_{P} \left(\frac{D}{P}\right);$$

therefore when it is omitted and a contains a factor k'^2 our result for s_a is too great by $\sum_P (D/P)$. Then S′ should be diminished by this amount for each value of a containing a k'^2. The a's which are multiples of $k_i'^2$ are

$$k_i'^2, \quad 2k_i'^2, \quad 3k_i'^2, \quad \cdots, \quad a'k_i'^2, \quad \cdots, \quad \left(\frac{A}{k_i'^2}\right)k_i'^2.$$

By expanding $\sum_{P}(D/P)$ into its separate terms we see that its value for $a = a'k_i'^2$ is equal to its value for $a = a'$, independently of whether or not a' is itself a multiple of the squares of one or more values of k'. Therefore the excess introduced into S' from these multiples of $k_i'^2$ is the summation of $\sum_{P}(D/P)$ for

$$a = 1, 2, 3, \cdots, \left(\frac{A}{k_i'^2}\right),$$

which is the above formula for S' with A replaced by $[A/k_i'^2]$. Denote this by

$$c_i^{(1)} = \sum_{P_o}\left(\frac{A}{k_i'^2 P_o}\right)\left(\frac{D}{P_o}\right) + \sum_{P_e}\left(\frac{A + k_i'^2 \cdot P_e}{2k_i'^2 \cdot P_e}\right)\left(\frac{D}{P_e}\right).$$

Then if, as we sum $c_i^{(1)}$ with respect to i, λ values of $k_i'^2$ are factors of the same a our result includes the correction from this a λ times when it should contain it but once. This may be remedied by applying the principle of cross-classification. Let $c_i^{(\nu)}$ be the sum of the corrections for all a's which are multiples of the product of the squares of ν values of k', the subscript i denoting the particular combination of k's just as when $\nu = 1$ it denotes the particular k'. Then

$$\sum_i c_i^{(1)} - \sum_i c_i^{(2)} + \sum_i c_i^{(3)} - \cdots + (-1)^{\lambda+1} \cdot \sum_i c_i^{(\lambda)}$$

represents the correction for all a's which are multiples of no more than λ factors of the type k'^2, and letting λ be the largest number of such factors in any a this becomes the total correction to be applied to S'. Then, denoting the number of reduced forms in the complete periods for a given value of Δ by S we have

$$\text{(9)} \qquad S = S' - \sum_i c_i^{(1)} + \sum_i c_i^{(2)} - \sum_i c_i^{(3)} + \cdots + (-1)^{\lambda}\sum_i c_i^{(\lambda)}$$

But

$$S' = \sum_i c_i^{(0)} = c^{(0)}$$

and we may write

$$\text{(10)} \qquad S = \sum_{\nu=0}^{\lambda}\{(-1)^{\nu}\sum_i c_i^{(\nu)}\}.$$

Let $l_i^{(\nu)}$ = the ith combination of ν factors k'^2, then

$$\text{(11)} \qquad c_i^{(\nu)} = \sum_{P_o}\left(\frac{A}{l_i^{(\nu)}P_o}\right)\left(\frac{D}{P_o}\right) + \sum_{P_e}\left(\frac{A + l_i^{(\nu)}P_e}{2 \cdot l_i^{(\nu)} \cdot P_e}\right)\left(\frac{D}{P_e}\right).$$

Put (11) in (10) and get

$$\text{(12)} \quad S = \sum_{\nu=0}^{\lambda}(-1)^{\nu} \cdot \sum_i\left[\sum_{P_o}\left(\frac{A}{l_i^{(\nu)}P_o}\right)\left(\frac{D}{P_o}\right) + \sum_{P_e}\left(\frac{A + l_i^{(\nu)}P_e}{2 \cdot l_i^{(\nu)} \cdot P_e}\right)\left(\frac{D}{P_e}\right)\right].$$

In finding S by this formula first compute the terms for $\nu = 0$. In most cases this completes the computation. If there are values of k', they are each odd factors of Δ and must be less than or equal to $\Delta^{\frac{1}{4}}$ for k'^2 is a factor of one or more values of a and $a \leq \Delta^{\frac{1}{2}}$. There will not be a large number of such factors and the number of products of the squares of two k's which divide one or more a's is much less. e. g. the smallest possible k' is 3 for which $\Delta \geqq 81$; if $k' = 5$, $\Delta \geqq 625$; the smallest possible value of $l_i^{(2)}$ is $3^2 \cdot 5^2 = 225$ for which $\Delta \geqq 50625$. Moreover when $\nu > 0$ the summations with respect to P_o and P_e contain very few terms because of the rapid decrease in size of the greatest integer coefficients. The value of (D/P_o) must be obtained where P_o runs through the odd numbers which contain no square and are less than or equal to A. (D/P_e) may then be evaluated from (D/P_o) for

$$\left(\frac{D}{P_e}\right) = \left(\frac{D}{2^{a_0}}\right) \cdot \left(\frac{D}{P_o'}\right)$$

where P_o' runs through the smaller values of P_o. All terms containing (D/P_e) where $P_e \equiv 2 \pmod 4$ may be omitted for $(D/2) = 0$, and similarily all terms containing (D/P_o) may be omitted where $[D, P_o] \neq 1$.

We have derived (12) with the understanding that S is the number of reduced forms for D on the values of a from 1 to $A = [\sqrt{\Delta}]$ inclusive, but A may be given any value less than $[\sqrt{\Delta}]$ and the resulting value of S is the number of reduced forms over the corresponding range in a.

We now consider the number of reduced forms on those values of a lying in the incomplete periods. Here a runs from $[\sqrt{\Delta}] + 1$ to $[\sqrt{4\Delta/3}]$ or to $\left(\dfrac{-1 + 2\sqrt{1 + 3\Delta}}{3}\right)$ inclusive, according as the greatest value is even or odd. In computing use $[\sqrt{4\Delta/3}]$ first; if this is odd use $\left(\dfrac{-1 + 2\sqrt{1 + 3\Delta}}{3}\right)$. The number of a's in the incomplete periods is approximately $3/20\sqrt{\Delta}$.

The number of solutions of the congruence $b^2 \equiv D \pmod a$ is not in general the number of reduced forms for a and Δ in these periods for here $\Delta < a^2$ which, as seen from the equation $b^2 = ac - \Delta$, makes it possible for c to be less than or equal to a. Therefore the summation made over the complete periods cannot be extended over the incomplete periods. Theoretically all we can say is that the number of forms for a given a and Δ, where $a > \sqrt{\Delta}$, may be determined from the number of solutions of $x^2 \equiv D \pmod a$ whose squares are greater than or equal to $a^2 - \Delta$; for if c is greater than or equal to a, b^2 or $ac - \Delta$ is greater than or equal to $a^2 - \Delta$.

Practically the computing of the number of forms for Δ in these periods is comparatively easy. Assume that we have first computed the number for the complete periods. Of the values of a in the incomplete periods those which have one or more factors (including a itself) which are values of k'^2 or of which D is a non-residue will have no forms. Such are readily found, principally by the use of the previously evaluated quadratic symbols. The remaining values of a will have forms belonging to them unless ruled out by the condition that $c \geqq a$. For each of these a's we must find the number of values of b such that $b^2 = ac - \Delta$, where $c \geqq a$ and $b^2 \leqq a^2/4$. To do this form the numbers

$$a \cdot a - \Delta, \quad a(a+1) - \Delta, \quad a(a+2) - \Delta, \quad \cdots$$

or

$$a^2 - \Delta, \quad a^2 - \Delta + a, \quad a^2 - \Delta + 2a, \quad \cdots, \quad a^2 - \Delta + i \cdot a, \quad \cdots$$

so long as

$$a^2 - \Delta + i \cdot a \leqq a^2/4.$$

(Note that $i \leqq 1/a(\Delta - 3a^2/4)$ which shows i to decrease rapidly as a increases.) The perfect squares among these numbers are the values of b^2, each of which yields two forms excepting $b^2 = a^2 - \Delta$ or $a^4/4$ in which cases there are no forms for the negative values of b.

Example I.—Using (12) to find the number of reduced forms and from this the class number, h for

$$-\mathrm{D} = \Delta = 2010 = 2 \cdot 3 \cdot 5 \cdot 67 \qquad \mathrm{A} = 44.$$

(1) terms for $\nu = 0$:

$$\sum_{P_o} \left(\frac{\mathrm{A}}{\mathrm{P}_o}\right)\left(\frac{\mathrm{D}}{\mathrm{P}_o}\right) = 44\left(\frac{\mathrm{D}}{1}\right) + 6\left(\frac{\mathrm{D}}{7}\right) + 4\left(\frac{\mathrm{D}}{11}\right) + 3\left(\frac{\mathrm{D}}{13}\right) + 2\left(\frac{\mathrm{D}}{17}\right)$$
$$+ 2\left(\frac{\mathrm{D}}{19}\right) + \left(\frac{\mathrm{D}}{23}\right) + \left(\frac{\mathrm{D}}{29}\right) + \left(\frac{\mathrm{D}}{31}\right) + \left(\frac{\mathrm{D}}{37}\right) + \left(\frac{\mathrm{D}}{41}\right) + \left(\frac{\mathrm{D}}{43}\right)$$
$$= 44 - 6 + 4 - 3 + 2 + 2 - 1 + 1 + 1 + 1 + 1 + 1 = 47$$

$$\sum_{P_e} \left(\frac{\mathrm{A} + \mathrm{P}_e}{2\mathrm{P}_e}\right)\left(\frac{\mathrm{D}}{\mathrm{P}_e}\right) = 6\left(\frac{\mathrm{D}}{4}\right) + 3\left(\frac{\mathrm{D}}{8}\right) + \left(\frac{\mathrm{D}}{16}\right) + \left(\frac{\mathrm{D}}{28}\right) + \left(\frac{\mathrm{D}}{32}\right) + \left(\frac{\mathrm{D}}{44}\right)$$
$$= 6\left(\frac{\mathrm{D}}{4}\right) + 3\left(\frac{\mathrm{D}}{8}\right) + \left(\frac{\mathrm{D}}{16}\right) + \left(\frac{\mathrm{D}}{4}\right)\left(\frac{\mathrm{D}}{7}\right) + \left(\frac{\mathrm{D}}{32}\right)$$
$$+ \left(\frac{\mathrm{D}}{4}\right)\left(\frac{\mathrm{D}}{11}\right)$$

but for Δ even and $\alpha_0 > 1$ $(\mathrm{D}/2^{\alpha_0}) = -1$ $\therefore$ this becomes

$$= -6 - 3 - 1 + 1 - 1 - 1 = -11.$$

(2) terms for $\nu = 1$; $l_i^{(1)} = 3$ or 5:
when $l_i^{(1)} = 3$, put $[\mathrm{A}/9] = 4$ for A and repeating computation as in (1) get for $\mathrm{P}_o : -4\left(\frac{\mathrm{D}}{1}\right) = -4$ for $\mathrm{P}_e : -\left(\frac{\mathrm{D}}{4}\right) = 1$.
when $l_i^{(1)} = 5$, put $[\mathrm{A}/25] = 1$ for A, and get for P_o: $-(\mathrm{D}/1) = -1$.
There are no further terms for $\nu = 1$ and none for $\nu > 1$. $\therefore$ collecting the above results we have

$$\mathrm{S} = 47 - 11 - 4 + 1 - 1 = 32.$$

To find the number of reduced forms in the incomplete periods: here

$$a = 45,\ 46,\ 47,\ 48,\ 49,\ 50,\ 51.$$

By (6) and the values of (D/P) already found we readily see that none of these a's, except possibly 51, can have any reduced forms. For $a = 51$ form the numbers

$$a^2 - \Delta, \quad a^2 - \Delta + a, \quad \text{etc., which are } 591,\ 642.$$

neither of which are squares, therefore there are no reduced forms on $a = 51$ and hence none in the incomplete periods,

$$\therefore h = \mathrm{S} = 32.$$

The computation of this result by the known formula—see ¶106 of Dirichlet-Dedekind, *Zahlentheorie*—

$$h = 2\sum_{1}^{3}(\alpha/\Delta) = 2\sum(\alpha/\Delta)$$

where the limits of α are defined by

$$\Delta/8 < \alpha < 3\Delta/8,$$

would be much more difficult as one may see from the following considerations. The range of α is from 252 to 753 inclusive, i. e. over 502 values. Then excluding all α's which are not relatively prime to Δ we have left about 152 values of α such that $(\alpha/\Delta) = 1$ or -1. Also $\Delta = 2010$, a comparatively large number, which complicates the evaluating of (α/Δ).

Example II.—To find h when $\Delta = 827$, a prime.

$$\mathrm{A} = [\sqrt{827}] = 28$$

terms for $\nu = 0$:

$$\sum_{P_o}\left(\frac{\mathrm{A}}{\mathrm{P}_o}\right)\left(\frac{\mathrm{D}}{\mathrm{P}_o}\right) = 28\left(\frac{\mathrm{D}}{1}\right) + 9\left(\frac{\mathrm{D}}{3}\right) + 5\left(\frac{\mathrm{D}}{5}\right) + 4\left(\frac{\mathrm{D}}{7}\right) + 2\left(\frac{\mathrm{D}}{11}\right) + 2\left(\frac{\mathrm{D}}{13}\right)$$

$$+\left(\frac{\mathrm{D}}{15}\right) + \left(\frac{\mathrm{D}}{17}\right) + \left(\frac{\mathrm{D}}{19}\right) + \left(\frac{\mathrm{D}}{21}\right) + \left(\frac{\mathrm{D}}{23}\right)$$

$$= 28 + 9 - 5 - 4 + 2 - 2 - 1 - 1 + 1 - 1 + 1 = 27$$

$$\sum_{P_e}\left(\frac{\mathrm{A} + \mathrm{P}_e}{2\mathrm{P}_e}\right)\left(\frac{\mathrm{D}}{\mathrm{P}_e}\right) = 4\left(\frac{\mathrm{D}}{4}\right) + 2\left(\frac{\mathrm{D}}{8}\right) + \left(\frac{\mathrm{D}}{12}\right) + \left(\frac{\mathrm{D}}{16}\right) + \left(\frac{\mathrm{D}}{20}\right) + \left(\frac{\mathrm{D}}{24}\right)$$

$$+\left(\frac{\mathrm{D}}{28}\right) = 4 - 2 + 1 - 1 - 1 - 1 - 1 = -1.$$

Since Δ is a prime there are no values of k' and therefore no terms for $\nu > 0$

$$\therefore \mathrm{S} = 27 - 1 = 26.$$

In the incomplete periods:

$$\left[\sqrt{\frac{4\Delta}{3}}\right] = 33$$

which is odd therefore we find

$$\left[\frac{-1 + 2\sqrt{1 + 3\Delta}}{3}\right] = 32,$$

and the values of a in these periods are 29, 30, 31 and 32.

We see, by formula (2), that there are no forms for $a = 29$, 30, or 32 for

$$\left(\frac{\mathrm{D}}{29}\right) = -1, \quad \left(\frac{\mathrm{D}}{5}\right) = -1 \quad \text{and} \quad \left(\frac{\mathrm{D}}{32}\right) = -1$$

But $(\mathrm{D}/31) = 1$. Then we form the numbers $a^2 - \Delta + i \cdot a$ which are: 134, 165, 196, 227, of which 196 is a perfect square giving $b = \pm 14$.

$\therefore$ we get 2 reduced forms in the incomplete periods. Adding this to S we get 28 as the entire number of reduced forms for $\Delta = 827$.

But $827 = 8n - 5 \therefore h = 3/4 \cdot 28 = 21$.

To get this result by the known formula we must compute

$$h = \sum_{0}^{4} (\alpha/\Delta)$$

where α runs from 0 to 413 which makes 413 symbols, $(\alpha/827)$, to evaluate.

Part III.

We define single reduced forms as reduced forms which have no opposites. Such forms belong to the three types (1) (a, b, a), (2) $(a, a/2, c)$, (3) $(a, 0, c)$. It is our purpose now to determine the number of reduced forms of this kind occurring on a given Δ.

Theorem I.—The number of single reduced forms of type (1) on Δ is the number of times Δ can be expressed as the product of two factors, the same in parity, the smaller of which is greater than or equal to one third of the larger.

The form is (a, b, a). $\Delta = a^2 - b^2 = (a - b)(a + b)$ let $a - b = \gamma$ and $a + b = \delta$ from which

$$a = \frac{\gamma + \delta}{2} \quad \text{and} \quad b = \frac{\delta - \gamma}{2}.$$

But $b \leqq a/2$ then

$$\frac{\delta - \gamma}{2} \leqq \frac{\gamma + \delta}{4} \quad \text{and} \quad \gamma \geqq \frac{\delta}{3}.$$

And conversely given values of γ and δ, the same in parity, and satisfying these conditions, an a and b may be found giving a form (a, b, a). γ and δ must be alike even or odd to give integers for

$$a = \frac{\gamma + \delta}{2} \quad \text{and} \quad b = \frac{\delta - \gamma}{2}.$$

Theorem II.—The number of single reduced forms of type (2) on Δ is the number of times Δ can be expressed as the product of two factors, the same in parity, the smaller of which is less than or equal to one third of the larger.

The form is $(a, a/2, c)$ where $a = 2a'$

$$\Delta = ac - \frac{a^2}{4} = 2a'c - a'^2 = a'(2c - a')$$

$$c \geqq a \therefore 2c - a' > a'.$$

Let $a' = \gamma$ and $2c - a' = \delta$ then

$$c = \frac{\gamma + \delta}{2} \quad \text{and} \quad a = 2\gamma.$$

But $c \geqq a$ then

$$\frac{\gamma + \delta}{2} \geqq 2\gamma \quad \text{or} \quad \gamma \leqq \frac{\delta}{3}.$$

Conversely, given factors γ and δ, the same in parity, and satisfying these conditions we can get a form $(a, a/2, c)$.

Theorem III.—The number of single reduced forms of type (3) on Δ is the number of times Δ can be expressed as the product of two factors.

The form is $(a, 0, c)$ and $\Delta = a \cdot c$ from which the theorem follows.

We need to note the possibility of one form belonging to two types. This can happen only in cases of (1) and (2), and of (1) and (3). If a form belongs to both (1) and (2) $\gamma = \delta/3$ and $\Delta = 3\gamma^2$; if it belongs to both (1) and (3), $\gamma = \delta = a$ and $\Delta = \gamma^2 = a^2$. Thus we have one of these special cases only if Δ is a square or 3 times a square.

Theorem IV.—If

$$\Delta = \prod_{i=1}^{n} p_i^{\alpha_i}$$

and is odd, the number of single reduced forms is $\prod_1^n (1 + \alpha_i)$.

The number of factors of Δ is $\prod_1^n (1 + \alpha_i)$. Then the number of times Δ can be expressed as the product of two factors is $\frac{1}{2} \cdot \prod_1^n (1 + \alpha_i)$, except when $\Delta = a^2$. By theorems I and II we get $\frac{1}{2} \cdot \prod_1^n (1 + \alpha_i)$ forms each belonging to one of the first two types. By theorem III we get $\frac{1}{2} \cdot \prod_1^n (1 + \alpha_i)$ forms of type (3). Then if no form of (1) or (2) belongs to (3) we have in all $\prod_1^n (1 + \alpha_i)$ single forms.

As already noted the only time a form of (1) or (2) belongs to (3) is in case of $(a, 0, a)$. Here $\Delta = a^2$ and therefore the number of times it may be expressed as the product of two factors, counting the case of $\Delta = a \cdot a$, is $\frac{1}{2}\left[\prod_1^n (1 + \alpha_i) + 1\right]$. We then have $\frac{1}{2}\left[\prod_1^n (1 + \alpha_i) + 1\right]$ forms of types (1) and (2) and the same number of type (3), making $\prod_1^n (1 + \alpha_i) + 1$ in all. But since $(a, 0, a)$ belongs to both (1) and (3) we have counted it twice. Therefore subtracting one we get $\prod_1^n (1 + \alpha_i)$ single forms as before.

Theorem V.—If

$$\Delta = \prod_0^n p_i^{\alpha_i}$$

(where $p_0 = 2$), the number of single reduced forms is $\alpha_0 \prod_1^n (1 + \alpha_i)$.

First assume $\Delta \neq a^2$. The number of factors of Δ is $(1 + \alpha_0) \prod_1^n (1 + \alpha_i)$ and therefore the number of forms of type (3) is

$$\frac{1+\alpha_0}{2}\prod_1^n (1+\alpha_i).$$

The number of times we can write $\Delta = \gamma \cdot \delta$ where γ and δ are the same in parity is

$$\frac{\alpha_0 - 1}{2}\prod_1^n (1+\alpha_i).$$

For γ and δ must each be even and not divisible by 2^{α_0}, and the number of factors of Δ of this kind is

$$(\alpha_0 - 1)\prod_1^n (1+\alpha_i).$$

Then there are

$$\frac{\alpha_0 - 1}{2}\prod_1^n (1+\alpha_i)$$

forms of the two types (1) and (2). Therefore the entire number of single forms is

$$\left(\frac{\alpha_0+1}{2}+\frac{\alpha_0-1}{2}\right)\prod_1^n (1+\alpha_i) = \alpha_0\prod_1^n (1+\alpha_i).$$

If $\Delta = a^2$ it may be shown, as in the proof of theorem IV, that the result is unchanged.

In particular when Δ contains no square factor the expression for the number of single forms becomes 2^n where n is the number of odd primes in Δ.

Part IV.

All reduced forms on a value of Δ which contains no square factor are primitive, i. e. $[a, b, c] = 1$. Primitive forms are said to be of the first or second sort according as $[a, 2b, c] = 1$ or 2. We will now determine the distribution of these two classes of forms in the tabulation. We know that primitive forms of the second sort can occur only on values of Δ where $\Delta = 8n - 1$ or $8n - 5$. Therefore we will confine our discussion to values of Δ of these types.

Theorem I.—When $\Delta = 8n - 5$ and contains no square factor, and $a \equiv 2 \pmod 4$ all forms are of the second sort and for $a \not\equiv 2 \pmod 4$ all forms are of the first sort.

If (a, b, c) is of the second sort a and c are even and b is odd.

$$a \cdot c = \Delta + b^2 \quad \text{but} \quad b^2 = 8m + 1$$

then

$$a \cdot c = 8n - 5 + 8m + 1$$
$$= 8(n + m) - 4$$

and $ac \equiv 4 \pmod 8$ which may be only when:

(1) $a \equiv 4 \pmod 8$ and c is odd

or

(2) $$a \equiv 2 \pmod 4 \quad \text{and} \quad c \equiv 2 \pmod 4$$

or

(3) $$a \text{ is odd} \quad \text{and} \quad c \equiv 4 \pmod 8.$$

But we can have forms of the second sort only under (2), since a and c must both be even.

Conversely whenever $a \equiv 2 \pmod 4$ we have forms of the second sort. For if $a \equiv 2 \pmod 4$ and $ac \equiv 4 \pmod 8$ c must be even. And since Δ is odd b must be odd.

Theorem II.—When $\Delta = 8n - 1$ and contains no square factor, one half the forms on $a \equiv 0 \pmod 8$ and all forms on other even a's are of the second sort.

$$\begin{aligned} a \cdot c &= \Delta + b^2 \\ &= 8n - 1 + 8m + 1 \\ &= 8(n + m) \end{aligned}$$

which can be only when:

(1) $$a \equiv 0 \pmod 8 \quad \text{and} \quad c \text{ odd or even}$$

or

(2) $$a \equiv 4 \pmod 8 \quad \text{and} \quad c \equiv 0 \pmod 2$$

or

(3) $$a \equiv 2 \pmod 4 \quad \text{and} \quad c \equiv 0 \pmod 4.$$

It is evident that all forms on values of a under (2) and (3) are of the second sort. We investigate forms on values of $a \equiv 0 \pmod 8$ as follows:

If (a, b_1, c_1) and (a, b_2, c_2) occur on the same Δ, then

$$ac_1 - b_1^2 = ac_2 - b_2^2$$

and

$$b_2^2 - b_1^2 \equiv 0 \pmod a;$$

and conversely if

$$b_2^2 - b_1^2 \equiv 0 \pmod a$$

and (a, b_1, c_1) occurs on Δ, (a, b_2, c_2) occurs on the same Δ.

Let (a, b_1, c_1) be on a given Δ and let α be a factor of a. Then if $a/\alpha - b_1 = b_2$ the form (a, b_2, c_2) is on the same Δ when

$$\left(\frac{a}{\alpha} - b_1\right)^2 - b_1^2 \equiv 0 \pmod a$$

$$\frac{a^2}{\alpha^2} - 2\frac{a}{\alpha}b_1 \equiv 0 \pmod a$$

$$a\left(\frac{a/\alpha - 2b_1}{\alpha}\right) \equiv 0 \pmod a$$

which is true when

$$\frac{a}{\alpha} - 2b_1 \equiv 0 \pmod \alpha.$$

Now all values of b on $a \equiv 0 \pmod 8$ are odd. Put $b_1 = 4h_1 \pm 1$ and $a = 8k$. Let $\alpha = 2$ then

$$\frac{a}{\alpha} = \frac{a}{2} \equiv 0 \pmod 4$$

and

$$\frac{a}{2} - 2b_1 \equiv 0 \pmod 2.$$

Therefore b_1 and $a/2 - b_1 = b_2$ occur on the same values of Δ and a.

$$b_2 = a/2 - b_1 = 4k - (4h_1 \pm 1) = 4(k - h_1) \mp 1 = 4h_2 \mp 1$$

in which $h_2 = k - h_1$ or $h_1 + h_2 = k$. Now

$$b_2^2 - ac_2 = b_1^2 - ac_1$$

$$b_2^2 - b_1^2 = a(c_2 - c_1)$$

$$= 8k(c_2 - c_1).$$

Also

$$b_2^2 - b_1^2 = (4h_2 \pm 1)^2 - (4h_1 \mp 1)^2$$

$$= 16(h_2^2 - h_1^2) \pm 8(h_2 + h_1)$$

$$= 8(h_2 + h_1)[2(h_2 - h_1) \pm 1].$$

Therefore

$$c_2 - c_1 = 2(h_2 - h_1) \pm 1$$

and c_2 and c_1 are different modulus 2. Then one of the forms (a, b_1, c_1) and (a, b_2, c_2) is of the first sort and the other is of the second sort, i. e., for each form of the first sort on a given $\Delta = 8n - 1$ and a, where $a \equiv 0 \pmod 8$, there is a form of the second sort on the same Δ and a and conversely.

It is known* (1) that for $\Delta = 8n - 1$ and for $\Delta = 3$ the number of forms of the first sort equals the number of those of the second sort; and (2) that for $\Delta = 8n - 5$ (excepting $\Delta = 3$) the number of forms of the second sort is one-third of the number of those of the first sort. Attempts to prove these relationships by the methods of this paper have been unsuccessful, but the following theorem may be proved and is of interest in this connection.

Theorem III.—If $\Delta = 8n - 1$ or $8n - 5$ and contains no square factor the number of single forms of the first sort equals the number of single forms of the second sort.

From Part III recall the three types of single forms, viz: (1) when $a = c$, (2) when $b = a/2$, (3) when $b = 0$.

Now a form of type (1), (a, b, a) is second sort, for if a and b were both odd Δ would be even and if a were odd and b even we would have

$$\Delta = a^2 - b^2 = 8m + 1 - 4k = 8n - 3 \quad \text{or} \quad 8n - 7.$$

Therefore a is even and b is odd.

* Dirichlet-Dedekind, *Zahlentheorie*, § 97.

A form of type (2) is of the second sort, for: $(a, a/2, c)$ may be written $(2b, b, c)$. Then

$$c = \frac{\Delta + b^2}{2b}$$

in which b is odd and

$$b^2 = 8m + 1.$$

Then

$$c = \frac{8n - 1 + 8m + 1}{2b} \quad \text{or} \quad \frac{8n - 5 + 8m + 1}{2b}$$

$$= 4 \cdot \frac{n + m}{b} \quad \text{or} \quad 2 \cdot \frac{2(n + m) - 1}{b}$$

Therefore c is even.

A form of type (3), $(a, 0, c)$ is first sort for otherwise Δ would contain a square factor. Therefore all forms of types (1) and (2) are of the second sort and those of type (3) are of the first sort. But in Part III we saw that the number of forms of types (1) and (2) equals the number of those of type (3), from which the theorem follows.

Δ \ a	1.	2.	3.	4.	4.	5.	6.	7.	8.
1	0 1								
2	0 2								
3	0 3	1 2							
4	0 4	0 2							
5	0 5	1 3							
6	0 6	0 3							
7	0 7	1 4							
8	0 8	0 4	1 3						
9	0 9	1 5	0 3						
10	0 10	0 5							
11	0 11	1 6	±1 4						
12	0 12	0 6	0 4		2 4				
13	0 13	1 7							
14	0 14	0 7	±1 5						
15	0 15	1 8	0 5	1 4					
16	0 16	0 8		0 4	2 5				
17	0 17	1 9	±1 6						
18	0 18	0 9	0 6						
19	0 19	1 10		±1 5					
20	0 20	0 10	±1 7	0 5	2 6				
21	0 21	1 11	0 7			2 5			
22	0 22	0 11							
23	0 23	1 12	±1 8	±1 6					
24	0 24	0 12	0 8	0 6	2 7	1 5			
25	0 25	1 13				0 5			
26	0 26	0 13	±1 9			±2 6			
27	0 27	1 14	0 9	±1 7			3 6		
28	0 28	0 14		0 7	2 8				
29	0 29	1 15	±1 10			±1 6			
30	0 30	0 15	0 10			0 6			
31	0 31	1 16		±1 8		±2 7			
32	0 32	0 16	±1 11	0 8	2 9		2 6		
33	0 33	1 17	0 11				3 7		
34	0 34	0 17				±1 7			
35	0 35	1 18	±1 12	±1 9		0 7	1 6		
36	0 36	0 18	0 12	0 9	2 10	±2 8	0 6		
37	0 37	1 19							
38	0 38	0 19	±1 13				±2 7		
39	0 39	1 20	0 13	±1 10		±1 8	3 8		
40	0 40	0 20		0 10	2 11	0 8		3 7	
41	0 41	1 21	±1 14			±2 9	±1 7		
42	0 42	0 21	0 14				0 7		
43	0 43	1 22		±1 11					
44	0 44	0 22	±1 15	0 11	2 12	±1 9	±2 8		
45	0 45	1 23	0 15			0 9	3 9	2 7	
46	0 46	0 23				±2 10			
47	0 47	1 24	±1 16	±1 12			±1 8	±3 8	
48	0 48	0 24	0 16	0 12	2 13		0 8	1 7	4 8
49	0 49	1 25				±1 10		0 7	
50	0 50	0 25	±1 17			0 10	±2 9		

Δ \ a	1.	2.	3.	4.	4.	5.	6.	7.	8.	8.	9.	9.	10.	11.
51	0 51	1 26	0 17	±1 13		±2 11	3 10							
52	0 52	0 26		0 13	2 14			±2 8						
53	0 53	1 27	±1 18				±1 9							
54	0 54	0 27	0 18			±1 11	0 9	±3 9						
55	0 55	1 28		±1 14		0 11		±1 8		3 8				
56	0 56	0 28	±1 19	0 14	2 15	±2 12	±2 10	0 8		4 9				
57	0 57	1 29	0 19				3 11							
58	0 58	0 29												
59	0 59	1 30	±1 20	±1 15		±1 12	±1 10	±2 9						
60	0 60	0 30	0 20	0 15	2 16	0 12	0 10		2 8					
61	0 61	1 31				±2 13		±3 10						
62	0 62	0 31	±1 21				±2 11	±1 9						
63	0 63	1 32	0 21	±1 16			3 12	0 9	1 8	±3 9				
64	0 64	0 32		0 16	2 17	±1 13			0 8	4 10				
65	0 65	1 33	±1 22			0 13	±1 11				4 9			
66	0 66	0 33	0 22			±2 14	0 11	±2 10						
67	0 67	1 34		±1 17										
68	0 68	0 34	±1 23	0 17	2 18		±2 12	±3 11	±2 9					
69	0 69	1 35	0 23			±1 14	3 13	±1 10						
70	0 70	0 35				0 14		0 10						
71	0 71	1 36	±1 24	±1 18		±2 15	±1 12		±1 9	±3 10				
72	0 72	0 36	0 24	0 18	2 19		0 12		0 9	4 11		3 9		
73	0 73	1 37						±2 11						
74	0 74	0 37	±1 25			±1 15	±2 13				±4 10			
75	0 75	1 38	0 25	±1 19		0 15	3 14	±3 12					5 10	
76	0 76	0 38		0 19	2 20	±2 16		±1 11	±2 10					
77	0 77	1 39	±1 26				±1 13	0 11			2 9			
78	0 78	0 39	0 26				0 13							
79	0 79	1 40		±1 20		±1 16			±1 10	±3 11				
80	0 80	0 40	±1 27	0 20	2 21	0 16	±2 14	±2 12	0 10	4 12	1 9			
81	0 81	1 41	0 27			±2 17	3 15				0 9	±3 10		
82	0 82	0 41						±3 13						
83	0 83	1 42	±1 28	±1 21			±1 14	±1 12			±4 11			
84	0 84	0 42	0 28	0 21	2 22	±1 17	0 14	0 12	±2 11				4 10	
85	0 85	1 43				0 17							5 11	
86	0 86	0 43	±1 29			±2 18	±2 15				±2 10			
87	0 87	1 44	0 29	±1 22			3 16	±2 13	±1 11	±3 12				
88	0 88	0 44		0 22	2 23				0 11	4 13				
89	0 89	1 45	±1 30			±1 18	±1 15	±3 14			±1 10			
90	0 90	0 45	0 30			0 18	0 15	±1 13			0 10	±3 11		
91	0 91	1 46		±1 23		±2 19		0 13					3 10	
92	0 92	0 46	±1 31	0 23	2 24		±2 16		±2 12		±4 12			
93	0 93	1 47	0 31				3 17							
94	0 94	0 47				±1 19		±2 14					±4 11	
95	0 95	1 48	±1 32	±1 24		0 19	±1 16		±1 12	±3 13	±2 11		5 12	
96	0 96	0 48	0 32	0 24	2 25	±2 20	0 16	±3 15	0 12	4 14			2 10	5 11
97	0 97	1 49						±1 14						
98	0 98	0 49	±1 33				±2 17	0 14			±1 11			
99	0 99	1 50	0 33	±1 25		±1 20	3 18				0 11	±3 12	1 10	
100	0 100	0 50		0 25	2 26	0 20			±2 13				0 10	

Δ \ a	1.	2.	3.	4.	4.	5.	6.	7.	8.	8.	9.	9.	10.	11.	12.	12.	13.	14.
101	0 101	1 51	±1 34			±2 21	±1 17	±2 15			±4 13		±3 11					
102	0 102	0 51	0 34				0 17											
103	0 103	1 52		±1 26				±3 16	±1 13	±3 14								
104	0 104	0 52	±1 35	0 26	2 27	±1 21	±2 18	±1 15	0 13	4 15	±2 12		±4 12					
105	0 105	1 53	0 35			0 21	3 19	0 15					5 13	4 11				
106	0 106	0 53				±2 22							±2 11					
107	0 107	1 54	±1 36	±1 27			±1 18				±1 12			±5 12				
108	0 108	0 54	0 36	0 27	2 28		0 18	±2 16	±2 14		0 12	±3 13				6 12		
109	0 109	1 55				±1 22							±1 11					
110	0 110	0 55	±1 37			0 22	±2 19	±3 17			±4 14		0 11					
111	0 111	1 56	0 37	±1 28		±2 23	3 20	±1 16	±1 14	±3 15			±3 12					
112	0 112	0 56		0 28	2 29			0 16	0 14	4 16				3 11				
113	0 113	1 57	±1 38				±1 19				±2 13							
114	0 114	0 57	0 38			±1 23	0 19						±4 13					
115	0 115	1 58		±1 29		0 23		±2 17					5 14					
116	0 116	0 58	±1 39	0 29	2 30	±2 24	±2 20		±2 15		±1 13		±2 12	±4 12				
117	0 117	1 59	0 39				3 21	±3 18			0 13	±3 14		2 11				
118	0 118	0 59						±1 17						±5 13				
119	0 119	1 60	±1 40	±1 30		±1 24	±1 20	0 17	±1 15	±3 16	±4 15		±1 12			5 12		
120	0 120	0 60	0 40	0 30	2 31	0 24	0 20		0 15	4 17			0 12	1 11		6 13		
121	0 121	1 61				±2 25							±3 13	0 11				
122	0 122	0 61	±1 41				±2 21	±2 18			±2 14							
123	0 123	1 62	0 41	±1 31			3 22							±3 12				
124	0 124	0 62		0 31	2 32	±1 25		±3 19	±2 16				±4 14					
125	0 125	1 63	±1 42			0 25	±1 21	±1 18			±1 14		5 15					
126	0 126	0 63	0 42			±2 26	0 21	0 18			0 14	±3 15	±2 13					
127	0 127	1 64		±1 32					±1 16	±3 17				±4 13				
128	0 128	0 64	±1 43	0 32	2 33		±2 22		0 16	4 18	±4 16			±2 12		4 12		
129	0 129	1 65	0 43			±1 26	3 23	±2 19					±1 13	±5 14				
130	0 130	0 65				0 26							0 13					
131	0 131	1 66	±1 44	±1 33		±2 27	±1 22	±3 20			±2 15		±3 14	±1 12		±5 13		
132	0 132	0 66	0 44	0 33	2 34		0 22	±1 19	±2 17					0 12		6 14		
133	0 133	1 67						0 19									6 13	
134	0 134	0 67	±1 45			±1 27	±2 23				±1 15		±4 15	±3 13				
135	0 135	1 68	0 45	±1 34		0 27	3 24		±1 17	±3 18	0 15	±3 16	5 16		3 12			
136	0 136	0 68		0 34	2 35	±2 28		±2 20	0 17	4 19			±2 14					
137	0 137	1 69	±1 46				±1 23				±4 17							
138	0 138	0 69	0 46				0 23	±3 21						±4 14				
139	0 139	1 70		±1 35		±1 28		±1 20					±1 14	±2 13				
140	0 140	0 70	±1 47	0 35	2 36	0 28	±2 24	0 20	±2 18		±2 16		0 14	±5 15	2 12	±4 13		
141	0 141	1 71	0 47			±2 29	3 25						±3 15					
142	0 142	0 71												±1 13				
143	0 143	1 72	±1 48	±1 36			±1 24	±2 21	±1 18	±3 19	±1 16			0 13	1 12	±5 14		
144	0 144	0 72	0 48	0 36	2 37	±1 29	0 24		0 18	4 20	0 16	±3 17	±4 16		0 12	6 15	5 13	
145	0 145	1 73				0 29		±3 22					5 17	±3 14				
146	0 146	0 73	±1 49			±2 30	±2 25	±1 21			±4 18		±2 15				±6 14	
147	0 147	1 74	0 49	±1 37			3 26	0 21							±3 13			7 14
148	0 148	0 74		0 37	2 38				±2 19									
149	0 149	1 75	±1 50			±1 30	±1 25				±2 17		±1 15	±4 15				
150	0 150	0 75	0 50			0 30	0 25	±2 22					0 15	±2 14				

UNIVERSITY OF CALIFORNIA PUBLICATIONS
IN
MATHEMATICS

Vol. 1, No. 6, pp. 115–162 July 6, 1914

THE ABELIAN EQUATIONS OF THE TENTH DEGREE, IRREDUCIBLE IN A GIVEN DOMAIN OF RATIONALITY.

BY

CHARLES GUSTAVE PAUL KUSCHKE.

CONTENTS.

INTRODUCTION.

In the following pages is found the number of types of Abelian equations of the tenth degree irreducible in a given domain of rationality. The various types are discussed with respect to their construction and the character of their roots. The rational domain does not need to be the natural domain; in fact it will appear that there is no irreducible Abelian equation in the natural domain of rationality R (1).

If $f(x) = 0$ is an irreducible equation of the tenth degree, then all the ten roots

x_i $(i = 1$ to $10)$ must be distinct, for otherwise

$$f(x) = 0 \quad \text{and} \quad \frac{d}{dx} f(x) = 0$$

would have a common root, which is impossible, as $\frac{d}{dx} f(x)$ is only of the ninth degree in x.* From the irreducibility of $f(x)$ follows further that its Galois group is transitive† and as $10 = 2 \cdot 5$ this group must be imprimitive.‡ Our first question therefore is: What are all the imprimitive groups of the tenth degree? These groups are found by Cole§ and divided into three types.

1. The groups which contain only 5 systems of intransitivity.
2. The groups which contain only 2 systems of intransitivity.
3. The groups which contain both 5 and 2 systems of intransitivity.

To type 1 belong thirteen groups, the largest of which is of order 3840 containing all the other twelve as subgroups

To type 2 belong fifteen groups. The largest group is of order 28 800 and contains all the other fourteen groups as subgroups.

Type 3 includes eight groups, of which the largest is of order 240.

Netto‖ showed that the group of an Abelian equation is an Abelian group and vice versa. Our first problem is, therefore, to search for all Abelian groups among the above thirty-six imprimitive groups. Expressing the group of order 240 in type 3 as a substitution group, we see the following relation between the eight groups of type 3, which we may call

$$G_{240}; \quad G_{120}; \quad G'_{120}; \quad G_{40}; \quad G_{20}; \quad G'_{20}; \quad G_{10}; \quad G'_{10}$$

where the subscripts indicate the order:

G_{240} contains all the remaining seven groups as subgroups.

G'_{20} is a subgroup of G'_{120} whereas G_{20} is a subgroup of G_{120} as well as of G_{40} and contains both G'_{10} and G_{10} as subgroups. Writing this relation in a better form we have the following scheme

$$G_{240}\begin{cases} \left.\begin{matrix} G_{120} \\ G_{40} \end{matrix}\right\} G_{20} \begin{cases} G_{10} \\ G'_{10} \end{cases} \\ G'_{120}\, G'_{20} \end{cases}$$

As G'_{10} is the dihedral group of order 10 and G'_{20} contains five subgroups of order 4 which are conjugate by Sylow's theorem, it follows that the cyclical group G_{10} is the only Abelian group of type 3.

We shall use as G_{10} the ten powers of

$$s = (x_1 x_7 x_3 x_9 x_5 x_6 x_2 x_8 x_4 x_{10})$$

* Weber, *Algebra*, vol. I, p. 454.

† Weber, *Algebra*, vol. I, p. 482.

‡ Netto, *Substitutionentheorie*, §237.

§ Cole, *Quart. Journ. Math.*, vol. 27, 1895, p. 39; but also Miller, *Bull. Amer. Math. Soc.*, vol. 1, pp. 67–72.

‖ *Loc. cit.*, §180, and C. Jordan, *Traité d'Analyse*, §402.

and in dealing with substitutions we shall write only the subscripts of the x's, replacing the subscript ten by zero.

Whenever we can show that a given equation $f(x) = 0$ has a Galois group with two as well as five systems of imprimitivity, but which cannot contain G'_{10} or G'_{20}, $f(x) = 0$ must be, according to the above scheme, an equation belonging to G_{10} if $f(x)$ is irreducible. For the proof of the fact that G'_{10} and G'_{20} is not a subgroup of the Galois group of $f(x) = 0$ we shall need the two substitutions

$$\sigma_1 = (16)(20)(39)(48)(57) \text{ which is in } G'_{10}$$

and

$$\sigma_2 = (1748)(2936)(50) \text{ which is in } G'_{20}$$

These three substitutions s, σ_1, σ_2 already show that we have chosen as our two systems of imprimitivity

$$[x_1x_2x_3x_4x_5], \quad [x_6x_7x_8x_9x_{10}]$$

and as our five systems of imprimitivity

$$[x_1x_6], \quad [x_2x_7], \quad [x_3x_8], \quad [x_4x_9], \quad [x_5x_{10}]$$

Looking over types 1 and 2 for Abelian groups we may easily show that there is no other Abelian group.

Regarding type 2, the order of an Abelian group of degree 10 cannot exceed the order $2 \cdot 5^2 = 50$,* so that there is only one group to be considered, namely the group of order 50, which Cole† indicates by

$$G_{50} \equiv (x_1x_2x_3x_4x_5)_5(x_6x_7x_8x_9x_{10})_5(x_1x_6 \cdot x_2x_7 \cdot x_3x_8 \cdot x_4x_9 \cdot x_5x_{10})$$

This group, however, contains the two substitutions

$$(16)(27)(38)(49)(50)$$

and

$$[(12345)(16)(27)(38)(49)(50)]^5 = (19)(74)(20)(85)(36)$$

so that G_{50} would contain more than one substitution of order 2, which is impossible here by Sylow's theorem, hence there is no Abelian group of type 2.

Regarding type 1 we note that the only bases for the construction of the thirteen groups in question are

$$(16)(27)(38)(49)(50)$$

and

$$\{(16)(27)(38)(49)(50)\}_{\text{pos}}$$

so that every group of type 1 contains the substitution

$$a = (16)(27).$$

Further, an examination of Cole's list shows that every group contains the substitution

$$b = (13579)(68024),$$

hence every group of type 1 contains

$$a^{-1}ba = (63529)(18074) \neq b,$$

and hence no group of type 1 is Abelian.

* Netto, *loc. cit.*, §170.

† *Loc. cit.*, p. 41, and Miller, *loc. cit.*

We have then the result:

All irreducible Abelian equations of the tenth degree belong to the cyclical group of order ten, which we have chosen to be

$$G_{10} \equiv \{s = (x_1 x_7 x_3 x_9 x_5 x_6 x_2 x_8 x_4 x_{10})\}.$$

Throughout the whole investigation we shall assume that there is given a certain rational domain, for which the necessary and sufficient conditions for all the irreducible Abelian equations are to be derived, as the reducibility of a polynomial depends on the rational domain employed. We shall not distinguish at all between the quantities lying in the natural domain, usually denoted by $R(1)$, and quantities lying in the given domain. All quantities lying in the given rational domain shall be called "rationally known" or simply "rational quantities," for we do not restrict our investigation to the natural rational domain and have therefore to give to the expression "rational" a broader meaning than is done usually in mathematics. Whenever we mean specially a quantity lying in the natural rational domain we shall always write "quantity belonging to $R(1)$."

THE IRREDUCIBLE ABELIAN EQUATIONS OF THE TENTH DEGREE.

The corresponding group is

$$G_{10} \equiv \{s = (1739562840)\}.$$

Throughout these investigations let ϵ be a primitive root of

$$x^{10} - 1 = 0,$$

and ω be a primitive root of

$$x^5 - 1 = 0.$$

Then we may write

$$\begin{aligned}
\omega &= \epsilon^2 &&= -\epsilon^7 = \tfrac{1}{4}[-1 + \sqrt{5} + \sqrt{-10 - 2\sqrt{5}}],\\
\omega^2 &= \epsilon^4 &&= -\epsilon^9 = \tfrac{1}{4}[-1 - \sqrt{5} + \sqrt{-10 + 2\sqrt{5}}],\\
\omega^3 &= \epsilon^6 &&= -\epsilon \;\,= \tfrac{1}{4}[-1 - \sqrt{5} - \sqrt{-10 + 2\sqrt{5}}],\\
\omega^4 &= \epsilon^8 &&= -\epsilon^3 = \tfrac{1}{4}[-1 + \sqrt{5} - \sqrt{-10 - 2\sqrt{5}}],\\
\omega^5 &= \epsilon^{10} &&= +1,\\
& \;\;\;\;\epsilon^5 &&= -1.
\end{aligned}$$

Let us first consider the five systems of intransitivity and form the functions

$$\varphi_i = x_i + x_{i+5} \qquad (i = 1 \text{ to } 5)$$

These five functions, while formally distinct, may be numerically equal. Assuming then that at least two of these five φ_i are equal, we have by applying our group G_{10} to the relation

$$\varphi_i = \varphi_{i'} \qquad (i \neq i')$$

the result that they are all equal to one another and whenever in our equation $f(x) = 0$ we have

$$\sum_{j=1}^{j=10} x_j = 0,$$

then

$$\varphi_i = 0 \text{ and hence } x_i = -x_{i+5} \qquad (i = 1 \text{ to } 5),$$

so that our equation $f(x) = 0$ is a quintic in x^2. We note here that the transformation of $f(x) = 0$ into an equation in which

$$\sum_1^{10} x_j = 0$$

is a very simple matter and does not change the character of $f(x) = 0$. Therefore without loss of generality we may assume that in our $f(x) = 0$ the sum of the roots vanishes, as we are then gaining much simplification in our investigations.

This leads us to a division of our subject into the two parts

$$\left.\begin{array}{ll} \text{A.} & \varphi_i = 0 \\ \text{B.} & \varphi_i \neq 0 \end{array}\right\} i = 1 \text{ to } 5.$$

A. $\varphi_i = 0 \quad (i = 1 \text{ to } 5)$.

Here our given equation is a quintic in x^2 with the roots x_i $(i = 1 \text{ to } 5)$. This quintic can be taken again in the reduced form and hence we may write throughout this section

$$\sum_{i=1}^{i=5} x_i^2 = 0.$$

The following Lagrange's system is the skeleton for our investigations in this section

$$\begin{aligned} \psi_1 &= x_1 + \omega^3 x_2 + \omega x_3 + \omega^4 x_4 + \omega^2 x_5, \\ \psi_2 &= x_1 + \omega^4 x_2 + \omega^3 x_3 + \omega^2 x_4 + \omega x_5, \\ \psi_3 &= x_1 + \omega x_2 + \omega^2 x_3 + \omega^3 x_4 + \omega^4 x_5, \\ \psi_4 &= x_1 + \omega^2 x_2 + \omega^4 x_3 + \omega x_4 + \omega^3 x_5, \\ \psi_5 &= x_1 + x_2 + x_3 + x_4 + x_5. \end{aligned}$$

Consider now for a moment the two systems of intransitivity by forming the function

$$y = \prod_{i=1}^{i=5} x_i,$$

which is changed by G_{10} into

$$-y = \prod_{i=1}^{i=5} x_{i+5}.$$

It follows then that y^2 is rationally known, and we have, if $y^2 = c$, the condition *c is rational, but not a perfect square* and $-c$ is of course the constant term in $f(x) = 0$ throughout part A.

Just as $\sqrt{c}$, so is

$$\psi_5 = \sum_{i=1}^{i=5} x_i$$

a function which is changed only by the odd substitutions of G_{10}; it follows then if

$$\frac{\psi_5}{\sqrt{c}} = 5r,$$

then *r is rationally known.*

Similarly ψ_i^5 ($i = 1$ to 4) takes only two values under G_{10}, so that

$$\frac{\psi_i^5}{\sqrt{c}}$$

is a quantity in our given rational domain if we adjoin to it ω, and we must have, for instance,

$$\psi_1^5 = \sqrt{c}[k + l\sqrt{5} + m\sqrt{-10+2\sqrt{5}} + n\sqrt{-10-2\sqrt{5}}],$$

where k, l, m, n are necessarily rational quantities.

Replacing finally k, l, m, n respectively by 5^5c^2K, $5^5\cdot c^2L$, 5^5c^2M, 5^5c^2N, then K, L, M, N are rationally known, and Lagrange's system becomes

$$\psi_i = 5\sqrt{c}\,\rho_i \qquad (i = 1 \text{ to } 4),$$

$$\rho_1 = \sqrt[5]{K + L\sqrt{5} + M\sqrt{-10+2\sqrt{5}} + N\sqrt{-10-2\sqrt{5}}},$$

$$\rho_2 = \sqrt[5]{K - L\sqrt{5} - N\sqrt{-10+2\sqrt{5}} + M\sqrt{-10-2\sqrt{5}}},$$

$$\rho_3 = \sqrt[5]{K - L\sqrt{5} + N\sqrt{-10+2\sqrt{5}} - M\sqrt{-10-2\sqrt{5}}},$$

$$\rho_4 = \sqrt[5]{K + L\sqrt{5} - M\sqrt{-10+2\sqrt{5}} - N\sqrt{-10-2\sqrt{5}}},$$

Combining the two forms of Lagrange's system, we get by addition the roots of $f(x) = 0$, namely

$$\begin{aligned}
x_1 &= -x_6 = +\sqrt{c}[\rho_1 + \rho_2 + \rho_3 + \rho_4 + r],\\
x_2 &= -x_7 = +\sqrt{c}[\omega^2\rho_1 + \omega\rho_2 + \omega^4\rho_3 + \omega^3\rho_4 + r],\\
x_3 &= -x_8 = +\sqrt{c}[\omega^4\rho_1 + \omega^2\rho_2 + \omega^3\rho_3 + \omega\rho_4 + r],\\
x_4 &= -x_9 = +\sqrt{c}[\omega\rho_1 + \omega^3\rho_2 + \omega^2\rho_3 + \omega^4\rho_4 + r],\\
x_5 &= -x_{10} = +\sqrt{c}[\omega^3\rho_1 + \omega^4\rho_2 + \omega\rho_3 + \omega^2\rho_4 + r].
\end{aligned}$$

It is seen now by direct computation that

$$\psi_1\psi_4 = 25c\sqrt[5]{P + Q\sqrt{5}} = (\omega + \omega^4)f_1 + (\omega^2 + \omega^3)f_2 = 25c\rho_1\rho_4,$$

and

$$\psi_2\psi_3 = 25c\sqrt[5]{P - Q\sqrt{5}} = (\omega^2 + \omega^3)f_1 + (\omega + \omega^4)f_2 = 25c\rho_2\rho_3,$$

where

$$\begin{aligned}
P &= K^2 + 5L^2 + 10(M^2 + N^2),\\
Q &= 2KL - 2(M^2 - N^2) - 8MN,\\
f_1 &= x_1x_3 + x_1x_4 + x_2x_4 + x_2x_5 + x_3x_5,\\
f_2 &= x_1x_2 + x_1x_5 + x_2x_3 + x_3x_4 + x_4x_5.
\end{aligned}$$

Hence P and Q are rational and, as f_1 and f_2 are unchanged by our G_{10}, both f_1 and f_2 are rational too. It follows then that $\rho_1\rho_4$ and $\rho_2\rho_3$, $\psi_1\psi_4$ and $\psi_2\psi_3$ are rationally known if we adjoin $\sqrt{5}$ to our rational domain; for $(\omega^2 + \omega^3)$ as well as $(\omega + \omega^4)$ are quantities in $R(\sqrt{5})$. Therefore $P \pm Q\sqrt{5}$ is a perfect fifth power of a rational quantity in this extended domain and hence we may write

$$\rho_1\rho_4 = R + S\sqrt{5} \quad \text{and} \quad \rho_2\rho_3 = R - S\sqrt{5},$$

where then R and S are rationally known and defined by

$$P = R^5 + 50R^3S^2 + 10RS^4,$$
$$Q = 5R^4S + 50R^2S^3 + 25S^5.$$

From our condition

$$\sum_1^5 x_i^2 = 0$$

follows by direct computation

$$r^2 + 2(\rho_1\rho_4 + \rho_2\rho_3) = 0,$$

hence

$$R = -\frac{r^2}{4}.$$

Whenever our given rational domain would not include any imaginary quantity, then $R = r = 0$, for if $R < 0$ then under our condition $P < 0$ but P is the sum of squares.

We take up this natural division here and write

A. I. $r = 0$. A. II. $r \neq 0$.

A. I. $r = R = P = \psi_5 = 0 \therefore \rho_1\rho_4 = -\rho_2\rho_3$

This case will be subdivided again, namely into

1. $S = 0$. 2. $S \neq 0$.

1. If $S = 0$, then $\rho_1\rho_4 = \rho_2\rho_3 = 0$ and we may take without losing any generality

$$\rho_1 = \rho_2 = 0.$$

From $\rho_1 = 0$ follows:

$$K + L\sqrt{5} + M\sqrt{-10 + 2\sqrt{5}} + N\sqrt{-10 - 2\sqrt{5}} = 0,$$

and, from $\rho_2 = 0$

$$K - L\sqrt{5} - N\sqrt{-10 + 2\sqrt{5}} + M\sqrt{-10 - 2\sqrt{5}} = 0,$$

hence

$$\psi_3 = 5\sqrt{c}\sqrt[5]{2(K - L\sqrt{5})} \quad \text{and} \quad \psi_4 = 5\sqrt{c}\sqrt[5]{2(K + L\sqrt{5})}.$$

By addition we get further

$$2K + (M - N)\sqrt{-10 + 2\sqrt{5}} + (M + N)\sqrt{-10 - 2\sqrt{5}} = 0,$$

or

$$2K + \sqrt{-10 + 2\sqrt{5}}\left[M - N + (M + N)\frac{4\sqrt{5}}{-10 + 2\sqrt{5}}\right] = 0.$$

The condition

$$\prod_{i=1}^{i=5} x_i = \sqrt{c}$$

shows

$$K = \left(\frac{1}{2c}\right)^2,$$

and hence $K \neq 0$, so that the coefficient of $\sqrt{-10+2\sqrt{5}}$ cannot vanish and ω *belongs to our rational domain* for with $\sqrt{-10+2\sqrt{5}}$ is $\sqrt{5}$ rationally known and hence also

$$\sqrt{-10-2\sqrt{5}} = \frac{4\sqrt{5}}{\sqrt{-10+2\sqrt{5}}}$$

As $K \neq 0$, both ψ_3 and ψ_4 cannot vanish; in fact, if

$$\rho_1 = \rho_2 = \rho_3 = \rho_4 = 0$$

then

$$x_1 = x_2 = x_3 = x_4 = x_5$$

and $f(x)$ is reducible. We may, however, very well have that three ρ's vanish and we again take a subdivision without losing any generality by writing

$$\alpha.\ \rho_3 = 0;\quad \rho_4 \neq 0. \qquad \beta.\ \rho_3 \neq 0;\quad \rho_4 \neq 0.$$

α. If $\rho_3 = 0$, then

$$K = L\sqrt{5} = \frac{1}{4c^2}$$

and

$$\rho_4 = \sqrt[5]{1/c^2}.$$

In this case the ten roots of $f(x) = 0$ are

$$\begin{aligned}
x_1 &= -x_6 = \sqrt{c}\,\rho_4 = \sqrt[10]{c},\\
x_2 &= -x_7 = \omega^2\sqrt[10]{c},\\
x_3 &= -x_8 = \omega^4\sqrt[10]{c},\\
x_4 &= -x_9 = \omega\sqrt[10]{c},\\
x_5 &= -x_{10} = \omega^3\sqrt[10]{c}.
\end{aligned}$$

The corresponding equation is

$$f(x) = x^{10} - c = 0,$$

and hence *c is not a perfect fifth power in our domain,* for then $f(x)$ splits up into five rational factors each of the second degree.

Our next question is: Are the derived conditions also sufficient that $f(x) = 0$ represents an irreducible Abelian equation? In this simple case it is readily seen from the above roots that $x^{10} - c = 0$ is irreducible under the derived conditions. Adjoining $\sqrt{c}$ to the given rational domain, then $x^{10} - c$ splits up into two rational factors $(x^5 - \sqrt{c})$ and $(x^5 + \sqrt{c})$; adjoining $\sqrt[5]{c}$ to our given domain, then $x^{10} - c$ splits up into five rational factors, each of the type $(x^2 - \omega^i\sqrt[5]{c})$. We see therefore that the Galois group of $f(x) = 0$ can only be a group of type 3 in Cole's list.* In fact the group is G_{10}, for the ten roots in our cycle

$$s = (1739562840)$$

satisfy the condition, that every root is the same rational function of the preceding one in the cycle, namely

$$x_1 = -\omega^3 x_{10};\quad x_7 = -\omega^3 x_1;\quad x_3 = -\omega^3 x_7,\quad \text{etc.,}$$

ω being rational by condition.

---

* *Loc. cit.*

Type I: If c is neither a perfect fifth nor a perfect second power of any quantity in the given rational domain and if ω, c, are in this domain, then $x^{10} - c = 0$ is an irreducible Abelian equation having the above roots.

An example is furnished by taking $c = 2$. Here $\sqrt{-10 + 2\sqrt{5}}$ must be adjoined to $R(1)$.

$$f(x) \equiv x^{10} - 2 \equiv (x^5 - \sqrt{2})(x^5 + \sqrt{2})$$
$$= (x^2 - \sqrt[5]{2})(x^2 - \omega\sqrt[5]{2})(x^2 - \omega^2\sqrt[5]{2})(x^2 - \omega^3\sqrt[5]{2})(x^2 - \omega^4\sqrt[5]{2}) = 0$$

is an irreducible Abelian equation under the above conditions.

$$\beta. \quad \rho_3 \neq 0; \quad \rho_4 \neq 0.$$

The function

$$\frac{\psi_3^2}{\psi_4} = \frac{(x_1 + \omega x_2 + \omega^2 x_3 + \omega^3 x_4 + \omega^4 x_5)^2}{x_1 + \omega^2 x_2 + \omega^4 x_3 + \omega x_4 + \omega^3 x_5}$$

is a two-valued function under G_{10} just as $\sqrt{c}$; hence, writing

$$\frac{\psi_3^2}{\psi_4\sqrt{c}} = 5e,$$

then *e is rationally known but not zero,* and we have

$$25c\sqrt[5]{4(K - L\sqrt{5})^2} = 25e\cdot c\sqrt[5]{2(K + L\sqrt{5})},$$

or, as

$$K = \frac{1}{4c^2},$$

$$2^4\cdot 5L^2c^4 - 2^3L\sqrt{5}c^2(1 + e^5c^2) + (1 - 2e^5c^2) = 0,$$

which determines L for $L \neq 0$, as then

$$e = \sqrt[5]{2K} = \sqrt[5]{1/2c^2}$$

and

$$(x - x_1)(x - x_6) = x^2 - 4ce^2$$

would be a rational factor of $f(x) = 0$.

Solving for L we get

$$L = \frac{\sqrt{5}}{20c^2}[1 + e^5c^2 \pm e^2c\sqrt{e^6c^2 + 4e}];$$

as L is rationally known it follows $e^6c^2 + 4e$ *is a perfect square of a rational quantity.* Observing that

$$\rho_3^2 = \sqrt[5]{2^2(+K - L\sqrt{5})^2},$$

we have to choose the negative sign in the expression for L and hence

$$L = \frac{\sqrt{5}}{20c^2}[1 + e^5c^2 - e^2c\sqrt{e^6c^2 + 4e}].$$

So we have

$$\rho_3 = \sqrt[5]{-\frac{e^5}{2} + \frac{e^2}{2c}\sqrt{e^6c^2 + 4e}},$$

$$\rho_4 = \sqrt[5]{\frac{1}{c^2} + \frac{e^5}{2} - \frac{e^2}{2c}\sqrt{e^6c^2 + 4e}},$$

and $f(x) = 0$ becomes

$$f(x) \equiv x^{10} - 10c^2\rho_3{}^3\rho_4x^6 - 25\rho_3{}^2\rho_4{}^4c^3x^4 + (15\rho_3{}^6\rho_4{}^2 - 10\rho_3\rho_4{}^7)c^4x^2 - c = 0.$$

All coefficients of $f(x)$ are rational at the same time when

$$e = \frac{\rho_3{}^2}{\rho_4}$$

is rational, for as ω is in our domain $\rho_3{}^5$ and $\rho_4{}^5$ are rationally known; hence $\frac{\rho_3{}^2}{\rho_4} \cdot \rho_4{}^5$ and therefore $\rho_3{}^3\rho_4$ is rationally known; but then all powers of $\rho_3{}^3\rho_4$ are rational, so that $\rho_3\rho_4{}^2$; $\rho_3{}^4\rho_4{}^3$; $\rho_3{}^2\rho_4{}^4$ are in our given domain. Further $\rho_3 \cdot \rho_4$ is irrational, for if it were rational

$$\frac{\rho_3\rho_4{}^2}{\rho_3\rho_4} = \rho_4,$$

and hence ρ_3 are rationally known, but then $(x - x_1)(x - x_6)$ would be a rational factor of $f(x)$. It follows then that

$$-\frac{e^5}{2} + \frac{e^2}{2c}\sqrt{e^6c^2 + 4e}$$

is not a perfect fifth power of a rational quantity.

The roots of $f(x) = 0$ are now

$$\begin{aligned} x_1 &= -x_6 = \sqrt{c}(\rho_3 + \rho_4), \\ x_2 &= -x_7 = \sqrt{c}(\omega^4\rho_3 + \omega^3\rho_4), \\ x_3 &= -x_8 = \sqrt{c}(\omega^3\rho_3 + \omega\rho_4), \\ x_4 &= -x_9 = \sqrt{c}(\omega^2\rho_3 + \omega^4\rho_4), \\ x_5 &= -x_{10} = \sqrt{c}(\omega\rho_3 + \omega^2\rho_4), \end{aligned}$$

where

$$\rho_3 = \sqrt[5]{-\frac{e^5}{2} + \frac{e^2}{2c}\sqrt{e^6c^2 + 4e}},$$

$$\rho_4 = \sqrt[5]{\frac{1}{c^2} + \frac{e^5}{2} - \frac{e^2}{2c}\sqrt{e^6c^2 + 4e}}.$$

Regarding the sufficiency of the derived conditions, we note first that the sum of any two roots is either zero or irrational as ω is rational and both ρ_3 and ρ_4 are irrational. The sum of two roots can be zero only if their subscripts differ by 5, in which case however the product $x_i \cdot x_{i+5}$ ($i = 1$ to 5) is irrational, for it is of the form

$$c(\omega^i\rho_3 + \omega^j\rho_4)^2,$$

and must contain the irrationalities ρ_3 and ρ_4 by conditions. It follows then, that $f(x)$ cannot contain a rational factor of the second degree. Similarly it is seen that by means of our conditions there is no rational factor of the fourth degree, for the product of any four roots is a polynomial of the fourth degree in the ρ's with rational coefficients, and this cannot be rationally known, as the ρ's satisfy an irreducible quintic. There is certainly no rational factor of odd degree, for the constant term would contain $\sqrt{c}$ as a factor, which cannot be expressed rationally in terms of the ρ's. It follows then that $f(x)$ is irreducible and its Galois group is transitive. Furthermore $f(x)$ contains two as well as five systems of intransitivity, for the factors $(x - x_i)(x - x_{i+5})$ are free from $\sqrt{c}$ and the two factors

$$\prod_{i=1}^{i=5} (x + x_i) \quad \text{and} \quad \prod_{i=1}^{i=5} (x - x_{i+5})$$

are free from ρ_3 and ρ_4 and contain only those combinations of ρ_3 and ρ_4 or those powers of ρ_3 and ρ_4 which have been seen to be rationally known; for we have

$$\sum_{i=1}^{i=5} x_i = 0,$$

$$\sum_{\substack{i=1\\j=2}}^{\substack{i=4\\j=5}} x_i x_j = 0 \quad (i < j),$$

$$_{i,\,j,\,k}\sum_{\substack{1\\i<j<k}}^{5} x_i x_j x_k = 5c\sqrt{c}\rho_4{}^2\rho_3,$$

$$_{i,\,j,\,k,\,l}\sum_{\substack{1\\i<j<k<l}}^{5} x_i x_j x_k x_l = -5c^2\rho_3{}^3\rho_4,$$

$$\prod_{i=1}^{i=5} x_i = \sqrt{c}.$$

The Galois group of our above $f(x) = 0$ is therefore a group of type 3. Next we show that neither G_{10}' nor G_{20}' can be a subgroup of this Galois group. Applying

$$\sigma_1 = (16)(20)(39)(48)(57)$$

as well as

$$\sigma_2 = (1748)(2936)(50)$$

to the function

$$5e = \frac{(x_1 + \omega x_2 + \omega^2 x_3 + \omega^3 x_4 + \omega^4 x_5)^2}{\sqrt{c}(x_1 + \omega^2 x_2 + \omega^4 x_3 + \omega x_4 + \omega^3 x_5)} \neq 0$$

we change $5e$ to zero, which is contrary to condition. σ_1 as well as σ_2 cannot belong to the group of $f(x) = 0$ under our derived conditions and hence the group of $f(x) = 0$ is G_{10}.

Type II. If c, e, ω, $\sqrt{e^6c^2 + 4e}$ *but not*

$$\sqrt[5]{-\frac{e^5}{2} + \frac{e^2}{2c}\sqrt{e^6c^2 + 4e}} \quad \textit{and} \quad \sqrt{c}$$

are quantities in our given domain and if $e \neq 0$, *then the above equation* $f(x) = 0$ *with the roots on page* 124 *is an irreducible Abelian equation.*

An example is furnished by taking $e = 1$, $c = \frac{3}{2}$; then

$$4e + c^2e^6 = \left(\frac{5}{2}\right)^2 \qquad \rho_3 = \sqrt[5]{1/3} \qquad \rho_4 = \sqrt[5]{1/9}$$

$$f(x) = x^{10} - \frac{15}{2}x^6 - \frac{75}{8}x^4 + \frac{105}{16} - x^2\,\frac{3}{2} = 0.$$

The roots are

$$x_1 = -x_6 = \sqrt{3/2}[\sqrt[5]{1/3} + \sqrt[5]{1/9}],$$
$$x_2 = -x_7 = \sqrt{3/2}[\omega^4\sqrt[5]{1/3} + \omega^3\sqrt[5]{1/9}],$$
$$\cdot\quad\cdot\quad\cdot\quad \text{etc.}\quad\cdot\quad\cdot\quad\cdot$$

As given domain may be taken $R(\omega)$.

2. If $S \neq 0$ and hence $\rho_i \neq 0$ $(i = 1 \text{ to } 4)$, we shall find that there is no Abelian type.

Actual computation shows

$$\psi_1^2\psi_2 + \psi_4^2\psi_3 = \sum_{i=1}^{i=5} x_i^3 - \frac{3+\sqrt{5}}{2}F_1 - \frac{3-\sqrt{5}}{2}F_2 + 2(1+\sqrt{5})F_3 + 2(1-\sqrt{5})F_4$$

and

$$\psi_3^2\psi_1 + \psi_2^2\psi_4 = \sum_{i=1}^{i=5} x_i^3 - \frac{3-\sqrt{5}}{2}F_1 - \frac{3+\sqrt{5}}{2}F_2 + 2(1-\sqrt{5})F_3 + 2(1+\sqrt{5})F_4,$$

where

$$F_1 = x_1^2x_2 + x_2^2x_1 + x_1^2x_5 + x_5^2x_1 + x_2^2x_3 + x_3^2x_2 + x_4^2x_3 + x_3^2x_4 + x_4^2x_5 + x_5^2x_4,$$
$$F_2 = x_1^2x_3 + x_3^2x_1 + x_1^2x_4 + x_4^2x_1 + x_2^2x_4 + x_4^2x_2 + x_2^2x_5 + x_5^2x_2 + x_3^2x_5 + x_5^2x_3,$$
$$F_3 = x_1x_2x_3 + x_1x_2x_5 + x_1x_4x_5 + x_2x_3x_4 + x_3x_4x_5,$$
$$F_4 = x_1x_2x_4 + x_1x_3x_4 + x_1x_3x_5 + x_2x_3x_5 + x_2x_4x_5.$$

All these four functions are double-valued under G_{10} just as $\sqrt{c}$. It follows, then, if λ_1 and λ_2 are in our given rational domain, that we may write

$$\rho_1^2\rho_2 + \rho_4^2\rho_3 = \lambda_1 + \lambda_2\sqrt{5}, \tag{1}$$
$$\rho_2^2\rho_4 + \rho_3^2\rho_1 = \lambda_1 - \lambda_2\sqrt{5}. \tag{2}$$

λ_1 and λ_2 cannot both vanish at once, for then, observing that

$$\rho_1\rho_4 = +S\sqrt{5}, \tag{3}$$
$$\rho_2\rho_3 = -S\sqrt{5}, \tag{4}$$

we would have

$$-\rho_4^4 = \rho_2^3\rho_3 \quad\text{and}\quad \frac{1}{\rho_4^2} = \frac{\rho_2}{\rho_3^3};$$

hence, by proper multiplication of the last two,

$$-1 = \frac{\rho_2^5}{\rho_3^5} \quad\text{or}\quad \rho_2 = -\rho_3,$$

and hence (2) appears in the form $\rho_1 + \rho_4 = 0$ so that $f(x)$ would contain the rational factor

$$(x - x_1)(x - x_6) = x^2.$$

Using equations (3) and (4) in order to eliminate ρ_2 and ρ_4 in (1) and (2) and eliminating finally ρ_3, we get

$$\frac{\rho_1^5G^2}{100S^4} = \tfrac{1}{2}[(\lambda_1 - \lambda_2\sqrt{5}) \pm \sqrt{(\lambda_1 - \lambda_2\sqrt{5})^2 - 20S^3\sqrt{5}}], \tag{5}$$

where

$$G = \lambda_1 + \lambda_2\sqrt{5} \pm \sqrt{(\lambda_1 + \lambda_2\sqrt{5})^2 + 20S^3\sqrt{5}}.$$

Eliminating ρ_3 and ρ_4 in (1) and (2) by means of (3) and (4) and finally between these two new equations (1*a*) and (1*b*) eliminating ρ_2 we get

$$\frac{4\rho_1^5}{G^2} = \frac{\lambda_1 - \lambda_2\sqrt{5} \pm \sqrt{(\lambda_1 - \lambda_2\sqrt{5})^2 - 20S^3\sqrt{5}}}{10S^2}. \tag{6}$$

Comparing (5) and (6) we have

$$G^4 = 2000S^6, \tag{7}$$

and hence there are two possibilities. If

$$(\lambda_1 + \lambda_2\sqrt{5})[\lambda_1 + \lambda_2\sqrt{5} \pm \sqrt{(\lambda_1 + \lambda_2\sqrt{5})^2 + 20S^3\sqrt{5}}] = 0,$$

then

$$\lambda_1 + \lambda_2\sqrt{5} = 0, \tag{8}$$

for if the other factor were zero we would have $S = 0$. If

$$(\lambda_1 + \lambda_2\sqrt{5})[\lambda_1 + \lambda_2\sqrt{5} \pm \sqrt{(\lambda_1 + \lambda_2\sqrt{5})^2 + 20S^3\sqrt{5}}] = -\,20S^3\sqrt{5},$$

then

$$(\lambda_1 + \lambda_2\sqrt{5})^2 = -\,20S^3\sqrt{5}. \tag{9}$$

Doing with ρ_2 exactly as has just been done with ρ_1 we are led to the following result: We must have
either

$$\lambda_1 - \lambda_2\sqrt{5} = 0 \tag{10}$$

or

$$(\lambda_1 - \lambda_2\sqrt{5})^2 = +\,20S^3\sqrt{5}. \tag{11}$$

One of the two relations (8) and (9) must exist with one of the two relations (10) and (11). As we saw, $\lambda_1 = \lambda_2 = 0$ is impossible, hence we cannot pair (8) and (10). Also relation (9) cannot exist with relation (11), for we would get

$$\lambda_1^2 + 5\lambda_2^2 = 0,$$

which leads to $S = 0$. It follows that either (8) and (11) or (9) and (10) exist together. Both cases are identical in the abstract; they differ only in the sign of $\sqrt{5}$. Assuming then

$$\lambda_1 + \lambda_2\sqrt{5} = 0, \tag{8}$$

we have from (11)

$$\lambda_1^2 = 5S^3\sqrt{5}, \tag{11a}$$

and $\sqrt{5}$ is rational and further $\sqrt{S\sqrt{5}}$ is in our domain. Equations (1) and (2) appear now in the form

$$\rho_1^2\rho_2 = -\,\rho_4^2\rho_3, \tag{12}$$

and

$$\rho_2^2\rho_4 + \rho_3^2\rho_1 = 2\lambda_1, \tag{13}$$

which may be written, by observing (3) and (4),

$$\rho_2^2\rho_4 + \frac{5S^3\sqrt{5}}{\rho_2^2\rho_4} = 2\lambda_1,$$

or, by means of (11*a*),

$$\lambda_1 = \rho_2^2\rho_4;$$

hence from (13) we have

$$\lambda_1 = \rho_2^2\rho_4 = \rho_3^2\rho_1;$$

hence

$$\rho_1 = \frac{\rho_2^2\rho_4}{\rho_3^2}. \tag{14}$$

From (12) we get, by means of (3), (4), and (11*a*)

$$\rho_1^2\rho_2 = +\frac{\lambda_1^2}{\rho_1^2\rho_2};$$

hence

$$\rho_1^2\rho_2 = \pm\lambda_1,$$

and we have finally

$$\lambda_1 = \rho_2^2\rho_4 = \rho_3^2\rho_1 = \pm\rho_1^2\rho_2 = \mp\rho_4^2\rho_3, \tag{15}$$

so that

$$\rho_1 = \pm\frac{\rho_3^2}{\rho_2} \tag{16}$$

and

$$\rho_1 = \mp\frac{\rho_4^2}{\rho_3}. \tag{17}$$

Observing (14), (16), (17) we may write

$$\pm\rho_1^2 = \frac{\rho_2^2\rho_4}{\rho_3^2}\cdot\frac{\rho_3^2}{\rho_2} = \rho_2\rho_4,$$

$$\pm\rho_1^3 = \left(\frac{\rho_4^2}{\rho_3}\right)^2\cdot\frac{\rho_3^2}{\rho_2} = \frac{\rho_4^4}{\rho_2}.$$

Multiplying the last two relations together we obtain

$$\rho_1^5 = \rho_4^5 \quad \text{or} \quad \rho_1 = \rho_4, \tag{18}$$

so that

$$\sqrt{\rho_1\rho_4} = \rho_1 = \rho_4 = \sqrt{S\sqrt{5}}$$

is rationally known, as we saw above. With ρ_1 and ρ_4 are ρ_2 and ρ_3 rationally known by means of (15) and hence

$$(x - x_1)(x - x_6)$$

is a rational factor of $f(x)$. There is no irreducible Abelian type under A. I. 2.

A. II. $r \neq 0$.

We had the relation

$$r^2 + 2(\rho_1\rho_4 + \rho_2\rho_3) = 0,$$

so that at most two ρ's can vanish, either $\rho_1 = \rho_4 = 0$ or $\rho_2 = \rho_3 = 0$. Without loss of generality we may take the following division of case II, $\rho_1 \neq 0$; $\rho_4 \neq 0$ and

$$(1)\ \rho_2 = \rho_3 = 0, \qquad (2)\ \rho_2 = 0;\ \ \rho_3 \neq 0, \qquad (3)\ \rho_2 \neq 0;\ \ \rho_3 \neq 0.$$

1. If $\rho_2 = \rho_3 = 0$ then

$$K = L\sqrt{5}$$

and

$$N = -M\frac{1+\sqrt{5}}{2}.$$

Suppose for a moment $K = L = 0$ then $\rho_1 = -\rho_4$ so that $f(x)$ contains the rational factor $(x^2 - cr)$. It follows then

$$K = L\sqrt{5} \neq 0$$

and hence $\sqrt{5}$ *is in our given domain of rationality.*

We may now write

$$\rho_1^5 = 2K + M'\sqrt{-10 + 2\sqrt{5}},$$
$$\rho_4^5 = 2K - M'\sqrt{-10 + 2\sqrt{5}},$$

where

$$M' = \frac{5+\sqrt{5}}{2}M$$

is necessarily rationally known.

$$(\rho_1\rho_4)^5 = -\left(\frac{r^2}{2}\right)^5 = 4K^2 - M'^2(-10 + 2\sqrt{5}),$$

so that M' can be determined by c and r as K is found from the condition

$$\prod_{i=1}^{i=5} x_i = \sqrt{c},$$

which leads to the relation

$$K = \frac{1 - \frac{19}{4}r^5c^2}{4c^2};$$

hence $r^5c^2 \neq \frac{4}{19}$ as $K \neq 0$ and $c \neq 0$.

$$M' = \frac{1}{8c^2}\sqrt{\frac{16 - 152r^5c^2 + 363r^{10}c^4}{-10 + 2\sqrt{5}}} \neq 0,$$

for if $M' = 0$, then $M = N = 0$ and $\rho_1 = -\rho_4$, so that $\rho_1\rho_4 = -\rho_1^2 = -\rho_4^2$ and therefore ρ_1, ρ_4 would be rational and hence $f(x)$ is reducible. It follows then

$$\sqrt{\frac{16 - 152r^5c^2 + 363r^{10}c^4}{-10 + 2\sqrt{5}}} \neq 0,$$

but is in our rational domain.

The roots of $f(x) = 0$ are given on page 120, where the ρ's are defined by

$$\rho_1 = \sqrt[5]{\frac{1 - \frac{19}{4} r^5c^2}{2c^2} + \frac{1}{8c^2}\sqrt{16 - 152r^5c^2 + 363r^{10}c^4}},$$

$$\rho_2 = \rho_3 = 0,$$

$$\rho_4 = \sqrt[5]{\frac{1 - \frac{19}{4} r^5c^2}{2c^2} - \frac{1}{8c^2}\sqrt{16 - 152r^5c^2 + 363r^{10}c^4}}.$$

The corresponding equation becomes

$$f(x) = x^{10} + \frac{115}{4}r^4c^2x^6 + 5\left(\frac{15}{2}r^6 - 2\frac{r}{c^2}\right)c^3x^4 + \left(\frac{541}{16}r^8 - \frac{35}{c^2}r^3\right)c^4x^2 - c = 0.$$

We ask again: Are the derived conditions also sufficient, that $f(x) = 0$ is an irreducible Abelian equation?

Our $f(x)$ cannot have any rational factor of odd degree, for the constant term would contain $\sqrt{c}$ as a factor which is irrational by conditions. $\sqrt{c}$ and ρ_i cannot be expressed rationally by each other. Further, as every root is of the form

$$\pm\sqrt{c}[\omega^k\rho_1 + \omega^{-k}\rho_4 + r] = \pm\sqrt{c}\left[\omega^k\rho_1 - \frac{r^2}{2\omega^k\rho_1} + r\right],$$

the sum of any two roots assumes one of the two forms either

$$\text{(1)} \qquad \pm\sqrt{c}\left[\rho_1(\omega^k + \omega^j) - \frac{r^2(\omega^k + \omega^j)}{2\rho_1\omega^{k+j}} + 2r\right],$$

or

$$\text{(2)} \qquad \pm\sqrt{c}\left[\rho_1 - \frac{r^2}{2\rho_1\omega^{k-j}}\right](\omega^k - \omega^j).$$

The type (2) shows that, if $k = j$, then the sum of the two roots vanishes, which can happen only if the subscripts differ by five. In this case, however, the product x_ix_{i+5} contains the irrationality ρ_1, for we have

$$x_ix_{i+5} = -c\left[\omega^k\rho_1 - \frac{r^2}{2\omega^k\rho_1} + r\right]^2 \neq 0.$$

In all other cases the sum of any two roots does not vanish and hence both irrationalities remain in the expression of the sum, for if in (2)

$$\rho_1 = \frac{r^2}{2\rho_1\omega^{k-j}},$$

then

$$r = \rho_1\omega^\mu\sqrt{2},$$

which cannot happen, as ρ_1 satisfies an irreducible quintic, $\sqrt{2}$ a quadratic and ω an irreducible quadratic for $\sqrt{5}$ is rational. If in (1)

$$\rho_1(\omega^k + \omega^j) - \frac{r^2(\omega^k + \omega^j)}{2\rho_1\omega^{k+j}} = 0,$$

then

$$\rho_1^2 = \left(\frac{r}{\omega^\lambda}\right)^2 \cdot \frac{1}{2},$$

which would be again impossible, as ρ_1 is irrational and satisfies an irreducible quintic.

Exactly the same thing can be shown in the case of a factor of the fourth degree, having now instead of $\rho_1(\omega^k + \omega^j)$ and ω^{k+j}, always $\rho_1(\omega^k + \omega^j + \omega^l + \omega^m)$ and $\omega^{k+j+l+m}$ respectively. If

$$\omega^k + \omega^l + \omega^j + \omega^m = -1,$$

then

$$\omega^{k+l+j+m} = +1,$$

and the sum of the four roots vanishes; but then the product of the four roots is irrational for the same reason as in the case $x_i x_{i+5}$.

Whenever, then, ρ_1 is irrational, we see, that the above conditions are sufficient for the irreducibility of $f(x)$. Hence we write

$$\sqrt[5]{\frac{1 - \frac{19}{4} r^5 c^2}{2c^2} + \frac{1}{8c^2}\sqrt{\frac{16 - 152c^2r^5 + 363c^4r^{10}}{-10 + 2\sqrt{5}}}}$$

is irrational.

Under the above conditions, then, the Galois group is transitive and must be again of type 3 in Cole's list for the five factors

$$(x - x_i)(x - x_{i+5}) \quad (i = 1 \text{ to } 5)$$

are rational after the adjunction of ρ_1 to the rational domain and the two factors

$$\prod_{i=1}^{i=5} (x - x_i) \quad \text{and} \quad \prod_{i=1}^{i=5} (x - x_{i+5})$$

are rational after the adjunction of $\sqrt{c}$ to the given domain of rationality for we find:

$$\sum_{i=1}^{=5} x_i = 0,$$

$$\sum_{\substack{1 \\ i<j}}^{5} {}_{i,j}\, x_i x_j = \frac{25}{2} r^2 c,$$

$$\sum_{\substack{1 \\ i<j<k}}^{5} {}_{i,j,k}\, x_i x_j x_k = \frac{35}{2} r^3 c \sqrt{c},$$

$$\sum_{\substack{1 \\ i<j<k<l}}^{5} {}_{i,j,k,l}\, x_i x_j x_k x_l = \frac{45}{4} r^4 c^2,$$

$$\sum_{i=1}^{i=5} x_i = +\sqrt{c}.$$

Now σ_1 interchanges $\pm\psi_1$ and $\mp\psi_4$, from which would follow that $\rho_1 = \pm\rho_4$, if σ_1 were in our Galois group, and hence $K = 0$ or $M' = 0$, which is contrary to our conditions.

σ_2 changes $\psi_1\psi_4 \neq 0$ to zero which is impossible for any substitution in our

Galois group, as $\psi_1\psi_4$ is rationally known. It follows then that our $f(x) = 0$ is an irreducible Abelian equation.

Type III. If c, $\sqrt{5}$, r *but not* $\sqrt{c}$ *are in our given rational domain, and if*

$$\sqrt{\frac{16 - 152r^5c^2 + 363r^{10}c^4}{-10 + 2\sqrt{5}}} \neq 0$$

but rational and

$$\sqrt[5]{\frac{1 - \frac{19}{4}r^5c^2}{2c^2} + \frac{1}{8c^2}\sqrt{\frac{16 - 152r^5c^2 + 363c^4r^{10}}{-10 + 2\sqrt{5}}}}$$

is irrational and $r \neq 0$, $r^5c^2 \neq \frac{4}{19}$ *then the above* $f(x)$ *is an irreducible Abelian equation with roots as given above.*

An example is furnished by taking $r = 1$; $c = 2$. Then

$$\frac{16 - 152c^2r^5 + 363c^4r^{10}}{-10 + 2\sqrt{5}} = \frac{16 \cdot 163}{-5 + \sqrt{5}};$$

hence our rational domain must contain

$$\sqrt{\frac{163}{-5 + \sqrt{5}}}$$

$$f(x) = x^{10} + 115x^6 + 280x^4 + 401x^2 - 2 = 0$$

The roots are

$$x_1 = -x_6 = +\sqrt{2}\left[\sqrt[5]{-\frac{9}{4} + \frac{1}{8}\sqrt{326}} + \sqrt[5]{-\frac{9}{4} - \frac{1}{8}\sqrt{326}} + 1\right],$$

$$x_2 = -x_7 = +\sqrt{2}\left[\omega^2\sqrt[5]{-\frac{9}{4} + \frac{1}{8}\sqrt{326}} + \omega^3\sqrt[5]{-\frac{9}{4} - \frac{1}{8}\sqrt{326}} + 1\right],$$

. . . .

etc.

2. If $\rho_2 = 0$, $\rho_i \neq 0$ $(i = 1, 3, 4)$, we have

$$\rho_3 = \sqrt[5]{2(K - L(\sqrt{5})} \neq 0,$$

which follows from $\rho_2 = 0$, that is, from

$$K - L\sqrt{5} = N\sqrt{-10 + 2\sqrt{5}} - M\sqrt{-10 - 2\sqrt{5}},$$

so that ω *is rationally known.*

The functions

$$\frac{\psi_3^2}{\psi_4\sqrt{c}} \quad \text{and} \quad \frac{\psi_1^2}{\psi_3\sqrt{c}}$$

are seen to be rationally known, if applied to by G_{10}. Calling them $5e$ and $5d$

respectively we have

$$e = \frac{\rho_3^2}{\rho_4} \quad \text{and} \quad d = \frac{\rho_1^2}{\rho_3},$$
$$\rho_1^5 = -\frac{ed^2r^2}{2},$$
$$\rho_3^5 = +\frac{e^2r^4}{4d},$$
$$\rho_4^5 = +\frac{r^8}{2^4ed^2}.$$

If ρ_1 were rational, then ρ_4 and ρ_3 are rational and vice versa; hence we have the condition e, d, *are rationally known and* $2^4ed^2r^2$ *is not a perfect fifth power.* The last condition excludes $e = 0$, and $d = 0$ as well as $r = 0$. The corresponding equation becomes

$$f(x) = x^{10} + 5A_1r^2c^2x^6 + 5A_2r^3c^3x^4 + \frac{5}{2}A_3r^4c^4x^2 - c = 0,$$

where

$$A_1 = \frac{23}{4}r^2 + de + 3re - \frac{1}{2}\frac{r^2e}{d} - \frac{3}{2}\frac{r^3}{d},$$
$$A_2 = -2r^3 + d^2e - \frac{e^2r^2}{d} - \frac{r^6}{8d^2e} - 5r^2e + \frac{15}{2}rde - \frac{5}{2}\frac{r^3e}{d} - \frac{5}{4}re^2 - \frac{5}{16}\frac{r^5}{d^2},$$
$$A_3 = \frac{71}{8}r^4 + \frac{1}{2}\frac{r^2e^3}{d} - \frac{r^8}{16d^3e} + \frac{3d^2e^2}{2} + \frac{3r^4e^2}{8d^2} + \frac{3r^6}{4d^2} - 7r^2de - \frac{5}{4}\frac{r^4e}{d} - \frac{7}{8}\frac{r^7}{d^2e},$$
$$+ 7rd^2e - \frac{r^7e^2}{d} - \frac{7}{2}\frac{r^5}{d} - \frac{51}{2}r^3e + 5r^2e^2 - 5re^2d - \frac{5}{4}\frac{r^5e}{d^2}.$$

The roots are—

$$x_1 = -x_6 = +\sqrt{c}\left[-\sqrt[5]{\frac{ed^2r^2}{2}} + \sqrt[5]{\frac{e^2r^4}{4d}} + \sqrt[5]{\frac{r^8}{16ed^2}} + r\right],$$
$$x_2 = -x_7 = +\sqrt{c}\left[-\omega^2\sqrt[5]{\frac{ed^2r^2}{2}} + \omega^4\sqrt[5]{\frac{e^2r^4}{4d}} + \omega^3\sqrt[5]{\frac{r^8}{16ed^2}} + r\right],$$

. . . .

etc.

r, c, e, d are not independent of each other, for they have to satisfy the identity

$$\frac{1}{c^2} = \frac{19}{4}r^5 + \frac{r^8}{16ed^2} + \frac{r^4e^2}{4d} - \frac{ed^2r^2}{2} + \frac{15}{8}\frac{r^6}{d} + \frac{15}{4}r^4e + \frac{5}{2}r^3de - \frac{5}{4}\frac{r^5e}{d},$$

which follows from the condition

$$\prod_{i=1}^{i=5} x_i = \sqrt{c}.$$

Regarding the sufficiency of the derived conditions, we first note that $f(x)$ cannot have a rational factor of odd degree, for the constant term would contain $\sqrt{c}$ as a factor and $\sqrt{c}$ cannot be expressed rationally by ρ_i ($i = 1, 3, 4$). Whenever the sum of any two or four roots vanishes, then these sums must be of the form

$$x_i + x_{i+5} \quad \text{or} \quad x_i + x_{i+5} + x_j + x_{j+5} \quad (i \neq j),$$

in which cases the product of the two or four roots is of the form

$$-c[\omega^k\rho_1+\omega^l\rho_3+\omega^m\rho_4+r]^2 \quad \text{or}$$
$$+c^2[\omega^r\rho_1+\omega^s\rho_3+\omega^t\rho_4+r]^2\cdot[\omega^k\rho_1+\omega^l\rho_3+\omega^m\rho_4+r]^2.$$

In each bracket stands an expression satisfying an irreducible quintic and hence only a power of $s\equiv 0$ (mod. 5) can make the product rational. Both products are of lower degree than five in the ρ's and hence they are irrational. Whenever the sum of any two or four roots does not vanish, it must contain $\sqrt{c}$ as a factor. It follows, then, that $f(x)$ is irreducible; therefore its Galois group is transitive. This group must contain again two as well as five systems of intransitivity, for the five products

$$(x-x_i)(x-x_{i+5}) \quad (i=1 \text{ to } 5).$$

do not show the irrationality $\sqrt{c}$, and the two products

$$\prod_{i=1}^{i=5}(x-x_i) \quad \text{and} \quad \prod_{i=1}^{i=5}(x-x_{i+5})$$

do not show the irrationality ρ, for we have

$$\sum_{i=1}^{i=5}x_i=5r\sqrt{c},$$
$$\sum_{\substack{i,j=1\\i<j}}^{i,j=5}x_ix_j=+\frac{25}{2}r^2c,$$
$$\sum_{\substack{i,j,k=1\\i<j<k}}^{i,j,k=5}x_ix_jx_k=\frac{5cr^2\sqrt{c}}{4}\left[14r-2e+\frac{r^2}{d}\right],$$
$$\sum_{\substack{i,j,k,l=1\\i<j<k<l}}^{i,j,k,l=5}x_ix_jx_kx_l=\frac{5c^2r^2}{4}\left[11r^2+2de+2\frac{r^3}{d}-\frac{er^2}{d}-2re\right],$$
$$\prod_{i=1}^{i=5}x_i=\sqrt{c}.$$

Applying σ_1 and σ_2 to $e=\dfrac{\rho_3{}^2}{\rho_4}$, we come to the *reductio ad absurdum* $e=0$.

It follows then that $f(x)=0$ is an irreducible Abelian equation under the derived conditions.

Type IV. If c, ω, d, r, e but not $\sqrt[5]{2^4ed^2r^2}$ and $\sqrt{c}$ are in the given rational domain, and if the identity exists

$$\frac{1}{c^2}=\frac{19}{4}r^5+\frac{r^8}{16ed^2}+\frac{r^4e^2}{4d}-\frac{ed^2r^2}{2}+\frac{15}{8}\frac{r^6}{d}+\frac{15}{4}r^4e+\frac{5}{2}r^3de-\frac{5}{4}\frac{r^5e}{d},$$

then the above $f(x)=0$ with the given roots is an irreducible Abelian equation of the tenth degree.

An example is furnished by supposing

$$r=d=e=1, \qquad c=\frac{4}{\sqrt{183}}.$$

Here $\sqrt{183}$ must be in our rational domain as well as ω, but not $\sqrt[4]{183}$; further, the identity between r, d, e, c exists as given above and $2^4ed^2r^2$ is not a perfect fifth power; hence $\sqrt[5]{2}$ must be a quantity not in the given domain. The corresponding equation is

$$f(x) = x^{10} + \frac{619}{183}x^6 - \frac{1019,808}{183\sqrt{183}}x^4 + \frac{13714,88}{183^2}x^2 - \frac{4}{\sqrt{183}} = 0.$$

The roots are:

$$x_1 = -x_6 = \frac{2}{\sqrt[4]{183}}\left[-\sqrt[5]{\frac{1}{2}} + \sqrt[5]{\frac{1}{4}} + \sqrt[5]{\frac{1}{16}} + 1\right],$$

$$x_2 = -x_7 = \frac{2}{\sqrt[4]{183}}\left[-\omega^2\sqrt[5]{\frac{1}{2}} + \omega^4\sqrt[5]{\frac{1}{4}} + \omega^3\sqrt[5]{\frac{1}{16}} + 1\right],$$

. . . etc. . . .

3. $\rho_i \neq 0$ $(i = 1 \text{ to } 4)$.

We proceed to show, that this case is impossible here just as case A. I, 2.

We saw above that, if λ_1 and λ_2 are in our given rational domain, we may put

$$\rho_1^2\rho_2 + \rho_4^2\rho_3 = \lambda_1 + \lambda_2\sqrt{5}, \tag{1}$$

$$\rho_2^2\rho_4 + \rho_3^2\rho_1 = \lambda_1 - \lambda_2\sqrt{5}, \tag{2}$$

$$\rho_1\rho_4 = +R + S\sqrt{5} \equiv T \neq 0, \tag{3}$$

$$\rho_2\rho_3 = +R - S\sqrt{5} \equiv Q \neq 0. \tag{4}$$

Eliminating ρ_2 and ρ_4 in (1) and (2) by means of (3) and (4), and finally removing ρ_3 in the new equation (2), we come to the result

$$\frac{A^2\rho_1^5}{4T^4} = \frac{\lambda_1 - \lambda_2\sqrt{5} \pm \sqrt{(\lambda_1 - \lambda_2\sqrt{5})^2 - 4Q^2T}}{2}, \tag{5}$$

where

$$A = \lambda_1 + \lambda_2\sqrt{5} \pm \sqrt{(\lambda_1 + \lambda_2\sqrt{5})^2 - 4T^2Q}.$$

Removing ρ_3 and ρ_4 in (1) and (2) by means of (3) and (4), and finally ρ_2, then equation (2) can be written

$$\frac{4\rho_1^5}{A^2} = \frac{\lambda_1 - \lambda_2\sqrt{5} \pm \sqrt{(\lambda_1 - \lambda_2\sqrt{5})^2 - 4Q^2T}}{2Q^2}. \tag{6}$$

Comparing (5) and (6) we have

$$A^2 = \pm 2^2QT^2. \tag{7}$$

The last relation may be written

$$(\lambda_1 + \lambda_2\sqrt{5})[\lambda_1 + \lambda_2\sqrt{5} \pm \sqrt{(\lambda_1 + \lambda_2\sqrt{5})^2 - 4QT^2}] = 4QT^2 \text{ or } = 0, \tag{8}$$

hence there are only two possibilities, namely

$$\lambda_1 + \lambda_2\sqrt{5} = 0, \tag{9}$$

or

$$(\lambda_1 + \lambda_2\sqrt{5})^2 = 4QT^2 \neq 0. \tag{10}$$

Doing the same thing with ρ_2, we find, that one of the following two relations must exist:

$$\lambda_1 - \lambda_2\sqrt{5} = 0 \tag{11}$$

or

$$(\lambda_1 - \lambda_2 \sqrt{5})^2 = 4Q^2T. \tag{12}$$

Pairing equations (9) and (11), then $\lambda_1 = \lambda_2 = 0$; hence we get from

$$\text{equation (2) that} + \frac{1}{\rho_4^2} = -\frac{\rho_2^2}{\rho_3^2(2R - \rho_2\rho_3)},$$

$$\text{and equation (1) that} + \rho_4^4 = -\frac{\rho_2(2R - \rho_2\rho_3)^2}{\rho_3},$$

$$\therefore + 1 = \rho_4^4 \cdot \left(\frac{1}{\rho_4^2}\right)^2 = \frac{+\rho_2^5}{\rho_3^5}, \quad \text{or} \quad \rho_2 = \rho_3$$

and from (2) again we get, if $\rho_2 = \rho_3$, that $\rho_1 = -\rho_4$.

From $\rho_2 = \rho_3$ we have

$$N\sqrt{-10 + 2\sqrt{5}} - M\sqrt{-10 - 2\sqrt{5}} = 0,$$

and from $\rho_1 = -\rho_4$ we have

$$K + L\sqrt{5} = 0.$$

As now $\rho_i \neq 0$, it follows $\sqrt{5}$ is rationally known but then $\rho_1\rho_4 = -\rho_1^2 = -\rho_4^2$ as well as $\rho_2\rho_3 = \rho_2^2 = \rho_3^2$, and hence ρ_i is rationally known, so that $f(x)$ contains the rational factor $(x - x_1)(x - x_6)$. It follows then that $\lambda_1 = \lambda_2 = 0$ is excluded.

Pairing (9) and (12), then $\sqrt{5}$ is rationally known, as

$$\lambda_1 = -\lambda_2\sqrt{5} \neq 0$$

and we have from (12)

$$\lambda_1^2 = Q^2T,$$

or

$$\lambda_1 = (R - S\sqrt{5})\sqrt{R + S\sqrt{5}}.$$

From (2) we have

$$\rho_2^2\rho_4 = 2\lambda_1 - \rho_3^2\rho_1 = 2\lambda_1 - \frac{Q^2T}{\rho_2^2\rho_4},$$

or

$$(\rho_2^2\rho_4)^2 - 2\lambda_1\rho_2^2\rho_4 + \lambda_1^2 = 0,$$

$$\rho_2^2\rho_4 = \lambda_1.$$

It follows again by (2)

$$\rho_2^2\rho_4 = \rho_1\rho_3^2 = \lambda_1$$

and hence

$$\rho_1 = \frac{\rho_2^2\rho_4}{\rho_3^2}. \tag{13}$$

From (1) we have

$$\rho_1^2\rho_2 = -\rho_4^2\rho_3 \tag{14}$$

(13), (14): eliminating ρ_1

$$\therefore \frac{\rho_2^4\rho_4^2\rho_2}{\rho_3^4} = -\rho_4^2\rho_3,$$

or

$$\rho_2 = -\rho_3 \tag{15}$$

and hence $\rho_1 = \rho_4$ by (13). We have arrived at a similar impossible result as in the case $\lambda_1 = \lambda_2 = 0$.

The case of pairing (10) and (11) is identical to the case just discussed; the only difference lies in the other sign of $\sqrt{5}$.

It remains then to show that relation (10) and relation (12) cannot exist together. From (2) we have

$$\rho_2^2\rho_4 = \lambda_1 - \lambda_2\sqrt{5} - \rho_3^2\rho_1 = \lambda_1 - \lambda_2\sqrt{5} - \frac{Q^2T}{\rho_2^2\rho_4}$$

or

$$(\rho_2^2\rho_4)^2 - (\lambda_1 - \lambda_2\sqrt{5})\rho_2\rho_4 + Q^2T = 0,$$

$$\therefore \text{by (12)} \quad \rho_2^2\rho_4 = +\frac{\lambda_1 - \lambda_2\sqrt{5}}{2}$$

or by (2) again

$$\rho_2^2\rho_4 = \rho_3^2\rho_1 \quad \text{or} \quad \rho_1 = \frac{\rho_2^2\rho_4}{\rho_3^2}. \tag{16}$$

Similarly from (1) follows exactly in the same way by means of (10)

$$\rho_1^2\rho_2 = \frac{\lambda_1 + \lambda_2\sqrt{5}}{2} = \rho_4^2\rho_3. \tag{17}$$

Eliminating ρ_1 between (16) and (17), we are led to the relation

$$\rho_2 = \rho_3$$

and hence, by (16),

$$\rho_1 = \rho_4.$$

Now $\rho_1^2\rho_2 + \rho_2^2\rho_4 = \lambda_1$ and as $\rho_2 = \rho_3$

$$\rho_1^2\rho_3 + \rho_2^2\rho_4 = \lambda_1 = \text{rationally known.}$$

Applying however G_{10} to $\rho_1^2\rho_3 + \rho_2^2\rho_4$, we see it is changed into $\omega\rho_1^2\rho_3 + \rho_2^2\rho_4$ and, as $\rho_i \neq 0$, we have the result:

There is no Abelian equation corresponding to case A. II, 3.

B. ALL $\varphi_i = x_i + x_{i+5}$ $(i = 1$ TO $5)$ ARE DISTINCT.

The irreducible Abelian equations, left for consideration, do not reduce to quintics in x^2.

Let us consider first the five systems of imprimitivity by taking the function

$$\prod_{i=1}^{i=5}(x_i - x_{i+5}), \quad (i = 1 \text{ to } 5)$$

This function is a double-valued function under G_{10}; hence its square is a quantity in our given rational domain. Putting then

$$\gamma = \prod_{i=1}^{i=5}(x_i - x_{i+5})^2 \quad (i = 1 \text{ to } 5),$$

γ is rationally known, but not a perfect square.

The Lagrange resolvent

$$\psi_1 = x_1 + \epsilon x_7 + \epsilon^2 x_3 + \epsilon^3 x_9 + \epsilon^4 x_5 + \epsilon^5 x_6 + \epsilon^6 x_2 + \epsilon^7 x_8 + \epsilon^8 x_4 + \epsilon^9 x_{10}$$

can be written in the form

$$\psi_1 = x_1 - x_6 + \omega^3(x_2 - x_7) + \omega(x_3 - x_8) + \omega^4(x_4 - x_9) + \omega^2(x_5 - x_{10}).$$

Similarly we may write the other values of ψ_1 under G_{10} in the form:

$$\psi_2 = x_1 + x_6 + \omega(x_2 + x_7) + \omega^2(x_3 + x_8) + \omega^3(x_4 + x_9) + \omega^4(x_5 + x_{10}),$$
$$\psi_3 = x_1 - x_6 + \omega^4(x_2 - x_7) + \omega^3(x_3 - x_8) + \omega^2(x_4 - x_9) + \omega(x_5 - x_{10}),$$
$$\psi_4 = x_1 + x_6 + \omega^2(x_2 + x_7) + \omega^4(x_3 + x_8) + \omega(x_4 + x_9) + \omega^3(x_5 + x_{10}),$$
$$\psi_5 = x_1 - x_6 + (x_2 - x_7) + (x_3 - x_8) + (x_4 - x_9) + (x_5 - x_{10}),$$
$$\psi_6 = x_1 + x_6 + \omega^3(x_2 + x_7) + \omega(x_3 + x_8) + \omega^4(x_4 + x_9) + \omega^2(x_5 + x_{10}),$$
$$\psi_7 = x_1 - x_6 + \omega(x_2 - x_7) + \omega^2(x_3 - x_8) + \omega^3(x_4 - x_9) + \omega^4(x_5 - x_{10}),$$
$$\psi_8 = x_1 + x_6 + \omega^4(x_2 + x_7) + \omega^3(x_3 + x_8) + \omega^2(x_4 + x_9) + \omega(x_5 + x_{10}),$$
$$\psi_9 = x_1 - x_6 + \omega^2(x_2 - x_7) + \omega^4(x_3 - x_8) + \omega(x_4 - x_9) + \omega^3(x_5 - x_{10}),$$
$$\psi_{10} = x_1 + x_6 + (x_2 + x_7) + (x_3 + x_8) + x_4 + x_9) + (x_5 + x_{10}),$$

Taking our $f(x) = 0$ again in the reduced form, then

$$\sum_{j=1}^{j=10} x_j = \psi_{10} = 0 \quad \text{and} \quad \sum_{i=1}^{i=5} x_i = \frac{\psi_5}{2}.$$

Both functions, $\sum_{i=1}^{i=5} x_i$ and $\prod_{i=1}^{i=5} (x_i - x_{i+5})$, are double-valued functions under G_{10}; hence calling

$$5r = \frac{\sum_{i=1}^{i=5} x_i}{\prod_{i=1}^{i=5} (x_i - x_{i+5})},$$

then *r is in our given domain* and $\psi_5 = 10\sqrt{\gamma} \cdot r$.

All $\psi_a{}^5$ are unchanged by G_{10} if a is even; and if a is odd then $\frac{\psi_a{}^5}{\sqrt{c}}$ must be rationally known if we adjoin ω to the rational domain. Analogy to case A shows that we may write

$$\psi_1 = 10\sqrt{\gamma}\sqrt[5]{K_1 + L_1\sqrt{5} + M_1\sqrt{-10 + 2\sqrt{5}} + N_1\sqrt{-10 - 2\sqrt{5}}} = 10\rho_1\sqrt{\gamma},$$

$$\psi_2 = 10 \cdot \sqrt[5]{K_2 - L_2\sqrt{5} + N_2\sqrt{-10 + 2\sqrt{5}} - M_2\sqrt{-10 - 2\sqrt{5}}} = 10\rho_2,$$

$$\psi_3 = 10\sqrt{\gamma}\sqrt[5]{K_1 - L_1\sqrt{5} - N_1\sqrt{-10 + 2\sqrt{5}} + M_1\sqrt{-10 - 2\sqrt{5}}} = 10\sqrt{\gamma}\rho_3,$$

$$\psi_4 = 10\sqrt[5]{K_2 + L_2\sqrt{5} - M_2\sqrt{-10 + 2\sqrt{5}} - N_2\sqrt{-10 - 2\sqrt{5}}} = 10\rho_4,$$

$$\psi_6 = 10\sqrt[5]{K_2 + L_2\sqrt{5} + M_2\sqrt{-10 + 2\sqrt{5}} + N_2\sqrt{-10 - 2\sqrt{5}}} = 10\rho_6,$$

$$\psi_7 = 10\sqrt{\gamma}\sqrt[5]{K_1 - L_1\sqrt{5} + N_1\sqrt{-10 + 2\sqrt{5}} - M_1\sqrt{-10 - 2\sqrt{5}}} = 10\sqrt{\gamma}\rho_7,$$

$$\psi_8 = 10\sqrt[5]{K_2 - L_2\sqrt{5} - N_2\sqrt{-10 + 2\sqrt{5}} + M_2\sqrt{-10 - 2\sqrt{5}}} = 10\rho_8,$$

$$\psi_9 = 10\sqrt{\gamma}\sqrt[5]{K_1 + L\sqrt{5} - M_1\sqrt{-10 + 2\sqrt{5}} - N_1\sqrt{-10 - 2\sqrt{5}}} = 10\sqrt{\gamma}\rho_9,$$

$$\psi_5 = 5r\sqrt{\gamma}, \qquad \psi_{10} = 0.$$

As our $\varphi_1 = x_1 + x_6$ takes five distinct values under G_{10}, the quintic resolvent equation

$$\prod_{i=1}^{i=5} (\varphi - \varphi_i) = 0$$

is an irreducible Abelian equation of the fifth degree in φ and the corresponding group in the x's is G_{10}, which either leaves every φ_i fixed or interchanges them cyclically.

We have now from our Lagrange system by addition

$$\begin{aligned}
\varphi_1 &= x_1 + x_6 = 2[\rho_6 + \rho_8 + \rho_2 + \rho_4],\\
\varphi_2 &= x_2 + x_7 = 2[\omega^2\rho_6 + \omega\rho_8 + \omega^4\rho_2 + \omega^3\rho_4],\\
\varphi_3 &= x_3 + x_8 = 2[\omega^4\rho_6 + \omega^2\rho_8 + \omega^3\rho_2 + \omega\rho_4],\\
\varphi_4 &= x_4 + x_9 = 2[\omega\rho_6 + \omega^3\rho_8 + \omega^2\rho_2 + \omega^4\rho_4],\\
\varphi_5 &= x_5 + x_{10} = 2[\omega^3\rho_6 + \omega^4\rho_8 + \omega\rho_2 + \omega^2\rho_4],
\end{aligned}$$

and, defining

$$\varphi_{i+5} = x_i - x_{i+5} \qquad (i = 1 \text{ to } 5),$$

then

$$\begin{aligned}
\varphi_6 &= x_1 - x_6 = 2\sqrt{\gamma}[+\rho_1 + \rho_3 + \rho_7 + \rho_9 + r],\\
\varphi_7 &= x_2 - x_7 = 2\sqrt{\gamma}[\omega^2\rho_1 + \omega\rho_3 + \omega^4\rho_7 + \omega^3\rho_9 + r],\\
\varphi_8 &= x_3 - x_8 = 2\sqrt{\gamma}[\omega^4\rho_1 + \omega^2\rho_3 + \omega^3\rho_7 + \omega\rho_9 + r],\\
\varphi_9 &= x_4 - x_9 = 2\sqrt{\gamma}[\omega\rho_1 + \omega^3\rho_3 + \omega^2\rho_7 + \omega^4\rho_9 + r],\\
\varphi_{10} &= x_5 - x_{10} = 2\sqrt{\gamma}[\omega^3\rho_1 + \omega^4\rho_3 + \omega\rho_7 + \omega^2\rho_9 + r].
\end{aligned}$$

Knowing φ_i and φ_{i+5}, we know of course x_j $(j = 1 \text{ to } 10)$. The roots are:

$$\begin{aligned}
x_1, x_6 &= \rho_6 + \rho_8 + \rho_2 + \rho_4 \pm \sqrt{\gamma}[\rho_1 + \rho_3 + \rho_7 + \rho_9 + r],\\
x_2, x_7 &= \omega^2\rho_6 + \omega\rho_8 + \omega^4\rho_2 + \omega^3\rho_4 \pm \sqrt{\gamma}[\omega^2\rho_1 + \omega\rho_3 + \omega^4\rho_7 + \omega^3\rho_9 + r],\\
x_3, x_8 &= \omega^4\rho_6 + \omega^2\rho_8 + \omega^3\rho_2 + \omega\rho_4 \pm \sqrt{\gamma}[\omega^4\rho_1 + \omega^2\rho_3 + \omega^3\rho_7 + \omega\rho_9 + r],\\
x_4, x_9 &= \omega\rho_6 + \omega^3\rho_8 + \omega^2\rho_2 + \omega^4\rho_4 \pm \sqrt{\gamma}[\omega\rho_1 + \omega^3\rho_3 + \omega^2\rho_7 + \omega^4\rho_9 + r],\\
x_5, x_{10} &= \omega^3\rho_6 + \omega^4\rho_8 + \omega\rho_2 + \omega^2\rho_4 \pm \sqrt{\gamma}[\omega^3\rho_1 + \omega^4\rho_3 + \omega\rho_7 + \omega^2\rho_9 + r],
\end{aligned}$$

where $+\sqrt{\gamma}$ belongs to x_i and $-\sqrt{\gamma}$ belongs to x_{i+5} $\Big\}$ $(i = 1 \text{ to } 5)$.

Actual computation shows

$$\psi_4\psi_6 = 10^2\sqrt[5]{P_2 + Q_2\sqrt{5}} = (\omega + \omega^4)F_1 + (\omega^2 + \omega^3)F_2 + \sum_{i=1}^{i=5}\varphi_i^2$$

and

$$\psi_2\psi_8 = 10^2\sqrt[5]{P_2 - Q_2\sqrt{5}} = (\omega^2 + \omega^3)F_1 + (\omega + \omega^4)F_2 + \sum_{i=1}^{i=5}\varphi_i^2,$$

where

$$\begin{aligned}
P_2 &= K_2^2 + 5L_2^2 + 10(M_2^2 + N_2^2),\\
Q_2 &= 2K_2L_2 - 2(M_2^2 + N_2^2) - 8M_2N_2,\\
F_1 &= \varphi_1\varphi_3 + \varphi_2\varphi_4 + \varphi_1\varphi_4 + \varphi_2\varphi_5 + \varphi_3\varphi_5,\\
F_2 &= \varphi_1\varphi_2 + \varphi_1\varphi_5 + \varphi_2\varphi_3 + \varphi_3\varphi_4 + \varphi_4\varphi_5.
\end{aligned}$$

Both F_1 and F_2 are unchanged by G_{10} and hence they are rationally known. We may therefore write

$$\psi_4\psi_6 = 100\rho_4\rho_6 = 100(R_2 + S_2\sqrt{5}),$$

$$\psi_2\psi_8 = 100\rho_2\rho_8 = 100(R_2 - S_2\sqrt{5}),$$

where R_2 and S_2 are defined by

$$P_2 = R^5 + 50R_2^3S_2 + 10R_2S^4,$$
$$Q_2 = 5R_2^4S_2 + 50R_2^2S_2^3 + 25S_2^5.$$

We have the relations:

$$\rho_4\rho_6 + \rho_2\rho_8 = 2R_2,$$

$$\rho_4\rho_6 - \rho_2\rho_8 = 2S_2\sqrt{5}.$$

We shall divide case B into (I.) $R_2 = 0$; (II.) $R_2 \neq 0$. The case $R_2 = 0$ will have two parts, (1) $S_2 = 0$; (2) $S_2 \neq 0$.

B. I. (1) $R_2 = S_2 = 0$.

Without loss of generality we may assume $\rho_6 = \rho_8 = 0$; hence we have here again the similar relation as under (A.)

$$2K_2 + \sqrt{-10 + 2\sqrt{5}}\left[M_2 - N_2 + (M_2 + N_2)\frac{4\sqrt{5}}{-10 + 2\sqrt{5}}\right] = 0.$$

Further from $\rho_6 = 0$ follows

$$\rho_4 = \sqrt[5]{2(K_2 + L_2\sqrt{5})}$$

and from $\rho_8 = 0$ follows

$$\rho_2 = \sqrt[5]{2(K_2 - L\sqrt{5})}.$$

It follows then $K_2 \neq 0$, for then $\rho_2 = -\rho_4$ and φ_1 would be zero; hence the expression in the bracket cannot vanish and therefore ω *is in our rational domain.*

For the same reason, at most three ρ_i can vanish and we shall divide the discussion into

(α) $\rho_2 = 0$; $\rho_4 \neq 0$ (β) $\rho_2 \neq 0$; $\rho_4 \neq 0$.

(α) $\rho_2 = 0$; hence

$$K_2 = L\sqrt{5} \neq 0 \quad \text{and} \quad \rho_4 = \sqrt[5]{4K_2}$$

$4K_2$ is not a perfect fifth power, for then φ_1 would be rationally known. Putting

$$\frac{\psi_9}{\psi_4\sqrt{\gamma}} = \delta'; \qquad \frac{\psi_3}{\psi_4^2\sqrt{\gamma}} = \vartheta'; \qquad \frac{\psi_7}{\psi_4^3\sqrt{\gamma}} = \zeta'; \qquad \frac{\psi_1}{\psi_4^4\sqrt{\gamma}} = \eta'$$

then δ', ϑ', ζ', η' are quantities in our given domain, for numerators and denominators change at the same time the sign, or are left fixed by G_{10}. We get then

$$\varphi_6 = x_1 - x_6 = 2\sqrt{\gamma}[10^3\eta'\sqrt[5]{(4K_2)^4} + 10^2\zeta'\sqrt[5]{(4K_2)^3} + 10\vartheta'\sqrt[5]{(4K_2)^2} + \delta'\sqrt[5]{4K_2} + r].$$

Putting

$$10^3\eta' = \eta; \qquad 10\vartheta' = \vartheta; \qquad 10^2\zeta' = \zeta$$
$$\delta' = \delta$$

$4K_2 = \kappa$ *is not a perfect fifth power, but rationally known*
$\delta, \eta, \vartheta, \zeta$ *are rationally known*

we have

$$x_1, x_6 = \sqrt[5]{\kappa} \pm \sqrt{\gamma}[r + \delta\sqrt[5]{\kappa} + \vartheta\sqrt[5]{\kappa^2} + \zeta\sqrt[5]{\kappa^3} + \eta\sqrt[5]{\kappa^4}],$$

$$x_2, x_7 = \omega^3\sqrt[5]{\kappa} \pm \sqrt{\gamma}[r + \omega^3\delta\sqrt[5]{\kappa} + \omega\vartheta\sqrt[5]{\kappa^2} + \omega^4\zeta\sqrt[5]{\kappa^3} + \omega^2\eta\sqrt[5]{\kappa}],$$

$$\cdot \quad \cdot \quad \cdot \quad \text{etc.} \quad \cdot \quad \cdot \quad \cdot$$

The corresponding equation is

$$f(x) = [x^5 - A_1x^4 + A_2x^3 - A_3x^2 + A_4x - A_5] \times [x^5 - B_1x^4 + B_2x^3 - B_3x^2 + B_4x - B_5] = 0,$$

where

$$A_1, B_1 = \pm 5r\sqrt{\gamma},$$

$$A_2, B_2 = +10\gamma r^2 - 5\gamma\kappa(\delta\eta + \vartheta\zeta) \mp 5\eta\kappa\sqrt{\gamma},$$

$$A_3, B_3 = -15r\gamma\eta\kappa + 10\gamma\delta\zeta\kappa + 5\gamma\kappa\vartheta^2 \pm 5\sqrt{\gamma}[2\gamma r^3 - 3r\gamma\kappa(\eta\delta + \vartheta\zeta) + \zeta\kappa + \gamma\kappa^2\eta(\zeta^2 + \vartheta\eta) + \gamma\kappa\delta(\vartheta^2 + \delta\zeta)],$$

$$A_4, B_4 = 5\gamma^2r^4 + 5\gamma^2\kappa^2(\delta^2\eta^2 + \vartheta^2\zeta^2) - 5\gamma^2\kappa(\zeta^3\delta\kappa + \delta^3\vartheta + \eta^3\zeta\kappa^2 + \vartheta^3\eta\kappa) + 5\gamma\eta^2\kappa^2 - 15r^2\gamma^2(\delta\eta + \vartheta\zeta)\kappa + 10\gamma^2r\kappa(\zeta^2\eta\kappa + \delta^2\zeta + \eta^2\vartheta\kappa + \vartheta^2\delta) + 10\gamma r\zeta\kappa - 5\gamma^2\eta\delta\vartheta\zeta\kappa^2 - 15\gamma\delta\vartheta\kappa \pm 5\sqrt{\gamma}[-\gamma\zeta^3\kappa^2 - \vartheta\kappa - 3r^2\gamma\eta\kappa + 2\gamma r\vartheta^2\kappa - \gamma\eta\vartheta\zeta\kappa^2 - 3\gamma\delta^2\vartheta\kappa + 2\gamma\eta^2\delta\kappa^2 + 8r\gamma\delta\zeta\kappa],$$

$$A_5, B_5 = \kappa + 5\gamma^2\delta^4\kappa + 10\gamma\delta^2\kappa - 5\gamma^2\zeta\vartheta^3\kappa^2 - 5\gamma^2\eta^3\vartheta\kappa^3 - 5\gamma\zeta\eta\kappa^2 - 15\gamma^2\kappa^2\zeta\eta\delta^2 + 5\gamma^2\zeta^2\eta^2\kappa^3 - 5\gamma^2r^3\eta\kappa + 5\gamma^2r^2\vartheta^2\kappa + 10r^2\gamma^2\delta\zeta\kappa - 5r\gamma^2\zeta^3\kappa^2 - 5r\gamma\vartheta\kappa - 5r\gamma^2\eta\vartheta\zeta\kappa^2 - 15r\gamma\vartheta\delta\kappa - 15r\gamma^2\vartheta\delta^2\kappa + 10r\gamma^2\delta\eta^2\kappa^2 \pm \sqrt{\gamma}[\gamma^2\eta^5\kappa^4 + \gamma^2\vartheta^5\kappa^2 + \gamma^2\zeta^5\kappa^3 + \gamma^2\delta^5\kappa + \gamma^2r + 5\delta\kappa + 10\gamma\delta^3\kappa - 5\gamma^2\zeta\delta^3\eta\kappa^2 - 5\gamma^2\zeta\vartheta^3\delta\kappa^2 - 5\gamma^2\eta\zeta^3\vartheta\kappa^3 - 5\gamma^2\delta\eta^3\vartheta\kappa^3 - 15\gamma\eta\zeta\delta\kappa^2 + 5\gamma^2\kappa^2(\zeta^2\eta^2\delta\kappa + \zeta^2\vartheta\delta^2 + \eta^2\zeta\vartheta^2\kappa + \delta^2\eta\vartheta^2) + 5\gamma\zeta^2\vartheta\kappa^2 + 5\gamma\eta\vartheta^2\kappa^2 - 5r^3\gamma^2\kappa(\zeta\vartheta + \eta\delta) + 5\gamma^2\kappa r^2(\zeta^2\eta\kappa + \delta^2\zeta + \eta^2\vartheta\kappa + \vartheta^2\delta) + 5\gamma r^2\zeta\kappa - 5r\gamma^2\kappa(\zeta^3\delta\kappa + \delta^3\vartheta + \eta^3\zeta\kappa^2 + \vartheta^3\eta\kappa) + 5r\gamma^2\kappa^2(\eta^2\delta^2 + \zeta^2\vartheta^2) - 5\gamma^2r\eta\delta\zeta\vartheta\kappa^2 + 5r\gamma\eta^2\kappa^2].$$

$\gamma, r, \delta, \eta, \vartheta, \zeta$ are not independent of each other, for the condition

$$\prod_{i=1}^{i=5}(x_i - x_{i+5}) = \sqrt{\gamma}$$

leads to *the identity*

$$\frac{1}{2^5\gamma^2} = \eta^5\kappa^4 + \zeta^5\kappa^3 + \vartheta^5\kappa^2 + \delta^5\kappa + r^5 - 5[\eta^3\vartheta\delta\kappa^3 + \vartheta^3\zeta\delta\kappa^2 + \zeta^3\eta\vartheta\kappa^3 + \delta^3\eta\zeta\kappa^2] + 5[\eta\vartheta^2\delta^2\kappa^2 + \vartheta\zeta^2\delta^2\kappa^2 + \zeta\eta^2\vartheta^2\kappa^3 + \delta\eta^2\zeta^2\kappa^3] + 5r\kappa^2(\eta^2\delta^2 + \vartheta^2\zeta^2) + 5r^2(\eta^2\vartheta\kappa^2 + \vartheta^2\delta\kappa + \zeta^2\eta\kappa^2 + \delta^2\zeta\kappa) - 5r^3(\delta\eta + \vartheta\zeta)\kappa - 5r(\vartheta^3\zeta\kappa^3 + \vartheta^3\eta\kappa^2 + \zeta^3\delta\kappa^2 + \delta^3\vartheta\kappa) - 5\kappa^2\delta\eta\zeta\vartheta.$$

There is obviously another condition to be imposed upon r, δ, ζ, ϑ, η namely *not all five can vanish* at once, for if

$$r = \delta = \zeta = \vartheta = \eta,$$

then the irrationality $\sqrt{\gamma}$ disappears entirely and $f(x)$ splits up into the two rational factors

$$\prod_{i=1}^{i=5} (x - x_i) \quad \text{and} \quad \prod_{i=1}^{i=5} (x - x_{i+5}),$$

as can be seen from the above A_i and B_i.

We ask again: Are the derived conditions also sufficient, that the above $f(x) = 0$ is an irreducible Abelian equation?

$f(x)$ cannot contain a linear, rationally known factor, for $\sqrt{\gamma}$ and $\sqrt[5]{\kappa}$ cannot be expressed rationally by each other; the former satisfies an irreducible quadratic, the latter an irreducible quintic. The sum of any two, three, or four roots is of the forms respectively

$$(\omega^\lambda + \omega^\mu)\sqrt[5]{\kappa} + \sqrt{\gamma} f_1(\omega, \sqrt[5]{\kappa}),$$

$$(\omega^\lambda + \omega^\mu + \omega^\nu)\sqrt[5]{\kappa} + \sqrt{\gamma} f_2(\omega, \sqrt[5]{\kappa}),$$

$$(\omega^\lambda + \omega^\mu + \omega^\nu + \omega^\pi)\sqrt[5]{\kappa^3} + \sqrt{\gamma} f_2(\omega, \sqrt[5]{\kappa}).$$

Hence, for the same reason that $\sqrt{\gamma}$ and $\sqrt[5]{\kappa}$ cannot be expressed rationally by each other, the last three forms must be irrational, even if

$$f_i(\omega, \sqrt[5]{\kappa}) = 0$$

$\sqrt[5]{\kappa}$ can disappear in a factor of the fifth degree, only if the roots of this factor are x_i or x_{i+5} ($i = 1$ to 5) and in fact the above symmetric functions of x_i and x_{i+5}, which we called A_i and B_i, show $\sqrt{\gamma}$ as the only irrationality. Our question is then simply: Can $\sqrt{\gamma}$ disappear in all A_i ($i = 1$ to 5) without contradicting the derived necessary conditions? A_1 and A_2 show that we must have (a) $r = \eta = 0$. Then we must have further, if $\sqrt{\gamma}$ shall disappear,

$$(b)\ \zeta + \gamma\zeta\delta^2 + \gamma\delta\vartheta^2 = 0 \quad (\text{by } A_3)$$

and

$$(c)\ \vartheta + 3\gamma\vartheta\delta^2 + \gamma\zeta^3\kappa = 0 \quad (\text{by } A_4).$$

Let us first see whether one of the quantities ζ, ϑ, δ, $1 + 3\gamma\delta^2$, can vanish without contradicting the necessary conditions derived.

(1) If $\delta = 0$ then by (b) $\zeta = 0$ and hence by (c) $\vartheta = 0$.

This case would require

$$r = \eta = \delta = \zeta = \vartheta = 0$$

and contradicts our conditions.

(2) If $\zeta = 0$ then by (b) $\delta = 0$ or $\vartheta = 0$ as $\gamma \neq 0$.

$\zeta = \delta = 0$ requires $\vartheta = 0$ by means of (c) and is therefore again impossible.

If then

$$r = \eta = \zeta = \vartheta = 0 \qquad \delta \neq 0$$

then A_5 shows that

$$\gamma^2\delta^4 + 10\gamma\delta^2 + 5 = 0$$

if $\sqrt{\gamma}$ shall disappear in A_5. Hence

$$\gamma\delta^2 \neq -5 \pm 2\sqrt{5}$$

and, as our identity leads to

$$\frac{1}{2^5\gamma^2} = \delta^5\kappa,$$

we must have

$$\gamma \textit{ is not a perfect fifth power}$$

for otherwise κ is a perfect fifth power which contradicts the conditions.

(3) If $\vartheta = 0$ then $\zeta = 0$, which is again case (2).

(4) If $1 + 3\gamma\delta^2 = 0$ then $\zeta = 0$, which is again case (2).

Assuming then that (*a*), (*b*) and (*c*) are true and that $\zeta \neq 0$, $\vartheta \neq 0$, $\delta \neq 0$, $1 + 3\gamma\delta^2 \neq 0$, we must have

$$\vartheta = -\frac{\gamma\zeta^3\kappa}{1 + 3\gamma\delta^2} \text{ by } (c)$$

and hence by (*b*)

$$\zeta^5 = -\frac{(1 + \gamma\delta^2)(1 + 3\gamma\delta^2)^2}{\gamma^3\delta\kappa^2}$$

so that from A_5 it must follow

$$1 + 15\gamma\delta^2 + 60\gamma^2\delta^4 + 75\gamma^3\delta^6 + 25\gamma^4\delta^8 \neq 0.$$

In this last case, that is

$$r = \eta = 0, \qquad \vartheta = -\frac{\gamma\zeta^3\kappa}{1 + 3\gamma\delta^2}, \qquad \zeta^5 = \frac{(1 + \gamma\delta^2)(1 + 3\gamma\delta^2)^2}{\gamma^3\delta\kappa^2}$$

our identity reduces to

$$\kappa = \frac{\gamma^2\delta^3}{2^5[1 + \gamma\delta^2 + 14\gamma^2\delta^4 + 48\gamma^3\delta^6 + 20\gamma^4\delta^8]}.$$

Under these conditions $\prod_{i=1}^{i=5}(x - x_i)$ must contain coefficients involving $\sqrt{\gamma}$.

This proves, that $f(x)$ is irreducible, if all conditions are observed. Furthermore the group of $f(x) = 0$ must have two as well as five systems of intransitivity; for A_i and B_i show that $f(x)$ splits up into two rational factors after having adjoined $\sqrt{\gamma}$ to the rational domain; and having adjoined ρ_4 to the rational domain then $f(x) = 0$ splits up into five equations with rational coefficients, namely into the five equations

$$(x - x_i)(x - x_{i+5}) = 0.$$

It follows then that the group of $f(x) = 0$ is a subgroup of the G_{240} above and it must be G_{10}, as σ_1 as well as σ_2 change $\kappa = \psi_4^5 \neq 0$ into a quantity which is zero.

Type V. If

$$\omega,\ r,\ \gamma,\ \delta,\ \vartheta,\ \zeta,\ \eta,\ \kappa$$

are quantities of the given rational domain, but not

$$\sqrt{\gamma}, \quad \sqrt[5]{\kappa}$$

and if the identity exists as given on page 141 *then* $f(x) = 0$ *with the roots on page* 141 *is an irreducible Abelian equation, if not all of the five quantities* r, ϑ, η, ζ, δ *are zero—*

with two special exceptions, namely

(1) *If* $r = \eta = \vartheta = \zeta = 0$ *then* $\gamma\delta^2 \neq -5 \pm 2\sqrt{5}$, $[\therefore \sqrt[5]{\gamma} = $ *irrational*$]$

(2) *If* $r = \eta = 0, \quad \zeta^5 = -\dfrac{(1+\gamma\delta^2)(1+3\gamma\delta^2)^2}{\gamma^3\delta\kappa^2}, \quad \vartheta = -\dfrac{\gamma\zeta^3\kappa}{1+3\gamma\delta^2}$

then

$$1 + 15\gamma\delta^2 + 60\gamma^2\delta^4 + 75\gamma^3\delta^6 + 25\gamma^4\delta^8 \neq 0.$$

Illustrations:

(1) $$\gamma = 2, \quad \eta = 2, \quad r = \vartheta = \zeta = \delta = 0;$$

then from the identity we have $\kappa = \dfrac{1}{8}$.

Hence the roots are:

$$x_1, x_6 = \sqrt[5]{\frac{1}{8}} \pm \sqrt{2}\sqrt[5]{\frac{1}{8^4}},$$

$$x_2, x_7 = \omega^3\sqrt[5]{\frac{1}{8}} \pm \omega^2\sqrt{2}\sqrt[5]{\frac{1}{8^4}},$$

. . . etc. . . .

The corresponding equation is

$$f(x) = 16x^{10} - 60x^6 - 4x^5 - 20x^3 + 25x^2 - 5x - 0,75 = 0.$$

Our domain must include $R(\omega)$.

(2) $$\zeta = \vartheta = \delta = \eta = 0, \quad \kappa = 2, \quad r = 2$$

then by the identity we get

$$\gamma = 2^{-5}.$$

Our domain must include $R(\omega)$.

$$f(x) = x^{10} + \frac{55}{2^5}x^6 - 4x^5 + \frac{45}{2^8}x^4 + \frac{5}{2^{12}}x^2 - \frac{5}{2^4}x + \frac{2^{10}-1}{2^{15}} = 0.$$

The roots are:

$$x_1, x_5 = \sqrt[5]{2} \pm \sqrt{\frac{1}{2^3}},$$

$$x_2, x_7 = \omega^3\sqrt[5]{2} \pm \sqrt{\frac{1}{2^3}}.$$

. . . etc. . . .

(β) $\rho_2 \neq 0, \ \rho_4 \neq 0.$

As ω is rationally known, we may put

$$\rho_2{}^5 = \lambda, \quad \rho_4{}^5 = \kappa$$

where λ, κ *are in our given domain.* Similarly, as under (α), we may write, if ν', μ', δ', ϑ' are in the given domain

$$\frac{\psi_7}{\sqrt{\gamma}\psi_2} = \nu', \qquad \frac{\psi_9}{\sqrt{\gamma}\psi_4} = \delta',$$

$$\frac{\psi_1}{\sqrt{\gamma}\psi_2{}^3} = \mu', \qquad \frac{\psi_3}{\sqrt{\gamma}\psi_4{}^2} = \vartheta'.$$

Hence

$$\varphi_6 = x_1 - x_6 = 2\sqrt{\gamma}[10^2\mu' \sqrt[5]{\lambda^3} + 10\vartheta' \sqrt[5]{\kappa^2} + \nu' \sqrt[5]{\lambda} + \delta' \sqrt[5]{\kappa} + r]$$

or, putting

$$10^2\mu' = \mu; \quad 10\vartheta' = \vartheta; \quad \nu' = \nu; \quad \delta' = \delta,$$

then ϑ, δ, μ, ν *are rationally known* and *not all five quantities*, ϑ, δ, μ, ν, r *can vanish*, as then $\varphi_6 = 0$. Our roots become

$$x_1, x_6 = \sqrt[5]{\lambda} + \sqrt[5]{\kappa} \pm \sqrt{\gamma}[\mu \sqrt[5]{\lambda^3} + \vartheta \sqrt[5]{\kappa^2} + \nu \sqrt[5]{\lambda} + \delta \sqrt[5]{\kappa} + r],$$

$$x_2, x_7 = \omega^4 \sqrt[5]{\lambda} + \omega^3 \sqrt[5]{\kappa} \pm \sqrt{\gamma}[\omega^2\mu \sqrt[5]{\lambda^3} + \omega\vartheta \sqrt[5]{\kappa^2} + \omega^4\nu \sqrt[5]{\lambda} + \omega^3\delta \sqrt[5]{\kappa} + r].$$

. . . . etc. . . .

Both $\sqrt[5]{\lambda}$ *and* $\sqrt[5]{\kappa}$ *are not in our given domain*, for otherwise $(x - x_i)(x - x_{i+5})$ is a rational factor of $f(x)$ and as both $\sqrt[5]{\lambda}$ and $\sqrt[5]{\kappa}$ are rational or irrational at the same time. Further, $\lambda \neq \pm \kappa$, for then

$$\nu'\vartheta' = \frac{\psi_3\psi_7}{\gamma\psi_2\psi_4^2}$$

would become

$$\frac{\psi_3\psi_7}{\gamma\psi_2^3} = \nu'\vartheta'.$$

Applying G_{10} to $\psi_3\psi_7$ we see that it is a rationally known function and hence ψ_2^3 and therefore $\rho_2 = \sqrt[5]{\lambda}$ would be rational.

$\psi_2\psi_4^2$ is unchanged by G_{10}; hence if $\tau = \sqrt[5]{\lambda\kappa^2}$ then τ *is a quantity in our rational domain* and we have the relations:

$$\lambda = \frac{\tau^5}{\kappa^2}; \qquad \sqrt[5]{\lambda^3\kappa} = \frac{\tau^3}{\kappa}; \qquad \sqrt[5]{\lambda^4\kappa^3} = \frac{\tau^4}{\kappa}.$$

The quantities

$$\tau, \ \kappa, \ \gamma, \ r, \ \vartheta, \ \delta, \ \mu, \ \nu$$

are connected by *the identity*, based upon

$$\prod_{i=1}^{i=5} (x_i - x_{i+5}) = \sqrt{\gamma}$$

$$\begin{aligned}\frac{1}{2^5\gamma^2} = {} & \frac{\mu^5\tau^{15}}{\kappa^6} + \vartheta^5\kappa^2 + \frac{\nu^5\tau^5}{\kappa^2} + \delta^5\kappa + r^5 - 5\left[\mu^3\vartheta\delta\frac{\tau^9}{\kappa^3} + \vartheta^3\nu\delta\kappa\tau + \nu^3\mu\vartheta\frac{\tau^6}{\kappa^2} + \delta^3\mu\nu\frac{\tau^4}{\kappa}\right] \\ & + 5\left[\mu\vartheta^2\delta^2\tau^3 + \vartheta\nu^2\delta^2\tau^2 + \nu\mu^2\vartheta^2\frac{\tau^7}{\kappa^2} + \delta\mu^2\nu^2\frac{\tau^8}{\kappa^3}\right] + 5r\left[\mu^2\delta^2\frac{\tau^6}{\kappa^2} + \vartheta^2\nu^2\tau^2\right] \\ & + 5r\left[\mu^2\vartheta\frac{\tau^6}{\kappa^2} + \vartheta^2\delta\kappa + \nu^2\mu\frac{\tau^5}{\kappa^2} + \delta^2\nu\tau\right] - 5r\left[\mu^3\nu\frac{\tau^{10}}{\kappa^4} + \vartheta^3\mu\tau^3 + \nu^3\delta\frac{\tau^3}{\kappa} + \delta^3\vartheta\kappa\right] \\ & - 5r^3\left[\mu\delta\frac{\tau^3}{\kappa} + \vartheta\nu\tau\right] - 5\mu\nu\delta\vartheta r\frac{\tau^4}{\kappa}.\end{aligned}$$

The roots may now be written in the form

$$x_1, x_6 = \frac{\tau}{\sqrt[5]{\kappa^2}} + \sqrt[5]{\kappa} \pm \sqrt{\gamma}\left[\frac{\mu\tau^3}{\kappa\sqrt[5]{\kappa}} + \vartheta\sqrt[5]{\kappa^2} + \frac{\nu\tau}{\sqrt[5]{\kappa^2}} + \delta\sqrt[5]{\kappa} + r\right],$$

$$x_2, x_7 = \frac{\omega^4\tau}{\sqrt[5]{\kappa^2}} + \omega^3\sqrt[5]{\kappa} \pm \sqrt{\gamma}\left[\frac{\omega^2\mu\tau^3}{\kappa\sqrt[5]{\kappa}} + \omega\vartheta\sqrt[5]{\kappa^2} + \frac{\omega^4\nu\tau}{\sqrt[5]{\kappa^2}} + \omega^3\delta\sqrt[5]{\kappa} + r\right],$$

. . . . etc. . . .

The corresponding equation is

$$f(x) = (x^5 - A_1x^4 + A_2x^3 - A_3x^2 + A_4x - A_5) \times (x^5 - B_1x^4 + B_2x^3 - B_3x^2 + B_4x - B_5) = 0,$$

where

$$A_1, B_1 = \pm 5r\sqrt{\gamma},$$

$$A_2, B_2 = 10r^2\gamma - 5\gamma\tau\left(\mu\delta\frac{\tau^2}{\kappa} + \vartheta\nu\right) \mp 5\tau\sqrt{\gamma}\left(\mu\frac{\tau^2}{\kappa} + \vartheta\right),$$

$$\begin{aligned} A_3, B_3 = {} & 5\tau + 5\gamma(\delta^2\tau + \vartheta^2\kappa) - 15r\gamma\tau\left(\mu\frac{\tau^2}{\kappa} + \vartheta\right) + 10\gamma\left(\mu\nu\frac{\tau^5}{\kappa^2} + \nu\delta\tau\right) \\ & \pm 5\sqrt{\gamma}\left[2\gamma r^3 + \mu\frac{\tau^5}{\kappa^2} + \nu\tau + 2\delta\tau + \nu^2\mu\gamma\frac{\tau^5}{\kappa^2} + \gamma\delta^2\nu\tau + \gamma\mu^2\vartheta\frac{\tau^6}{\kappa^2} + \vartheta^2\delta\kappa \right. \\ & \left. - 3\gamma r\mu\delta\frac{\tau^3}{\kappa} - 3\gamma r\vartheta\nu\tau\right], \end{aligned}$$

$$\begin{aligned} A_4, B_4 = {} & -5\frac{\tau^3}{\kappa} + 5\gamma^2r^4 + 5\gamma^2\tau^2\left(\mu^2\delta^2\frac{\tau^4}{\kappa^2} + \vartheta^2\nu^2\right) - 5\gamma^2\left(\nu^3\delta\frac{\tau^3}{\kappa} + \delta^3\vartheta\kappa\right. \\ & \left. + \mu^3\nu\frac{\tau^{10}}{\kappa^4} + \vartheta^3\mu\tau^3\right) + 5\gamma\tau^2(\mu^2\tau + \vartheta^2) - 15r^2\gamma^2\tau\left(\mu\delta\frac{\tau^2}{\kappa} + \vartheta\nu\right) \\ & + 10\gamma^2r\left(\nu^2\mu\frac{\tau^5}{\kappa^2} + \delta^2\nu\tau + \mu^2\vartheta\frac{\tau^6}{\kappa^2} + \vartheta^2\delta\kappa\right) + 10\gamma r\left(\mu\frac{\tau^5}{\kappa^2} + \nu\tau\right) \\ & - 5\gamma^2\mu\vartheta\delta\nu\frac{\tau^4}{\kappa} - 5\gamma\mu\vartheta\frac{\tau^4}{\kappa} - 15\gamma\nu\frac{\tau^3}{\kappa}(\delta + \nu) - 15\gamma\delta\vartheta\kappa \\ & + 40r\gamma\delta\tau \pm 5\sqrt{\gamma}\left[2r\tau - \gamma\tau^3\left(\frac{\nu^3}{\kappa} + \frac{\mu^3\tau^7}{\kappa^4}\right) - \frac{\delta\tau^3}{\kappa} - \vartheta\kappa\right. \\ & - 3r^2\gamma\tau\left(\mu\frac{\tau^2}{\kappa} + \vartheta\right) + 2\gamma r(\delta^2\tau + \vartheta^2\kappa) - \gamma\vartheta\mu\frac{\tau^4}{\kappa}(\nu + \delta) - 3\gamma\nu^2\delta\frac{\tau^3}{\kappa} \\ & \left. - 3\nu\frac{\tau^3}{\kappa} - 3\gamma\delta^2\vartheta\kappa + 2\gamma\mu^2\delta\frac{\tau^6}{\kappa^2} + 2\gamma\vartheta^2\nu\tau^2 + 8r\gamma\nu\tau\left(\mu\frac{\tau^4}{\kappa^2} + \delta\right)\right], \end{aligned}$$

$$\begin{aligned} A_5, B_5 = {} & \frac{\tau^5}{\kappa^2} + \kappa + 5\gamma^2\left(\frac{\nu^4\tau^5}{\kappa^2} + \delta^4\kappa\right) + 10\gamma\left(\frac{\nu^2\tau^5}{\kappa^2} + \delta^2\kappa\right) \\ & - 5\gamma^2\nu\vartheta^3\kappa\tau - 5\gamma^2\tau\left(\delta^3\mu\frac{\tau^3}{\kappa} + \vartheta^3\delta\kappa + \mu^3\vartheta\frac{\tau^8}{\kappa^3}\right) \\ & - 5\gamma\mu\tau\left(\nu\frac{\tau^3}{\kappa} + \vartheta\frac{\tau^5}{\kappa^2}\right) - 15\gamma\mu\delta\frac{\tau^4}{\kappa} - 15\gamma^2\mu\nu\tau\left(\delta^2\frac{\tau^3}{\kappa} + \vartheta\nu\frac{\tau^5}{\kappa^2}\right) \\ & + 5\gamma\mu^2\frac{\tau^9}{\kappa^3} + 5\gamma^2\mu^2\frac{\tau^7}{\kappa^2}\left(\nu^2\frac{\tau}{\kappa} + \vartheta^2\right) + 10\gamma\vartheta\tau^2(\delta + \nu) \\ & - 5\gamma^2r^3\tau\left(\vartheta + \mu\frac{\tau^2}{\kappa}\right) + 5\gamma r^2\tau + 5\gamma^2r^2(\delta^2\tau + \vartheta^2\kappa) \\ & + 10r^2\gamma^2\nu\tau\left(\mu\frac{\tau^4}{\kappa^2} + \delta\right) - 5r\gamma^2\tau^3\left(\frac{\nu^3}{\kappa} + \mu^3\frac{\tau^7}{\kappa^4}\right) - 5r\gamma\left(\delta\frac{\tau^3}{\kappa} + \vartheta\kappa\right) \\ & - 5r\gamma^2\mu\vartheta\frac{\tau^4}{\kappa}(\nu + \delta) - 15r\gamma\nu\frac{\tau^3}{\kappa} - 15r\gamma\delta\left(\nu\frac{\tau^3}{\kappa} + \vartheta\kappa\right) \\ & - 15r\gamma^2\delta\left(\nu^2\frac{\tau^3}{\kappa} + \vartheta\delta\kappa\right) + 10r\gamma^2\tau^2\left(\nu\vartheta^2 + \delta\mu^2\frac{\tau^4}{\kappa^2}\right) \\ & \pm\sqrt{\gamma}\left[5\nu\frac{\tau^5}{\kappa^2} + 5\delta\kappa + \gamma^2\left(\mu^5\frac{\tau^{15}}{\kappa^6} + \vartheta^5\kappa^2 + \frac{\nu^5\tau^5}{\kappa^2} + \delta^5\kappa + r\right)\right. \end{aligned}$$

$$+ 10\gamma\left(\frac{\nu^3\tau^5}{\kappa^2} + \delta^3\kappa\right) - 5\gamma^2\tau\left(\mu\nu\delta^3\frac{\tau^3}{\kappa} + \nu\vartheta^3\delta\kappa + \mu\nu^3\vartheta\frac{\tau^5}{\kappa^2} + \delta\mu^3\vartheta\frac{\tau^8}{\kappa}\right)$$
$$- 5\gamma\vartheta^3\kappa\tau - 5\mu\frac{\tau^4}{\kappa} - 15\gamma\mu\nu\tau\left(\delta\frac{\tau^3}{\kappa} + \vartheta\frac{\tau^5}{\kappa^2}\right) - 15\gamma\mu\delta^2\frac{\tau^4}{\kappa}$$
$$- 5\gamma^2\tau^2\left(\nu^2\mu^2\delta\frac{\tau^6}{\kappa^3} + \nu^2\vartheta\delta^2 + \mu^2\vartheta^2\nu\frac{\tau^5}{\kappa^2} + \delta^2\vartheta^2\mu\tau\right) + 5\vartheta\tau^2(1 + \gamma\nu^2)$$
$$+ 5\gamma\tau^2\left(\delta\mu^2\frac{\tau^6}{\kappa} + \vartheta\delta^2 + \mu\vartheta^2\tau\right) + 20\gamma\vartheta\nu\delta\tau^2 - 5\gamma^2r^3\tau\left(\nu\vartheta + \mu\delta\frac{\tau^2}{\kappa}\right)$$
$$+ 5\gamma^2r^2\left(\nu^2\mu\frac{\tau^5}{\kappa^2} + \delta^2\nu\tau + \mu^2\vartheta\frac{\tau^6}{\kappa^2} + \vartheta^2\delta\kappa\right) + 5\gamma r^2\tau\left(\mu\frac{\tau^4}{\kappa^2} + \nu\right)$$
$$+ 10r^2\gamma\delta\tau - 5r\frac{\tau^3}{\kappa} - 5r\gamma^2\left(\nu^3\delta\frac{\tau^3}{\kappa} + \delta^3\vartheta\kappa + \mu^3\nu\frac{\tau^{10}}{\kappa^4} + \vartheta^3\mu\tau^3\right)$$
$$+ 5r\gamma^2\tau^2\left(\nu^2\vartheta^2 + \delta^2\mu^2\frac{\tau^4}{\kappa^2}\right) - 5r\gamma^2\mu\vartheta\nu\delta\frac{\tau^4}{\kappa} + 5r\gamma\tau^2\left(\vartheta^2 + \mu^2\frac{\tau^4}{\kappa^2}\right)$$
$$- 5r\gamma\mu\vartheta\frac{\tau^4}{\kappa} - 15r\gamma\nu^2\frac{\tau^3}{\kappa}\Bigg].$$

Regarding the sufficiency of the derived conditions, we first note that the group of the above $f(x) = 0$ must contain two as well as five systems of intransitivity; for, adjoining $\sqrt[5]{\kappa}$ to the rational domain, $f(x)$ splits up into the five rational factors $(x - x_i)(x - x_{i+5})$, as seen directly from the roots of $f(x) = 0$. The above A_i and B_i show, that $f(x)$ is the product of two rational factors, if $\sqrt{\gamma}$ is adjoint to the rational domain. It follows then that the group of $f(x) = 0$ must be a subgroup of the above G_{240}, if $f(x)$ is irreducible.

Just as under (α) the sum of any two, three, or four roots can not be rational under the derived conditions, as $\sqrt{\gamma}$ and $\sqrt[5]{\kappa}$ cannot be expressed rationally by each other and as ω is rationally known. The same is true for any single root and hence $f(x)$ cannot contain any rational factor of the first, second, third or fourth degree. The sum of any five roots is either irrational for the same reason or it is zero, which can happen, only when we pick out the five roots belonging to the same system of intransitivity. It remains then to see when the coefficients of $\sqrt{\gamma}$ in all A_i may vanish. A_1 and A_2 can be rational only if respectively

$$(a)\quad r = 0 \qquad\qquad (b)\quad \mu\frac{\tau^2}{\kappa} + \vartheta = 0$$

as $\tau \neq 0$, $\gamma \neq 0$.

A_3 is rational under the assumption that (a) and (b) hold and if

$$(c)\qquad -\frac{\vartheta\tau^2}{\kappa} + \nu + 2\delta + \gamma\left(-\frac{\nu^2\vartheta\tau^2}{\kappa} + \delta^2\nu + \vartheta^3 + \frac{\vartheta^2\delta\kappa}{\tau}\right) = 0$$

Assuming the truth of (a) and (b), again A_4 is rational, if

$$(d)\qquad + \gamma\left(\frac{\vartheta^3\tau^2}{\kappa} - \nu^3\tau\right) - \left(\delta\tau + \frac{\vartheta\kappa}{\tau^2}\right) + \gamma\vartheta^2\kappa(\nu + \delta) - 3\gamma\nu^2\delta\tau - 3\nu\tau$$
$$- \frac{3\gamma\kappa\vartheta\delta^2}{\tau^2} + 2\gamma\delta\vartheta^2 + 2\gamma\nu\vartheta^2 = 0,$$

and A_5 is rational, if

$$
(e)\quad
\begin{aligned}
&\frac{1}{5}\gamma^2\left(-\frac{\vartheta^5\tau^3}{\kappa}+\frac{\vartheta^5\kappa^2}{\tau^2}+\frac{\nu^5\tau^3}{\kappa^2}+\frac{\delta^5\kappa}{\tau^2}\right)+\frac{\nu\tau^3}{\kappa^2}+\frac{\delta\kappa}{\tau^2}+2\gamma\left(\frac{\nu^3\tau^3}{\kappa^2}+\frac{\delta^3\kappa}{\tau^2}\right)\\
&\quad-\gamma^2\vartheta\left(\frac{\nu\delta\vartheta^2\kappa}{\tau}-\nu\delta^3-\nu^3\vartheta\tau-\vartheta^2\delta\tau\right)-\frac{\gamma\vartheta^3\kappa}{\tau}+2\vartheta+3\gamma\vartheta\nu\left(\delta+\frac{\vartheta\tau^2}{\kappa}\right)\\
&\quad+3\gamma\vartheta\delta^2\tau-\gamma^2\vartheta\left(\frac{\nu^2\vartheta\delta\tau^2}{\kappa^2}+\nu^2\delta^2+\nu\vartheta^3\tau-\frac{\delta^2\vartheta^2\kappa}{\tau}\right)+\vartheta\gamma\nu^2\\
&\quad+\gamma\vartheta\left(\frac{\delta\vartheta\tau^2}{\kappa}+\delta^2-\frac{\vartheta^2\kappa}{\tau}\right)+20\vartheta\nu\delta=0.
\end{aligned}
$$

We have then the following condition: if $r \neq 0$, $f(x)$ is irreducible and *if* $r = 0$ *and* $\mu = -\vartheta\kappa/\tau^2$ *then* τ, κ, ν, δ, ϑ *must not satisfy* at the same time all three relations (*c*), (*d*), (*e*).

If then all the above conditions are true $f(x)$ is irreducible and the group is a subgroup of G_{240}.

If $\mu = \vartheta = \delta = \nu = 0$, then our identity reduces to $2^5\gamma^2r^5 = 1$ and hence in this case γ is a perfect fifth power of a rational quantity.

σ_1 cannot be a substitution of the Galois group of $f(x) = 0$, for it produces, if applied to

$$\tau = \rho_2\rho_4^2 \neq 0,$$

$\rho_8\rho_6^2$ which is equal to zero.

σ_2 applied to $\tau = \rho_2\rho_4^2$ produces $\rho_6\rho_2^2 = 0$.

It follows then that the Galois group of $f(x) = 0$ can only be G_{10} and hence $f(x) = 0$ is an irreducible Abelian equation.

Type VI. If $\omega, \gamma, r, \tau, \kappa, \vartheta, \delta, \mu, \nu$ *but not* $\sqrt{\gamma}$, $\sqrt[5]{\kappa}$ *are in our given rational domain, and if not all five quantities* r, ϑ, δ, μ, ν *vanish, and if the identity of page* 145 *exists, then* $f(x) = 0$ *with the roots on page* 145 *is an irreducible Abelian equation with the special restriction, that, if* $r = 0$ *and* $\mu = -\vartheta\kappa/\tau^2$, *then* ν, δ, ϑ, τ, κ *must not satisfy all the three relations* (*c*), (*d*), (*e*) *of pages* 147–148.

Illustrations:

(1) $\mu = \vartheta = \delta = \nu = 0, \quad r = 2.$

To satisfy the given identity $\gamma = \frac{1}{2^5}$, τ, κ may be chosen arbitrarily according to the given conditions. The roots are:

$$x_1, x_6 = \frac{\tau}{\sqrt[5]{\kappa^2}} + \sqrt[5]{\kappa} \pm \frac{1}{2\sqrt{2}},$$

$$x_2, x_7 = \frac{\omega^4\tau}{\sqrt[5]{\kappa^2}} + \omega^3\sqrt[5]{\kappa} \pm \frac{1}{2\sqrt{2}},$$

. . . . etc. . . .

$$
\begin{aligned}
f(x) = x^{10} &- \frac{5}{8}x^8 - 10\tau x^7 + 5\left(\frac{1}{2^5}-\frac{2\tau^3}{\kappa}\right)x^6 - \left(2\kappa-\frac{5}{4}\tau+2\frac{\tau^5}{\kappa^2}\right)x^5\\
&+\left(25\tau^2-\frac{25}{8}\right)x^4+\left(\frac{125}{2^5}\tau+\frac{50\tau^4}{\kappa}-\frac{5}{2}\frac{\tau^5}{\kappa^2}-\frac{5}{2}\kappa\right)x^3
\end{aligned}
$$

$$+\left(\frac{25}{2^4}\frac{\tau^3}{\kappa}-\frac{75}{2^{12}}+10\tau\cdot\kappa+35\frac{\tau^6}{\kappa^2}-\frac{75}{2^2}\tau^2\right)x^2$$

$$+\left(\frac{75}{2^8}\tau^2-\frac{75}{2^4}\frac{\tau^4}{\kappa}-\frac{5}{2^5}\kappa-\frac{5}{2^5}\frac{\tau^5}{\kappa_2}+10\tau^3+10\frac{\tau^8}{\kappa^3}\right)x$$

$$+\frac{\tau^{10}}{\kappa^4}+\kappa^2+\frac{25}{2^6}\tau^2+2\frac{\tau^5}{\kappa}-\frac{5}{2^2}\kappa\tau-2^5\cdot 5^2+\frac{25}{2^8}\frac{\tau^3}{\kappa}-\frac{35}{2^8}\frac{\tau^6}{\kappa^2}=0.$$

Assuming $\tau = 1$, $\kappa = 2$ then

$$(x) = x^{10}-\frac{5}{8}x^8-10x^7-\frac{155}{32}x^6-\frac{13}{4}x^5-\frac{175}{8}x^4+\frac{725}{2^5}x^3 +\frac{44085}{2^{12}}x^2+\frac{465}{2^8}x+\frac{405457}{2^9}=0.$$

The roots are:

$$x_1,\ x_6 = \frac{1}{\sqrt[5]{4}}+\sqrt[5]{2}\pm\frac{1}{2\sqrt{2}},$$

$$x_2,\ x_7 = \frac{\omega^4}{\sqrt[5]{4}}+\omega^3\sqrt[5]{2}\pm\frac{1}{2\sqrt{2}},$$

. . . etc. . . .

The given domain must include $R(\omega)$ but not $\sqrt{2}$.

(2) $r = \mu = \vartheta = \delta = 0$, $\nu = 1$, $\gamma = 2$.

As this is the case,

$$r = 0 \quad\text{and}\quad \mu = -\frac{\vartheta\kappa}{\tau^2} = 0,$$

we first must see that not all three relations (*c*), (*d*), (*e*) on pages 147–148 are satisfied. In our case (*c*) reduces to $\nu = 0$ and hence the values chosen are admissible. Our identity reduces to

$$\frac{1}{2^7} = \frac{\nu^5\tau^5}{\kappa^2} = \frac{\tau^5}{\kappa^2}.$$

Choosing $\kappa = 2$, then $\sqrt[5]{2}$ must not be in our domain and $\sqrt{2}$ must not in the given domain, but ω must be rationally known. We have then $\tau = \frac{1}{2}$.

$$f(x) = x^{10}+5x^7-\frac{35}{8}x^6+\frac{29}{2^6}x^5-\frac{25}{4}x^4+\frac{1575}{2^4}x^3+\frac{1727}{2^8}x^2-\frac{2075}{2^{10}}x+\frac{1165}{2^{14}}=0.$$

The roots are:

$$x_1,\ x_6 = \frac{1}{2^5\sqrt[5]{4}}+\sqrt[5]{2}\pm\frac{\sqrt{2}}{2^5\sqrt[5]{4}},$$

$$x_2,\ x_7 = \frac{\omega^4}{2^5\sqrt[5]{4}}+\omega^3\sqrt[5]{2}\pm\frac{\omega^4\sqrt{2}}{2^5\sqrt[5]{4}},$$

. . . etc. . . .

B. I. (2) $R_2 = 0$, $S_2 \neq 0$.

There is no irreducible Abelian equation of this proposed type. The proof is here omitted as it is entirely identical to the proof given under A. I. 2; we need only to replace

$$\rho_1, \quad \rho_2, \quad \rho_3, \quad \rho_4, \quad x_i,$$

by, respectively,

$$\rho_6, \quad \rho_8, \quad \rho_2, \quad \rho_4, \quad \varphi_i.$$

B. II. $R_2 \neq 0$.

From the relation

$$\rho_6\rho_4 + \rho_2\rho_8 = 2R_2$$

we can have at most two of the four ρ's equal to zero: either $\rho_6 = \rho_4 = 0$ or $\rho_2 = \rho_8 = 0$. Both cases are identical in the abstract. We therefore cover all cases corresponding to B. II. by dividing B. II. into the three sub-cases

(1) $$\rho_8 = \rho_2 = 0,$$
(2) $$\rho_8 = 0; \quad \rho_2 \neq 0,$$
(3) $$\rho_8 \neq 0; \quad \rho_2 \neq 0.$$

B. II. (1) $\rho_8 = \rho_2 = 0, \quad \rho_4 \neq 0, \quad \rho_6 \neq 0.$

From the conditions:

$$\rho_2 = 0 \text{ follows } K_2 - L_2\sqrt{5} = -N_2\sqrt{-10+2\sqrt{5}} + M_2\sqrt{-10-2\sqrt{5}},$$
$$\rho_8 = 0 \text{ follows } K_2 - L_2\sqrt{5} = +N_2\sqrt{-10+2\sqrt{5}} - M_2\sqrt{-10-2\sqrt{5}},$$
$$\therefore K_2 - L_2\sqrt{5} = 0 \quad \text{or} \quad K_2 = L_2\sqrt{5}.$$

Suppose for a moment $K_2 = L_2 = 0$; then $\rho_6 = -\rho_4$ and $\varphi_1 = x_1 + x_6 = 0$, which is impossible; hence $K = L\sqrt{5} \neq 0$ and therefore $\sqrt{5}$ *is in our given domain.*

It further follows from the above, that

$$\frac{N_2}{M_2} = \frac{\sqrt{-10-2\sqrt{5}}}{\sqrt{-10+2\sqrt{5}}},$$

or

$$N_2 = -M_2\frac{1+\sqrt{5}}{2},$$

so that

$$\rho_6{}^5 = 2K_2 - \frac{20M_2}{\sqrt{-10+2\sqrt{5}}}$$

and

$$\rho_4{}^5 = 2K_2 + \frac{20M_2}{\sqrt{-10+2\sqrt{5}}};$$

hence

$$\rho_4\rho_6 = 2R_2 = \sqrt[5]{4K_2{}^2 - \frac{400}{-10+2\sqrt{5}}M_2{}^2},$$

or, putting

$$M_2' = \frac{20}{-10+2\sqrt{5}}M_2,$$

then M_2' *is rationally known* and

$$(2R_2)^5 = 4K_2{}^2 - M_2'(-10+2\sqrt{5}) \text{ *is a perfect fifth power.*}$$

The following four functions ψ_1/ψ_6, $\psi_3/\psi_4{}^2$, $\psi_7/\psi_6{}^2$, ψ_9/ψ_4 are seen to be double-valued functions under G_{10}; hence, if we write

$$\rho_1 = \pi\rho_6, \quad \rho_3 = \vartheta\rho_4^2, \quad \rho_7 = \sigma\rho_6^2, \quad \rho_9 = \delta\rho_4,$$

then π, ϑ, σ, δ are quantities in our given rational domain after the adjunction of ω.

By actual computation we have

$$\psi_1\psi_9 = 10^2\gamma\rho_1\rho_9 = \iota\sum_1^5 \varphi^2_{i+5} + (\omega + \omega^4)F_1 + (\omega^2 + \omega^3)F_2,$$

$$\psi_3\psi_7 = 10^2\gamma\rho_3\rho_7 = \iota\sum_1^5 \varphi^2_{i+5} + (\omega^2 + \omega^3)F_1 + (\omega + \omega^4)F_2,$$

where

$$F_1 = \varphi_6\varphi_8 + \varphi_6\varphi_9 + \varphi_7\varphi_9 + \varphi_7\varphi_{10} + \varphi_8\varphi_{10},$$
$$F_2 = \varphi_6\varphi_7 + \varphi_6\varphi_{10} + \varphi_7\varphi_8 + \varphi_8\varphi_9 + \varphi_9\varphi_{10}.$$

F_1, F_2 are seen to be rationally known, if applied to by G_{10} and as $\sqrt{5}$ is in our given domain, the two functions $\rho_1\rho_9$ and $\rho_3\rho_7$ are rationally known and hence from

$$\rho_1\rho_9 = \pi\delta\rho_4\rho_6 = 2\pi\delta R_2$$

and

$$\rho_3\rho_7 = \sigma\vartheta\rho_4^2\rho_6^2 = 4\sigma\vartheta R_2^2$$

it follows that $\pi\delta$ and $\sigma\vartheta$ are in our given domain.

Further,

$$\rho_1^5 + \rho_9^5 = (K_1 + L_1\sqrt{5})2 = \delta^5\rho_4^5 + \rho_6^5\pi^5$$
$$= 2K_2(\pi^5 + \delta^5) + M_2'\sqrt{-10 + 2\sqrt{5}}(\delta^5 - \pi^5).$$

Similarly

$$\rho_3^5 + \rho_7^5 = (K_1 - L_1\sqrt{5})2 = \vartheta^5\rho_4^{10} + \sigma^5\rho_6^{10}$$
$$= [4K_2^2 + M_2'^2(-10 + 2\sqrt{5})](\vartheta^5 + \sigma^5) + 2^2K_2M_2'\sqrt{-10 + 2\sqrt{5}}(\vartheta^5 - \sigma^5).$$

Summarizing now the facts that $\rho_1^5 + \rho_9^5$, $\rho_3^5 + \rho_7^5$, $\sigma\vartheta$, $\delta\pi$ are quantities in our given domain and that δ, π, σ, ϑ are also in our domain, if we adjoin ω to this domain, then it is seen that we may write

(1) $\pi = a + b\sqrt{-10 + 2\sqrt{5}}$, (2) $\delta = a - b\sqrt{-10 + 2\sqrt{5}}$,

(3) $\vartheta = \alpha + \beta\sqrt{-10 + 2\sqrt{5}}$, (4) $\sigma = \alpha - \beta\sqrt{-10 + 2\sqrt{5}}$,

where a, b, α, β *are in our given rational domain.* The relation

$$(2R_2)^5 = 4K_2^2 - M_2'^2(-10 + 2\sqrt{5}) \tag{5}$$

which we established above and the identity

$$\begin{aligned}\frac{1}{2^5\gamma^2} = {} & \pi^5\rho_6^5 + \delta^5\rho_4^5 + \vartheta^5\rho_4^{10} + \sigma^5\rho_6^{10} + r^5 + 5\cdot 2R_2(\vartheta\sigma 2R_2 - \pi\delta)[(\pi^2\vartheta \\ & + \delta^2\sigma)4R_2^4 - (\vartheta^2\delta\rho_4^5 + \sigma^2\pi\rho_6^5)] + 5r\cdot 4R_2^2(\pi^2\delta^2 + \vartheta^2\sigma^2) \\ & + 5r^2[(\pi^2\vartheta + \delta^2\sigma)4R_2^2 + (\vartheta^2\delta\rho_4^5 + \sigma^2\pi\rho_6^5)] - 5r\cdot 2R_2(\pi\delta + \vartheta\sigma\cdot 2R_2) \\ & - 5r\pi\delta\vartheta\sigma\cdot 8R_2^3 - 5r(\pi^3\sigma\rho_6^5 + \delta^3\vartheta\rho_4^5) - 5r\cdot 2R_2(\vartheta^3\pi\rho_4^5 + \sigma^3\delta\rho_6^5),\end{aligned} \tag{6}$$

which is based upon the condition

$$\prod_{i=1}^{i=5} (x_i - x_{i+5}) = \sqrt{\gamma},$$

enable us to express M_2' and K_2 in terms of R_2, σ, ϑ, π, δ, γ. The relation (6) does not involve any quantity lying outside of our rational domain, for we have

$$\pi\delta = a^2 - b^2(-10 + 2\sqrt{5}),$$

$$\vartheta\sigma = \alpha^2 - \beta^2(-10 + 2\sqrt{5}),$$

$$\pi^2\vartheta + \delta^2\sigma = 2\alpha[a^2 + b^2(-10 + 2\sqrt{5})] + 4\beta ab(-10 + 2\sqrt{5}),$$

$$\begin{aligned}\vartheta^2\delta\rho_4^5 + \sigma^2\pi\rho_6^5 &= 4K_2[a\alpha^2 + a\beta^2(-10 + 2\sqrt{5}) - 2\alpha\beta b(-10 + 2\sqrt{5})] \\ &\quad + 2M_2'[2a\alpha\beta - b\alpha^2 - b\beta^2(-10 + 2\sqrt{5})](-10 + 2\sqrt{5}),\end{aligned}$$

$$\begin{aligned}\pi^3\sigma\rho_6^5 + \delta^3\vartheta\rho_4^5 &= 4K_2\alpha a^3 + (-10 + 2\sqrt{5})2[\sigma ab\alpha K_2 \\ &\quad - (3a^2b + b^3[-10 + 2\sqrt{5}])(2\beta K_2 + M_2') \\ &\quad + \beta M_2'(a^3 + 3ab[-10 + 2\sqrt{5}])],\end{aligned}$$

$$\begin{aligned}\vartheta^3\pi\rho_4^5 + \sigma^3\delta\rho_6^5 &= 4K_2a\alpha + 2(-10 + 2\sqrt{5})[\sigma a\alpha\beta^2K_2 + 6b\alpha^2\beta K_2 \\ &\quad + 2K_2b\beta^2(-10 + 2\sqrt{5}) + M_2'(b\alpha - 3b\alpha\beta^2[-10 + 2\sqrt{5}] \\ &\quad + 3a\alpha^2\beta + a\beta^2(-10 + 2\sqrt{5})],\end{aligned}$$

$$\begin{aligned}\pi^5\rho_6^5 + \delta^5\rho_4^5 &= 4K_2[a^5 + 10a^3b[-10 + 2\sqrt{5}] + 5ab^4[-10 + 2\sqrt{5}]^2] \\ &\quad - 2M_2'(-10 + 2\sqrt{5})[5a^4b + 10a^2b^3[-10 + 2\sqrt{5}] \\ &\quad + b^5[-10 + 2\sqrt{5}]^2],\end{aligned}$$

$$\begin{aligned}\vartheta^5\rho_4^{10} + \sigma^5\rho_6^{10} &= 2[4K_2^2 + M_2'^2(-10 + 2\sqrt{5})][\alpha^5 + \alpha^3\beta^2(-10 + 2\sqrt{5}) \\ &\quad + \alpha\beta^4(-10 + 2\sqrt{5})^2] + 8K_2 + M_2'[\alpha^4\beta + \alpha^2\beta^3(-10 + 2\sqrt{5}) \\ &\quad + \beta^5(-10 + 2\sqrt{5})^2](-10 + 2\sqrt{5}).\end{aligned}$$

Having found K_2 and M_2' from (5) and (6), we know ρ_4 and ρ_6.

The roots of $f(x) = 0$ are

$$\begin{aligned}x_1, x_6 &= \rho_4 + \rho_6 \pm \sqrt{\gamma}[(a + b\sqrt{-10 + 2\sqrt{5}})\rho_6 + (\alpha + \beta\sqrt{-10 + 2\sqrt{5}})\rho_4^2 \\ &\quad + (\alpha - \beta\sqrt{-10 + 2\sqrt{5}})\rho_6^2 + (a - b\sqrt{-10 + 2\sqrt{5}})\rho_4 + r],\end{aligned}$$

$$\begin{aligned}x_2, x_7 &= \omega^2\rho_6 + \omega^3\rho_4 \pm \sqrt{\gamma}[\omega^2(a + b\sqrt{-10 + 2\sqrt{5}})\rho_6 + \omega(\alpha + \beta\sqrt{-10 + 2\sqrt{5}})\rho_4^2 \\ &\quad + (\alpha - \beta\sqrt{-10 + 2\sqrt{5}})\omega^4\rho_6^2 + \omega^3(a - b\sqrt{-10 + 2\sqrt{5}})\rho_4 + r],\end{aligned}$$

. . . etc., . . .

where

$$\rho_4 = 2K_2 + M_2'\sqrt{-10 + 2\sqrt{5}},$$

$$\rho_6 = 2K_2 - M_2'\sqrt{-10 + 2\sqrt{5}},$$

where K_2 and M_2' are found from the relations (5) and (6).

The corresponding equation is

$$f(x) = (x^5 - A_1x^4 + A_2x^3 - A_3x^3 + A_4x - A_5) \times (x^5 - B_1x^4 + B_2x^3 - B_3x^2 + B_4x - B_5) = 0,$$

where

$$A_1, B_1 = \pm 5r\sqrt{\gamma},$$

$$A_2, B_2 = 10\gamma r^2 - 10R_2 - 10R_2\gamma(\pi\delta + 2\sigma\vartheta R_2) \mp 10R_2\sqrt{\gamma}(\pi + \delta),$$

$$\begin{aligned} A_3, B_3 = {} & 5\gamma(\sigma^2\rho_6^5 + \vartheta^2\rho_4^5) - 15r\gamma \cdot 2R_2(\pi\delta + \vartheta\sigma) + 10\gamma(\pi\vartheta + \sigma\delta)4R_2^2 \\ & \pm \sqrt{\gamma}5[2\gamma r^3 + 4R_2^2(\sigma + \vartheta) + \gamma(\sigma^2\pi\rho_6^5 + \vartheta^2\delta\rho_4^5) + 4\gamma R_2^2(\delta^2\sigma + \pi^2\vartheta) \\ & - 6rR_2 - 6rR_2(\pi\delta + \sigma\vartheta 2R_2)], \end{aligned}$$

$$\begin{aligned} A_4, B_4 = {} & + 20R_2^2 + 5\gamma^2r^4 - 30\gamma r^2R_2 + 20\gamma^2R_2^2(\pi^2\delta^2 + 4\sigma^2\vartheta^2R_2^2) \\ & - 5\gamma^2[(\sigma^3\delta\rho_6^5 + \vartheta^3\pi\rho_4^5)2R_2 + \delta^3\vartheta\rho_4^5 + \pi^3\sigma\rho_6^5)] + 20\gamma R_2^2(\pi^2 + \delta^2) \\ & - 30r^2\gamma^2R_2(\pi\delta + 2\sigma\vartheta R_2) + 10\gamma^2r[\sigma^2\pi\rho_6^5 + \vartheta^2\delta\rho_4^5 + 4R_2^2(\delta^2\sigma + \pi^2\vartheta)] \\ & + 40\gamma rR_2^2(\sigma + \vartheta) - 40\gamma\sigma\vartheta R_2^3(1 + \gamma\pi\delta) - 15\gamma(\pi\sigma\rho_6^5 + \vartheta\delta\rho_4^5) \\ & + 80\gamma\pi\delta R_2^2 \pm 5\sqrt{\gamma}[- 2\gamma R_2(\sigma^3\rho_6^5 + \vartheta^3\rho_4^5) + 2\gamma r(\sigma^2\rho_6^5 + \vartheta^2\rho_4^5) \\ & - (\vartheta\rho_4^5 + \sigma\rho_6^5) - 6r^2\gamma R_2(\pi + \delta) - 8\gamma(\pi + \delta)\vartheta\sigma R_2^3 \\ & - 3\gamma(\pi^2\sigma\rho_6^5 + \delta^2\vartheta\rho_4^5) + 8\gamma R_2^2(\pi + \delta)(1 + \pi\delta) + 32r\gamma R_2^2(\pi\vartheta + \sigma\delta)], \end{aligned}$$

$$\begin{aligned} A_5, B_5 = {} & 4K_2 + 5\gamma^2(\pi^4\rho_6^5 + \delta^4\rho_4^5) + 10\gamma(\pi^2\rho_6^5 + \delta^2\rho_4^5) \\ & - 5\gamma^2[2R_2(\sigma\delta^3 + \pi^3\vartheta) + \sigma\vartheta(\vartheta^2\rho_4^5 + \sigma^2\rho_6^5)]4R_2^2 \\ & - 40\gamma R_2^3(\sigma\pi + \delta\vartheta) - 120\gamma R_2^3(\sigma\delta + \pi\vartheta) - 120\gamma^2R_2^3\pi\delta(\sigma\delta + \vartheta\pi) \\ & + 10\gamma R_2(\sigma^2\rho_6^5 + \vartheta^2\rho_4^5) + 10\gamma^2R_2(\sigma^2\pi^2\rho_6^5 + \delta^2\vartheta^2\rho_4^5) \\ & - 10\gamma^2rR_2(\pi + \delta) + 5\gamma^2r^2(\sigma^2\rho_6^5 + \vartheta^2\rho_4^5) + 40r^2\gamma^2R_2^2(\delta\sigma + \pi\vartheta) \\ & - 10r\gamma^2R_2(\sigma^3\rho_6^5 + \vartheta^3\rho_4^5) - 5r\gamma(\vartheta\rho_4^5 + \sigma\rho_6^5) - 40r\gamma^2\vartheta\sigma R_2^3(\pi + \delta) \\ & - 15r\gamma(\vartheta\delta\rho_4^5 + \sigma\pi\rho_6^5) - 15r\gamma^2(\vartheta\delta^2\rho_4^5 + \sigma\pi^2\rho_6^5) \\ & + 40r\gamma R_2^2(\pi + \delta) + 40r\gamma^2R_2^2\pi\delta(\pi + \delta) \\ & \pm \sqrt{\gamma}[\gamma^2(\pi^5\rho_6^5 + \vartheta^5\rho_4^{10} + \sigma^5\rho_6^{10} + \delta^5\rho_4^5 + r^5) + 5(\pi\rho_6^5 + \delta\rho_4^5) \\ & + 10\gamma(\pi^3\rho_6^5 + \delta^3\rho_4^5) - 20\gamma^2R_2^2[2R_2\pi\delta(\sigma\delta^2 + \vartheta\pi^2) + \vartheta\sigma(\delta\vartheta^2\rho_4^5 + \sigma^2\pi\rho_6^5)] \\ & - 40R_2^3(\sigma + \vartheta) - 120\gamma R_2^3\pi\delta(\sigma + \vartheta) - 120\gamma R_2^3(\sigma\delta^2 + \vartheta\pi^2) \\ & + 10\gamma^2R_2(\pi\delta(\sigma^2\pi\rho_6^5 + \vartheta^2\delta\rho_4^5) + 8R_2^3\sigma\vartheta(\sigma\delta^2 + \vartheta\pi^2)) \\ & + 10\gamma R_2(\sigma\vartheta 8R_2^3(\sigma + \vartheta) + \sigma^2\delta\rho_6^5 + \pi\vartheta^2\rho_4^5) - 10\gamma r^3R_2 \\ & - 10\gamma^2r^3R_2(\pi\delta + 2R_2\sigma\vartheta) + 5\gamma^2(\sigma^2\pi\rho_6^5 + \vartheta^2\delta\rho_4^5)r^2 \\ & + 20\gamma^2r^2R_2^2(\delta^2\sigma + \pi^2\vartheta) + 20\gamma r^2R_2^2(\sigma + \vartheta) + 20rR_2^2 \\ & - 5r\gamma^2(\delta^3\vartheta\rho_4^5 + \pi^3\sigma\rho_6^5) - 10r\gamma^2R_2(\sigma^3\delta\rho_6^5 + \vartheta^3\pi\rho_4^5) \\ & + 20\pi\gamma^2R_2^2(\pi^2\delta^2 + 4R_2^2\sigma^2\vartheta^2) - 40r\gamma^2R_2^3\sigma\vartheta\pi\delta \\ & + 20\gamma R_2^2(\pi^2 + \delta^2) - 40r\gamma R_2^3\sigma\vartheta + 80r\gamma R_2^2\pi\delta]. \end{aligned}$$

Our next question is: Are the derived conditions also sufficient, that $f(x) = 0$ with the above roots is an irreducible Abelian equation?

First of all we note that *not all five quantities a, b, α, β and r can be zero at the* same time, for then $f(x)$ would contain the two rational factors

$$x^5 - A_1x^4 + A_2x^3 - A_3x^2 + A_4x - A_5$$

and

$$x^5 - B_1x^4 + B_2x^3 - B_3x^2 + B_4x - B_5,$$

as seen from the above functions A_i, B_i. Further

$$\rho_4 \neq \pm \rho_6,$$

for then

$$\rho_4\rho_6 = \pm \rho_4^2 = \pm \rho_6^2 = 2R_2$$

and hence

$$\varphi_1 = x_1 + x_6$$

would be rationally known. It follows then that equations (5) and (6) must lead to

$$K_2 \neq 0 \quad \text{and} \quad M_2' \neq 0$$

otherwise we have no irreducible equation corresponding to case B. ω may be rationally known or not; it certainly cannot be expressed rationally by ρ_4 or ρ_6, for if ω is not in our domain, then ω satisfies an irreducible quadratic, as $\sqrt{5}$ is rational, whereas ρ_4 satisfies an irreducible quintic. It follows then, as just explained under B. I, that the sum of any two, three, or four roots must always contain the irrationality ρ_4 and ρ_6 respectively and that $f(x)$ cannot have a rational factor of the first, second, third, and fourth degree.

Whenever the sum of any five roots does not vanish, this sum is irrational for the same reason; the first case, however, can happen only if we pick out five roots belonging to the same system of intransitivity. Our question of the irreducibility of $f(x)$ simply amounts to the following question: Under what conditions can all five A_i be rationally known?

A_1 shows that we must have $r = 0$, if $\sqrt{\gamma}$ shall disappear. Further, from A_2, we must have $\pi + \delta = 0$ as $R_2 \neq 0$; hence

$$a = 0.$$

Under these conditions we get further the relations from A_3

$$\text{(I)} \qquad \frac{4\alpha R_2^2}{-10+2\sqrt{5}} = \frac{4K_2b\alpha\beta + M_2'b(\alpha^2 + \beta^2[-10+2\sqrt{5}])}{1 + b^2(-10+2\sqrt{5})},$$

from A_4:

$$\text{(II)} \qquad \begin{aligned} &4\gamma R_2K_2[\alpha^3 + 3\alpha\beta^2(-10+2\sqrt{5})] + 2\gamma R_2M_2'(-10+2\sqrt{5})[3\alpha^2\beta \\ &\quad + \beta^3(-10+2\sqrt{5})] + [2K_2\alpha + \beta M_2'(-10+2\sqrt{5})] \\ &\quad \times [1 + 3\gamma b^2(-10+2\sqrt{5})] = 0, \end{aligned}$$

and from A_5:

$$\text{(III)} \qquad \begin{aligned} &\gamma^2\{2b^5M_2'(-10+2\sqrt{5})^3 + [8K_2^2 + 2M_2'^2(-10+2\sqrt{5})] \\ &\times[\alpha^5 + 10\alpha^3\beta^2(-10+2\sqrt{5}) + 5\alpha\beta^4(-10+2\sqrt{5})^2] \\ &+ 8K_2M_2'(-10+2\sqrt{5})[5\alpha^4\beta + 10\alpha^2\beta^2(-10+2\sqrt{5}) \\ &+ \beta^5(-10+2\sqrt{5})\} - 10M_2'(-10+2\sqrt{5})(b + 2b^3\gamma) \\ &- 80R_2^3\alpha[\gamma^2b^4(-10+2\sqrt{5}) - 1] + 10R_2b(-10+2\sqrt{5})[8K_2^2\alpha\beta \\ &+ 2M_2'\alpha^2 + 2M_2'\beta^2(-10+2\sqrt{5})][2\gamma^2R_2(\alpha^2 - \beta^2[-10+2\sqrt{5}]) \\ &- b^2(-10+2\sqrt{5}) - \gamma] + 160R_2^4\gamma\alpha[\alpha^2 - \beta^2(-10+2\sqrt{5})] \\ &\times[1 + \gamma b^2(-10+2\sqrt{5})] = 0. \end{aligned}$$

If then $r = a = 0$ and equations I, II, III are true $f(x)$ is reducible, having two rational factors each of the fifth degree. If, however, not all these five equations are satisfied, $f(x)$ is irreducible and must have two as well as five systems of intransitivity, as seen directly from the A_i, B_i respectively from the roots. Its group, being therefore a subgroup of G_{240}, can only be G_{10}, as σ_1 and σ_2 are impossible in the Galois group of $f(x) = 0$ under our conditions, for we have

σ_1 interchanges ρ_4 and ρ_6, so that $\rho_4 = \pm \rho_6$, if σ_1 were in our Galois group and hence $M_2' = 0$ or $K_2 = 0$, but this contradicts our conditions.

Similarly σ_2 is impossible in the Galois group, as σ_2 changes ρ_4 to $\omega^4\rho_2 = 0$, which contradicts $R_2 \neq 0$. It follows then that our $f(x) = 0$ is an irreducible Abelian equation under the derived conditions.

Type VII. If $\sqrt{5}$, α, β, a, b, K_2, M_2', γ, r, R_2, *but not* $\sqrt{\gamma}$ *are quantities in our given rational domain and if not all five quantities* α, β, a, b, r *are zero and if relations* (5) *and* (6) *on page* 151 *exist and if* $R_2 \neq 0$, $K_2 \neq 0$, $M_2' \neq 0$*—then the above* $f(x) = 0$ *with the given roots is an irreducible Abelian equation, if*

$$\sqrt[5]{2K_2 \pm M_2' \sqrt{-10 + 2\sqrt{5}}}$$

is not in our domain and if not all five equations $r = a = 0$, (I), (II), (III), *are satisfied.*

Illustrations:

(1) $$a = b = \alpha = \beta = 0, \qquad r = 2, \qquad M_2' = 1.$$

Our identity (6) shows

$$\frac{1}{2^5\gamma^2} = r^5;$$

hence $\gamma = 2^{-5}$.

Equation (5) shows

$$(2R_2)^5 = 4K_2^2 - (-10 + 2\sqrt{5}),$$

hence, taking

$$K_2 = +\sqrt{\frac{\sqrt{5}}{2}},$$

we have

$$R_2 = \tfrac{1}{2}\sqrt[5]{10}.$$

Here our rational domain cannot include $\sqrt{2}$, and

$$\sqrt[5]{2\sqrt{\frac{\sqrt{5}}{2}} \pm \sqrt{-10+2\sqrt{5}}},$$

but must contain $\sqrt{5}$, $\sqrt[5]{10}$, $\sqrt{\dfrac{\sqrt{5}}{2}}$.

As $r \neq 0$, not all five equations $r = \alpha = 0$, (I), (II), (III) are satisfied and hence all conditions are satisfied. The roots are

$$x_1, x_6 = \sqrt[5]{\sqrt{2\sqrt{5}} + \sqrt{-10 + 2\sqrt{5}}} + \sqrt[5]{\sqrt{2\sqrt{5}} - \sqrt{-10 + 2\sqrt{5}}} \pm \sqrt{\frac{1}{2^3}},$$

$$x_2, x_7 = \omega^3 \sqrt[5]{\sqrt{2\sqrt{5}} + \sqrt{-10 + 2\sqrt{5}}} + \omega^2 \sqrt[5]{\sqrt{2\sqrt{5}} - \sqrt{-10 + 2\sqrt{5}}} \pm \sqrt{\frac{1}{2^3}}$$

. . . . etc. . . . .

$$f(x) = x^{10} - \left(\frac{1}{8} + 2\sqrt[5]{10}\right) 5x^8 + \left(\frac{9}{8} - 16\sqrt[5]{10} + 25\sqrt[5]{10^2}\right) x^6 - 4\sqrt{2\sqrt{5}}x^5$$

$$+ \frac{1}{8}\left(\frac{1}{20} - 308\sqrt[5]{10^2} + \frac{119}{2}\sqrt[5]{10}\right) x^4 + 5(1 - 4\sqrt[5]{10})\sqrt{2\sqrt{5}}x^3$$

$$+\left(\frac{189}{5\cdot 2^{12}}-\frac{1}{10}\sqrt[5]{10}+\frac{45}{16}\sqrt[5]{10^2}-\frac{743}{2^7}\sqrt[5]{10^3}\right)x^2-5\sqrt{2\sqrt{5}}\left(\sqrt[5]{10}+\frac{1}{16}\right)x$$

$$+8\sqrt{5}+\frac{1}{5^2\cdot 2^{12}}-\frac{1}{5\cdot 2^8}\sqrt[5]{10}+\frac{7}{5\cdot 2^6}\sqrt[5]{10^2}+\frac{1}{4}\sqrt[5]{10^3}+\sqrt[5]{10^4}=0.$$

(2) $$r=\alpha=\beta=b=0,\qquad a=1,\qquad M_2'=1,$$

$$R_2=-\frac{1}{2}\sqrt[5]{2\sqrt{5}},\quad K_2=\frac{1}{2}\sqrt{-10},\quad \gamma=\frac{1}{2^3\sqrt[4]{-10}}.$$

By these values chosen all conditions are satisfied.

The roots are:

$$x_1, x_6=\sqrt[5]{\sqrt{-10}+\sqrt{-10+2\sqrt{5}}}+\sqrt[5]{\sqrt{-10}-\sqrt{-10+2\sqrt{5}}}$$
$$\pm\frac{1}{2}\sqrt{\frac{1}{2\sqrt[4]{-10}}}\left[\sqrt[5]{\sqrt{-10}+\sqrt{-10+2\sqrt{5}}}+\sqrt[5]{\sqrt{-10}-\sqrt{-10+2\sqrt{5}}}\right],$$

$$x_2, x_7=\omega^3\sqrt[5]{\sqrt{-10}+\sqrt{-10+2\sqrt{5}}}+\omega^2\sqrt[5]{\sqrt{-10}-\sqrt{-10+2\sqrt{5}}}$$
$$\pm\frac{1}{2}\sqrt{\frac{1}{2\sqrt[4]{-10}}}\left[\omega^3\sqrt[5]{\sqrt{-10}+\sqrt{-10+2\sqrt{5}}}+\omega^2\sqrt[5]{\sqrt{-10}-\sqrt{-10+2\sqrt{5}}}\right],$$

. . . . etc. . . .

$$f(x)=x^{10}+2\sqrt[5]{2\sqrt{5}}\left(5+\frac{1}{8\sqrt[4]{-10}}\right)x^8+\sqrt[5]{20}\left[45+\frac{1}{2^6\sqrt{-10}}-\frac{30}{2^3\sqrt[4]{-10}}\right]x^6$$
$$+4\sqrt{-10}\left[1+\frac{5}{4\sqrt[4]{-10}}+\frac{5}{2^6\sqrt{-10}}\right]x^5+5\sqrt[4]{40\sqrt{5}}$$
$$\times\left[20+\frac{9}{\sqrt[4]{-10}}+\frac{55}{2^6\sqrt{-10}}\right]x^4+40\sqrt{-10}\sqrt[5]{2\sqrt{5}}\left[1+\frac{21}{8\sqrt[4]{-10}}\right.$$
$$\left.+\frac{27}{2^6\sqrt{-10}}+\frac{3}{2^9\sqrt[4]{-10^3}}\right]x^3+100\sqrt[5]{2^4\cdot 5^2}\left[1+\frac{5}{2^3\sqrt[4]{-10}}+\frac{7}{2^6\sqrt{-10}}\right.$$
$$\left.-\frac{1}{2^9\sqrt[4]{-10^3}}\right]x^2+40\sqrt{-10}\sqrt[5]{20}\left[1+\frac{1}{4\sqrt{-10}}+\frac{5}{2^4\sqrt{-10}}\right.$$
$$\left.+\frac{1}{2^7\sqrt[4]{-10^3}}+\frac{1}{5\cdot 2^{13}}\right]x+8\sqrt{-10}\left[26+\frac{15}{2\sqrt[4]{-10}}+\frac{55}{2^4\sqrt{-10}}\right.$$
$$\left.+\frac{5}{2^7\sqrt[4]{-10^3}}-\frac{13}{2^{12}\cdot 5}\right]=0.$$

Here the given rational domain must not include

$$\sqrt{2\sqrt[4]{-10}},\quad \sqrt[5]{\sqrt{-10}\pm\sqrt{-10+2\sqrt{5}}},$$

but it must contain $\sqrt[4]{-10}$, $\sqrt[5]{2\sqrt{5}}$.

B. II. (2) $\rho_8=0,\quad \rho_2\neq 0,\quad \rho_4\neq 0,\quad \rho_6\neq 0.$

From $\rho_8=0$ follows

$$K_2-L_2\sqrt{5}=+N_2\sqrt{-10+2\sqrt{5}}-M_2\sqrt{-10-2\sqrt{5}},$$

hence

$$\rho_2 = \sqrt[5]{2(K_2 - L_2\sqrt{5})} \neq 0.$$

It follows then ω *is rationally known.*

Putting then

$$\frac{\rho_1}{\rho_6} = \pi, \quad \frac{\rho_3}{\rho_4^2} = \zeta, \quad \frac{\rho_9}{\rho_4} = \delta, \quad \frac{\rho_7}{\rho_2} = \sigma,$$

then $\pi, \zeta, \delta, \sigma$ *are in the given domain.*

Here again *not all five quantities* $r, \pi, \zeta, \delta, \sigma$ *can be zero at once,* for then $f(x)$ would be reducible. As ω is rationally known, we may put $\rho_2^2/\rho_4 = g$ and $\rho_6^2/\rho_2 = h$, where g, h *are quantities in our given domain,* $g \neq 0$, $h \neq 0$.

We have then

$$\rho_6^5 = \frac{\rho_2^2}{\rho_4}\left(\frac{\rho_6^2}{\rho_2}\right)^2 \rho_6\rho_4 = +\, 2gh^2R_2,$$

$$\rho_2^5 = \left(\frac{\rho_2^2}{\rho_4}\right)^2 \frac{\rho_2}{\rho_6^2}\cdot(\rho_4\rho_6)^2 = +\,\frac{4g^2R_2^2}{h},$$

$$\rho_4^5 = \frac{(\rho_6\rho_4)^5}{\rho_6^5} = +\,\frac{2^4R_2^4}{gh^2}.$$

If ρ_6 were rationally known, all three ρ_i would be rational and if ρ_6 is not in our given rational domain, then ρ_2 and ρ_4 cannot be rationally known. Whenever ρ_i is in our given rational domain, $f(x)$ splits up into the two rational factors

$$\prod_{i=1}^{i=5}(x - x_i) \quad \text{and} \quad \prod_{i=1}^{i=5}(x - x_{i+5}).$$

It follows then that $2gh^2R_2$ *is not a perfect fifth power.*

The last condition includes the three conditions above, namely

$$R_2 \neq 0, \quad h \neq 0, \quad g \neq 0.$$

We have here again an *identity,* based upon

$$\prod_{i=1}^{i=5}(x_i - x_{i+5}) = \sqrt{\gamma}$$

namely

$$\begin{aligned}\frac{1}{2^5\gamma^2} = {} & 2\pi^5gh^2R_2 + \frac{4\sigma^5g^2R_2^4}{h} + \frac{2^8\zeta^5R_2^8}{g^2h^4} + \frac{2^4\delta^5R_2^4}{gh^2} \\ & - 40R_2^3\left(\pi^3\zeta\delta + \frac{8\zeta^3\sigma\delta R_2^3}{h^3g} + \frac{\sigma^3\pi\zeta g}{h} + \frac{\delta^3\pi\sigma}{h}\right) \\ & + 20R_2^2\left(\frac{2^3\pi\zeta^2\delta^2R_2^3}{gh^2} + \frac{4\zeta\sigma^2\delta^2R_2^2}{h^2} + \frac{2^2\sigma\pi^2\zeta^2R_2^2}{h} + \delta\pi^2\sigma^2g\right) \\ & + 20rR_2^2\left(\pi^2\delta^2 + \zeta^2\sigma^2\frac{4R_2^2}{h^2}\right) + 10rR_2\left(2\pi^2\zeta R_2 + \frac{8\zeta^2\delta R_2^3}{gh^2} + \sigma^2\pi g\right. \\ & \left. + \frac{2\delta^2\sigma R_2}{h}\right) - 10r^3R_2\left(\pi\delta + \frac{2\sigma\zeta R_2}{h}\right) - 10rR_2\left(\pi^3\zeta hg + \frac{2^4\zeta^3\pi R_2^4}{gh^2}\right. \\ & \left. + \frac{2\sigma^3\delta gR_2}{h} + \frac{2^3\delta^3\zeta R_2^3}{gh^2}\right) - \frac{40r\pi\delta\zeta\sigma R_2^3}{h} + r^5.\end{aligned}$$

The roots of $f(x) = 0$ are here

$$x_1, x_6 = \sqrt[5]{2gh^2R_2} + \sqrt[5]{\frac{4g^2R_2^2}{h}} + \sqrt[5]{\frac{2^4R_2^4}{gh^2}} \pm \sqrt{\gamma}\left[\pi\sqrt[5]{2gh^2R_2} + \zeta\sqrt[5]{\frac{2^8R_2^8}{g^2h^4}} + \sigma\sqrt[5]{\frac{4g^2R_2^2}{h}} + \delta\sqrt[5]{\frac{2^4R_2^4}{gh^2}} + r\right],$$

. . . . etc. . . . .

$$f(x) = (x^5 - A_1x^4 + A_2x^3 - A_3x^2 + A_4x - A_5) \times (x^5 - B_1x^4 + B_2x^3 - B_3x^2 + B_4x - B_5) = 0,$$

where

$$A_1, B_1 = \pm 5r\sqrt{\gamma},$$

$$A_2, B_2 = +10\gamma r^2 - 10R_2 - 10\gamma R_2(\pi\delta + \zeta\sigma \cdot 2R_2) \mp \sqrt{\gamma}10R_2\left(\pi + \delta + \frac{\zeta 2R_2}{h}\right),$$

$$\begin{aligned} A_3, B_3 = {} & 10R_2\left(g + \frac{2R_2}{h}\right) + 10\gamma R_2\left(\sigma^2 g + \frac{\delta^2 2R_2}{h} + \frac{8R_2^3\zeta^2}{gh^2}\right) \\ & - 30\gamma rR_2\left(\pi + \frac{2R_2\zeta}{h} + \delta\right) + 20\gamma R_2\left(\pi\zeta 2R_2 + \pi\sigma g + \frac{\sigma\delta 2R_2}{h}\right) \\ & \pm \sqrt{\gamma}\left[10\gamma r^3 + 10R_2\left(\pi g + \frac{2\sigma R_2}{h} + 2\zeta R_2\right)\right. \\ & + 10\gamma R_2\left(\sigma^2\pi g + \frac{\delta^2\sigma 2R_2}{h} + \pi^2\zeta 2R_2 + \frac{8R_2^3\zeta^2\delta}{gh^2}\right) \\ & \left. - 30rR_2\left(1 + \gamma\pi\delta + \gamma\frac{\zeta\sigma 2R_2}{h}\right) + 20R_2\left(\frac{2R_2\delta}{h} + \sigma g\right)\right], \end{aligned}$$

$$\begin{aligned} A_4, B_4 = {} & -10R_2\left(\frac{2R_2 g}{h} + hg\right) + 20R_2^2 + 5\gamma^2r^4 - 30\gamma r^2R_2 \\ & + 20\gamma^2R_2^2\left(\pi^2\delta^2 + \frac{4R_2^2\zeta^2\sigma^2}{h^2}\right) - 10\gamma^2R_2\left(\frac{\sigma^3\delta 2R_2 g}{h} + \frac{8R_2^3\delta^3\zeta}{gh^2}\right. \\ & \left. + \pi^3\sigma hg + \frac{2^4R_2^4\zeta^3\pi}{gh^2}\right) + 20\gamma R_2^2\left(\pi^2 + \frac{4R_2^2\zeta^2}{h^2}\right) + 20\gamma R_2^2\delta^2 \\ & - 30r^2\gamma^2R_2\left(\pi\delta + \frac{2R_2\zeta\sigma}{h}\right) + 20\gamma^2 r\left(\sigma^2\pi g + \frac{2\delta^2\sigma R_2}{h}\right. \\ & \left. + 2\pi^2\zeta R_2 + \frac{2^3\zeta^2\delta R_2^3}{gh^2}\right) + 20\gamma rR_2\left(\pi g + \frac{2R_2\sigma}{h} + 2\zeta R_2\right) \\ & - \frac{40\gamma^2R_2^3\pi\sigma\delta\zeta}{h} - \frac{40\zeta R_2^3}{h}(\pi + \sigma + \delta) - 30\gamma R_2 hg\pi(\sigma + \pi) \\ & - \frac{60}{h}R_2^2 g\sigma(\delta + \sigma) - \frac{240\gamma\delta\zeta R_2^4}{gh^2} + 80\gamma R_2^2\pi\delta + 80r\gamma R_2\left(\sigma g + \frac{2R_2\delta}{h}\right) \\ & \pm 5\sqrt{\gamma}\left[4rR_2\left(g + \frac{2R_2}{h}\right) - 2\gamma R_2\left(\frac{\sigma^3 2R_2 g}{h} + \pi^3 gh + \frac{2^4R_2^4\zeta^3}{gh^2}\right)\right. \\ & - 2R_2\left(\frac{2R_2\delta g}{h} + \frac{8R_2^3\zeta}{gh^2} + \sigma gh\right) - 30r^2\gamma\delta R_2 - 30r^2\gamma R_2\left(\pi + \frac{2R_2\zeta}{h}\right) \\ & + 4R_2\left(\sigma^2 g + \frac{2\delta^2R_2}{h} + \frac{8R_2^3\zeta^2}{gh^2}\right) - \frac{8\zeta R_2^3}{h} - \frac{8\gamma R_2^3\zeta}{h}(\pi\sigma + \pi\delta + \sigma\delta) \end{aligned}$$

$$- 6\pi R_2 hg(\gamma\pi\sigma + 1) - 12R_2^2\frac{\sigma g}{h}(\sigma\delta\gamma + 1) - \frac{48\gamma\delta^2\zeta R_2^4}{gh^2}$$

$$+ 8\delta\pi\gamma R_2^2(\pi + \delta) + 8R_2^2(\pi + \delta) + \frac{32\gamma\sigma\zeta^2 R_2^4}{h^2}$$

$$+ 16r\gamma R_2\left(2\pi\zeta R_2 + \pi\sigma g + \frac{2\sigma\delta R_2}{h}\right)\Bigg],$$

$$A_5, B_5 = 2R_2\left(gh^2 + \frac{2g^2R_2}{h} + \frac{2^3R_2^3}{gh^2}\right) + 10\gamma^2R_2\left(\pi^4gh^2 + \frac{2\sigma^4g^2R_2}{h} + \frac{2^3R_2^3\delta^4}{gh^2}\right)$$

$$+ 20\gamma R_2\left(\pi^2gh^2 + \frac{2R_2\sigma^2g^2}{h} + \frac{2^3R_2^3\delta^2}{gh^2}\right) - \frac{40R_2^3}{h} - \frac{40\gamma^2R_2^3\delta^3}{h}(\sigma + \pi)$$

$$- \frac{320\gamma^2R_2^6\zeta^3}{gh^3}(\sigma + \delta) - \frac{40R_2^3\zeta\gamma^2}{h}(\sigma^3g + \pi^3h)$$

$$- \frac{40\gamma R_2^3}{h}(\sigma\pi + \pi\zeta g + \delta\zeta h) - \frac{120\gamma R_2^3}{h}(\sigma\delta + \pi\delta + \zeta\sigma g + \pi\zeta h + \delta^2)$$

$$- \frac{120\gamma^2R_2^3\pi}{h}(\sigma\delta^2 + \zeta\sigma^2g + \delta\pi\zeta h) + 20R_2^2g(1 + \sigma^2\gamma + \pi^2\gamma)$$

$$+ \frac{80\gamma R_2^4\zeta^2}{h}\left(1 + \frac{2R_2}{gh^2}\right) + 20\gamma^2R_2^2\left(\sigma^2\pi^2g + \frac{\pi^2\zeta^2\cdot 2^2R_2^2}{h} + \frac{2^3R_2^3\zeta^2\delta^2}{gh^2}\right)$$

$$+ 40g\gamma R_2^2\delta(\pi + \sigma) + \frac{160R_2^4\gamma\zeta}{h^2}(\delta + \sigma) + 80\gamma\pi\sigma R_2^2g - 10\gamma^2r^3R_2$$

$$- 10\gamma^2r^3R_2\left(\pi + \frac{2R_2\zeta}{h}\right) + 10\gamma r^2gR_2(1 + \sigma^2)$$

$$+ \frac{20\gamma r^2R_2^2}{h}\left(1 + \delta^2 + \frac{2^2R_2^2\zeta^2}{gh}\right) + 20\gamma^2r^2R_2\left(\pi\sigma g + \frac{2R_2\delta\sigma}{h} + 2R_2\pi\zeta\right)$$

$$- 10r\gamma^2R_2\left(\frac{2R_2\sigma^3g}{h} + \pi^3gh + \frac{2^4R_2^4\zeta^3}{gh^2}\right) - 10r\gamma R_2\left(\frac{2R_2\delta g}{h}\right.$$

$$\left. + \frac{2^3R_2^3\zeta}{gh^2} + \sigma gh\right) - \frac{40r\gamma^2R_2^3\zeta}{h}(\pi\sigma + \pi\delta + \sigma\delta) - \frac{40r\gamma\zeta R_2^3}{h}$$

$$- 30\gamma rR_2g\left(\frac{2R_2\sigma}{h} + \pi h\right) - 30r\gamma^2R_2\left(\frac{2R_2\delta\sigma^2g}{h} + \frac{2^3R_2^3\zeta\delta^2}{gh^2} + \pi^2\sigma gh\right)$$

$$+ 40r\gamma R_2^2(\pi + \delta) + 40r\gamma^2R_2^2\left(\pi\delta^2 + \frac{2^2\sigma\zeta^2R_2^2}{h^2} + \delta\pi^2\right)$$

$$\pm \sqrt{\gamma}\left\{\gamma^2\left(2\pi^5gh^2R_2 + \frac{2^8\zeta^5R_2^8}{g^2h^4} + \frac{2^2\sigma^5R_2^2g^2}{h} + \frac{2^4\delta^5R_2^4}{gh^2} + r^5\right)\right.$$

$$+ 10R_2\left(\pi gh^2 + \frac{2\sigma g^2R_2}{h} + \frac{2^3\delta R_2^3}{gh^2}\right) + 20\gamma R_2\left(\pi^3gh^2 + \frac{2\sigma^3g^3R_2}{h}\right.$$

$$\left. + \frac{2^3\delta^3R_2^3}{gh^2}\right) - 40\gamma^2R_2^3\left(\frac{\sigma\delta^3\pi}{h} + \frac{\sigma\delta\zeta^3 2^3R_2^3}{gh^3} + \frac{\pi\sigma^3\zeta g}{h} + \delta\pi^3\zeta\right)$$

$$- \frac{40\gamma R_2^3}{h}\left(\delta^3 + \frac{2^3R_2^3\zeta^3}{gh^2}\right) - \frac{40R_2^3}{h}(\sigma + \pi + \zeta g + \zeta h) - \frac{120\delta R_2^3}{h}$$

$$- \frac{120\gamma R_2^3\pi}{h}\left(\sigma\delta + \zeta\sigma g + \delta\zeta h + \frac{\sigma\delta^2}{\pi} + \delta^2 + \frac{\zeta\sigma^2g}{\pi} + \zeta\pi h\right)$$

$$+ 20\gamma^2R_2^2\left(\sigma^2\delta\pi^2g + 2^2R_2^2\cdot\frac{\sigma^2\zeta\delta^2}{h^2} + 2^2R_2^2\cdot\frac{\pi^2\sigma\zeta^2}{h} + 2^3R_2^3\cdot\frac{\delta^2\pi\zeta^2}{gh^2}\right)$$

$$+ 20R_2^2\left(\delta g + \frac{2^2R_2^2\zeta}{h^2}\right) + 20\gamma R_2^2g\delta(\sigma^2+\pi^2) + \frac{80R_2^4\zeta\gamma}{h^2}(\sigma^2+\delta^2)$$

$$+ \frac{80\gamma R_2^4\zeta}{h}\left(\sigma + \frac{2\pi\zeta R_2}{gh}\right) + 40R_2^2g(\pi+\sigma) + 80\sigma\delta R_2^2\left(\pi g + \frac{2^2R_2^2\zeta}{h^2}\right)$$

$$- 10\gamma r^3R_2\left(1 + \frac{2R_2\sigma\zeta}{h} + \pi\delta\right) + 10\gamma^2r^2R_2\pi g(1+\sigma^2)$$

$$+ \frac{20\gamma r^2R_2^2}{h}\left(\sigma\delta^2 + \pi^2\zeta h + \sigma + \zeta h + \frac{2^2\zeta^2\delta R_2^2}{g\cdot h}\right)$$

$$+ 20r^2\gamma R_2\left(\sigma g + \frac{2\delta R_2}{h}\right) - 10rgR_2\left(\frac{2R_2}{h} + h\right) + 20rR_2^2$$

$$- 10r\gamma^2R_2\left(\frac{2R_2\sigma^3\delta g}{h} + \frac{2^3R_2^3\delta^3\zeta}{gh^2} + \pi^3\sigma hg + \frac{2^4R_2^4\zeta^3\pi}{gh^2}\right)$$

$$+ 20r\gamma^2R_2^2\left(\frac{2^2R_2^2\sigma^2\zeta^2}{h^2} + \delta^2\pi^2\right) - \frac{40r\gamma R_2^3\zeta}{h}(\pi\sigma\delta\gamma + \pi + \delta + \sigma)$$

$$+ 20\gamma rR_2^2\left(1 + \pi^2\delta^2 + \frac{2^2R_2^2\zeta^2}{h^2}\right) - 30r\gamma R_2g\left(\frac{2R_2\sigma^2}{h} + \pi^2h\right)$$

$$+ 80r\gamma\pi\delta R_2^2\Big\}.$$

Regarding the sufficiency of the derived conditions we first decide whether $f(x)$ can be reducible under our conditions obtained so far. As ρ_2, ρ_4, ρ_6 each satisfy an irreducible quintic, $\sqrt{\gamma}$ an irreducible quadratic, $\sqrt{\gamma}$ cannot be expressed rationally by ρ_2, ρ_4 or ρ_6, and hence no root can be rational, for we may write $\omega^j\rho_2 + \omega^k\rho_4 + \omega^{-k}\rho_6$ in the form $\frac{\omega^j\rho_6^2}{h} + \frac{\omega^k 2R_2}{\rho_6} + \omega^{-k}\rho_6$. If the last expression were rational, we must have necessarily that ρ_6 is a rational quantity. It follows then, that no root is here rationally known and $f(x)$ has no rational linear factor. The same theory holds for the sum of any two, three or four roots, as ω is rationally known and we have here again, that the sum of any two, three, four or five roots must be irrational except when it is zero which happens only for

$$_i\sum_1^5 x_i \quad\text{and}\quad _i\sum_1^5 x_{i+5}.$$

Looking into the above A_i and B_i, we see that they contain as only the irrationality $\sqrt{\gamma}$. This irrationality drops out in A_1 and B_1 if

(1) $$r = 0,$$

in A_2 and B_2 if

(2) $$\pi + \delta + \frac{\zeta 2R_2}{h} = 0,$$

and hence in A_3 and B_3 if

(3) $$\left.\begin{array}{l}\pi g(1+\gamma\sigma^2) + 2\sigma g - \dfrac{(\pi+\delta)^3h\zeta\delta\gamma}{g} \\ -\dfrac{\pi+\delta}{\zeta}[\sigma + \zeta h + \gamma\delta^2\sigma + \gamma\pi^2\zeta h + 2\delta]\end{array}\right\} = 0.$$

in A_4 and B_4 if, in addition to (1) and (2),

$$(4)\quad \begin{aligned} &+\frac{\pi+\delta}{\zeta}[\gamma\sigma^3 g+\delta g-2\delta^2+3\sigma g(1+\sigma\delta\gamma)] \\ &+2\sigma^2 g-\pi^3 hg\gamma-\sigma hg-3\pi hg(1+\gamma\pi\sigma) \\ &+\frac{(\pi+\delta)^3 h}{\zeta^2 g}[\gamma\zeta^2+3\gamma\delta^2-2\gamma\sigma g\zeta+1-2\zeta] \\ &+\frac{(\pi+\delta)^2 h}{\zeta}[+1-\gamma(\pi\sigma+\sigma\delta+3\pi\delta)]. \end{aligned}$$

Finally, assuming the truth of (1) and (2), A_5 and B_5 are rationally known, if

$$(5)\quad \begin{aligned} &\pi^5 gh^2\gamma^2+5\pi gh^2+10\pi^3 gh^2\gamma-5\frac{\pi+\delta}{\zeta}\Big[\frac{\sigma^5 g^2}{5}\gamma^2+\sigma g^2+2\gamma\sigma^3 g^2 \\ &+\delta gh(\gamma^2\sigma^2\pi^2+1+\gamma\sigma^2+\gamma\pi^2+4\sigma\pi)+2gh(\pi+\sigma)\Big] \\ &-5\frac{(\pi+\delta)^2 h}{\zeta^2}\Big[\gamma^2\sigma\delta^3\pi+\gamma^2\sigma^3\pi\zeta g+\delta\pi^3\zeta h\gamma^2+\gamma\delta^3+\zeta \\ &+\frac{1}{h}(\sigma+\pi+\zeta g)+3\delta+\frac{3\gamma\pi}{h}\Big(\delta\sigma+\zeta\sigma g+\frac{\sigma\delta^2}{\pi}+\delta^2+\frac{\zeta\sigma^2 g}{\pi}\Big)\Big] \\ &-5\frac{(\pi+\delta)^3 h}{g\zeta^3}\Big[+3\gamma\pi\zeta^2 g+\frac{\delta^5}{5}\gamma^2+\delta+2\gamma\delta^3+\gamma^2\sigma^2\zeta\delta^2 g \\ &+\gamma^2 g\pi^2\sigma\zeta^2 h+\zeta g+\zeta g\gamma(\sigma^2+\delta^2)+\gamma\zeta\sigma gh+4\sigma\delta\zeta g\Big] \\ &+5\frac{(\pi+\delta)^4}{g\zeta^2}h^2\pi\gamma(\delta^2\gamma+1)+5\frac{(\pi+\delta)^5 h^2}{g\zeta^2}\gamma(1+\gamma\sigma\delta)-\frac{(\pi+\delta)^7}{g^2\zeta^2}h^3\gamma^2. \end{aligned}$$

Hence, to assure the irreducibility of $f(x)$, we must have: *Not all of the above five equations must be satisfied.*

If this condition is observed in addition to the conditions already found, then $f(x)$ is irreducible, its group being therefore transitive, must be imprimitive with two as well as five systems of intransitivity, for $f(x)$ contains the five rational factors $(x-x_i)(x-x_{i+5})$, if we adjoin ρ_2, ρ_4 or ρ_6 to the given rational domain; and adjoining $\sqrt{\gamma}$ then the above functions A_i, B_i show that $f(x)$ splits up into the two rational factors

$$\prod_{i=1}^{i=5}(x-x_i) \quad\text{and}\quad \prod_{i=1}^{i=5}(x-x_{i+5}).$$

It follows then that the Galois group of $f(x)=0$ is a subgroup of G_{240} and it can only be G_{10} under our derived conditions, as σ_1 and σ_2 are impossible substitutions in the group of $f(x)=0$. σ_1 changes $\rho_2{}^5\neq 0$ to $\rho_8{}^5=0$ and σ_2 applied to $\rho_6{}^5$ produces $\rho_8{}^5=0$, so that under the conditions derived $f(x)=0$ is an irreducible Abelian equation.

Type VIII. If ω, π, σ, δ, ζ, g, h, γ, r, R_2 *are quantities in our given rational domain and if* $\sqrt{\gamma}$, $\sqrt[5]{2gh^2R_2}$ *are not in our given domain, and if* r, π, ζ, δ, σ *are not all at the same time zero and if the identity on page* 157 *exists and if not all five equations on pages* 160–161 *are satisfied, then* $f(x) = 0$ *of page* 158 *with the roots on page* 158 *is an irreducible Abelian equation.*

An example may be furnished by taking $r = \zeta = \sigma = \delta = 0$.

Here our identity reduces to

$$\frac{1}{2^5\gamma^2} = 2gh^2R_2\pi.$$

Assuming then

$$g = h = R_2 = \pi = 1,$$

we have

$$\gamma^2 = \frac{1}{2^6}; \quad \text{hence} \quad \gamma = \frac{1}{2^3}.$$

$\sqrt{2}$ and $\sqrt[5]{2}$ cannot be in our rational domain, as $\sqrt{\gamma}$ as well as $\sqrt[5]{2gh^2R_2}$ are irrational. Further equation (2) on page 160 is not satisfied.

$$\left.\begin{aligned} f(x) = x^{10} - 20x^8 - 60x^7 + 65x^6 + \frac{8971}{2^4}x^5 + \frac{9075}{2^3}x^4 \\ + \frac{51755}{2^6}x^3 + \frac{65925}{2^7}x^2 + \frac{96475}{2^9}x + \frac{433563}{2^{13}} \end{aligned}\right\} = 0.$$

The roots are:

$$x_2, x_7 = \omega^4\sqrt[5]{4} + \omega^2\sqrt[5]{2} + \omega^3\sqrt[5]{16} \pm \frac{\omega^2\sqrt[5]{2}}{2\sqrt{2}},$$

$$x_1, x_6 = \sqrt[5]{4} + \sqrt[5]{2} + \sqrt[5]{16} \pm \frac{\sqrt[5]{2}}{2\sqrt{2}},$$

. . . etc. . . .

The given rational domain must contain ω.

B. II. (3) $\rho_i \neq 0 \quad (i = 2, 4, 6, 8)$.

There is no irreducible Abelian equation of the proposed type. The proof, which is here omitted for brevity, is identical to the proof given under A. II. (3) if we replace there

$$\rho_1, \quad \rho_2, \quad \rho_3, \quad \rho_4, \quad x_i,$$

respectively by

$$\rho_6, \quad \rho_8, \quad \rho_2, \quad \rho_4, \quad \varphi_i.$$

C. CONCLUSION.

There are eight types of irreducible Abelian equations of the tenth degree. Types I to IV are quintics in x^2. *There is no such equation in the natural rational domain* $R(1)$, *for in all types* $\sqrt{5}$ *must be a quantity of the given domain. Six types are possible, only if* ω *belongs to the rational domain.*

UNIVERSITY OF CALIFORNIA PUBLICATIONS
IN
MATHEMATICS

Vol. 1, No. 7, pp. 163-169 February 16, 1915

ABRIDGED TABLES OF HYPERBOLIC FUNCTIONS.

BY

F. E. PERNOT

In the calculation of the operating characteristics of long transmission circuits the most convenient and direct solution is afforded by the use of hyperbolic functions of complex variables. For lines consisting of large conductors in which the losses are small it is desirable to make computations to the degree of accuracy afforded by six-place tables of logarithms. Also, in such cases, the argument to which the hyperbolic functions are taken from the tables is small; for power circuits usually below 0.5.

The most convenient table of these functions is that of the Smithsonian Institution, in which are tabulated both the logarithms and natural values for arguments from zero to 6.0, five decimal places being given. A table of Gudermannian functions is also appended, which makes it possible to find the values of functions (hyperbolic) to six places; not, however, without involving a double interpolation, first to the Gudermannian and then from a table of trigonometric functions.

For small arguments the hyperbolic cosine is not varying rapidly, hence easy interpolations are assured. The hyperbolic sine, on the other hand, is varying with the same rapidity as the argument, and therefore the interpolations for log sinh x are cumbersome unless the tabular interval is very small. These considerations are immediately seen from the series expressing the two functions:

$$\left.\begin{aligned} \cosh x &= 1 + \frac{x^2}{\underline{|2}} + \frac{x^4}{\underline{|4}} + \frac{x^6}{\underline{|6}} + \ldots\ldots \\ \sinh x &= x + \frac{x^3}{\underline{|3}} + \frac{x^5}{\underline{|5}} + \frac{x^7}{\underline{|7}} + \ldots\ldots \end{aligned}\right\} (1)$$

In all cases the hyperbolic tangent is most easily derived from sinh x and cosh x by

$$\tanh x = \frac{\sinh x}{\cosh x} \quad (2)$$

The series for sinh x in (1) can be written

$$\sinh x = x\left(1 + \frac{x^2}{\lfloor 3} + \frac{x^4}{\lfloor 5} + \frac{x^6}{\lfloor 7} + \ldots\right) \quad (3)$$

which immediately suggests the advisability of tabulating values of the quantity in brackets,

$$y = 1 + \frac{x^2}{\lfloor 3} + \frac{x^4}{\lfloor 5} + \frac{x^6}{\lfloor 7} + \ldots \quad (4)$$

for this quantity is seen to have an even smaller rate of change than cosh x. The value of the hyperbolic sine is then given by

$$\left.\begin{aligned} \sinh x &= xy \\ y &= \frac{\sinh x}{x} \end{aligned}\right\} \quad (5)$$

Since extended computations are most conveniently done by logarithms, the tabulation was made of log y. Log x naturally being at hand, the value of log sinh x is immediately found by simple addition of log x and log y, and, further, it is found without having to make any inconvenient interpolations. A glance at the values of log sinh x as tabulated in any table will immediately impress one with the impracticability of interpolating directly for log sinh x for values of x between 0 and 0.10, in which region a large proportion of the values of x fall for the particular work above referred to. This scheme of tabulating the ratio of a function of a variable to the variable itself is the same as is used in the "S and T" tables found in Bremiker's logarithm tables, where for small angles the ratios of the trigonometric sine and tangent to the angle are given.

CONSTRUCTION OF TABLES

The following table contains values to base 10 of log $\frac{\sinh x}{x}$ and log cosh x, together with the differences to be used in interpolating. The arguments progress in steps of 0.005 from zero to 0.600, giving 121 entries. Logarithms are given to seven places.

The first column, $\log_{10} \frac{\sinh x}{x}$, was computed by first evaluating the expression for y in (4) and then taking out log y from tables. For the small values of x used, this series is very convergent, hence the labor involved was not excessive. The evaluation of the series was made for alternate entries, and the intermediate values obtained by interpolation.

Using this value of log $\frac{\sinh x}{x}$ the value of log $\frac{1}{\sinh^2 x}$ was formed. Using this as argument in Zech's table of addition logarithms, the value of $\log_{10} \cosh^2 x$ is immediately obtained, from the relation $\cosh^2 x = \sinh^2 x + 1$. Interpolations in the addition logarithm tables were made to the nearest even number in order that the resulting value of log cosh x might appear to the nearest unit in the last place.

The complete set of values was checked by differences, and in a few cases the last unit was changed by one in order to give uniformity in the second differences, which in both tabulations are practically constant.

In addition to the check by differences, every tenth entry was checked independently by calculating directly the values of

$$y = \frac{\sinh x}{x} = \frac{\epsilon^x - \epsilon^{-x}}{2x} \text{ and } \cosh x = \frac{\epsilon^x + \epsilon^{-x}}{2},$$

using the tabulated values of the exponentials as given to nine decimals in tables of the exponential function by J. W. L. Glaisher, F. R. S., published in the *Transactions of the Cambridge Philosophical Society,* vol. XIII. From the above values, the logarithm to seven places was taken from tables, which agreed with the previously tabulated values to the nearest unit of the last place, except in two or three cases where a difference of one was noted.

To facilitate interpolation, the values of the first derivatives of the function multiplied by the tabular interval ($\omega = 0.005$) are tabulated in units of the last place in the tabulated seven-place logarithm of the function. The second differences are also tabulated.

For the first three columns:

$$\left.\begin{aligned} f(x) &= \log_{10}\frac{\sinh x}{x} = \log_{10} y \\ \omega f'(x) &= \omega\frac{d}{dx}\log_{10}\frac{\sinh x}{x} = \omega\frac{d}{dx}\log_{10} y \\ &= \omega\frac{d}{dx}\log_{10}\left(1 + \frac{x^2}{\underline{|3}} + \frac{x^4}{\underline{|5}} + \ldots\right) \\ &= 2\omega M\frac{x}{y}\left(\frac{1}{\underline{|3}} + \frac{2x^2}{\underline{|5}} + \frac{3x^4}{\underline{|7}} + \frac{4x^6}{\underline{|9}} + \ldots\right) \end{aligned}\right\} \quad (6)$$

M = modulus of the common logarithm system = 0.43429448.

This series is very convergent, and was used in the form for the value of $\omega = 0.005$,

$$\omega f'(x) = 0.01M\frac{x}{y}\left(\frac{1}{\underline{|3}} + \frac{2x^2}{\underline{|5}} + \frac{3x^4}{\underline{|7}} + \ldots\right) \quad (7)$$

Of course the above value was multiplied by 10^7 to reduce to units of the last place in the tabulation of log y. Values of log y to be used in the computation were taken directly from the table.

Δ_2 is the average of the differences in $\omega f'(x)$ immediately preceding and following a tabular value of log y.

For the last three columns:

$$\left.\begin{aligned} f(x) &= \log_{10}\cosh x \\ \omega f'(x) &= \omega\frac{d}{dx}\log_{10}\cosh x \\ &= \omega M\frac{\sinh x}{\cosh x} \end{aligned}\right\} \quad (8)$$

Thus

$$\omega f'(x) = 0.005M \tanh x \quad (9)$$

Δ_2 is the same as defined for the previous case.

It is to be noted that the second differences, or accelerations if x be considered equicrescent, are practically constant. Easy and accurate interpolations are thus assured, even to seven decimals and for tabular intervals in the argument as great as 0.005, which is five times as great an interval as is used in the Smithsonian Tables for the functions tabulated to five decimals only.

TO INTERPOLATE USING SECOND DIFFERENCES

Let it be required to determine the value of $f(x)$ for an argument

$$x = x_0 \pm a$$

x_0 is the value of the argument nearest that of x. a, then, is the distance over which the interpolation has to be made, and $\frac{a}{\omega} = n =$ the fraction of the tabular interval. Using the second differences, we have

$$f(x) = f(x_0 \pm n\omega) = f(x_0) \pm n\,\omega f'(x) + \frac{n^2}{2}\Delta_2 \quad (10)$$

$f(x_0)$ is the value of the function corresponding to the argument x_0.

Illustration:

Required, log sinh 0.1163740

log cosh 0.1163740

$x =$ 0.1163740

$x_0 =$ 0.115

$a = +0.0013740$

$n = \frac{a}{.005} = +\ 0.27480$

		for sinh x	for cosh x
$f(x_0)$	=	0.0009569	0.0028655
$n\,\omega f'(x)$	=	228.6	683.2
$\frac{1}{2}n^2\Delta_2$	=	1.4	4.1
$f(x)$	=	0.0009799	0.0029342
log x	=	9.0658560—10	

log sinh x = 9.0668359—10 log cosh x = 0.0029342

log tanh x = 9.0639017—10

LOGARITHMS OF HYPERBOLIC FUNCTIONS OF A REAL VARIABLE

From $x=0$ to $x=0.600$

x	$\log_{10}\frac{\sinh x}{x}$	$\omega f'(x)$	Δ_2	$\log_{10}\cosh x$	$\omega f'(x)$	Δ_2
0.000	0.0000000	000.0	36.2	0.0000000	000.0	108.6
.005	.0000018	36.2	36.2	.0000054	108.6	108.6
.010	.0000072	72.4	36.2	.0000217	217.1	108.6
.015	.0000162	108.6	36.2	.0000489	325.7	108.6
.020	.0000289	144.8	36.2	.0000869	434.2	108.5
.025	.0000452	180.9	36.2	.0001357	542.8	108.5
.030	.0000651	217.1	36.2	.0001954	651.2	108.5
.035	.0000886	253.3	36.2	.0002659	759.7	108.4
.040	.0001158	289.5	36.2	.0003473	868.1	108.4
.045	.0001466	325.7	36.2	.0004396	976.5	108.4
.050	.0001809	361.8	36.2	.0005426	1084.8	108.3
.055	.0002189	398.0	36.2	.0006565	1193.1	108.2
.060	.0002605	434.2	36.2	.0007813	1301.3	108.2
.065	.0003057	470.4	36.2	.0009168	1409.5	108.1
.070	.0003546	506.5	36.2	.0010632	1517.5	108.0
.075	.0004071	542.7	36.2	.0012203	1625.6	108.0
.080	.0004632	578.8	36.1	.0013882	1733.5	107.9
.085	.0005229	615.0	36.1	.0015670	1841.3	107.8
.090	.0005862	651.1	36.1	.0017565	1949.1	107.7
.095	.0006531	687.2	36.1	.0019568	2056.7	107.6
.100	.0007236	723.3	36.1	.0021679	2164.3	107.5
.105	.0007977	759.4	36.1	.0023897	2271.7	107.4
.110	.0008755	795.6	36.1	.0026222	2379.0	107.3
.115	.0009569	831.7	36.1	.0028655	2486.2	107.2
.120	.0010418	867.8	36.1	.0031194	2593.3	107.0
.125	.0011304	903.8	36.1	.0033841	2700.3	106.9
.130	.0012226	939.9	36.1	.0036595	2807.1	106.8
.135	.0013184	976.0	36.0	.0039456	2913.8	106.6
.140	.0014178	1012.0	36.0	.0042423	3020.4	106.5
.145	.0015208	1048.1	36.0	.0045496	3126.8	106.3
.150	.0016274	1084.1	36.0	.0048676	3233.0	106.2
.155	.0017376	1120.1	36.0	.0051962	3339.1	106.0
.160	.0018514	1156.1	36.0	.0055354	3445.0	105.8
.165	.0019688	1192.1	36.0	.0058852	3550.8	105.7
.170	.0020899	1228.1	36.0	.0062456	3656.3	105.5
.175	.0022145	1264.1	36.0	.0066165	3761.8	105.3
.180	.0023427	1300.1	36.0	.0069979	3867.0	105.1
.185	.0024745	1336.0	36.0	.0073899	3972.0	104.9
.190	.0026099	1372.0	36.0	.0077923	4076.9	104.7
.195	.0027489	1407.9	35.9	.0082052	4181.5	104.5
.200	.0028915	1443.8	35.9	.0086286	4286.0	104.3

x	$\log_{10}\frac{\sinh x}{x}$	$\omega f'(x)$	Δ_2	$\log_{10}\cosh x$	$\omega f'(x)$	Δ_2
.200	.0028915	1443.8	35.9	.0086286	4286.0	104.3
.205	.0030377	1479.7	35.9	.0090624	4390.2	104.1
.210	.0031874	1515.6	35.9	.0095066	4494.2	103.9
.215	.0033407	1551.4	35.9	.0099612	4598.0	103.7
.220	.0034977	1587.3	35.8	.0104262	4701.6	103.5
.225	.0036582	1623.1	35.8	.0109015	4805.0	103.3
.230	.0038223	1658.9	35.8	.0113872	4908.1	103.0
.235	.0039900	1694.7	35.8	.0118832	5011.1	102.8
.240	.0041612	1730.5	35.8	.0123894	5113.7	102.5
.245	.0043361	1766.3	35.8	.0129059	5216.2	102.3
.250	.0045145	1802.1	35.7	.0134326	5318.3	102.1
.255	.0046965	1837.8	35.7	.0139696	5420.3	101.8
.260	.0048821	1873.5	35.7	.0145167	5522.0	101.6
.265	.0050712	1909.2	35.7	.0150739	5623.4	101.3
.270	.0052639	1944.9	35.7	.0156413	5724.5	101.0
.275	.0054602	1980.5	35.6	.0162188	5825.4	100.8
.280	.0056600	2016.2	35.6	.0168064	5926.1	100.5
.285	.0058634	2051.8	35.6	.0174040	6026.4	100.2
.290	.0060704	2087.4	35.6	.0180117	6126.5	99.9
.295	.0062809	2123.0	35.6	.0186293	6226.3	99.7
.300	.0064950	2158.5	35.5	.0192569	6325.8	99.4
.305	.0067126	2194.0	35.5	.0198945	6425.0	99.1
.310	.0069338	2229.6	35.5	.0205419	6523.9	98.8
.315	.0071586	2265.1	35.5	.0211993	6622.5	98.5
.320	.0073869	2300.6	35.5	.0218665	6720.9	98.2
.325	.0076187	2336.0	35.4	.0225435	6818.9	97.9
.330	.0078541	2371.5	35.4	.0232302	6916.6	97.6
.335	.0080930	2406.9	35.4	.0239267	7014.0	97.2
.340	.0083354	2442.2	35.4	.0246330	7111.1	96.9
.345	.0085814	2477.6	35.3	.0253490	7207.8	96.6
.350	.0088309	2512.9	35.3	.0260746	7304.3	96.3
.355	.0090839	2548.2	35.3	.0268098	7400.4	96.0
.360	.0093405	2583.5	35.3	.0275546	7496.2	95.6
.365	.0096006	2618.8	35.2	.0283090	7591.7	95.3
.370	.0098643	2654.0	35.2	.0290730	7686.8	95.0
.375	.0101315	2689.2	35.2	.0298464	7781.6	94.6
.380	.0104022	2724.4	35.2	.0306293	7876.1	94.3
.385	.0106764	2759.5	35.1	.0314216	7970.2	93.9
.390	.0109541	2794.7	35.1	.0322233	8064.0	93.6
.395	.0112353	2829.8	35.1	.0330344	8157.4	93.3
.400	.0115201	2864.9	35.1	.0338548	8250.5	92.9
.405	.0118083	2899.9	35.0	.0346845	8343.2	92.6
.410	.0121000	2934.9	35.0	.0355234	8435.6	92.2
.415	.0123952	2969.9	35.0	.0363716	8527.6	91.8
.420	.0126940	3004.9	35.0	.0372289	8619.2	91.5
.425	0.0129963	3039.8	34.9	.0380954	8710.5	91.1

x	$\log_{10} \frac{\sinh x}{x}$	$\omega f'(x)$	Δ_2	$\log_{10} \cosh x$	$\omega f'(x)$	Δ_2
.425	0.0129963	3039.8	34.9	.0380954	8710.5	91.1
.430	.0133020	3074.7	34.9	.0389710	8801.4	90.7
.435	.0136112	3109.6	34.9	.0398557	8892.0	90.4
.440	.0139239	3144.5	34.8	.0407494	8982.2	90.0
.445	0.142400	3179.3	34.8	.0416521	9072.0	89.6
.450	.0145597	3214.1	34.8	.0425638	9161.4	89.3
.455	.0148828	3248.8	34.7	.0434844	9250.5	88.9
.460	.0152095	3283.5	34.7	.0444139	9339.2	88.5
.465	.0155396	3318.2	34.7	.0453522	9427.5	88.1
.470	.0158732	3352.9	34.6	.0462993	9515.4	87.7
.475	.0162102	3387.5	34.6	.0472552	9602.9	87.3
.480	.0165507	3422.1	34.6	.0482199	9690.1	86.9
.485	.0168946	3456.7	34.6	.0491932	9776.8	86.6
.490	.0172420	3491.2	34.5	.0501752	9863.2	86.2
.495	.0175929	3525.7	34.5	.0511659	9949.2	85.8
.500	.0179472	3560.2	34.5	.0521651	10034.8	85.4
.505	.0183049	3594.6	34.4	.0531728	10119.9	85.0
.510	.0186661	3629.0	34.4	.0541890	10204.7	84.6
.515	.0190307	3663.4	34.3	.0552137	10289.1	84.2
.520	.0193988	3697.7	34.3	.0562468	10373.1	83.8
.525	.0197703	3732.0	34.3	.0572883	10456.7	83.4
.530	.0201452	3766.3	34.2	.0583382	10539.9	83.0
.535	.0205235	3800.5	34.2	.0593963	10622.7	82.6
.540	.0209053	3834.7	34.2	.0604627	10705.1	82.2
.545	.0212905	3868.8	34.1	.0615373	10787.1	81.8
.550	.0216791	3903.0	34.1	.0626201	10868.6	81.4
.555	.0220711	3937.0	34.1	.0637111	10949.8	81.0
.560	.0224665	3971.1	34.0	.0648101	11030.6	80.6
.565	.0228653	4005.1	34.0	.0659171	11111.0	80.2
.570	.0232675	4039.1	34.0	.0670322	11190.9	79.7
.575	.0236731	4073.0	33.9	.0681553	11270.4	79.3
.580	.0240821	4106.9	33.9	.0692863	11349.5	78.9
.585	.0244945	4140.8	33.8	.0704252	11428.3	78.5
.590	.0249103	4174.6	33.8	.0715720	11506.5	78.1
.595	.0253294	4208.4	33.8	.0727265	11584.4	77.7
.600	.0257519	4242.2	33.7	.0738888	11661.9	77.3

UNIVERSITY OF CALIFORNIA PUBLICATIONS
IN
MATHEMATICS

Vol. 1, No. 8, pp. 171-186 February 18, 1920

A LIST OF OUGHTRED'S MATHEMATICAL SYMBOLS, WITH HISTORICAL NOTES

BY

FLORIAN CAJORI

There are two considerations which have induced the present writer to exhibit William Oughtred's symbols in a table: (1) The extraordinary emphasis laid by Oughtred upon the invention and use of mathematical symbols at a period when the notation of modern elementary algebra was in a state of formation, and (2) the inaccessibility of Oughtred's books to most mathematicians of the present time. By means of this table it is hoped that some of the desired information is placed within easier reach and that some of the erroneous statements relating to his notation may be avoided hereafter. Omitted from the list are all astronomical symbols used by Oughtred, also a few very specialized and tentative symbols in mechanics which seem devoid of historical interest.

The texts referred to are placed at the head of the table; the symbols, in the column at the extreme left. Each number in the table indicates the page of the text cited at the head of the column containing the number on which the symbol on the left occurs. Thus the notation 0|500, 5 occurs on page 3 of the *Clavis* of 1652. The page assigned is not always the first on which the symbol occurs in that volume. Some desirable information which could not be inserted in the table itself is found in the accompanying notes. The table affords an interesting view of the growth and changes in notation in Oughtred's works during half a century. The following is a key to the abbreviated book titles used in the table:

Clavis of 1631 = Arithmeticae in|numeris et speci-|ebvs institvtio:|Qvae tvm logisticae, tvm analyti|cae, atqve adeo|totivs mathematicae, qvasi|clavis|est.| — Ad nobilissimvm spe|-ctatissimumque invenem Dn. Gvilel|mvm Howard, Ordinis, qui dici|tur, Balnei Equitem, honoratissimi Dn.| Thomae, Comitis Arvndeliae & | Svrriae, Comitis Mareschal|li Angliae, &c. filium. — |Londini,| Apud Thomam Harpervm,| M. DC. XXXI.

Clavis of 1647 = The Key | of the | Mathematicks | New Forged and Filed:| Together with | A Treatise of the Resolution | of all kinde of Affected Aequa|tions in Numbers.| With the Rule of Compound Usury; And demonstration of the | Rule of false Position.| And a most easie Art of delineating all | manner of Plaine Sun-Dyalls. Geome|trically taught | By | Will. Oughtred.| London,| Printed by Tho. Harper, for Rich.| Whitaker, and are to be sold at his | shop in Pauls Church-yard 1647.

Clavis of 1648 = Clavis mathematica | denuò limata,| sive potius | fabricata.| Cui accedit | Tractatus de Resolutione Aequatio | num qualitercunque adfectarum | in numeris:| Et Declaratio tum decimi elementi Euclidis | de lateribus incommensurabilibus: tum | decimi tertii & decimi quarti ele|menti de quinque solidis | Regularibus.| Atque hic passim | Logisticae decimalis, & Logarithmorum | Doctrina intexitur.| Autore Gulielmo Oughtredo Anglo.| Londini,|Excudebat Thomas Harper, sumptibus Thomae | Whitakeri, apud quem venales sunt in | Cœmiterio D. Pauli. 1648.

Clavis of 1652 = Guilelmi Oughtred | AEtonensis,| quondam Collegii Regalis in Cantabrigia Socii, Clavis mathematicae | denvo limata,| Sive potius | fabricata.| Cum aliis quibusdam ejusdem | Commentationibus, quae in se|quenti pagina recensentur.|Editio tertia auctior & emendatior.| Oxoniae, | Excudebat Leon. Lichfield, Veneunt | apud Tho. Robinson. 1652.|

Clavis of 1667 = Guilelmi Oughtred | AEtonensis,| quondam Collegii Regalis | in Cantabrigia Socii,| Clavis mathematicae | Denvo Limata,| Sive potius | Fabricata.| Cum aliis quibusdam ejusdem | Commentationibus, quae in se|quenti pagina recensentur.| Editio quarta auctior & emendatior.| Oxoniae,| Typis Lichfieldianis, Acad. Typog.| Veneunt apud Joh. Crosley, &.| Amos Curteyne. 1667.|

Clavis of 1693 = Guilelmi Oughtred | AEtonensis,| Quondam Collegii Regalis in | Cantabrigia Socii,| Clavis Mathematicae | denuo Limata,| Sive potius | Fabricata.| Cum aliis quibusdam ejusdem | Commentationibus, quae in | sequenti pagina recensentur.| Editio Quinta auctior & emendatior.| Oxoniae,| Excudebat Leon. Lichfield.| Anno Dom. MDCXCIII.

A second impression of the fifth Latin edition appeared in 1698. The changes in the title-page come after the word "emendatior." In 1698 the closing part of the title-page is as follows:

> . . . emendatior.| Ex Recognitione D. Johannis Wallis, S.T.D.| Geometriae Professoris Saviliani.| Oxoniae:| Typis Leon. Lichfield: Impensis Tho. Leigh | ad Insigne Pavonis juxta Ecclesiam S. Dun-|stani, Lond. 1698.

Clavis of 1694 = Mr. William Oughtred's | Key | of the | Mathematicks.| Newly Translated from the Best Edition.| With Notes,| Rendring it Easie and Intelligble to | less Skilful Readers.| In which also,| Some Problems Left Unanswer'd by the Author are Resolv'd.| Absolutely necessary | For all Gagers, Surveyors, Gunners, Military-|Officers, Mariners, &c.| Recommended by Mr. E. Halley,| Fellow of the Royal Society.| London:| Printed for John Salusbury, at the | Rising-Sun in Cornhill. MDCXCIV.

An impression bearing the date 1702 is identical with the publication of 1694. No attempt was made to correct even the grossest oversights in proof reading. The title-page of 1702 is the only part set in type anew. Its wording is the same as in 1694, except for the part after the word "London;" in 1702 we read:

> . . . London;| Printed for Ralph Smith at the Bible | under the Piazza of the Royal-Exchange,| Cornhil. 1702.

We shall use the following abbreviations for the designation of tracts which were added to one or another of the different editions of the *Clavis Mathematicae:*

Eq. = De Aequationum affectarvm resolvtione in numeris.
Eu. = Elementi decimi Euclidis declaratio.
So. = De Solidis regularibus, tractatus
An. = De Anatocismo, sive usura composita.
Fa. = Regula falsæ positionis.
Ar. = Theorematum in libris Archimedis de sphaera & cylindro declaratio
Ho. = Horologia scioterica in plano, Geometricè delineandi modus.

List of the tracts added to the CLAVIS MATHEMATICAE of 1631 in its later editions, given in the order in which the tracts appear in each edition, the numbers in parentheses indicating the pagination:

Clavis of 1647: Eq. (121–169), An. (169–172), Fa. (173, 174), Ho. (1–30).

Clavis of 1648: Eq. (113–160), An. (161–164), Fa. (164, 165), Eu. (166–187), So. (188–207).

Clavis of 1652: Eq. (110–151), Eu. (1–22), So. (23–41), An. (42–44), Fa. (45, 46), Ar. (1–10), Ho. (1–41).

Clavis of 1667: Eq. (110–151), Eu. (1–21), So. (23–41), An. (42–44), Fa. (45, 46), Ar. (1–10), Ho. (1–41).

Clavis of 1693: Eq. (110–150), Eu. (1–20), So. (21–39), An. (40–42), Fa. (43, 44), Ar. (1–10), Ho. (1–43).

Clavis of 1694: Eq. (158–208).

I have seen a pamphlet, Oxford, 1662, containing the Eu., So., and Fa. I have seen also Ar. published as a separate pamphlet, Oxford, 1663.

Cir. of Prop. = The | Circle | of | Proportion,| and | The Horizontall | Instrument.| Both invented, and the vses of both | written in Latine by that learned Mathe-|matician Mr. W. O.| Bvt | Translated into English: and set forth for | the publique benefit by William Forster, louer | and practizer of the Mathematicall Sciences.| London | Printed by Avg. Mathewes,| dwelling in the Parsonage Court, neere | St. Brides. 1632.|

+ An Addition | vnto the Vse | of the Instrvment | called the circles of | Proportion, For the Working | of Nauticall Questions.| Together with certaine necessary Consi-| derations and Advertisements touching | Navigation.| All which, as also the former Rules concerning | this Instrument are to bee wrought not onely | Instrumentally, but with the penne, by Arith-|meticke, and the Canon of | Triangles.| Hereunto is also annexed the excellent Vse of two | Rulers for Calculation.| And is to follow after the 111 Page | of the first Part.| London, |Printed by Avgvstine Mathewes,| 1633.

In our tables "Ad." stands for *Addition vnto the Vse of the Instrument etc.*

Trigono. (Lat.) = Trigonometria:| Hoc est,| Modus computandi Triangulorum | Latera & Angulos, ex Canone | Mathematico traditus & demonstratus.| Collectus ex Chartis Clarissimi Domini | Willelmi Oughtred | AEtonensis.| Pèr | Richardum Stokesium Collegii Regalis in| Cantabrigia Socium.| et | Arthurum Haughton, Generosum.| Unâ cum Tabulis Sinuum, Tangent: & Secant, &c.| Londini,| Typis R. & L. W. Leybourn, Impensis Thomae Johnson, apud quem | vaeneunt sub signo Clavis Aureae in Coemeterio | S. Pauli, 1657.

In our tables, "Ca." stands for *Canones sinuum tangentium, secantium et logarithmorum pro sinubus et tangentibus,* which is the title for the tables in the above work.

Trigono. (Eng.) = Trigonometrie,| or,| The manner of calculating the Sides | and Angles of Triangles, by the | Mathematical Canon,| demonstrated.| By William Oughtred Etonens.| And published by Richard Stokes Fellow of Kings Colledge in | Cambridge, and Arthur Haughton Gentleman.| London,| Printed by R. and W. Leybourn, for Thomas Johnson at the | Golden Key in St. Pauls Church-yard.| M.DC.LVII.

Opu. Posth. = Guilelmi Oughtred | AEtonensis,| Quondam collegii Regalis | in | Cantabrigia Socii.| Opuscula Mathematica hactenus | inedita.| Oxonii,| E Theatro Sheldoniano,| Anno 1677.|

Opu. Posth. contains the following nine tracts:

Institutiones Mechanicae (pp. 1–54).
De variis corporum generibus gravitate & magnitudine comparatis (pp. 55–67).
Automata (pp. 68–86).
Quaestiones Diophanti Alexandrini Lib. 3. (pp. 87–129 .
De Triangulis planis rectangulis. (pp. 130–138).
De Divisione Superficierum. (pp. 139–153).
Musicae Elementa. (pp. 154–170).
De Propugnaculorum Munitionibus. (pp. 171–194).
Sectiones Angulares. (pp. 195–222+207–212, irregular pagination).

Ought. Explic. = Oughtredus explicatus,| sive | Commentarius | in | Ejus Clavem Mathematicam.| Cui additae sunt | Planetarum Observationes | & | Horologiorum Constructio.| Authore | Gilberto Clark.| Londini,| Typis Milonis Flesher; Veneunt apud Ric.| Davis, Bibliopolam Oxoniensem, 1682.|[1]

Symbols	Meanings of Symbols	Clavis Mathematicae							Circ. of Prop.[1] 1632, 1633	Trigono. (Latin) 1657	Opusc. Posth. 1677	Oughtr. Explic. 1682
		1631	1647	1648	1652	1667	1693	1694				
=	equal to	38	34	53	30	15	16	73	20	3	3	29
0⌊56	separatrix[2]	1	1	1	1	1	1	2	3	13	63	1
0.56	separatrix	....	....	....	....	....	....	....	....	235	....	....
.⌊56	separatrix[3]	....	....	....	....	....	....	17	....	....	....	5
0, 56	separatrix	....	....	....	....	....	....	....	....	....	221	....
0⌊500	same value as .5	....	....	3	3	3	3	....	....	....	....	2
a.b	ratio $a:b$, or $a-b$ [4]	5	8	7	12	7	7	25	7	3	3	27
2.314	separating[5]	....	....	....	....	....	....	....	4	235	....	....
2,314	the mantissa	....	Eq.136	158	150	113	113	175	....	....	....	....
$\overline{2}$.314	– characteristic	....	Eq.167	158	150	150	150	207	....	....	....	....
:	arithm. proportion[6]	....	....	22	21	21	21	32	....	....	....	....
:	a : b, ratio[7]	....	....	An.162	....	....	....	24	10	36	140	....
R.S	given ratio	21	28	33	32	25	25	49	19	....	87	42
::	geomet. proportion[8]	5	8	7	7	7	7	11	7	3	3	27
∺	contin. proportion	13	18	16	16	16	16	25	....	34	142	29
∵∴	contin. proportion	....	....	....	....	....	....	....	....	....	....	114
=	geom.[9] proportion	....	....	....	....	....	....	....	....	....	....	89
: :	( )[10]	45	57	107	104	52	53	149	96	....	101	....
:	(	40	58	99	92	56	92	119	35	32	101	75
: .	( )	....	115	106	104	104	104	95	....	....	102	53
:	( )	....	58	....	95	95	63	122	....	....	....	98
. :	( )	....	65	58	57	57	63	95	....	....	....	97
.	( )[11]	....	....	....	....	....	An. 42	97	....	....	....	116
. .	( )[11]	....	....	....	....	....	....	89	....	....	....	101
( )	( )[12]	....	....	....	....	....	....	....	....	....	....	81
∵	therefore	....	....	....	....	....	....	....	....	....	151	....
+	addition[13]	2	3	3	57	3	3	4	99	3	3	5
pl	addition	49	3	3	57	3	3	4	96	....	112	5
mo	addition[14]	....	....	....	....	....	....	4	....	....	....	....
–	subtraction	2	3	3	10	3	3	4	21	3	4	5
±	plus or minus	51	57	106	56	53	17	140	....	16	....	97
mi	subtraction	....	3	66	57	3	3	4	96	....	130	....
le	less[14]	....	....	....	....	....	....	4	....	....	....	....
$\overline{2}$	negative 2	....	1	9	1	1	5	16	....	....	....	8
×	multiplication[15]	7	10	10	10	10	10	13	37	32	143	....

Symbols	Meanings of Symbols	Clavis Mathematicae							Circ. of Prop. 1632, 1633	Trigono. (Latin) 1657	Opusc. Posth. 1677	Oughtr. Explic. 1682
		1631	1647	1648	1652	1667	1693	1694				
Hq bq	× by juxtaposition	7	11	10	11	10	37	13	....	5	87	17
in	multiplication[16]	7	10	10	10	10	10	....	96	....	219	59
$\frac{8}{4}$	fraction, division	8	12	11	11	11	11	23	21	16	9	5
a)b(c	b ÷ a = c	10	14	13	14	14	13	21	....	....	99	50
$\frac{4}{3}\rbrack\frac{3}{2}\lbrack\frac{9}{8}$	$\frac{3}{2} \div \frac{4}{3} = \frac{9}{8}$	....	....	....	....	....	....	....	....	....	156	....
Aq	AA	7	11	10	10	10	10	14	....	....	104	17
Ac	AAA	7	11	10	10	10	10	14	....	....	105	25
Aqq	AAAA	7	11	10	10	10	10	14	....	....	106	41
Aqc	AAAAA	7	11	10	10	10	10	55	....	....	....	67
Acc	AAAAAA	7	11	10	10	10	10	55	....	....	....	41
ABq	$\overline{AB}^2$ [17]	....	11	11	11	11	11	15	....	....	....	....
$\boxed{4}\cdot\cdot\boxed{10}$	4th . . 10th power	23	37	35	34	34	....	53	....	....	....	55
[4] · · [10]	4th . . 10th power	....	....	....	35	35	34	52	....	....	....	....
$a^2 \,.\,.\, a^7$	$a^2 \,.\,.\, a^7$	....	....	....	....	....	....	....	....	....	205	24
Q	qaesitum	....	17	16	16	16	16	25	....	....	....	....
Q	square[18]	38	33	31	57	30	30	47	28	5	100	75
Qu	square	....	....	....	....	....	....	....	....	....	....	105
C	cube	38	33	31	30	30	30	47	28	....	62	53
Cu	cube	....	136	128	123	123	123	175	....	....	....	....
QQ	4th power	45	33	61	30	30	30	47	....	....	210	....
QC	5th power	....	33	31	30	30	30	47	....	....	210	....
D	diameter	....	....	187	Eu. 21	Eu. 21	Eu. 20	....	37	....	....	....
L, l	latus, radix[19]	....	121	113	110	110	110	158	37	....	....	139
∠	angle	....	....	....	....	....	....	....	....	19	192	....
∠∠	angles	....	....	....	....	....	....	....	....	16	....	....
P	perimeter	....	....	....	....	....	....	....	37	....	....	....
P	ZA − Aq	41	....	....	....	....	....	....	....	....	....	....
R	radius	....	120	111	109	109	109	154	37	32	211	....
℞, R	remainder	....	152	134	126	128	142	....	....	....	....	45
R	rational	....	....	166	Eu. 1	Eu. 1	Eu. 1	....	....	....	....	....
Ω	superficies curva	....	....	....	Ar. I	Ar. 1	Ar. 1	....	....	....	....	....
√	root	....	33	31	30	30	30	47	....	....	102	....
√	square root	....	53	48	47	47	47	70	....	....	134	....
√q	square root	35	49	48	46	46	46	65	96	....	....	....
√b	latus binomii	....	33	31	30	30	30	47	....	....	....	....

Symbols	Meanings of Symbols	Clavis Mathematicae							Circ. of Prop. 1632, 1633	Trigono. (Latin) 1657	Opusc. Posth. 1677	Oughtr. Explic. 1682
		1631	1647	1648	1652	1667	1693	1694				
√r	latus residui	....	34	31	30	30	30	47	....	....	....	....
√u	sq. rt. of polyno.[20]	....	55	53	53	52	52	96	....	....	....	....
√qq	fourth root	35	52	47	46	46	48	69	....	....	....	....
√c	cube root	35	52	49	46	46	46	69	....	....	....	....
√qc	fifth root	35	49	47	46	46	46	65	....	....	....	....
√cc	sixth root	37	52	49	48	48	48	69	....	....	....	....
√ccc	ninth root	....	....	....	....	....	....	....	....	....	....	....
√cccc	twelfth root	37	52	49	48	48	48	69	....	....	....	....
√qu	square root	49	....	....	....	....	....	....	....	....	....	....
√[12] or √⌈12⌉	twelfth root	37	52	50	49	49	49	69	....	....	....	....
rq, rc	$\sqrt{\ }$, $\sqrt[3]{\ }$	....	....	....	....	....	....	....	....	....	....	73
r, ru	square root	....	....	....	....	....	....	....	....	....	....	74, 96
A, E	nos., $A > E$	21	33	31	30	30	30	47	....	....	87	53
Z	$A+E$	21	33	31	30	30	30	47	19	16	87	53
X	$A-E$	21	33	31	30	30	30	47	....	16	87	53
Z	A^2+E^2	41	33	31	30	30	30	47	....	....	98	54
X	A^2-E^2	41	33	31	30	30	30	47	....	....	99	54
Z	A^3+E^3	44	33	31	30	30	30	47	....	....	....	94
X	A^3-E^3	44	33	31	30	30	30	47	....	....	....	94
ʒ	$a+e$	....	....	167	Eu. 1	Eu. 1	Eu. I	....	....	....	....	....
ɛ	$a-e$	....	....	....	Eu. 1	Eu. 1	Eu. I	....	....	....	....	....
ʒ	a^2+b^2	....	....	167	Eu. 2	Eu. 2	Eu. I	....	....	....	....	....
ɛ	a^2-b^2	....	....	167	Eu. 2	Eu. 2	Eu. I	....	....	....	....	....
⊏‾	majus[21]	....	Ho. 17	166	145	Eu. 1	Eu. I	....	....	....	....	....
_⊐	minus	....	Ho. 17	166	Eu. 1	Eu. 1	Eu. I	....	....	....	....	....
⊏‾	non majus	....	....	166	Eu. 1	Eu. 1	Eu. I	....	....	....	....	....
_⊐	non minus	....	....	166	Eu. 1	Eu. 1	Eu. I	....	....	....	....	....
⊏_	minus[22]	....	Ho. 30	....	....	....	....	....	....	....	....	....
⊏‾	minus[22]	....	....	....	Ho. 31	Ho. 29	....	....	....	....	....	....
∵	major ratio	....	....	166	Eu. 1	Eu. 1	Eu. I	....	....	....	11	....
∴	minor ratio	....	....	166	Eu. 1	Eu. 1	Eu. I	....	....	....	6	....
<	less than[21]	....	....	....	....	....	....	....	....	....	4	....
>	greater than	....	....	....	....	....	....	....	....	....	4	....
⊓	commensurabilia	....	....	166	Eu. 1	Eu. 1	Eu. I	....	....	....	....	....
⊓	incommensurabilia	....	....	166	Eu. 1	Eu. 1	Eu. I	....	....	....	....	....

Symbols	Meanings of Symbols	Clavis Mathematicae 1631	1647	1648	1652	1667	1693	1694	Circ. of Prop. 1632, 1633	Trigono. (Latin) 1657	Opusc. Posth. 1677	Oughtr. Explic. 1682
[symbol]	commens. potentia	....	....	166	Eu. 1	Eu. 1	Eu. I	....	....	....	....	....
[symbol]	incommens. potentia	....	....	166	Eu. 1	Eu. 1	Eu. I	....	....	....	....	....
[symbol]	rationale	....	....	166	Eu. 1	Eu. 1	Eu. I	....	....	....	....	....
[symbol]	irrationale	....	....	166	Eu. 1	Eu. 1	Eu. I	....	....	....	....	....
[symbol]	medium	....	....	166	Eu. 1	Eu. 1	Eu. I	....	....	....	....	....
ς	line, cut extr. & mean ratio.	....	....	166	Eu. 1	Eu. 1	Eu. I	....	....	....	....	....
σ	major ejus portio	....	....	166	Eu. 1	Eu. 1	Eu. I	....	....	....	....	....
τ	minor ejus portio	....	....	166	Eu. 1	Eu. 1	Eu. I	....	....	....	....	....
sim	simile	....	....	166	Eu. 1	Eu. 1	Eu. I	....	....	33	....	....
[symbol]	proxime majus	....	....	166	Eu. 1	Eu. 1	Eu. I	....	....	....	....	....
[symbol]	proxime minus	....	....	166	Eu. 1	Eu. 1	Eu. I	....	....	....	....	....
[symbol]	aequale vel minus	....	....	166	Eu. 1	Eu. 1	Eu. I	....	....	....	....	....
[symbol]	aequale vel majus	....	....	166	Eu. 1	Eu. 1	Eu. I	....	....	....	....	....
▯	rectangulum	51	....	167	Eu. 2	Eu. 2	Eu. I	....	....	17	149	....
□	quadratum	....	....	167	Eu. 2	Eu. 2	Eu. I	....	....	....	....	....
△	triangulum	....	....	167	Eu. 2	Eu. 2	Eu. I	....	....	....	147	....
[symbol]	latus, radix	....	....	167	Eu. 2	Eu. 2	Eu. I	....	....	....	....	....
[symbol]	media proportion.	....	....	167	Eu. 2	Eu. 2	Eu. I	....	....	....	....	....
[symbol]	differentia[23]	....	....	....	Eu. 2	Eu. 2	Eu. I	....	....	....	....	....
‖	parallel	....	....	....	....	....	....	....	....	....	197	....
log	logarithm	....	172	158	150	150	122	207	....	17	....	....
log : Q:	log. of square	....	135	127	122	122	122	174	....	....	....	....
S	sine[24]	....	Ho. 29	....	....	....	....	....	96	5	172	....
t	tangent	....	Ho. 29	....	....	....	....	....	96	3	174	....
se	secant	....	....	....	....	....	....	....	....	14	....	....
sv	sinus versus	76	107	99	98	98	98	140	....	....	....	....
s ver	sinus versus[25]	....	....	....	....	....	....	....	....	5	....	....
sin : com	sine complement	....	....	....	....	....	....	....	Ad. 69	....	....	....
s co	cosine	....	....	....	....	....	....	....	96	3	174	....
t co	cotangent	....	....	....	....	....	....	....	96	3	....	....
se co	cosecant	....	....	....	....	....	....	....	....	4	....	....
sin	sine	....	....	....	Ho. 41	Ho. 41	Ho. 42	....	Ad. 69	35	....	37
tan	tangent	....	....	....	....	....	....	....	Ad. 69	Ca. 3	....	....
sec	secant	....	....	....	....	....	....	....	Ad. 41	....	....	....
sec:parall	sum of secants	....	....	....	....	....	....	....	Ad. 41	....	....	....

Symbols	Meanings of Symbols	Clavis Mathematicae 1631	1647	1648	1652	1667	1693	1694	Circ. of Prop. 1632, 1633	Trigono. (Latin) 1657	Opusc. Posth. 1677	Oughtr. Explic. 1682
tang	tangent	....	Ho. 29	....	Ho. 41	Ho. 41	Ho. 42	....	12	235	....	....
C	.01 of a degree	....	....	....	....	....	....	....	....	236	....	....
Cent	.01 of a degree	....	....	....	....	....	....	....	....	235	....	....
° ' ''	degr., min., sec.	....	21	20	21	20	21	32	66	....	....	36
Ho. ' ''	hours, min., sec.	....	....	....	....	....	....	....	67	....	....	....
,	$180^{\circ} -$ angle	....	....	....	....	....	....	....	....	2	....	....
$\bar{V}$	equal in no. of degr.	....	....	....	....	....	....	....	....	6	....	....
$\frac{\pi}{\delta}$	$\pi = 3.1416$	....	72	69	66	66	66	99	....	....	....	....
——	cancelled[26]	68	100	94	90	90	90	131	....	....	....	....
M	mean proportion	....	....	....	Ar. 1	Ar. 1	Ar. 1	....	....	....	....	....
m	minus	2	....	....	....	....	....	....	....	....	....	....
3 2 4 3 / 12 8 6 / 9	$\frac{3}{2} \times \frac{4}{3} = 2$, $\frac{3}{2} \div \frac{4}{3} = \frac{9}{8}$ [27]	20	32	30	29	29	29	45	....	....	....	....
Gr.	degree	....	20	19	Ho. 23	Ho. 23	19	29	....	235	....	....
min.	minute	....	....	....	....	....	....	....	Ad. 19	....	....	....
⅃⁄|	differentia	....	....	....	....	....	....	....	....	....	134	....
|—|	aequalia tempore	....	....	....	....	....	....	....	....	....	68	....
Lo	logarithm	....	....	....	....	....	....	....	....	Ca. 2	....	....
⌐|	separatrix	....	....	....	....	....	....	....	....	244	....	....
D	differentia	....	....	....	....	....	....	....	19	237	....	....
Tri, tri	triangle	....	76	191	Eu. 26	70	69	....	....	24	....	....
M	Cent. minute of arc	....	....	....	....	....	....	....	....	Ca. 2	....	....
X	multiplication[28]	....	....	....	....	....	....	....	....	5	101	16
Z cru	Z sum, X diff.	....	....	....	....	....	....	....	....	17	....	....
Z crur	of sides of	....	....	....	....	....	....	....	....	16	....	....
X cru	rectangle[29]	....	....	....	....	....	....	....	....	17	....	....
X crur	or triangle	....	....	....	....	....	....	....	....	16	....	....
A	unknown	38	53	51	50	50	50	72	....	....	113	84
L	altit. frust. of pyramid or cone	77	109	101	99	99	99	141	....	....	....	....
T	altit. of part cut off	77	109	101	99	99	99	142	....	....	....	....
α	first term (of progressions)	13	85, 18	80, 17	78, 16	78, 16	78, 16	116, 26	19	....	....	30, 116
ω	last term (of progressions)	....	85, 18	80, 17	78, 16	78, 16	78,16	116, 26	....	....	....	30, 116
T	no. of terms (of progressions)	....	85	80	78	78	78	116	....	....	....	116
X	common differ. (of progressions)	....	85	80	78	78	78	116	....	....	....	116
Z	sum of all terms (of progressions)	....	85, 18	80, 17	78, 16	78, 16	78, 16	116, 26	19	....	....	30, 116

HISTORICAL NOTES[2]

1. All the symbols, except "Log," which we saw in the 1660 edition of the *Circles of Proportion* are given in the editions of 1632 and 1633.

2. In the first half of the seventeenth century the notation for decimal fractions engaged the attention of mathematicians in England as it did elsewhere. In 1608 an English translation of Stevin's well-known tract was brought out, with some additions, in London by Robert Norton, under the title, *Disme: The Art of Tenths, or, De imall Arithmetike.* Stevin's notation is followed also by Henry Lyte in his *Art of Tens or Decimall Arithmetique, London*, 1619, and in *Johnsons Arithmetick*, 2.ed., London, 1633, where 3576.725 is written $3576|\overset{1}{7}\overset{2}{2}\overset{3}{5}$. William Purser in his *Compount Interest and Annuities*, London, 1634, p. 8, uses: as the separator, as did Adrianus Metius in his *Geometriae practicae pars I et II*, Lvgd. 1625, p. 149, and Rich. Balam in his *Algebra*, London, 1653, p. 4., Jonas Moore uses the inverted comma, 3‘571, in his *Arithmetick in two Books*, London, first part, p. 211. The decimal point first appears in John Napier's *Descriptio*, (Edward Wright's translation into English, London, 1616) and Napier's *Rabdologia*, Edinburgh, 1617. Oughtred's notation for decimals must have delayed the general adoption of the decimal point. E. Wells's *Elementa arithmetica universalis*, Oxford 1698, uses Oughtred's notation.

3. This mixture of the old and the new decimal notation occurs in the *Key* of 1694 (*Notes*) and in Gilbert Clark's *Oughtredus explicatus* only once; no reference is made to it in the table of errata of either book. In Oughtred's *Opuscula mathematica hactenus inedita*, the mixed notation $128,\underline{57}$ occurs on page 193 fourteen times. Oughtred's regular notation $128|\underline{57}$ hardly ever occurs in this book. We have seen similar mixed notations in the *Miscellanies: or Mathematical Lucubrations, of Mr. Samuel Foster, Sometime publike Professor of Astronomie in Gresham Colledge, in London* by John Twysden, London, 1659, page 13 of the "Observationes Eclipsium;" we find there $32.\underline{466}$, $31.\underline{008}$

4. The dot . used to indicate ratio is not, as claimed by some writers, used by Oughtred for *division.* Oughtred does not state in his book that. signifies division. We quote from an early and a late edition of the *Clavis.* He says in the *Clavis* of 1694, p. 45, and in the one of 1648, p. 30, "to continue ratios is to multiply them *as if they were* fractions." Fractions, as well as divisions, are indicated by a horizontal line. Nor does the statement from the *Clavis* of 1694, p. 20, and the edition of 1648, p. 12, "In Division, as the Divisor is to Unity, so is the Dividend to the Quotient" prove that he looked upon ratio as an indicated division. It does not do so any more than the sentence from the *Clavis* of 1694, and the one of 1648, p. 7, "In Multiplication, as 1 is to either of the factors, so is the other to the Product" proves that he considered ratio an indicated multiplication. Oughtred says (*Clavis* of 1694, p. 19, and the one of 1631, p. 8): 'If Two Numbers stand one above another with a Line drawn between them, 'tis as much as to say, that the upper is to be divided by the under; as $\frac{12}{4}$, and $\frac{5}{12}$."

In further confirmation of our view we quote from Oughtred's letter to W. Robinson:[3] "Division is wrought by setting the divisor under the dividend with a line between them."

---

[1] This is not a book written by Oughtred, but merely a commentary on the *Clavis.* Nevertheless, it seemed desirable to refer to its notation, which helps to show the changes then in progress.

[2] Note 1 refers to the *Circles of Proportion.* The other notes apply to the superscripts found in the column "Meanings of Symbols."

[3] Rigaud, *Correspondence of Scientific Men of the Seventeenth Century*, Vol. I, 1841, Letter VI, p. 8.

5. In Gilbert Clark's *Oughtredus explicatus* there is no mark whatever to separate the characteristic and mantissa. This is a step backward.

6. Oughtred's language (*Clavis* of 1652, p. 21) is: "Ut 7.4: 12.9 vel 7.7 −3: 12.12 −3. Arithmeticè proportionales sunt." As later in his work he does not use arithmetical proportion in symbolic analysis, it is not easy to decide whether the symbols just quoted were intended by Oughtred as part of his algebraic symbolism or merely as punctuation marks in ordinary writing. John Newton (*Institutio mathematica or, a mathematical institution*, London, 1654, p. 129) says: "As 8,5:6,3. Here 8 exceeds 5, as much as 6 exceeds 3."

John Wallis (*Mathesis universalis*, Oxonii, 1657, p. 229) says: "Et pariter 5,3; 11,9; 7,5; 19,17. sunt in eadem progressione arithmetica." In *Pavlini Chelvccii Institutiones anaylticae, editio post tertiam Romanam prima in Germania*, Viennae, 1761, p. 3, arithmetical proportion is indicated thus, 6.8 ∵ 10.12. Oughtred's notation is followed in the article "caractere" of the *Encyclopedie methodique* (Mathématiques) Paris, Liège, 1784. R. P. Bernard Lamy (*Elemens des mathematiques ou traite de la grandeur en general*, 3 éd., Amsterdam, 1692, p. 155) says: "Proportion arithmetique,

5,7 ∵ 10,12.

c'est à dire qu'il y a même difference entre 5 et 7, qu'entre 10 et 12."

In the *Nouveaux elemens de geometrie*, 2e éd., a La Haye, 1690, also in the edition issued at Paris, 1667, p. II., the same symbols are used for arithmetical progression as for geometrical progression, as in 7.3 :: 13.9 and 6.2 :: 12.4.

7. In the publications referred to in the table, of the years 1648 and 1694, the use of : to signify ratio has been found to occur only once in each copy; hence we are inclined to look upon this notation in these copies as printer's errors. W. W. Beman has pointed out in *L'Intermediaire des mathematiciens*, Vol. 9, 1902, p. 229, that Oughtred's Latin edition of his *Trigonometria*, 1657, contains in the explanations of the use of the tables, near the end, the use of : for *ratio*. It is improbable that the colon occurring in those tables was inserted by Oughtred himself.

In the *Trigonometria* proper, the colon does not occur and Oughtred's regular notation for ratio and proportion A.B :: C.D, is followed throughout. Moreover, in the English edition of Oughtred's trigonometry, printed in the same year, 1657, but subsequent to the Latin edition, the passage of the Latin edition containing the : is recast, the new notation for ratio is abandoned and Oughtred's regular notation introduced. The : used to designate ratio in Oughtred's *Opuscula mathematica hactenus inedita*, 1677, may have been introduced by the editor of the book.

We are able to show that the colon : was used to designate geometric ratio some years before 1657, by at least two authors, Vincent Wing the astronomer, and a schoolmaster who hides himself behind the initials "R.B." Wing wrote several works. In 1649 he published in London his *Urania Practica*, which, however, exhibits no special symbolism for proportion. But his *Harmonicon Coeleste*, London, 1651, contains many times Oughtred's notation A.B :: C.D, and many times also the new notation A : B :: C : D, the two notations being used interchangeably. Later there appeared from his pen, in London, three books in one volume, *Logistica astronomica* (1656), *Doctrina spherica* (1655) and *Doctrina theorica* (1655), each of which uses the notation, A : B :: C : D.

The book by the second author is entitled, *An Idea of Arithmetick at first designed for the use of the Free Schoole at Thurlow in Suffolk . . . by R.B., Schoolmaster there*, London, 1655. One finds in this text 1.6 :: 4.24 and also A : a :: C : c.

It is worthy of note also, that in a text entitled, *Johnsons Arithmetick; In two Bookes*, second edition, London, 1633, the colon : is used to designate a *fraction*. Thus $\frac{3}{4}$ is written 3 : 4. If a fraction be considered as an indicated division, then we have here the use of : for division at a period 51 years before Leibniz first employed it for that purpose. However, dissociated from the idea of a fraction, division is not designated by any symbol in Johnson's text. In dividing 8976 by 15 he writes the quotient 598 6 : 15.

8. Oughtred's notation A.B :: C.D, is the earliest serviceable symbolism for proportion. Before that proportions were either stated in words as was customary in rhetorical modes of

exposition, or else was expressed by writing the terms of the proportion in a line with dashes or dots to separate them. This practice was inadequate for the needs of the new symbolic algebra. Hence Oughtred's notation met with ready acceptance. It was used in Wingate's *Arithmetick made easie*, edited by John Kersey, London, 1650, p. 378, in Seth Ward's *In Ismaelis Bullialdi astr. philol. fundamenta inquisitio brevis*, Oxford, 1653, p. 7; in Isaac Barrow's *Euclidis data*, Cantabrigiae, 1657, p. 2; in John Wallis's *Operum mathematicorum pars altera* Oxonii, 1656, p. 181; in Samuel Foster's *Miscellanies* (posthumous), London, 1659, p. 1; in Jonas Moore's *Arithmetick in two Books*, London, 1660, second part, p. 61; in N. Mercator's *Logarithmotechnia*, London, 1668, p. 29; in Thomas Brancker's *Introduction to Algebra* (trans. of Rhonius), London, 1668, p. 37; in R. P. Bernard Lamy's *Elemens des mathematiques*, Amsterdam, 1692, p. 208. John Collins in his *Mariners Plain Scale New Plain'd*, London, 1659, and James Gregory in his *Appendicula ad veram circuli & hyperbolae quadraturam*, 1668, use :: for proportion and : for ratio. See Note 7.

9. We have seen this notation only once in this book, namely in the expression R.S. $=3.2$.

10. Oughtred says, *Clavis* of 1694, p. 47, in connection with the radical sign, "if the Power be included between two Points at both ends, it signifies the universal Root of all that Quantity so included; which is sometimes also signified by *b* and *r*, as the $\sqrt{}$b is the Binomial Root, the $\sqrt{}$r the Residual Root." This notation is in no edition strictly adhered to; the second : is often omitted when all the terms to the end of the polynomial are affected by the radical sign or by the sign for a power. In later editions still greater tendency to a departure from the original notation is evident. Sometimes one dot takes the place of the two dots at the end; sometimes the two end dots are given, but the first two are omitted; in a few instances one dot at both ends is used, or one dot at the beginning and no symbol at the end; however, these cases are very rare and are perhaps only printer's errors. We copy the following illustrations:

Q : A$-$E : est Aq$-$2AE$+$Eq, for $(A-E)^2=A^2-2AE+E^2$. From *Clavis of* 1631, p. 45.

$\frac{1}{2}$ BCq$\pm\sqrt{}$q : $\frac{1}{4}$ BCqq$-$CMqq. $=\left.\begin{matrix}\text{BAq}\\ \text{CAq}\end{matrix}\right\}$, for $\frac{1}{2}\,\overline{BC}^2\pm\sqrt{(\frac{1}{4}\overline{BC}^4-\overline{CM}^4)}=\overline{BA}^2$ or $\overline{CA}^2$. From *Clavis* of 1648, p. 106.

$\sqrt{}$q : BA$+$CA$=$BC$+$D, for $\sqrt{(BA+\overline{CA})}=BC+D$. From *Clavis* of 1631, p. 40.

$\frac{AB}{2}+\sqrt{}\text{q}\ \frac{ABq}{4}-\frac{C\times S}{R}: = A.$, for $\frac{\overline{AB}}{2}+\sqrt{\left(\frac{\overline{AB}^2}{4}-\frac{\overline{C\times S}}{R}\right)}=A$. From *Clavis* of 1652, p. 95.

Q.Hc$+$Ch : for $(Hc+Ch)^2$. From *Clavis* of 1652, p. 57.

Q.A$-$X$=$, for $(A-X)^2=$. From *Clavis* of 1694, p. 97.

$\frac{B}{2}+\text{r.u.}\ \frac{Bq}{4}-\text{CD}.=A$, for $\frac{B}{2}+\sqrt{\left(\frac{B^2}{4}-CD\right)}=A$. From *Oughtredus explicatus*, 1682, p. 101.

Oughtred's notation to designate aggregation did not originate with him. In a slightly different form it occurs in writings from the Netherlands. In 1603, *C. Dibvadii in geometriam Evclidis demonstratio numeralis*, Leyden, contains many expressions of this sort $\sqrt{}.136+\sqrt{}2048$, signifying $\sqrt{(136+\sqrt{2048})}$. The dot is used to indicate that the root of the binomial (not of 136 alone) is called for. This notation is used extensively in *Ludolphi à Cevlen de circvlo*, Leyden, 1619, and in *Willebrordi Snellii De circuli dimensione*, Leyden, 1621. In place of the single dot, Oughtred used the colon : , probably to avoid confusion with his notation for ratio. To avoid further possibility of uncertainty, he usually placed the colon both before and after the algebraic expression that was being aggregated.

We have noticed the use of the colon to express aggregation, in the manner of Oughtred, in the *Arithmetique made easie*, by Edmund Wingate, 2. edition by John Kersey, London, 1650, p. 387; in John Wallis' *Operum mathematicorum pars altera*, Oxonii, 1656, p. 186, as well as in the various parts of Wallis' *Treatise of Algebra*, London, 1685, and also in Jonas Moore's *Arith-*

metick in two Books, London, 1660, second part, p. 14. The 1630 edition of Wingate's book does not contain the part on algebra, nor the symbolism in question; these were probably added by John Kersey.

11. These notations to signify aggregation occur very seldom in the texts referred to and may be simply printer's errors.

12. Mathematical parentheses occur also on pages 75, 80 and 117 of G. Clark's *Oughtredus explicatus*.

Perhaps the earliest use of parentheses to indicate aggregation has been found by G. Eneström in a work of Tartaglia. In *Bibliotheca mathematica* S. 3, Vol. 7, 1906–1907, p. 296, Eneström says:

"Als Ergänzung einer früheren Notiz (BM. 6_3, 1905, S. 405) über Klammern als Zeichen der Zusammengehörigkeit verschiedener Ausdrücke zum Zwecke der Ausführung einer neuen Operation, bemerke ich, dass meines Wissens gewöhnliche (runde) Klammern für diesen Zweck zuerst von TARTAGLIA im 2. Bande (1556) seines *General trattato di numeri e misure* angewendet worden sind. Sehr oft kommen solche Klammern bei Wurzeln aus zusammengesetzten Ausdrücken vor (vgl. Bl. 167^b, 169^b, 170^b, 174^b, 177^a usw.); so z. B. drückt TARTAGLIA (Bl. 167^b) $\sqrt{\sqrt{28}-\sqrt{10}}$ auf folgende Weise aus: ℞ v. (℞ 28 men R 10); ℞ v. bedeutet „radiz universalis". Ausnahmsweise benu zt TARTAGLIA (Bl. 168^b, 169^a) die erste Klammernhälfte um zu bezeichnen, dass zwei vor einem Minuszeichen stehende Monome als ein einziger Term betrachtet werden sollen; so z. B. bedeutet (Bl. 168^b): men (22 men ℞ 6 nicht $-22-\sqrt{6}$ sondern $-(22-\sqrt{6})$. Dagegen hat TARTAGLIA meines Wissens die Klammern nie als Multiplikationszeichen gebraucht."

Parentheses had been used by Clavius, as appears from his *Operum mathematicorum tomus secundus*, Mogvntiae, M.DC.XI., algebra, p. 39, and by Albert Girard in his *Invention nouvelle en l'algebre*, Amsterdam, 1629, p. 17, but did not come into general use until after the time of Leibniz and the Bernoullis. However, they were used before the time of Leibniz in some elementary books, as for instance, in *La geometrie et practique generale d'icelle* par I. Errard de Bat-le-Duc, Ingenieur ordinaire de se Majesté, 3e éd., reveuë par D.H.P.E.M., Paris, 1619, p. 216; and in the *Novae geometriae clavis algebra*, authore P. Jacobo de Billy, Paris, 1643, p. 157; as also in an *Abridgement of the Precepts of Algebra, Written in French by James de Billy*, London, 1659, p. 346, and in the *Miscellanies: or Mathematical Lucrubations, of Mr. Samuel Foster, Sometime publike Professor of Astronomie in Gresham Colledge in London*, London, 1659, p. 7.

13. In the *Clavis* of 1631, p. 2, it says "Signum additionis siue affirmationis, est + plus" and "Signum subductionis, siue negationis est − minus." In the edition of 1694 it says simply "The Sign of Addition is + more" and "The Sign of Subtraction is − less," thereby ignoring, in the definition, the double function played by these symbols.

14. In the "Errata," following the preface of the 1694 edition it says, for "*more* or *mo.* r. [ead] *plus* or *pl.*", for *less* or *le.* r.[ead] *minus* or *mi.*"

15. Oughtred's *Clavis mathematicae* of 1631 is not the first appearance of × as a symbol for multiplication. In Edward Wright's translation of John Napier's *Descriptio*, entitled, *A Description of the Admirable Table of Logarithms*, London, 1618, the letter X is given as the sign of multiplication in the part of the book called "An Appendix to the Logarithmes, shewing the practise of the calculation of Triangles, etc." The authorship of this "Appendix" is not indicated. It has been surmised by Augustus De Morgan, as Dr. J. W. L. Glaisher has informed me, that this "Appendix" was written by Oughtred. The fact that it contains this symbol for multiplication and the then unusual word "cathetus" for the perpendicular leg of a right triangle (a term occurring in Oughtred's later works) are named by Glaisher in support of this conjecture. This first printed appearance of the symbol hardly explains its real origin. The author of the "Appendix" may have written the St. Andrew's Cross ×, while the printer may have substituted the letter X as the nearest available type for it. Studies as to the origin of × have been made by C. Le Paige, "Sur l'origine de certains signes d'operation," in *Annales de la societe scientifique de*

Bruxelles, seizième année, 1891–1892, second partie, pp. 79–82, and Gravelaar, "Over den oorsprong van ons maalteeken (×)" in *Wiskundig Tijdschrift*, 6e année. In 1551 E. O. Schreckenfuss placed the cross between two factors, one above the other.

The use of the letters x and X for multiplication is not uncommon during the seventeenth and beginning of the eighteenth centuries. We note the following instances: Vincent Wing, *Doctrina theorica*, London, 1656, p. 63; John Wallis, *Arithmetica infinitorum*, Oxford, 1655, pp. 115, 172; *Moore's Arithmetick in two Books*, by Jonas Moore, London, 1660, p. 108; the *Nouveaux elemens de geometrie*, Paris, 1667 ["par M.D.M.G.B."], p. 6; Lord Brounker in *Philosophical Transactions*, II, p. 466 (London, 1668); *Exercitatio geometrica, auctore Laurentio Lorenzinio, Vincentii Viviani discipulo*, Florence, 1721. The *Dioptrica nova* by William Molyneux of Dublin, London, 1692, uses * as the symbol of multiplication. John Wallis used the ⋈ in his *Elenchus geometriae Hobbianae*, Oxoniae, 1655, p. 23.

16. *in* as a symbol of multiplication carries with it also a *collective* meaning: For example, the *Clavis* of 1652 has on page 77, "Erit $\frac{1}{2}Z+\frac{1}{2}B$ in $\frac{1}{2}Z-\frac{1}{2}B=\frac{1}{4}Zq-\frac{1}{4}Bq$".

17. That is, the line AB squared.

18. These capital letters *precede* the expression to be raised to a power. Seldom are they used to indicate powers of monomials. From the *Clavis* of 1652, p. 65 we quote:

"Q : A+E : +Eq=2Q : $\frac{1}{2}$A+E : +2Q.$\frac{1}{2}$A."

i.e., $(A+E)^2+E^2=2(\frac{1}{2}A+E)^2+2\left(\frac{A}{2}\right)^2$

19. L and *l* stand for the same thing: *side* or *root*, *l* being used usually when the coefficients of the unknown quantity are given in Hindu-Arabic numerals, so that all the letters in the equation, viz., *l*, *q*, *c*, *qq*, *qc*, etc. are small letters. The *Clavis* of 1694, p. 158, uses L in a place where the Latin editions use *l*.

20. The symbol √u does not occur in the *Clavis* of 1631 and is not defined in the later editions. The following throws light upon its significance. In the 1631 edition, chapt. XVI, sect. 8, p. 40, the author takes √q BA+B=CA, gets from it √q BA=CA−B, then squares both sides and solves for the unknown A. He passes next to a radical involving *two* terms, and says:
"Item √q vniuers : BA+CA : −D=BC : vel per transpositionem √q : BA+CA=BC+D;" he squares both sides and solves for A. In the later editions he writes "√u" in place of "√q vniuers : " The *u* evidently stands or "vniuers," a word meaning *collective;* i.e. √u means the square root of all terms taken collectively.

21. Harriot's symbols > for greater and < for less were far superior to the corresponding symbols used by Oughtred. While Harriot's symbols are symmetric to a horizontal axis and asymmetric only to a vertical, Oughtred's symbols are asymmetric to both axes and therefore harder to remember. Indeed some confusion in their use occurred in Oughtred's own works, as is shown in the table. Isaac Barrow used ⊏ for majus and ⊐ for minus in his *Euclidis Data*, Cantabrigiae, 1657, p. 1. Seth Ward, a pupil of Oughtred, writes in his *In Ismaelis Bullialdi astronomiae philolaicae fundamenta inquisitio brevis*, Oxoniae, 1653, p. 1, ⊏ for majus and ⊏ for minus. Another pupil, John Wallis, writes in his *Treatise of Algebra*, London, 1685, p. 127 ⊐ for majus and ⊐ for minus. For further notices of discrepancy in the use of these symbols, see *Bibliotheca Mathematica*, 12_3, 1911–12, p. 64. Harriot's > and < easily won out over Oughtred's notation. Wallis follows Harriot almost exclusively; so does Thomas Gibson in his *Syntaxis mathematica*, London, 1655, p. 20; as does also Thomas Brancker in his *Introduction to Algebra* (translation of Rhonius' algebra), London, 1668, p. 76.

22. This notation for "less than" in the Ho. occurs only in the explanation of "Fig. EE" In the text (Chapt. IX) the regular notation explained in Eu. is used.

23. The symbol ↘ so closely resembles the symbol ∽ which was used by John Wallis in his *Operum mathematicorum pars prima*, Oxford, 1657, pp. 208, 247, 334, 335, that the two symbols were probably intended to be one and the same. It is difficult to assign a good reason why Wallis

who greatly admired Oughtred and was editor of the later Latin editions of his *Clavis mathematicae* should purposely reject Oughtred's ∽ and intentionally introduce ∾ as a substitute symbol. In fact, the symbol ∾ occurs twice, in the place of ∽, in the *Clavis* of 1698 (edited by Wallis), page 18 of the *Elementi decimi Euclidis declaratio,* while in the list o: symbols collected at the beginning of this *Declaratio* Oughtred's usual ∾ is given.

24. Von Braunmühl, in his *Geschichte der Trigonometrie,* 2. Teil, Leipzig, 1903, pp. 42, 91, refers to Oughtred's *Trigonometria* of 1657 as containing the earliest use of abbreviations of trigonometric functions and points out that half a century later the army of writers on trigonometry had hardly yet reached the standard set by Oughtred. This statement must be supplemented in several respects. Our table shows that Oughtred used such abbreviations a quarter of a century before the publication of his *Trigonometria.* Moreover, as early as 1624, the contractions *sin* and *tan* appear on the drawing representing Gunter's scale, but Gunter did not use them in his book, the *Description and Vse of the Sector, the Crosse-staffe and other Instruments,* London, 1624, except in the drawings of his scale, as in "the second book," p. 31. A competitor for the honour of first using abbreviations of the trigonometric functions is Richard Norwood, who, in his *Trigonometrie,* London, 1631, First Book, p. 20, and also in the second edition of it, 1651, p. 18, declares that "in these examples *s* stands for *sine*: *t* for *tangent*: *sc* for *sine complement*: *tc* for *tangent complement*: *sec*: for *secant.*" Another trigonometry that was published before Oughtred's and uses abbreviations is the *Idea Trigonometriae demonstratae,* Oxoniae, 1654, by Seth Ward, Savilian Professor of Astronomy at Oxford and at one time a pupil of Oughtred. On page 1 Ward writes:

s.	Sinus		Complementum
S'	Sinus Complementi	Z	Summa
t.	Tangens	X	Differentia
𝒯	Tangens Complementi	Cr	Crura

Anthony Wood, in his *Athenae Oxonienses* (Ed. Bliss) Vol. IV. London, 1820, p. 249, says in connection with Ward's trigonometry: "The method of which, mention'd in the preface to this book, Mr. Oughtred challenged for his."

Vincent Wing, in his *Logistica as'ronomica,* London, 1656, p. 8, uses *s* and *t* for "sine" and "tangent," and in his *Doctrina spherica,* London, 1655, p. 28, he uses *cs* for "cosine." John Ca well writes in his *Trigonometry* at the end of John Wallis's *Algebra,* London, 1685, *s* for sine Σ for cosine, ∫ for secant *T* for tangent, and 𝒯 for cotangent. For example, p. 3,

$\Sigma_{,}$A·ΣB :: ∫,B·∫,A/stands for cos A : cos B = sec B : sec A.

Mention should be made of abbreviations used even earlier than any of the preceding, given in "An Appendiz to the Logarithmes, shewing the practice of the Calculation of Triangles, etc." in Edward Wright's edition of Napier's *A Description of the Admirable Table of Logarithmes,* London, 1618, where we read on p. (4): "For the Logarithme of an arch or an angle, I set before it (s) for the antilogarithme or complement thereof (s*) and for the Differential (t)." In further explanation of this rather unsatisfactory passage, the author (Oughtred?) says, "As for example: s B+BC=CA. that is, the Logarithme of an angle B. at the Base of a plain right-angled triangle, increased by the addition of the Logarithm of BC. the hypothenuse thereof, is equall to the Logarithme of CA the cathetus." Here "logarithme of an angle B" evidently means "log. sin B," just as with Napier, "Logarithms of the arcs" signify really "Logarithms of the sines of the angles." In Napier's table, the numbers in the column marked "Differentiae" signify *og. sine* minus *log cosine,* that is the logarithms of the *tangents*; this explains the contraction (t) in the "Appendix," the notation (s*) for *cosine* resembles somewhat Seth Ward's notation. In view of the fact that Ward was an early pupil of Oughtred, we have here a new argument in favor of the hypothesis mentioned in Note 15 that Oughtred is the author of the "Appendix" mentioned above. Observe also that in the *Trigonometria* p. 2 Oughtred indicates 180^0- angle, by (').

25. This reference is to the *English* edition, the *Trigonometrie* of 1657. In the Latin edition there is printed on page 5, by mistake, *s* instead of *s versus*. The table of errata makes reference to this misprint.

26. The horizontal line was printed beneath the expression that was being crossed out. Thus, on page 68 of the *Clavis* of 1631 there is:

B Gqq$-\underline{B\ Gq\times 2\ BK\times BD}+BKq\times$BDq$=BGq\times$BDq$+$
BGq$\times$BKq$-\underline{BGq\times 2BK\times BD}+BGq\times$4 CAq.

27. This notation, says Oughtred, was used by ancient writers on music, who "are wont to connect the terms of ratios, either to be continued" as in $\frac{3}{2}\times\frac{4}{3}=2$, "or diminish'd" as in $\frac{3}{2}\div\frac{4}{3}=\frac{9}{8}$.

28. *See* Note 15.

29. *Cru* and *crur* are abbreviations for *crurum*, side of a rectangle or right triangle. Hence *Z cru* means the sum of the sides, *X cru*, the difference of the sides.

January 18, 1915.

UNIVERSITY OF CALIFORNIA PUBLICATIONS
IN
MATHEMATICS
Vol. 1, No. 9, pp. 187-209 February 17, 1920

ON THE HISTORY OF GUNTER'S SCALE AND THE SLIDE RULE DURING THE SEVENTEENTH CENTURY

BY

FLORIAN CAJORI

TABLE OF CONTENTS

I. INTRODUCTION

In my history of the slide rule[1], and my article on its invention[2] it is shewn that William Oughtred and not Edmund Wingate is the inventor, that Oughtred's circular rule was described in print in 1632, his rectilinear rule in 1633. Richard Delamain is referred to as having tried to appropriate the invention to himself[3] and as having written a scurrilous pamphlet against Oughtred. All our information

[1] F. Cajori, *History of the Logarithmic Slide Rule and Allied Instruments*, New York, 1909, pp. 7–14, also Addenda i–vi.

[2] F. Cajori, "On the Invention of the Slide Rule," in *Colorado College Publication*, Engineering Series Vol. 1, 1910. An abstract of this is given in *Nature* (London), Vol. 82, 1909, p. 267.

[3] F. Cajori, *History* etc., p. 14.

about Delamain was taken from De Morgan,[4] who, however, gives no evidence of having read any of Delamain's writings on the slide rule. Through Dr. Arthur Hutchinson of Pembroke College, Cambridge, I learned that Delamain's writings on the slide rule were available. In this article will be given: First, some details of the changes introduced during the seventeenth century in the design of Gunter's scale by Edmund Wingate, Milbourn, Thomas Brown, John Brown and William Leybourn; second, an account of Delamain's book of 1630 on the slide rule which antedates Oughtred's first *publication* (though Oughtred's date of *invention* is earlier than the date of Delamain's alleged invention) and of Delamain's later designs of slide rules; third, an account of the controversy between Delamain and Oughtred; fourth, an account of a later book on the slide rule written by William Oughtred, and of other seventeenth century books on the slide rule.

II. INNOVATIONS IN GUNTER'S SCALE

We begin with Anthony Wood's account of Wingate's introduction of Gunter's scale into France.[5]

> In 1624 he transported into France the rule of proportion, having a little before been invented by Edm. Gunter of Gresham Coll. and communicated it to most of the chiefest mathematicians then residing in Paris: who apprehending the great benefit that might accrue thereby, importun'd him to express the use thereof in the French tongue. Which being performed accordingly, he was advised by monsieur Alleawne the King's chief engineer to dedicate his book to monsieur the King's only brother, since duke of Orleans. Nevertheless the said work coming forth as an abortive (the publishing thereof being somewhat hastened, by reason an advocate of Dijon in Burgundy began to print some uses thereof, which Wingate had in a friendly way communicated to him) especially in regard Gunter himself had learnedly explained its use in a far larger volume.[6]

Gunter's scale, which Wingate calls the "rule of proportion," contained, as described in the French edition of 1624, four lines: (1) A single line of numbers; (2) a line of tangents; (3) a line of sines; (4) a line, one foot in length, divided into 12 inches and tenths of inches, also a line, one foot in length, divided into tenths and hundredths.

---

[4] Art. "Slide Rule" in the *Penny Cyclopaedia* and in the *English Cyclopaedia* [Arts and Sciences].

[5] Anthony Wood, *Athenae oxonienses* (Ed. P. Bliss), London, Vol. III, 1817, p. 423.

[6] The full title of the book which Wingate published on this subject in Paris is as follows:

L'Vsage|de la|Reigle de|Proportion|en l'Arithmetique &|Geometrie.|Par Edmond Vvingate,| Gentil-homme Anglois.|

E *ἀυῆsφιλεμα τῆs, ἔση ηολυμα τῆs.*

In tenui, sed nõ tenuis v∫u∫ve, laborne.|

A Paris,|Chez Melchior Mondiere,|demeurant en l'Isle du Palais,|à la|ruë de Harlay aux deux Viperes.|M. DC. XXIV.|Auec Priuilege du Roy.|

Back of the title page is the announcement:

> Notez que la Reigle de Proportion en toutes façons se vend à Paris chez Melchior Tauernier, Graueur & Imprimeur du Roy pour les Tailles douces, demeurant en l'Isle du Palais sur le Quay qui regarde la Megisserie à l'Espic d'or.

The English editions of this book which appeared in 1626 and 1628 are devoid of interest. The editions of 1645 and 1658 contain an important innovation.[7] In the preface the reasons why this instrument has not been used more are stated to be: (1) the difficulty of drawing the lines with exactness, (2) the trouble of working thereupon by reason (sometimes) of too large an extent of the compasses, (3) the fact that the instrument is not readily portable. The drawing of Wingate's arrangement of the scale in the editions of 1645 and 1658 is about 66 cm. (26.5 in.) long. It contains five parallel lines, about 66 cm. long, each having the divisions of one line marked on one side and of another line on the other side. Thus each line carries two graduations: (1) A single logarithmic line of numbers; (2) a logarithmic line of numbers thrice repeated; (3) the first scale repeated, but beginning with the graduations which are near the middle of the first scale, so that its graduation reads 4, 5, 6, 7, 8, 9, 1, 2, 3; (4) a logarithmic line of numbers twice repeated; (5) a logarithmic line of tangents; (6) a logarithmic line of sines; (7) the rule divided into 1000 equal parts; (8) the scale of latitudes; (9) a line of inches and tenths of inches; (10) a scale consisting of three kinds, viz., a gauge line, a line of chords, and a foot measure, divided into 1000 equal parts.

Important are the first and second scales, by which cube root extraction was possible "by inspection only, without the aid of pen or compass;" similarly the third and fourth scales, for square roots. This innovation is due to Wingate. The 1645 edition announces that the instrument was made in brass by Elias Allen, and in wood by John Thompson and Anthony Thompson in Hosier Lane.

William Leybourn, in his *The Line of Proportion or Numbers, Commonly called Gunter's Line, Made Easie*, London, 1673, says in his preface "To the Reader:"

> The Line of Proportion or Numbers, commonly called (by Artificers) Gunter's Line, hath been discoursed of by several persons, and variously applied to divers uses; for when Mr. Gunter had brought it from the Tables to a Line, and written some Uses thereof, Mr. Wingate added divers Lines of several lengths, thereby to extract the Square or Cube Roots, without doubling or trebling the distance of the Compasses: After him Mr. Milbourn, a Yorkshire Gentleman, disposed it in a Serpentine or Spiral Line, thereby enlarging the divisions of the Line.

On pages 127 and 128 Leybourn adds:

> Again, One T. Browne, a Maker of Mathematical Instruments, made it in a Serpentine or Spiral Line, composed of divers Concentrick Circles, thereby to enlarg the divisions, which was the contrivance of one Mr. Milburn a Yorkshire Gentleman, who writ thereof, and communicated his Uses to the aforesaid Brown, who (since his death) attributed it to himself: But whoever was the contriver of it, it is not without inconvenience; for it can in no wise be made portable; and besides (instead of compasses) an opening Joynt with thirds [threads] must be placed to move upon the Centre of the Instrument, without which no proportion can be wrought.

---

[7] The title-page of the edition of 1658 is as follows:

> The Use of the Rule of Proportion in Arithmetick & Geometrie. First published at Paris in the French tongue, and dedicated to Monsieur, the then king's onely Brother (now Duke of Orleance). By Edm. Wingate, an English Gent. And now translated into English by the Author. Whereinto is now also inserted the Construction of the same Rule, & a farther use thereof . . . 2nd edition inlarged and amended. London, 1658.

This Mr. Milburn is probably the person named in the diary of the antiquarian, Elias Ashmole, on August 13 [1646?]; "I bought of Mr. Milbourn all his Books and Mathematical Instruments."[8] Charles Hutton[9] says that Milburne of Yorkshire designed the spiral form about 1650. This date is doubtless wrong, for Thomas Browne who, according to Leybourn, got the spiral form of line from Milbourn, is repeatedly mentioned by William Oughtred in his *Epistle*[10] printed some time in 1632 or 1633. Oughtred does not mention Milbourn, and says (page 4) that the spiral form "was first hit upon by one Thomas Browne a Joyner, . . . the serpentine revolution being but two true semicircles described on severall centers."[11]

Thomas Brown did not publish any description of his instrument, but his son, John Brown, published in 1661 a small book,[12] in which he says (preface) that he had done "as Mr. Oughtred with Gunter's Rule, to a sliding and circular form; and as my father Thomas Brown into a Serpentine form; or as Mr. Windgate in his *Rule of Proportion.*" He says also that "this brief touch of the Serpentine-line I made bold to assert, to see if I could draw out a performance of that promise, that hath been so long unperformed by the promisers thereof." Accordingly in Chapter XX he gives a description of the serpentine line, "contrived in five (or rather 15) turn." Whether this description, printed in 1661, exactly fits the instrument as it was developed in 1632, we have no means of knowing. John Brown says:

1. First next the center is two circles divided one into 60, the other into 100 parts, for the reducing of minutes to 100 parts, and the contrary.

2. You have in seven turnes two inpricks, and five in divisions, the first Radius of the sines (or Tangents being neer the matter, alike to the first three degrees,) ending at 5 degrees and 44 minutes.

3. Thirdly, you have in 5 turns the lines of numbers, sines, Tangents, in three margents in divisions, and the line of versed sines in pricks, under the line of Tangents, according to Mr. *Gunter's* cross-staff: the sines and Tangents beginning at 5 degrees, and 44 minutes where

---

[8] *Memories of the Life of that Learned Antiquary, Elias Ashmole, Esq.; Drawn up by himself by way of Diary. With Appendix of original Letters.* Publish'd by Charles Burman, Esq., London, 1717, p. 23.

[9] *Mathematical Tables*, 1811, p. 36, and art. "Gunter's Line" in his *Phil. and Math. Dictionary*, London, 1815.

[10] *To the English Gentrie, and all others studious of the Mathematicks, which shall bee readers hereof. The just Apologie of Wil: Ovghtred, against the slaunderous insimulations of Richard Delamain, in a Pamphlet called Grammelogia, or the Mathematicall Ring, or Mirifica logarithmorum projectio circularis.* We shall refer to this document as *Epistle.* It was published without date in 32 unnumbered pages of fine print, and was bound in with Oughtred's *Circles of Proportion,* in the editions of 1633 and 1639. In the 1633 edition it is inserted at the end of the volume just after the *Addition vnto the Vse of the Instrument etc.*, and in that of 1639 immediately after the preface. It was omitted from the Oxford edition of 1660. The *Epistle* was also published separately. There is a separate copy in the British Museum, London. Aubrey, in his *Brief Lives*, edited by A. Clark, Vol. II, Oxford, 1898, p. 113, says quaintly, "He writt a stitch't pamphlet about 163(?4) against . . . Delamaine."

[11] Thomas Browne is mentioned by Stone in his *Mathematical Instruments*, London 1723, p. 16. See also Cajori, *History of the Slide Rule*, New York, 1909, p. 15.

[12] *The Description and Use of a Joynt-Rule: . . . also the use of Mr. White's Rule for measuring of Board and Timber, round and square; With the manner of Vsing the Serpentine-line of Numbers, Sines, Tangents, and Versed Sines.* By J. Brown, Philom., London, 1661.

the other ended, and proceeding to 90 in the sines, and 45 in the Tangents. And the line of numbers beginning at 10, and proceeding to 100, being one entire Radius, and graduated into as many divisions as the largeness of the instrument will admit, being 10 to 10 50 into 50 parts, and from 50 to 100 into 20 parts in one unit of increase, but the Tangents are divided into single minutes from the beginning to the end, both in the first, second and third Radiusses, and the sines into minutes; also from 30 minutes to 40 degrees, and from 40 to 60, into every two minutes, and from 60 to 80 in every 5th minute, and from 80 to 85 every 10th, and the rest as many as can be well discovered.

The versed sines are set after the manner of Mr. *Gunter's* Cross-staff, and divided into every 10th minutes beginning at 0, and proceeding to 156 going backwards under the line of Tangents.

4. Fourthly, beyond the Tangent of 45 in one single line, for one Turn is the secants to 51 degrees, being nothing else but the sines reitterated beyond 90.

5. Fifthly, you have the line of Tangents beyond 45, in 5 turnes to 85 degrees, whereby all trouble of backward working is avoided.

6. Sixthly, you have in one circle the 180 degrees of a Semicircle, and also a line of natural sines, for finding of differences in sines, for finding hour and Azimuth.

7. Seventhly, next the verge or outermost edge is a line of equal parts to get the Logarithm of any number, or the Logarithm sine and Tangent of any ark or angle to four figures besides the carracteristick.

8. Eightly and lastly, in the space place between the ending of the middle five turnes, and one half of the circle are three prickt lines fitted for reduction. The uppermost being for shillings, pence and farthings. The next for pounds, and ounces, and quarters of small *Averdupoies* weight. The last for pounds, shillings and pence, and to be used thus: If you would reduce 16s. 3d. 2q. to a decimal fraction, lay the hair or edge of one of the legs of the index on 16. 3½ in the line of 1. s. d. and the hair shall cut on the equal parts 81 16; and the contrary, if you have a decimal fraction, and would reduce it to a proper fraction, the like may you do for shillings, and pence, and pounds, and ounces.

The uses of the lines follow.

As to the use of these lines, I shall in this place say but little, and that for two reasons. First, because this instrument is so contrived, that the use is sooner learned then any other, I speak as to the manner, and way of using it, because by means of first second and third radiusses, in sines and Tangents, the work is always right on, one way or other, according to the Canon whatsoever it be, in any book that treats of the Logarithms, as *Gunter, Wells, Oughtred, Norwood*, or others, as in *Oughtred* from page 64 to 107.

Secondly, and more especially, because the more accurate, and large handling thereof is more then promised, if not already performed by more abler pens, and a large manuscript thereof by my *Sires* meanes, provided many years ago, though to this day not extant in print; so for his sake I claiming my interest therein, make bold to present you with these few lines, in order to the use of them: And first note,

1. Which soever of the two legs is set to the first term in the question, that I call the first leg always, and the other being set to the second term, I call the second leg . . .

The exact nature of the contrivance with the "two legs" is not described, but it was probably a flat pair of compasses, attached to the metallic surface on which the serpentine line was drawn. In that case the instrument was a slide rule, rather than a form of Gunter's line. In his publication of 1661, as also in later publications,[13] John Brown devoted more space to Gunter's scales, requiring the use of a separate pair of compasses, than to slide rules.

The same remark applies to William Leybourn who, after speaking of Seth Partridge's slide rule, returns to forms of Gunter's scale, saying:[14]

> There is yet another way of disposing of this Line of Proportion, by having one Line of the full length of the Ruler, and another Line of the same Radius broken in two parts between 3 and 4; so that in working your Compasses never go off of the Line: This is one of the best contrivances, but here Compasses must be used. These are all the Contrivances that I have hitherto seen of these Lines: That which I here speak of, and will shew how to use, is only two Lines of one and the same Radius, being set upon a plain Ruler of any length (the larger the better) having the beginning of one Line, at the end of the other, the divisions of each Line being set so close together, that if you find any number upon one of the Lines, you may easily see what number stands against it on the other Line. This is all the Variation. . . .
>
> *Example 1. If a Board be 1 Foot 64 parts broad, how much in length of that Board will make a Foot Square?* Look upon one of your Lines (it matters not which) for 1 Foot 64 parts, and right against it on the other Line you shall find 61; and so many parts of a Foot will make a Foot square of that Board.

This contrivance solves the equation $1.64x=1$, yielding centesimal parts of a foot.

James Atkinson[15] speaks of "Gunter's scale" as "usually of Boxwood . . . commonly 2 ft. long, $1\frac{1}{2}$ inch broad" and "of two kinds: *long Gunter* or *single Gunter*, and the *sliding Gunter*. It appears that during the seventeenth century (and long after) the Gunter's scale was a rival of the slide rule.

III. RICHARD DELAMAIN'S GRAMMELOGIA

We begin with a brief statement of the relations between Oughtred and Delamain. At one time Delamain, a teacher of mathematics in London, was assisted by Oughtred in his mathematical studies. In 1630 Delamain published the *Grammelogia*, a pamphlet describing a circular slide rule and its use. In 1631 he published another tract, on the *Horizontall Quadrant*.[16] In 1632 appeared Oughtred's *Circles of Proportion*[17] translated into English from Oughtred's Latin manuscript by another pupil, William Forster, in the preface of which Forster makes

[13] *A Collection of Centers and Useful Proportions on the Line of Numbers*, by John Brown, 1662(?), 16 pages; *Description and Use of the Triangular Quadrant*, by John Brown, London, 1671; *Wingate's Rule of Proportion in Arithmetick and Geometry: or Gunter's Line. Newly rectified by Mr. Brown and Mr. Atkinson, Teachers of the Mathematicks*, London, 1683; *The Description and Use of the Carpenter's-Rule: Together with the Use of the Line of Numbers commonly call'd Gunter's-Line*, by John Brown, London, 1704.

[14] William Leybourn, *op. cit.*, pp. 129, 130, 132, 133.

[15] James Atkinson's edition of Andrew Wakely's *The Mariners Compass Rectified*, London, 1694 [Wakely's preface dated 1664, Atkinson's preface, 1693]. Atkinson adds *An Appendix containing Use of Instruments most useful in Navigation.* Our quotation is from this *Appendix*, p. 199.

[16] R. Delamain, *The Making, Description, and Use of a small portable Instrument . . . called a Horizontall Quadrant*, etc., London, 1631.

[17] Oughtred's description of his circular slide rule of 1632 and his rectilinear slide rule of 1633, as well as a drawing of the circular slide rule, are reproduced in Cajori's *History of the Slide Rule*, Addenda, pp. ii–vi.

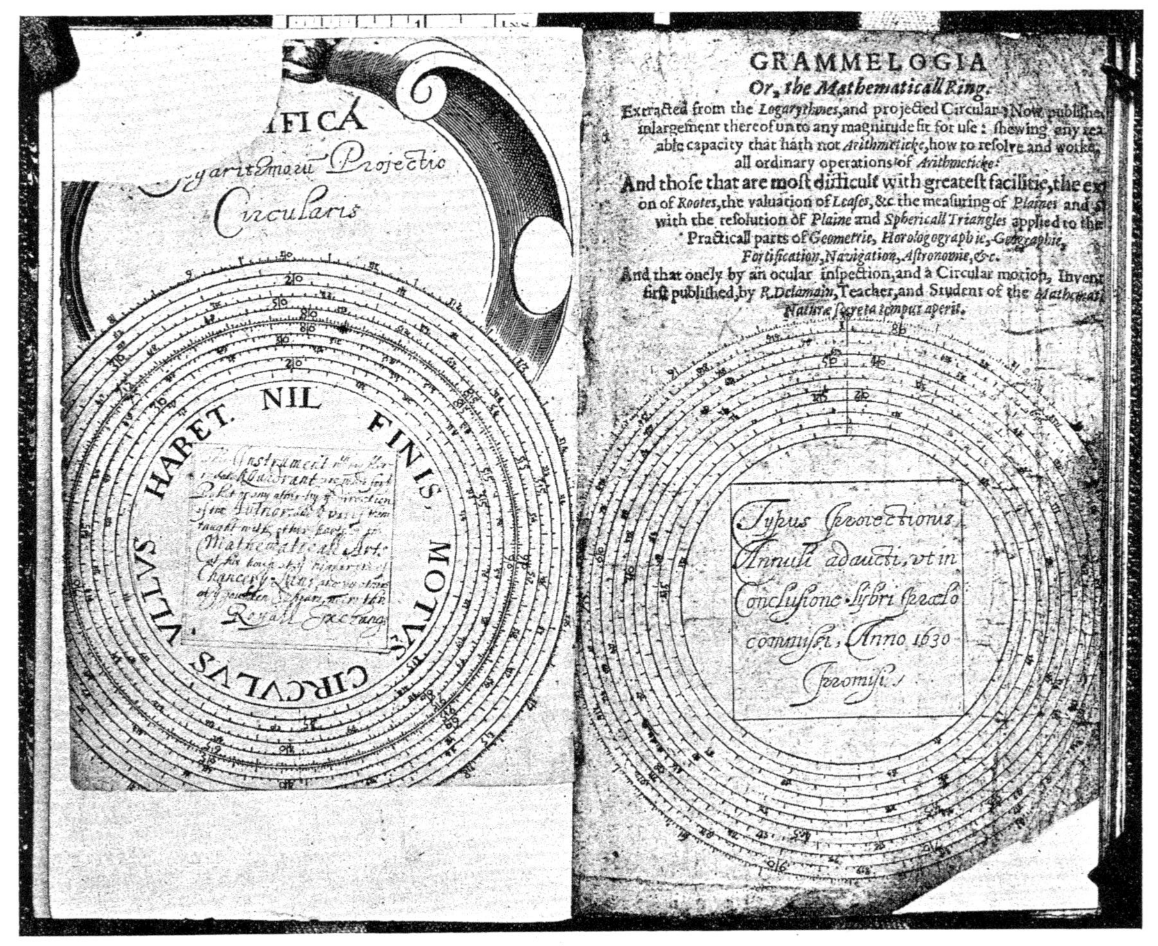

The two title-pages of the edition of the GRAMMELOGIA in the British Museum in London which we have called "Grammelogia IV."

the charge (without naming Delamain) that "another . . . went about to preocupate" the new invention. This led to verbal disputes and to the publication by Delamain of several additions to the *Grammelogia*, describing further designs of circular slide rules and also stating his side of the bitter controversy, but without giving the name of his antagonist. Oughtred's *Epistle* was published as a reply. Each combatant accuses the other of stealing the invention of the circular slide rule and the horizontal quadrant.

There are at least five different editions, or impressions, of the *Grammelogia* which we designate, for convenience, as follows:

Grammelogia I, 1630. One copy in the Cambridge University Library.[17]
Grammelogia II, I have not seen a copy of this.
Grammelogia III, One copy in the Cambridge University Library.[18]
Grammelogia IV, One copy in the British Museum, another in the Bodleian Library, Oxford.[19]
Grammelogia V, One copy in the British Museum.

In *Grammelogia I* the first three leaves and the last leaf are without pagination. The first leaf contains the title-page; the second leaf, the dedication to the King and the preface "To the Reader;" the third leaf, the description of the Mathe-

---

17 The full title of the *Grammelogia I* is as follows:

Gram̄elogia|or,|The Mathematicall Ring.|Shewing (any reasonable Capacity that hath| not Arithmeticke) how to resolve and worke|all ordinary operations of Arithmeticke.| And those which are most difficult with greatest|facilitie: The extraction of Roots, the valuation of|Leases, &c. The measuring of Plaines|and Solids.|With the resolution of Plaine and Sphericall|Triangles.|And that onely by an Ocular Inspection,|and a Circular Motion.|Naturae secreta tempus aperit.|London printed by John Haviland, 1630.

18 *Grammelogia III* is the same as *Grammelogia I*, except for the addition of an appendix, entitled:

De la Mains|Appendix|Vpon his|Mathematicall|Ring. Attribuit nullo (praescripto tempore) vitae|vsuram nobis ingeniique Deus.|London,|

. . . The next line or two of this title-page which probably contained the date of publication, were cut off by the binder in trimming the edges of this and several other pamphlets for binding into one volume.

19 *Grammelogia IV* has two title pages. The first is *Mirifica Logarithmoru' Projectio Circularis.* There follows a diagram of a circular slide rule, with the inscription within the innermost ring: *Nil Finis, Motvs, Circvlvs vllvs Habet.* The second title page is as follows:

Grammelogia|Or, the Mathematicall Ring.|Extracted from the Logarythmes, and projected Circular: Now published in the|inlargement thereof unto any magnitude fit for use: shewing any reason-|able capacity that hath not Arithmeticke how to resolve and worke,| all ordinary operations of Arithmeticke:|And those that are most difficult with greatest facilitie, the extracti-|on of Rootes, the valuation of Leases, &c. the measuring of Plaines and Solids,|with the resolution of Plaine and Sphericall Triangles applied to the|Practicall parts of Geometrie, Horologographie, Geographie|Fortification, Navigation, Astronomie, &c.|And that onely by an ocular inspection, and a Circular motion, Invented and first published, by R. Delamain, Teacher, and Student of the Mathematicks.| Naturae secreta tempus aperit.|

There is no date. There follows the diagram of a second circular slide rule, with the inscription within the innermost ring: *Typus proiectionis Annuli adaucti vt in Conslusione Lybri praelo commissi, Anno 1630 promisi.* There are numerous drawings in the *Grammelogia*, all of which, excepting the drawings of slide rules on the engraved title-pages of *Grammelogia IV* and *V*, were printed upon separate pieces of paper and then inserted by hand into the vacant spaces on the printed pages reserved for them. Some drawings are missing, so that the Bodleian *Grammelogia IV* differs in this respect slightly from the copy in the British Museum and from the British Museum copy of *Grammelogia V*.

matical Ring. Then follow 22 numbered pages. Counting the unnumbered pages, there are altogether 30 pages in the pamphlet. Only the first three leaves of this pamphlet are omitted in *Grammelogia IV* and *V*.

In *Grammelogia III* the *Appendix* begins with a page numbered 52 and bears the heading "Conclusion;" it ends with page 68, which contains the same two poems on the mathematical ring that are given on the last page of *Grammelogia I* but differs slightly in the spelling of some of the words. The 51 pages which must originally have preceded page 52, we have not seen. The edition containing these we have designated *Grammelogia II*. The reason for the omission of these 51 pages can only be conjectured. In Oughtred's *Epistle* (p. 24), it is stated that Delamain had given a copy of the *Grammelogia* to Thomas Brown, and that two days later Delamain asked for the return of the copy, "because he had found some things to be altered therein" and "rent out all the middle part." Delamain labored "to recall all the bookes he had given forth, (which were many) before the sight of *Brownes Lines*." These spiral lines Oughtred claimed that Delamain had stolen from Brown. The title-page and page 52 are the only parts of the *Appendix*, as given in *Grammelogia III*, that are missing in the *Grammelogia IV* and *V*.

Grammelogia IV answers fully to the description of Delamain's pamphlet contained in Oughtred's *Epistle*. It was brought out in 1632 or 1633, for what appears to be the latest part of it contains a reference (page 99) to the *Grammelogia I* (1630) as "being now more then two yeares past." Moreover, it refers to Oughtred's *Circles of Proportion*, 1632, and Oughtred's reply in the *Epistle* was bound in the *Circles of Proportion* having the *Addition* of 1633. For convenience of reference we number the two title-pages of *Grammelogia IV*, "page (1)" and "page (2)," as is done by Oughtred in his *Epistle*. *Grammelogia IV* contains, then, 113 pages. The page numbers which we assign will be placed in parentheses, to distinguish them from the page numbers which are *printed* in *Grammelogia IV*. The pages (44)–(65) are the same as the pages 1–22, and the pages (68)–(83) are the same as the pages 53–68. Thus only thirty-eight pages have page numbers printed on them. The pages (67) and (83) are identical in wording, except for some printer's errors; they contain verses in praise of the *Ring*, and have near the bottom the word "Finis." Also, pages (22) and (23) are together identical in wording with page (113), which is set up in finer type, containing an advertisement of a part of *Grammelogia IV* explaining the mode of graduating the circular rules. There are altogether six parts of *Grammelogia IV* which begin or end by an address to the reader, thus: "To the Reader," "Courteous Reader," or "To the courteous and benevolent Reader . . .," namely the pages (8), (22), (68), (89), (90), (108). In his *Epistle* (page 2), Oughtred characterizes the make up of the book in the following terms:

> In reading it . . . I met with such a patchery and confusion of disjoynted stuffe, that I was striken with a new wonder, that any man should be so simple, as to shame himselfe to the world with such a hotch-potch.

Grammelogia V differs from *Grammelogia IV* in having only the second title-page. The first title-page may have been torn off from the copy I have seen. A second difference is that the page with the printed numeral 22 in *Grammelogia IV* has after the word "Finis" the following notice:

> This instrument is made in Silver, or Brasse for the Pocket, or at any other bignesse, over against Saint Clements Church without Temple Barre, by Elias Allen.

This notice occurs also on page 22 of *Grammelogia I* and *III*, but is omitted from page 22 of *Grammelogia V*.

In his address to King Charles I, in his *Grammelogia I*, Delamain emphasizes the ease of operating with his slide rule by stating that it is "fit for use . . . as well on Horse backe as on Foot." Speaking "To the Reader," he states that he has "for many yeares taught the Mathematicks in this Towne," and made efforts to improve Gunter's scale "by some Motion, so that the whole body of Logarithmes might move proportionally the one to the other, as occasion required. This conceit in February last [1629] I struke upon, and so composed my *Grammelogia* or *Mathematicall Ring*; by which only with an *ocular inspection*, there is had at one instant all proportionalls through the said body of Numbers." He dates his preface "first of January, 1630." The fifth and sixth pages contain his "Description of the Grammelogia," the term *Grammelogia* being applied to the instrument, as well as to the book. His description is as follows:

> The parts of the Instrument are two Circles, the one moveable, and the other fixed; The moveable is that unto which is fastened a small pin to move it by; the other Circle may be conceived to be fixed; The circumference of the moveable Circle is divided into unequall parts, charactered with figures thus, 1. 2. 3. 4. 5. 6. 7. 8. 9. these figures doe represent themselves, or such numbers unto which a Cipher or Ciphers are added, and are varied as the occasion falls out in *the speech of Numbers*, so 1. stands for 1. or 10. or 100., &c. the 2. stands for 2. or 20. or 200. or 2000., &c. the 3. stands for 30. or 300. or 3000., &c.

After elaborating this last point and explaining the decimal subdivisions on the scales of the movable circle, he says that "the numbers and divisions on the fixed Circle, are the very same that the moveable are, . ." There is no drawing of the slide rule in this publication. The twenty-two numbered pages give explanations of the various uses to which the instrument can be put: "How to performe the Golden Rule" (pp. 1–3), "Further uses of the Golden Rule" (pp. 4–6), "Notions or Principles touching the disposing or ordering of the Numbers in the Golden Rule in their true places upon the Grammelogia" (pp. 7–11), "How to divide one number by another" (pp. 12, 13), "to multiply one Number by another" (pp. 14, 15), "To find Numbers in continuall proportion" (pp. 16, 17), "How to extract the Square Root," "How to extract the Cubicke Root" (pp. 18–21), "How to performe the Golden Rule" (the rule of proportion) is explained thus:

> Seeke the first number in the moveable, and bring it to the second number in the fixed, so right against the third number in the moveable, is the answer in the fixed.
>
> If the Interest of 100. li. be 8. li. in the yeare, what is the Interest of 65. li. for the same time.
>
> Bring 100. in the moveable to 8. in the fixed, so right against 65. in the moveable is 5.2. in the fixed, and so much is the Interest of 65. li. for the yeare at 8. li. for 100. li. *per annum*.

The *Instrument* not removed, you may at one instant right against any summe of money in the moveable, see the Interest thereof in the fixed: the reason of this is from the *Definition of Logarithmes.*

These are the earliest known printed instructions on the use of a slide rule. It will be noticed that the description of the instrument at the opening makes no references to logarithmic lines for the trigonometric functions; only the line of numbers is given. Yet the title-page promised the "resolution of Plaine and Sphericall Triangles." Page 22 throws light upon this matter:

If there be composed three Circles of equal thicknesse, A.B.C. so that the inner edge of D [should be B] and the outward egde of A bee answerably graduated with *Logarithmall signes* [sines], and the outward edge of B and the inner edge of A with *Logarithmes*; and then on the backside be graduated the *Logarithmall Tangents,* and againe the *Logarithmall signes* oppositly to the former graduations, it shall be fitted for the resolution of *Plaine* and *Sphericall Triangles.*

After twelve lines of further remarks on this point he adds:

Hence from the forme, I have called it a *Ring,* and *Grammelogia* by annoligie of a *Lineary speech;* which *Ring,* if it were projected in the *convex* unto two yards *Diameter,* or thereabouts, and the line *Decupled,* it would worke *Trigonometrie* unto seconds, and give *proportionall numbers* unto six places only by an *ocular inspection,* which would compendiate *Astronomicall calculations,* and be sufficient for the *Prosthaphaeresis* of the Motions: But of this as God shall give life and ability to health and time.

The unnumbered page following page 22 contains the patent and copyright on the instrument and book:

Whereas Richard Delamain, Teacher of Mathematicks, hath presented vnto Vs an Instrument called Grammelogia, or The Mathematicall Ring, together with a Booke so intituled, expressing the use thereof, being his owne Invention; we of our Gracious and Princely favour have granted unto the said Richard Delamain and his Assignes, Privilege, Licence, and Authority, for the sole Making, Printing and Selling of the said Instrument and Booke: straightly forbidding any other to Make, Imprint, or Sell, or cause to be Made, or Imprinted, or Sold, the said Instrument or Booke within any our Dominions, during the space of ten yeares next ensuing the date hereof, upon paine of Our high displeasure. Given under our hand and Signet at our Palace of Westminster, the fourth day of January, in the sixth yeare of our Raigne.

In the *Appendix* of *Grammelogia III,* on page 52 is given a description of an instrument promised near the end of *Grammelogia I*:

That which I have formerly delivered hath been onely upon one of the *Circles* of my *Ring,* simply concerning *Arithmeticall Propo tions,* I will by way of *Conclusion* touch upon some uses of the *Circles,* of *Logarithmall Sines,* and *Tangents,* which are placed on the edge of both the moveable and fixed *Circles* of the *Ring* in respect of *Geometricall Proportions,* but first of the description of these *Circles.*

First, upon the side that the *Circle of Numbers* is one, are graduated on the edge of the moveable, and also on the edge of the fixed the *Logarithmall Sines,* for if you bring 1. in the moveable amongst the *Numbers* to 1. in the fixed, you may on the other edge of the moveable and fixed see the *sines* noted thus 90. 90. 80. 80. 70. 70. 60. 60. &c. unto 6.6. and each degree subdivided, and then over the former divisions and figures 90. 90. 80. 80. 70. 70. &c. you have the other degrees, viz. 5. 4. 3. 2. 1. each of those divided by small points.

Secondly, (if the *Ring* is great) neere the outward edge of this side of the fixed against the *Numbers,* are the usuall divisions of a *Circle,* and the points of the *Compasse*: serving for observation in *Astronomy,* or *Geometry,* and the *sights* belonging to those divisions, may be placed on the moveable *Circle.*

Thirdly, opposite to those *Sines* on the other side are the *Logarithmall Tangents,* noted a like both in the moveable and fixed thus 6.6.7.7.8.8.9.9.10.10.15.15.20.20. &c. unto 45.45. which numbers or divisions serve also for their *Complements* to 90. so 40 gr. stands for 50. gr. 30. gr. for 60 gr. 20. gr. for 70. gr. &c. each degree here both in the moveable and fixed is also divided into parts. As for the degrees which are under 6. viz. 5.4.3.2.1. they are noted with small figures over this divided *Circle* from 45.40.35.30.25. &c. and each of those degrees divided into parts by small points both in the moveable and fixed.

Fourthly, on the other edge of the moveable on the same side is another graduation of *Tangents,* like that formerly described. And opposite unto it, in the fixed is a Graduation *of Logarithmall sines* in every thing answerable to the first descrition of Sines on the other side.

Fifthly, on the edge of the *Ring* is graduated a parte of the *Æquator,* numbered thus 10 20. 30. unto 100. and there unto is adjoyned the degrees of the *Meridian* inlarged, and numbred thus 10 20.30 unto 70. each degree both of the *Æquator,* and *Meridian* are subdivided into parts; these two graduated *Circles* serve to resolve such *Questions* which concerne *Latitude, Longitude, Rumb,* and *Distance,* in *Nauticall* operations.

Sixthly, to the concave of the *Ring* may be added a *Circle* to be elevated or depressed for any *Latitude,* representing the *Æquator,* and so divided into houres and parts with an *Axis,* to shew both the *houre,* and *Azimuth.* and within this *Circle* may be hanged a *Box,* and *Needle* with a Socket for a *staffe* to slide into it, and this accommodated with *scrue pines* to fasten it to the *Ring* and *staffe,* or to take it off at pleasure.

The pages bearing the printed numbers 53–68 in the *Grammelogia III, IV* and *V* make no reference to the dispute with Oughtred and may, therefore, be assumed to have been published before the appearance of Oughtred's *Circles of Proportion.* On page 53, "To the Reader," he says:

. . . you may make use of the Projection of the *Circles* of the *Ring* upon a *Plaine,* having the feet of a paire of *compasses* (but so that they be flat) to move on the *Center* of that *Plaine,* and those feet to open and shut as a paire of *Compasses* . . . now if the feet bee opened to any two termes or numbers in that *Projection,* then may you move the first foot to the third number, and the other foot shall give the *Answer*; . . . it hath pleased some to make use of this way. But in this there is a double labour in respect to that of the *Ring,* the one in fitting those feet unto the numbers assigned, and the other by moving them about, in which a man can hardly accommodate the *Instrument* with one hand, and expresse the *Proportionals* in writing with the other. By the *Ring* you need not but bring one number to another, and right against any other number is the *Answer* without any such motion. . . . upon that [the Ring] I write, shewing some uses of those *Circles* amongst themselves, and conjoyned with others . . . in *Astronomy, Horolographie,* in plaine *Triangles* applyed to *Dimensions, Navigation, Fortification,* etc. . . . But before I come to *Construction,* I have thought it convenient by way introduction, to examine the truth of the graduation of those *Circles* . . .

These are the words of a practical man, interested in the mechanical development of his instrument. He considers not only questions of convenience but also of accuracy. The instrument has, or may have now, also lines of sines and tangents. To test the accuracy of the circles of Numbers, "bring any number in the moveable to halfe of that number in the fixed: so any number or part in the fixed shall give his double in the moveable, and so may you trie of the thirds, fourths &c. of numbers, *vel contra*," (p. 54). On page 55 are given two small drawings, labelled, "A Type of the Ringe and Scheme of this Logarithmicall projection, the use followeth. These Instruments are made in Silver or Brasse by John Allen neare the Sauoy in the Strand."

IV. CONTROVERSY BETWEEN OUGHTRED AND DELAMAIN ON THE INVENTION OF THE CIRCULAR SLIDE RULE

Delamain's publication of 1630 on the 'Mathematicall Ring' does not appear at that time to have caused a rupture between him and Oughtred. When in 1631 Delamain brought out his *Horizontall Quadrant*, the invention of which Delamain was afterwards charged to have stolen from Oughtred, Delamain was still in close touch with Oughtred and was sending Oughtred in the Arundell House, London, the sheets as they were printed. Oughtred's reference to this in his *Epistle* (p. 20) written after the friendship was broken, is as follows:

> While he was printing his tractate of the Horizontall quadrant, although he could not but know that it was injurious to me in respect of my free gift to Master *Allen*, and of *William Forster*, whose translation of my rules was then about to come forth: yet such was my good nature, and his shamelessnesse, that every day, as any sheet was printed, hee sent, or brought the same to mee at my chamber in Arundell house to peruse which I lovingly and ingenuously did, and gave him my judgment of it.

Even after Forster's publication of Oughtred's *Circles of Proportion*, 1632, Oughtred had a book, *A canon of Sines Tangents and Secants*, which he had borrowed from Delamain and was then returning (*Epistle*, page (5)). The attacks which Forster, in the preface to the *Circles of Proportion*, made upon Delamain (though not naming Delamain) started the quarrel. Except for Forster and other pupils of Oughtred who urged him on to castigate Delamain, the controversy might never have arisen. Forster expressed himself in part as follows:

> . . . being in the time of the long vacation 1630, in the Country, at the house of the Reverend, and my most worthy friend, and Teacher, Mr. William Oughtred (to whose instruction I owe both my initiation, and whole progresse in these Sciences.) I vpon occasion of speech told him of a Ruler of Numbers, Sines, & Tangents, which one had be-spoken to be made (such as it vsually called Mr. Gunter's Ruler) 6 feet long, to be vsed with a payre of beame-compasses. "He answered that was a poore invention, and the performance very troublesome: But, said he, seeing you are taken with such mechanicall wayes of Instruments, I will shew you what deuises I have had by mee these many yeares." And first, hee brought to mee two Rulers of that sort, to be vsed by applying one to the other, without any compasses: and after that hee shewed mee those lines cast into a circle or Ring, with another moueable circle vpon it. I seeing the great expeditenesse of both those wayes; but especially, of the latter, wherein it farre excelleth any other Instrument which hath bin knowne; told him, I wondered that hee could so many yeares conceale such vseful inuentions, not onely from the world, but from my selfe, to whom in other parts and mysteries of Art, he had bin so liberall. He answered, "That the true way of Art is not by Instruments, but by Demonstration: and that it is a preposterous course of vulgar Teachers, to begin with Instruments, and not with the Sciences, and so in-stead of Artists, to make their Schollers only doers of tricks, and as it were Iuglers: to the despite of Art, losse of precious time, and betraying of willing and industrious wits,

vnto ignorance and idlenesse. That the vse of Instruments is indeed excellent, if a man be an Artist: but contemptible, being set and opposed to Art. And lastly, that he meant to commend to me, the skill of Instruments, but first he would haue me well instructed in the Sciences. He also shewed me many notes, and Rules for the vse of those circles, and of his Horizontall Instrument, (which he had proiected about 30 yeares before) the most part written in Latine. All which I obtained of him leaue to translate into English, and make publique, for the vse, and benefit of such as were studious, and louers of these excellent Sciences.

Which thing while I with mature, and diligent care (as my occasions would give me leaue) went about to doe: another to whom the Author in a louing confidence discouered this intent, using more hast then good speed, went about to preocupate; of which vntimely birth, and preuenting (if not circumuenting) forwardnesse, I say no more: but aduise the studious Reader, onely so farre to trust, as he shal be sure doth agree to truth & Art.

While in this dedication reference is made to a slide rule or "ring" with a "moveable circle," the instrument actually described in the *Circles of Proportion* consists of fixed circles "with an *index* to be opened after the manner of a paire of Compasses." Delamain, as we have seen, had decided preference for the moveable circle. To Oughtred, on the other hand, one design was about as good as the other; he was more of a theorist and repeatedly expressed his contempt for mathematical instruments. In his *Epistle* (page (25)), he says he had not "the one halfe of my intentions upon it" (the rule in his book), nor one with a "moveable circle and a thread, but with an opening Index at the centre (if so be that bee cause enough to make it to bee not the same, but another Instrument) for my part I disclaime it: it may go seeke another Master: which for ought I know, will prove to be *Elias Allen* himselfe: for at his request only I altered a little my rules from the use of the moveable circle and the thread, to the two armes of an Index."

All parts of Delamain's *Grammelogia IV*, except pages 1–22 and 53–68 considered above, were published after the *Circles of Proportion*, for they contain references to the ill treatment that Delamain felt or made believe that he felt, that he had received in the book published by Oughtred and Forster. Oughtred's reference to teachers whose scholars are "doers of tricks," "Iuglers," and Forster's allusion to "another to whom the Author in a loving confidence" explained the instrument and who "went about to preocupate" it, are repeatedly mentioned. Delamain says, (page (89)) that at first he did not intend to express himself in print, "but sought peace and my right by a private and friendly way." Oughtred's account of Delamain's course is that of an "ill-natured man" with a "virulent tongue," "sardonical laughter" and "malapert sawsiness." Contrasting Forster and Delamain, he says that, of the former he "had the very first moulding" and made him feel that "the way of Art" is "by demonstration." But Delamain was "already corrupted with doing upon Instruments, and quite lost from ever being made an Artist." (*Epistle* page (27)). Repeatedly does Oughtred assert Delamain's ignorance of mathematics. The two men were evidently of wholly different intellectual predilections. That Delamain loved instruments is quite evident, and we proceed to describe his efforts to improve the circular slide rule.

The *Grammelogia IV* is dedicated to King Charles I. Delamain says:

> . . . Everything hath his beginning, and curious *Arts* seldome come to the height at the first; It was my promise then to enlarge the *invention* by a way of *decuplating the Circles*, which I now present unto your *sacred Majestie* as the *quintessence* and *excellencie* there of . . .

His enlarged circular rules are illustrated in the Bodleian Library copy of *Grammelogia IV* by four diagrams, two of them being the two drawings on the two title-pages at the beginning of the *Grammelogia IV*, 4 inches in external diameter, and exhibiting eleven concentric circular lines carrying graduations of different sorts. In the second of these designs all circles are fixed. The other two drawings are each 10¾ inches in external diameter and exhibit 18 concentric circular lines; the folded sheet of the first of these drawings is inserted between pages (23) and (24), the second folded sheet between pages (83) and (84). All circles of this second instrument are fixed. Counting in the two small drawings in *Grammelogia III*, there are in all six drawings of slide rules in the Bodleian *Grammelogia IV*. On pages (24) to (43) Delamain explains the graduation of slide rules. He takes first a rule which has one circle of equal parts, divided into 1000 equal divisions. From a table of logarithms he gets $\log 2=0.301$; from the number 301 in the circle of equal parts he draws a line to the center of the circle and marks the intersection with the circles of numbers by the figure 2. Thus he proceeds with log 3, log 4, and so on; also with $\log \sin x$ and $\log \tan x$. For $\log \sin x$ he uses two circles, the first (see page (27)) for angles from $34' \, 24''$ to $5° \, 44' \, 22''$, the second circle from $5° \, 44' \, 22''$ to $90°$. The drawings do not show the seconds. He suggests many different designs of rules. On page (29) he says:

> For the single projection of the *Circles of my Ring*, and the dividing and graduating of them: which may bee so inserted upon the edges of *Circles* of mettle turned in the forme of a *Ring*, so that one *Circle* may moove betweene two fixed, by helpe of two stayes, then may there be graduated on the *face of the Ring*, upon the outer edge of the mooveable and inner edge of the fixed, the *Circle of Numbers*, then upon the inner edge of that mooveable *Circle*, and the outward edge of that inner fixed Circle may be inserted the *Circle of Sines*, and so according to the description of those that are usually made.

In addition to these lines he proceeds to mention the circle giving the ordinary division into degrees and minutes, and two circles of tangents on the other side of the rule.

Next Delamain explains an arrangement of all the graduation on one side of the rule by means of "a small channell in the innermost *fixed Circle*, in which may be placed a small single Index, which may have sufficient length to reach from the innermost edge of the *Mooveable Circle*, unto the outmost edge of the *fixed Circle*, which may be mooved to and fro at pleasure, in the channell, which Index may serve to shew the opposition of Numbers" (p. (31)). From this it is clear that the invention of the "runner" goes back to the very first writers on the slide rule.

After describing a modification of the above arrangement, he adds, "many other formes might be deliverd, about this single *projection*" (p. (32)).

Proceeding to the "enlarging" of the circles in the Ring, to, say, the "*Quadruple* to that which is single, that is, foure times greater," the "equall parts" are distributed over four circles instead of only one circle, but the general method of graduation is the same as before (p. (33)); there being now four circles carrying the logarithms of numbers, and so on. Next he points out "severall wayes how the Circles of the Mathematicall Ring (being inlarged) may be accommodated for practicall use:" (1) The Circles are all fixed in a plain and movable flat compasses (or better, a movable semicircle) are used for fixing any two positions; (2) There is a "double projection" of each logarithmic line "inlarged on a Plaine," one fixed, the other movable, as shown in his first figure on the title-page, a single index only being used; (3) use of "my great *Cylinder* which I have long proposed (in which all the Circles are of equall greatnesse,) and it may be made of any magnitude or capacity, but for a study (hee that will be at the charge) it may be of a yard diameter and of such an indifferent length that it may containe 100 or more Circles fixed parallel one to the other on the *Cylinder*, having a space betweene each of them, so that there may bee as many mooveable Circles, as there are fixed ones, and these of the mooveable linked, or fastened together, so that they may all moove together by the fixed ones in these spaces, whose edges both of the fixed, and mooveable being graduated by helpe of a single Index will shew the proportionalls by opposition in this double *Projection*, or by a double *Index* in a single *Projection*" (p. (36)).

Next follows the detailed description of his Ring "on a Plaine, according to the diagramme that was given the King (for a view of that projection) and afterwards the Ring it selve." The diagram is the large one which we mentioned as inserted between pages (23) and (24). The instrument has two circles, one moveable, upon each of which are described 13 distinct circular graduations. The lines on the fixed circle are: "The Circle of degrees and calendar," E. "Circle of equall parts, and part of the Equator, and Meridian," TT. "The Circle of Tangents," S. "The Circle of Sines," D. "The Circle of Decimals," N. "The Circle of Numbers." The lines on the movable circle are: N. "The Circle of Numbers," E. "The Circle of equated figures, and bodies," S. "The Circle of Sines," TT. "The Circle of Tangents," Y. "The Circle of time, yeares, and monethes."

On pages (84)–(88) Delamain explains an enlargement of his Ring for computations involving the sines of angles near to 90°. On page (86) he says:

> I have continued the *Sines* of the *Projection* unto two severall *revolutions*, the one beginning at 77.gr. 45.m. 6.s. and ends at 90.gr. (being the last *revolution* of the *decuplation* of the former, or the hundred part of that *Projection*) the other beginning at 86.gr. 6.m. 48.s. and ends at 90.gr. (being the last of a ternary of *decuplated revolutions*, or the thousand part of that *Projection*) and may bee thus used.

He explains the manner of using these extra graduations. Thus he claims to have attained degrees of accuracy which enabled him to do what "some one" had declared "could not bee done." It is hardly necessary to point out that Delamain's *Grammelogia IV* suggests designs of slide rules which inventors two hundred or more years later were endeavouring to produce. Which of Delamain's designs of rules were actually made and used, he does not state explicitly. He refers to a rule 18 inches in diameter as if it had been actually constructed (pages (86), (88)). Oughtred showed no appreciation of such study in designing and ridiculed Delamain's efforts, in his *Epistle.*

Additional elucidations of his designs of rules, along with explanations of the relations of his work to that of Gunter and Napier, and sallies directed against Oughtred and Forster, are contained on pages (8)–(21) of his *Grammelogia IV.*

V. INDEPENDENCE AND PRIORITY OF INVENTION

The question of independence and priority of invention is discussed by Delamain more specifically on pages (89)–(113); Oughtred devotes his entire *Epistle* to it. It is difficult to determine definitely which publication is the later, Delamain's *Grammelogia IV* or Oughtred's *Epistle.* Each seems to quote from the other. Probably the explanation is that the two publications contain arguments which were previously passed from one antagonist to the other by word of mouth or by private letter. Oughtred refers in his *Epistle* (p. (12)) to a letter from Delamain. We believe that the *Epistle* came after Delamain's *Grammelogia IV.* Delamain claims for himself the invention of the circular slide rule. He says in his *Grammelogia IV.* (p. (99)), "when I had a sight of it, which was in *February,* 1629 (as I specified in my *Epistle*) I could not conceale it longer, envying my selfe, that others did not tast of that which I found to carry with it so delightfull and pleasant a goate [taste] . . ." Delamain asserts (without proof) that Oughtred "never saw it as he now challengeth it to be his invention, untill it was so fitted to his hand, and that he made all his practise on it after the publishing of my *Booke* upon my *Ring,* and not before; so it was easie for him or some other to write some uses of it in Latin after Christmas, 1630 and not the *Sommer* before, as is falsely alledged by some one . . ." (p. (91)). Delamain's accusation of theft on the part of Oughtred cannot be seriously considered. Oughtred's reputation as a mathematician and his standing in his community go against such a supposition. Moreover, William Forster is a witness for Oughtred. The fact that Oughtred had the mastery of the rectilinear slide rule as well, while Delamain in 1630 speaks only of the circular rule, weighs in Oughtred's favour.

Oughtred says he invented the slide rule "above twelve yeares agoe," that is, about 1621, and "I with mine owne hand made me two such Circles, which I have used ever since, as my occasions required," (*Epistle* p. (22)). On the same page, he describes his mode of discovery thus:

> I found that it required many times too great a paire of Compasses [in using Gunter's line], which would bee hard to open, apt to slip, and troublesome for use. I therefore first devised to have another Ruler with the former: and so by setting and applying one to the other, I did not onely take away the use of Compasses, but also make the worke much more easy and expedite: when I should not at all need the motion of my hand, but onely the glancing at my sight: and with one position of the Rulers, and view of mine eye, see not one onely, but the manifold proportions incident unto the question intended. But yet this facility also

wanted not some difficulty especially in the line of tangents, when one arch was in the former mediety of the quadrant, and the other in the latter: for in this case it was needful that either one Ruler must bee as long againe as the other; or else that I must use an inversion of the Ruler, and regression. By this consideration I first of all saw that if those lines upon both Rulers were inflected into two circles, that of the tangents being in both doubled, and that those two Circles should move one upon another; they with a small thread in the center to direct the sight, would bee sufficient with incredible and wonderfull facility to worke all questions of Trigonometry . . .

Oughtred said that he had no desire to publish his invention, but in the vacation of 1630 finally promised William Forster to let him bring out a translation. Oughtred claims that Delamain got the invention from him at Alhallontide [November 1], 1630, when they met in London. The accounts of that meeting we proceed to give in double column.

Delamain's Statement

Grammelogia IV, page (98)

"... about Alhalontide 1630. (as our *Authors* reporteth) was the time he was *circumvented*, and then *his intent in a loving manner (as before) he opened unto me*, which particularly I will dismantle in the very naked truth: for, wee being walking together some few weekes before *Christmas*, upon *Fishstreet hill*, we discoursed upon sundry things *Mathematicall*, both *Theoreticall* and Practicall, and of the excellent inventions and helpes that in these dayes were produced, amongst which I was not a little taken with that of the *Logarythmes*, commending greatly the ingenuitie of Mr. Gunter in the *Projection*, and inventing of his *Ruler*, in the lines of proportion, extracted from these *Logarythmes* for ordinary *Practicall uses*; He replyed unto me (in these very words) What will yov say to an *Invention* that I have, which in a lesse extent of the *Compasses* shall worke truer then that of Mr. *Gunters Ruler*, I asked him then of what forme it was, he answered with some pause (which no doubt argued his suspition of mee that I might conceive it) that it was *Archingwise*, but now hee sayes that hee told mee then, it was *Circular* (but were I put to my oath to avoid the guilt of Conscience I would conclude in the former.) At which immediately I answered, I had the like my selfe, and so we discoursed not a word more touching that subject . . . Then after my coming home I sent him a sight of my *Projection* drawne in *Pastboard*: Now admit I had not the *Invention* of my *Ring* before I discoursed . . . it was not so facil for mee . . . to raise and compose so complete, and absolute an *Instrument* from so small a principle, or glimpse of light . . ."

Oughtred's Statement

Epistle, page (23)

"Shortly after my gift to *Elias Allen*, I chanced to meet with *Richard Delamain* in the street (it was at Alhallontide) and as we walked together I told him what an Instrument I had given to Master *Allen*, both of the Logarithmes projected into circles, which being lesse then one foot diameter would performe as much as one of Master *Gunters* Rulers of sixe feet long: and also of the Prostaphaereses of the Plannets and second motions. *Such an invention have I* said he: for now his *intentions* (that is his ambition) beganne to worke: . . . But he saith, *Then after my comming home I sent him a sight of my projection drawne in pastboard*. See how notoriously he jugleth without an Instrument. *Then after*: how long after? *a sight of my projection*: of how much? More then seven weekes after on December 23, he sent to mee the line of numbers onely set upon a circle: . . . and so much onely he presented to his Majesty: but as for Sine or tangent of his, there was not the least shew of any. Neither could he give to Master *Allen* any direction for the composure of the circles of his Ring, or for the division of them: as upon his oath Master *Allen* will testify how hee misled him, and made him labour in vain above three weeks together, until Master *Allen* himselfe found out his ignorance and mistaking, which is more cleare then is possible with any impudence to be outfaced."

Oughtred makes a further statement (*Epistle*, p. (24)) as follows:

> Delamain hearing that Brown with his *Serpentine* had *another line* by which he could worke to minutes in the 90 degree of sines . . . gave the [his] booke to Browne: who in thankfulnesse could not but gratify Delamain with his *Lines* also: and teach him the use of them, but especially of the *great Line*: with this caution on both sides, that one should not meddle with the others invention. Two dayes after *Delamain* . . . because he had found some things to be altered therin, . . . asked for the booke . . . but as soone as he had got it in his hands he rent out all the middle part with the two Schemes & put them up in his pocket & went his way . . . and . . . laboureth to recall all the bookes he had given forth . . . And shortly after this he got a new Printer (who was ignorant of his former Schemes) to print him new: giving him an especiall charge of the *outermost line newly graven* in the Plate, which indeed is *Brownes very line*: and then altering his book . . .

This and other statements made by Oughtred seem damaging to Delamain's reputation. But it is quite possible that Oughtred's guesses as to Delamain's motives are wrong. Moreover, some of Oughtred's statements are not first hand knowledge with him, but mere hearsay. One may accept his first hand facts and still clear Delamain of wrong doing. There is always danger that rival claimants of an invention or discovery will proceed on the assumption that no one else could possibly have come independently upon the same devices that they themselves did; the history of science proves the opposite. Seldom is an invention of any note made by only one man. We do not feel competent to judge Delamain's case. We know too little about him as a man. We incline to the opinion that the hypothesis of independent invention is the most plausible. At any rate, Delamain figures in the history of the slide rule as the publisher of the earliest book thereon and as an enthusiastic and skillful designer of slide rules.

The effect of this controversy upon interested friends was probably small. Doubtless few people read both sides. Oughtred says:[20] "this scandall . . . hath with them, to whom I am not knowne, wrought me much prejudice and disadvantage . ." Aubrey,[21] a friend of Oughtred, refers to Delamain "who was so sawcy to write against him" and remembers having seen "many yeares since, twenty or more good verses made" against Delamain. Another friend of Oughtred, William Robinson, who had seen some of Delamain's publications, but not his *Grammelogia IV*, wrote in a letter to Oughtred, shortly before the appearance of the latter's *Epistle*:

> I cannot but wonder at the indiscretion of Rich. Delamain, who being conscious to himself that he is but the pickpurse of another man's wit, would thus inconsiderately provoke and awake a sleeping lion . . . he hath so weakly (though in my judgment, vaingloriously enough) commended his own labour . . .[22]

Delamain presented King Charles I with one of his sun-dials, also with a manuscript and, later, with a printed copy of his book of 1630. A drawing of his improved slide rule was sent to the King and the *Grammelogia IV* is dedicated to him.

---

[20] *Epistle*, p. (8).

[21] Aubrey, *op cit.*, Vol. II., p. 111.

[22] Rigaud, *Correspondence of Scientific Men during the 17th Century*, Vol. I, Oxford, 1841, p. 11.

The King must have been favorably impressed, for Delamain was appointed tutor to the King in mathematics. His widow petitioned the House of Lords in 1645 for relief; he had ten children.[23]

Anthony Wood states that Charles I, on the day of his execution, commanded his friend Thomas Herbert "to give his son the duke of York his large ring-sundial of silver, a jewel his maj. much valued." Anthony Wood adds, "it was invented and made by Rich. Delamaine a very able mathematician, who projected it, and in a little printed book did shew its excellent use in resolving many questions in arithmetic and other rare operations to be wrought by it in the mathematics."[24]

VI. OUGHTRED'S GAUGING LINE, 1633

It has not been generally known, hitherto, that Oughtred designed a rectilinear slide rule for gauging and published a description thereof in 1633.[25] In his *Circles of Proportion*, chapter IX, Oughtred had offered a closer approximation than that of Gunter for the capacity of casks. The Gauger of London expostulated with Oughtred for presuming to question anything that Gunter had written. The ensuing discussion led to an invitation extended by the Company of Vintners to the instrument maker Elias Allen to request Oughtred to design a gauging rod.[26] This he did, and Allen received an order for "threescore" instruments. On page 19 Oughtred describes his 'Gauging Rod:'

> It consisteth of *two rulers of brasse* about 32 ynches of length, which also are halfe an ynch broad, and a quarter of an ynch thick . . . At one end of both those rulers are *two little sockets* of brasse fastened on strongly: by which the rulers are held together, and made to move one upon another, and to bee drawne out unto any length, as occasion shall require: and when you have them at the just length, there is upon one of the sockets *a long Scrue-pin* to scrue them fast.

There are graduations on three sides of the rulers, one graduation being the logarithmic line of numbers. He says (p. 39), "the maner of computing the *Gauge-divisions* I have concealed." W. Robinson, who was a friend of Oughtred, wrote him as follows:[27]

---

[23] *Dictionary of National Biography*, Art. "Delamain, Richard." See also Rev. Charles J. Robinson, *Taylors' School, from A.D. 1562 to 1874*, Vol. I, 1882, p. 151; *Journal of the House of Commons*, Vol. IV., p. 197*b*; *Sixth Report of the Royal Commission on Historical Manuscripts*, Part I, Report and Appendix, London, 1877. In this *Appendix*, p. 82, we read the following:

> Oct. 22 [1645] Petition of Sarah Delamain, relict of Richard Delamain. Petitioner's husband was servant to the King, and one of His Majesty's engineers for the fortification of the kingdom, and his tutor in mathematical arts; but upon the breaking out of the war he deserted the Court, and was called by the State to several employments, in fortifying the towns of Northampton, Newport, and Abingdon; and was also abroad with the armies as Quartermaster-General of the Foot, and therein died. Petitioner is left a disconsolate widow with ten children, the four least of whom are now afflicted with sickness, and petitioner has nothing left to support them. There are several considerable sums of money due to the petitioner, as well from the King as the State. Prays that she may have some relief amongst other widows. See L. J., VII. 6. 657.

[24] Anthony Wood, *Athenae Oxonienses* (Edition Bliss) Vol. IV., London, 1820, p. 34.

[25] *The New Artificial Gauging Line or Rod: together with rules concerning the use thereof: Invented and written by WILLIAM OUGHTRED*, etc., London, 1633. The copy we have seen is in the Bodleian Library, Oxford. The book is small sized and has 40 pages.

[26] Oughtred, *op. cit.*, p. 11.

[27] S. J. Rigaud, *Correspondence of Scientific Men of the 17th Century*, Oxford, Vol. I, 1841, p. 17.

I have light upon your little book of artificial gauging, wherewith I am much taken, but I want the rod, neither could I get a sight of one of them at the time, because Mr. Allen had none left . . . I forgot to ask Mr. Allen the price of one of them, which if not much I would have one of them." Oughtred annotated this passage thus: "Or in wood, if any be made in wood by Thompson or any other."

Another of Oughtred's admirers, Sir Charles Cavendish, wrote, on February 11, 1635 thus:[28]

I thank you for your little book, but especially for the way of calculating the divisions of your gauging rod. I wish, both for their own sakes and yours, that the citizens were as capable of the acuteness of this invention, as they are commonly greedy of gain, and then I doubt not but they would give you a better recompense than I doubt now they will.

On April 20, 1638, we find Oughtred giving Elias Allen directions[29] "about the making of the two rulers." As in 1633,[30] so now, Oughtred takes one ruler longer than the other. This 1633 instrument was used also as "a crosse-staffe to take the height of the Sunne, or any Starre above the Horizon, and also their distances." The longer ruler was called *staffe*, the shorter *transversarie*. While in 1633 he took the lengths of the two in the ratio "almost 3 to 2," in 1638, he took "the transversary three quarters of the staff's length, . . . that the divisions may be larger."

VII. OTHER SEVENTEENTH CENTURY SLIDE RULES

In my *History of the Slide Rule* I treat of Seth Partridge, Thomas Everard, Henry Coggeshall, W. Hunt and Sir Isaac Newton.[31] Of Partridge's *Double Scale of Proportion*, London, I have examined a copy dated 1661, which is the earliest date for this book that I have seen. As far as we know, 1661 is the earliest date of publications on the slide rule, since Oughtred and Delamain. But it would not be surprising if the intervening 28 years were found not so barren as they seem at present. The 1661 and 1662 impressions of Partridge are identical, except for the date on the title-page. William Leybourn, who printed Partridge's book, speaks in high appreciation of it in his own book.[32]

In 1661 was published also John Brown's first book, *Description and Use of a Joynt-Rule*, previously mentioned. In Chapter XVIII he describes the use of "Mr. Whites rule" for the measuring of board and timber, round and square. He calls this a "sliding rule." The existence, in 1661, of a "Whites rule" indicates activities in designing of which we know as yet very little. In his book of 1761, previously quoted, Brown gives a drawing of "White's sliding rule" (p. 193); also a special contrivance of his own, as indicated by him in these words:

A further improvement of the Triangular Quadrant, as I have made it several times, with a sliding Cover on the in-side, when made hollow, to carry Ink, Pens, and Compasses; then on the sliding Cover, and Edges, is put the Line of Numbers, according to Mr. White's first Contrivance for manner of operation; but much augmented, and made easie, by John Brown.

---

[28] Rigaud, *loc. cit.*, p. 22.
[29] Rigaud, *loc. cit.*, pp. 30, 31.
[30] Oughtred, *An Addition vnto the Vse of the Instrument called the Circles of Proportion*, London, 1633, p. 63.
[31] F. Cajori, *History of the Slide Rule*, New York, 1909, pp. 16–22, Addenda, pp. vi–ix
[32] W. Leybourn, *op. cit.*, 1673, Preface, and pp. 128–29.

He gives no drawing of his "triangular quadrant," hence his account of it is unsatisfactory. He explains the use of "gage-points." His placing logarithmic lines on the edges of instrument boxes was outdone in oddity later by Everard who placed them on tobacco-boxes.[33] In Brown's publication of 1704 the White slide rule is given again, "being as neat and ready a way as ever was used." He tells also of a "glasier's sliding rule." William Leybourn explains in 1673 how Wingate's double and triple lines for squaring and cubing, or square and cube root, can be used on slide rules.[34]

Beginning early in the history of the slide rule, when Oughtred designed his "gauging rod," we notice the designing of rules intended for very special purposes. Another such contrivance, which enjoyed long popularity, was the *Timber Measure by a Line*, by Hen. Coggeshall, Gent., London, 1677, a booklet of 35 pages. Coggeshall says in his preface:

> For what can be more ready and easie, then having set twelve to the length, to see the Content exactly against the Girt or Side of the Square. Whereas on Mr. Partridge's Scale the Content is the Sixth Number, which is far more troublesome then [even] with Compasses.

One line on Coggeshall's rule begins with 4 and extends to 40, these numbers being the "Girt" (a quarter of the circumference), which in ordinary practice of measuring round timber lies between 4 inches and 40 inches. This "Girt line" slides "against the line of Numbers in two Lengths, to which it is exactly equal." A second edition, 1682, shows some changes in the rule, as well as an enlargement and change of title of the book itself: *A Treatise of Measures, by a Two-foot Rule*, by H. C. Gent, London, 1682. In this, the description of the rule is given thus:

> There are four Lines on each flat of this Rule; two next the outward edges, which are Lines of Measure; and two next the inward edges, which are Lines of Proportion. On one flat, next the inward edges, is the Square-line [Girt-line in round timber measurement] with the Line of Numbers his fellow. Next the outward, a Line of Inches divided into Halfs, Quarters, and Half-Quarters; from 1 to 12 on one Rule; and from 12 to 24 on the other. On the other flat, next the inward edges, is the double Scale of Numbers [for solving proportions]. Next the outward on one Rule a Line of Inches divided each into ten parts; and this for gauging, etc. On the other a foot divided into 100 parts.

Later further changes were introduced in Coggeshall's rule.[34]

It is worthy of note that Coggeshall's slide rule book, *The Art of Practical Measuring*, was reviewed in the *Acta eruditorum*, anno 1691, p. 473; hence Leupold's description[35] of the rectilinear slide rule in his *Theatrum arithmetico-geometricum*, Leipzig, 1727, Cap. XIII, p. 71, is not the earliest reference to the rectilinear rule found in German publications. The above date is earlier even than Biler's reference to a circular slide rule in his *Descriptio instrumenti mathematici universalis* of 1696.

---

[33] Cajori *op. cit.*, Addenda, p. ix.

[34] William Leybourn, *op. cit.*, 1673, p. 35.

[34] See Cajori, *op. cit.*, pp. 20, 28, Addenda, p. ix.

[35] See F. Cajori, "A Note on the History of the Slide Rule," *Bibliotheca mathematica*, 3 F., Vol. 10, pp. 161–163.

Two noted slide rules for gauging were described by Tho. Everard, Philomath, in his *Stereometry made easie*, London, 1684. He designates his lines by the capital letters A, B, C, D, E. On the first instrument, *A* on the rule, and *B* and *C* on the slide, have each two radiuses of numbers, *D* has only one, while *E* has three. The second rule is described in an *Appendix*; it is one foot long, with two slides enabling the rule to be extended to 3 feet.

Everard's instruments were made in London by Isaac Carver who, soon after, himself wrote a sixteen-page *Description and Use of a New Sliding Rule, projected from the Tables in the Gauger's Magazine*, London, 1687, which was "printed for William Hunt" and bound in one volume with a book by Hunt, called *The Gauger's Magazine*, London, 1687. This appears to be the same William Hunt who later brought out descriptions of his own of slide rules. The instrument described by Carver "consists of three pieces, two whereof are moveable to be drawn out till the whole be 36 inches long." It has several non-logarithmic graduations, together with logarithmic lines marked A, B, C, D, of which A, B, C are "double lines," and D a "single line" used for squares and square roots. It is designed for the determination of the vacuity of a "spheroidal cask lying," a "spheroidal cask standing," and a "parabolical cask lying."

Another seventeenth century writer on the slide rule is John Atkinson, whom we have mentioned earlier. He says:[36] "The Lines of Numbers, Sines and Tangents, are set double, that is, one on each side, as the middle piece slides: which middle piece is so contrived, to slip to and fro easily, to slide out, and to be put in any side uppermost, in order to bring those Lines together (or against one another) most proper for solving the Question, wrought by *Sliding-Gunter*."

The data presented in this article show that, while the earliest slide rules were of the circular type, the later slide rules of the seventeenth century were of the rectilinear type.[37]

January 12, 1915.

[36] John Atkinson, *op. cit.*, 1694, p. 204.

[37] Probably the oldest slide rule now in existence is owned by St. John's College, Oxford, and is in the form of a brass disc, 1 ft. 6 in. in diameter. It was exhibited along with other instruments in May, 1919. According to the *Catalogue of a Loan Exhibition of Early Scientific Instruments* in Oxford, opened May 16, 1919, the instrument is inscribed with the name of the maker ("*Elias Allen fecit*") and with the name of the donor, Georgius Barkham. It is dated 1635, which is only three years after the first publication of Oughtred's description of his circular slide rule. It is stated in the *Catalogue:* "Unfortunately all the movable parts but the base-plate and a couple of thumb-screws are missing. The face of the instrument is engraved with Oughtred's *Horizontal Instrument.* The back is engraved with eleven Circles of Proportion as described in Arthur Haughton's book, a copy of which was presented to St. John's College by George Barkham, to explain the use of the instrument." As Arthur Haughton's Oxford edition of Oughtred's *Circles of Proportion* did not appear until 1660, it would seem that the instrument was probably not presented to the College before 1660. As far as is known, the next oldest slide rule is of the year 1654, kept in the South Kensington Museum, London, and is described in *Nature* of March 5, 1914. It is a rectilinear rule, "of boxwood, well made, and bound together with brass at the two ends. It is of the square type, a little more than 2 ft. in length, and bears the logarithmic lines first described by Edmund Gunter. Of these, the *num, sin* and *tan* lines are arranged in pairs, identical and contiguous, one line in each pair being on the fixed part, and the other on the slide." The instrument is inscribed, "Made by Robert Bissaker for T. W., 1654." Nowhere else have we seen reference to Robert Bissaker. His slide rule seems to antedate the "Whites rule." mentioned above. [This foot-note was added on October 15, 1919.]

UNIVERSITY OF CALIFORNIA PUBLICATIONS
IN
MATHEMATICS

Vol. 1, No. 10, pp. 211-222 February 17, 1920

ON A BIRATIONAL TRANSFORMATION CONNECTED WITH A PENCIL OF CUBICS

BY
ARTHUR ROBINSON WILLIAMS

PREFATORY NOTE

When I prepared the following paper I was acquainted only with the paper by Cayley, cited on page 212, and the articles therein mentioned. The problems of the construction of the cubic through nine points and the construction of the ninth point of intersection of all cubics through eight points, virtually the same problem, have been treated by Weddle, by Hart, by Chasles, by Cayley, and of course by later writers. Weddle's paper may be found in the *Cambridge and Dublin Mathematical Journal,* Vol. VI (1851), page 83, and Hart's in the same volume, page 181. Chasles' memoir appeared in *Comptes rendus lebdomadaire des seances de l'Academie des Sciences,* Vol. XXXVI (1853), page 942.

It does not seem to have occurred to these writers to inquire concerning the locus of the ninth point when seven remain fixed and the eighth moves in a given manner. However, Professor H. S. White, to whom this paper was referred, pointed out that the subject had been well covered by Geiser and Milinowski.

The paper by Geiser, "*Ueber Zwei Geometrische Probleme,*" is in *Crelle's Journal,* Vol. 67 (1867), page 78. Starting from the well-known fact, which follows from the elementary properties of the cubic, that if seven points be fixed and an eighth be taken on the line joining two of them, the ninth intersection of all cubics through the eight lies on the conic through the other five, and conversely; and making use of a theorem of Steiner's regarding the locus of the double points of all the cubics through seven points, he derives by synthetic and very simple methods practically all the results which I have obtained, including the special cases arising from relations among the seven fixed points. Naturally neither he nor Milinowski, whose method is also synthetic, mentions the system of lines which I consider in section 12.

Milinowski, whose paper on this subject is found in *Crelle's Journal,* Vol. 77 (1874), page 263, sets up a one-to-one correspondence between the cubics through seven points and the points of the plane. The same end is served by making correspond to a given cubic of the net its polar line with respect to a fixed point. To obtain the locus of the ninth point when seven are fixed and the eighth moves, I have expressed its co-ördinates directly in terms of those of the eighth. This would not be feasible of course for curves of higher order than the third. Thus in case of a net of quartics, if twelve points are fixed, there correspond to a thirteenth not one but three others. Milinowski's methods are equally applicable to this and to curve nets of higher order.

The second and longer portion of Geiser's paper is given to the discussion of an analogous problem in space. All quadric surfaces through seven fixed points have an eighth in common. Hesse, then *privatdocent* at Konigsberg, has given a construction for this eighth point in *Crelle's Journal*, Vol. 26 (1843), page 147. Geiser discusses the locus of the eighth point when six are fixed and the seventh moves in a given manner. The problem is rather more complex than that of the net of plane cubics, and the results obtained are very curious and interesting.

The theorem that all the cubics through eight points have a ninth in common, suggests the following transformation, which is evidently birational. Fix seven points, and make correspond to a variable eighth the remaining common point of all the cubics through the eight.

1. *Construction of the ninth intersection of the cubics through eight points.*

It follows from the properties of the cubic regarded as the locus of intersections of corresponding elements of a pencil of conics and a pencil of lines projectively related, that if any four of eight points be taken, the ninth point of intersection of all the cubics through the eight projects to the four selected points in four rays whose anharmonic ratio is the same as that of the four conics which have the four remaining points in common, and are determined respectively by the four chosen points.[1] Therefore 9 is seen to lie on a certain conic through the four chosen points. In particular, if we take two sets of four points which have three points in common, 9 will be determined as the fourth intersection of two conics three of whose intersections are known; or, in other words, as the point which projects simultaneously to two sets of four points (which have three points in common) in two sets of four rays, each four having a given anharmonic ratio. Thus $9\{4\,6,\,7\,8\} = (1\,2\,3\,5)\{4\,6,\,7\,8\}$, and $9\{5\,6,\,7\,8\} = (1\,2\,3\,4)\{5\,6,\,7\,8\}$; where $(1\,2\,3\,5)\{4\,6,\,7\,8\}$ means the anharmonic ratio of the four conics which have 1 2 3 5 in common, and which pass through 4 6 7 8 respectively. In these anharmonic ratios the order of elements is arbitrary, but once chosen must be consistently maintained. In the above notation, primary and secondary elements are separated by a comma. Cayley, in the paper cited, has given the following construction due to Dr. Hart for a ninth point so determined:

Let $(1\,2\,3\,5)\{4\,6,\,7\,8\}$ be denoted by p, and $(1\,2\,3\,4)\{5\,6,\,7\,8\}$ by s. The anharmonic ration of four conics of a pencil is given by that of their four points of inter-

[1] Cayley "On the Construction of the Ninth Point of Intersection of the Cubics which Pass through Eight Points." *Collected Papers*, Vol. 4, p. 495.

section with any line passing through any one of their four common points, or by that of the four tangents at one of the common points, or again by that of the po'ars of any point with respect to the four conics. After determining p and s, Dr. Hart proceeds thus: Let 74 and 65 meet in M, and on 74 take Q so that $\{4M, 7Q\} = p$. Let 85 meet 64 in N, and on 85 take R so that $\{5N, R8\} = s$. Let QR meet 65 in K and 64 in L. Then $7K$ and $8L$ give 9. The whole construction may be accomplished with the ruler.

2. *Analytic treatment of this construction shows that the co-ördinates of* 9, $\lambda'\mu'\nu'$, *are octavic functions of the co-ördinates of* 8, *of the form* $\lambda' = C_2C_3K_1$, $\mu' = C_3C_1K_2$, $\nu' = C_1C_2K_3$, *where the Cs are of degree three and the Ks of degree two.*

Now introducing coördinates, let 4 5 6 be the vertices of the triangle of reference; 6 being $(1 : 0 : 0)$, 5 being $(0 : 0 : 1)$, and 4 $(0 : 1 : 0)$. Then 4 5 is the line $x=0$, 56 is $y=0$, and 64 is $z=0$. Let 1 2 3 7 be $(\lambda_1 : \mu_1 : \nu_1)$, $(\lambda_2 : \mu_2 : \nu_2)$, etc., and let 8, which is to be regarded as variable, be $(\lambda : \mu : \nu)$. The quantities p and s are now themselves functions of $\lambda\ \mu\ \nu$. p is $(1\,2\,3\,5)\{4\,6,\ 7\,8\}$. The four conics all pass through 5, and it will therefore suffice to find the anharmonic ration of their four other points of intersection with the line 4 5, i.e., the line $x=0$. The conic $1\,2\,3\,5-4$ meets $x=0$ at 4, i.e. $(0 : 1 : 0)$. The equation of $1\,2\,3\,5-6$ may be written thus:

$$\begin{vmatrix} y^2 & xy & xz & yz \\ \mu_1^2 & \lambda_1\mu_1 & \lambda_1\nu_1 & \mu_1\nu_1 \\ \mu_2^2 & \lambda_2\mu_2 & \lambda_2\nu_2 & \mu_2\nu_2 \\ \mu_3^2 & \lambda_3\mu_3 & \lambda_3\nu_3 & \mu_3\nu_3 \end{vmatrix} = 0$$

Setting $x=0$, we have $y : z = a : b$, where a and b are the minors of yz and y^2 respectively. Similarly the equation of $1235-7$ may be written in the determinant form, and, setting $x=0$, we have $y : z = f : g$ where $f \equiv (\lambda_1^2 \cdot \mu_2^2 \cdot \lambda_3\mu_3 \cdot \lambda_7\nu_7)$ and $g \equiv (\lambda_1^2 \cdot \lambda_2\mu_2 \cdot \lambda_3\nu_3 \cdot \mu_7\nu_7)$. For $1\,2\,3\,5-8$ we have the same equation except that the coördinates of 7 are replaced by those of 8. Setting $x=0$, we have $y : z = F_8 : G_8$, where F_8 and G_8 are obtained on replacing $\lambda_7\mu_7\nu_7$ by $\lambda\mu\nu$ in f and g. The capital letters and the subscript will indicate the presence of the coordinates of the variable point. Hence the four points whose anharmonic ratio we desire are respectively $(0 : 1 : 0)$, $(0 : a : b)$, $(0 : f : g)$, $(0 : F_8 : G_8)$. Following the order of elements indicated above, we have the anharmonic ratio of the four conics $g\,(aG_8 - bF_8)/G_8(ag - bf) \equiv p$. In the same way the anharmonic ratio $(1\,2\,3\,4)\ \{5\,6,\ 7\,8\}$ is found to be $g(bH_8 - cG_8)/G_8(bh - cg)$. Only three new determinants have been introduced, $c \equiv (\nu_1^2 \cdot \lambda_2\mu_2 \cdot \lambda_3\nu_3)$, $h \equiv \lambda_1^2 \cdot \nu_2^2 \cdot \lambda_3\mu_3 \cdot \lambda_7\nu_7)$, and H_8, which is obtained on replacing $\lambda_7\mu_7\nu_7$ by $\lambda\mu\nu$ in h. Thus a, b, c are of the third order and involve only the coördinates of 123. The other determinants are of the fourth order and related as indicated.

In accordance with the construction outlined, we take Q on 7 4 so that $\{4M, 7Q\} = p$, where M is the intersection of 74 with 65, that is, of 7 4 with $y=0$. The coördinates of Q are thus found to be ;$(\lambda_7 : p\mu_7 : \nu_7)$. Similarly the intersection of 8 5 with 6 4 is N, and taking R on 8 5 so that $\{5N, R8\}=s$, R is $(s\lambda : \mu s : \nu)$. The intersection of QR with $y=0$ gives K, and with $z=0$ gives L. The intersection of $7K$ and $8L$ is 9. Thus for $\lambda'\mu'\nu'$, the coördinates of 9, we obtain

$$\left.\begin{aligned}\lambda' &= (\nu_7\lambda s-\lambda_7\nu)(\lambda_7\mu-\mu_7\lambda p)(p-s)\\ \mu' &= (\lambda_7\mu-\mu_7\lambda p)(\mu_7 p\nu-\nu_7\mu s)(s-1)\\ \nu' &= (\mu_7 p\nu-\nu_7\mu s)(\nu_7\lambda s-\lambda_7\nu)(1-p)\end{aligned}\right\}\ (1)$$

The following identities are at once evident

$$\left.\begin{aligned}&\lambda(\mu_7 p\nu-\nu_7\mu s)+\mu(\nu_7\lambda s-\lambda_7\nu)+\nu(\lambda_7\mu-\mu_7\lambda p)=0\\ &(p-s)+(s-1)+(1-p)=0\end{aligned}\right\}\ (a)$$

To obtain $\lambda'\mu'\nu'$ as homogeneous expressions in $\lambda\mu\nu$ it is necessary only to substitute the values of p and s in (1). After removal of a common denominator and the factor g from each numerator, we have

$$\left.\begin{aligned}\lambda' = &\{\nu_7\lambda(bH_8-cG_8)g-\lambda_7\nu(bh-cg)G_8\}\times\\ &\{\lambda_7\mu(ag-bf)G_8-\mu_7\lambda(aG_8-bF_8)g\}\times\\ &\{(bh-cg)(aG_8-bF_8)-(ag-bf)(bH_8-cG_8)\}.\\ \mu' = &\{\lambda_7\mu(ag-bf)G_8-\mu_7\lambda(aG_8-bF_8)g\}\times\\ &\{\mu_7\nu(bh-cg)(aG_8-bF_8)-\nu_7\mu(ag-bf)(bH_8-cG\ )\}\times\\ &\{(bH_8-cG_8)g-(bh-cg)G_8\}.\\ \nu' = &\{\mu_7\nu(bh-cg)(aG_8-bF_8)-\nu_7\mu(ag-bf)(bH_8-cG_8)\}\times\\ &\{\nu_7\lambda(bH_8-cG_8)g-\lambda_7\nu(bh-cg)G_8\}\times\\ &\{(ag-bf)G_8-(aG_8-bF_8)g\}.\end{aligned}\right\}\ (2)$$

Recalling that $F_8\ G_8\ H_8$ are determinants homogeneous and of degree two in $\lambda\ \mu\ \nu$, it is evident that $\lambda'\ \mu'\ \nu'$ are homogeneous of the eighth degree, and of the form

$$\left.\begin{aligned}\lambda' &= C_2C_3K_1\\ \mu' &= C_3C_1K_2\\ \nu' &= C_1C_2K_3\end{aligned}\right\}\ (2')$$

where the Cs are of the third degree and the Ks of the second. The identities connecting the Cs and Ks are

$$\left.\begin{aligned}&\lambda gC_1+\mu(ag-bf)C_2+\nu(bh-cg)C_3=0\\ &gK_1+(ag-bf)K_2+(bh-cg)K_3=0\end{aligned}\right\}\ (\beta)$$

3. *The factors C and K may be replaced by determinants proportional to them.*

From the form of $F_8\ G_8\ H_8$ it follows that they all vanish when the coördinates, $\lambda\ \mu\ \nu$, of 8 are replaced by those of any one of the points 1 2 3. Then, expanding in terms of $\lambda\ \mu\ \nu$, it is easily seen that $G_8=0$, $aG_8-bF_8=0$, $bH_8-cG_8=0$ represent respectively the conics 1 2 3 4 5, 1 2 3 5 6, 1 2 3 4 6. Noting that the left-hand members of these conics become g, $ag-bf$, $bh-cg$ when $\lambda_7\ \mu_7\ \nu_7$ are put for $\lambda\ \mu\ \nu$, it appears immediately that $C_2=0$, $C_3=0$, $C_1=0$ represent three nodal cubics, each of which passes through all the seven fixed points, and which have their double points at 4 5 6 respectively. Similarly $K_1=0$, $K_2=0$, $K_3=0$ represent three conics which have the points 1 2 3 7 in common and which pass through 6 4 5 respectively. The symmetry of $\lambda'\ \mu'\ \nu'$ is quite striking. Thus the two cubics that form part of λ' have their double points at (xy) and (xz), while the accompanying conic, K_1, passes through 6, i.e., the vertex opposite to $x=0$. Similarly for the factors of μ' and ν'.

Thus the form of each factor C and of each factor K is determined down to a constant multiplier by the character of the curve that it represents when equated to zero. For example $K_2=0$ gives the conic 1 2 3 7 4, and therefore K_2 differs at most by a constant factor from the determinant

$$\begin{vmatrix} \nu^2 & \lambda^2 & \mu\nu & \nu\lambda & \lambda\mu \\ \nu_1^2 & \lambda_1^2 & \mu_1\lambda_1 & \nu_1\lambda_1 & \lambda_1\mu_1 \\ \nu_2^2 & \lambda_2^2 & \mu_2\nu_2 & \nu_2\lambda_2 & \lambda_2\mu_2 \\ \nu_3^2 & \lambda_3^2 & \mu_3\nu_3 & \nu_3\lambda_3 & \lambda_3\mu_3 \\ \nu_7^2 & \lambda_7^2 & \mu_7\nu_7 & \nu_7\lambda_7 & \lambda_7\mu_7 \end{vmatrix}$$

Expanding this determinant by the elements of the first row, and expanding also $K_2\equiv(bH_8-cG_8)g-(bh-cg)G_8$, i.e. $b(gH_8-hG_8)$, comparison of the coefficients shows that K_2 is identical with the above determinant multiplied by $-bJ$. J is the determinant $(\lambda_1^2\cdot\lambda_2\mu_2\cdot\lambda_3\nu_3)\equiv\lambda_1\lambda_2\lambda_3(\lambda_1\cdot\mu_2\cdot\nu_3)$. Similar relations hold for the other Cs and Ks. Thus

$$\begin{aligned} K_1 &\equiv (\mu^2\cdot\nu_1^2\cdot\mu_2\nu_2\cdot\nu_3\lambda_3\cdot\lambda_7\nu_7)J^2b \\ K_2 &\equiv (\nu^2\cdot\lambda_1^2\cdot\mu_2\nu_2\cdot\nu_3\lambda_3\cdot\lambda_7\mu_7)(-Jb) \\ K_3 &\equiv (\lambda^2\cdot\mu_1^2\cdot\mu_2\nu_2\cdot\nu_3\lambda_3\cdot\lambda_7\nu_7)Jb \\ C_2 &\equiv (\lambda^2\mu\cdot\lambda_1^2\nu_1\cdot\lambda_2\mu_2\nu_2\cdot\lambda_3\nu_3^2\cdot\mu_7\nu_7^2)(-JJ_1) \\ C_3 &\equiv (\lambda^2\mu\cdot\lambda_1^2\nu_1\cdot\lambda_2\mu_2^2\cdot\lambda_3\mu_3\nu_3\cdot\mu_7^2\nu_7)JJ_1 \\ C_1 &\equiv (\lambda\mu^2\cdot\lambda_1\mu_1\nu_1\cdot\lambda_2\nu_2^2\cdot\mu_3^2\nu_3\cdot\mu_7\nu_7^2)J^2J_1 \end{aligned}$$

$J_1\equiv(\lambda_1\cdot\mu_2\cdot\nu_3)$. It is seen, therefore, that $\lambda'\ \mu'\ \nu'$ as given by (2) contain the common factor $-J_4J_1^2\equiv-\lambda_{14}\lambda_{24}\lambda_3 tJ_1$i.Removing this common factor, we may use for the Cs and Ks the determinants indicated in the last set of equations.

4. *Geometric character of the transformation.*

In (2) $\lambda'\ \mu'\ \nu'$ represent curves of the eighth order which have seven triple points in common, that is sixty-three $(=8^2-1)$ common intersections. Since the possession of a triple point at a given point imposes six conditions, it follows that a curve of order eight having seven fixed triple points is subjected to forty-two conditions, two less than the number sufficient to determine a curve of that order. This accords with the general theory of Cremona transformations which demands that in a birational transformation of degree n the expressions of the nth degree shall represent curves which have n^2-1 common intersections, while these intersections must represent

$$\frac{n\,(n+3)}{2}-2$$

conditions, two less than the number required to determine a curve of order n.

Therefore if 8 moves on a line not passing through any of the seven fixed points, 9 will describe a proper, i.e., undegenerate, curve of order eight having a triple point at each of the seven fixed points. If 8 moves on a conic which does not pass through any of the seven fixed points, 9 will describe a curve of order sixteen having a sextuple point at each of them. According to the general theory there will correspond to each of the seven fixed points a curve of order three. The nature of this correspondence appears immediately from (2). For example, if 8 is any point on C_1, 9 is $(1:0:0)$. But C_1 is the cubic which has its double point at $(1:0:0)$. For 8 on C_3, 9 is $(0:0:1)$, and for 8 on C_2, 9 is $(0:1:0)$, which are the double points of C_3 and C_2 respectively. Considerations of symmetry enable us to conclude immediately that in this transformation the curve of degree three corresponding to any one of the seven fixed points is the nodal cubic passing through all of them and having a double point at the one in question. These seven cubics taken together constitute the Jacobian of $L\ M\ N$; where $L\ M\ N$ are the functions of $\lambda'\ \mu'\ \nu'$ got by solving (2) for λ, for μ, and for ν respectively. Since in this transformation the coördinates of 8 are evidently the same functions of the coördinates of 9 as the coördinates of 9 are of those of 8, the same seven cubics constitute the Jacobian of $\lambda'\ \mu'\ \nu'$. If 8 is on K_1 (2) shows that 9 is on $x=0$. But K_1 is the conic 12376 and $x=0$ is the line 4 5. Similarly for the conics K_2 and K_3 and the lines $y=0$ and $z=0$. It thus appears from (2), as is obvious from the general properties of the transformation and from those of the cubic, that to the line joining two of the fixed points corresponds the conic through the other five. Or, in other words, the curve of order eight corresponding to such a line consists of this conic and the two cubics which correspond to the two fixed points. With the exception of the seven special cubics just mentioned, all cubics through the seven fixed points are self-corresponding. For the points of any one of them correspond in pairs. To a sextic which has a double point at each of the original points corresponds another sextic which itself has a double point at each of them. It will be seen that there is one such sextic all of whose points are invariant. It is in fact the locus of the invariant points of the transformation.

5. *There are infinitely many invariant points, and their locus is a sextic.*

To ascertain the invariant points, we may put $\lambda\ \mu\ \nu$ for $\lambda'\ \mu'\ \nu'$ in (1). Then dividing the first equation by the second, the first by the third, and the second by the third, and cross-multiplying, we have the three equations, two of which are independent,

$$\lambda(\mu_7 p\nu-\nu_7\mu s)(s-1)-\mu(\nu_7\lambda s-\lambda_7\nu)(p-s)=0$$
$$\mu(\nu_7\lambda s-\lambda_7\nu)(1-p)-\nu(\lambda_7\mu-\mu_7\lambda p)(s-1)=0$$
$$\nu(\lambda_7\mu-\mu_7\lambda p)(p-s)-\lambda(\mu_7 p\nu-\nu_7\mu s)(1-p)=0$$

Any invariant points must be common to the three curves represented by these equations. But all three reduce to

$$\lambda ps(\mu_7\nu-\nu_7\mu)+\mu s(\nu_7\lambda-\lambda_7\nu)+\nu p(\lambda_7\mu-\mu_7\lambda)=0 \tag{4}$$

Hence, instead of a finite number of invariant points there are infinitely many; in fact all the points of the curve (4). This curve is evidently a sextic, and, referring to (2′), may be written in any of the three forms

$$\begin{aligned}\lambda C_1K_2-\mu C_2K_1&=0,\\ \mu C_2K_3-\nu C_3K_2&=0,\\ \nu C_3K_1-\lambda C_1K_3&=0.\end{aligned}$$

Any of these equations shows that the sextic has a multiple point which is at least a double point at each of the seven fixed points. But no one of them can be of higher order than two, for in that case there would correspond to the sextic a curve of lower order than six.

An interesting property of the invariant points appears when they are considered in connection with the original construction. It will be recalled that 9 is the fourth intersection of two conics one of which is the locus of points that project to 4 6 7 8 in the anharmonic ratio p, while the points of the other project to 5 6 7 8 in the anharmonic ratio s. Let the tangents to these conics at 8 be l_p and l_s. Then the equations of these tangents may be easily obtained. For the anharmonic ratio of the four rays 84, 86, 87 and l_p is p, while that of 85, 96, 87, l_s is s. Making use of the fact that the anharmonic ratio of four lines of the form L, M, $L-rM$, $L-r_1M$ is r/r_1, and keeping the same order of elements as in calculating p and s, the equations of l_p and l_s are found to be respectively

$$\begin{aligned}p\nu x-r\nu y+(r\mu-\lambda p)z&=0\\ s\mu x-(k\nu+\lambda s)y+k\mu z&=0\end{aligned}$$

where
$$r\equiv-\frac{\nu_7\lambda-\lambda_7\nu}{\mu_7\nu-\nu_7\mu}$$

and
$$k\equiv\frac{\lambda_7\mu-\mu_7\lambda}{\mu_7\nu-\nu_7\mu}$$

The conics will have their fourth intersection at 8 if l_p and l_s coincide; that is if

$$\frac{p\nu}{s\mu}=\frac{r\nu}{k\nu+\lambda s}=\frac{r\mu-\lambda p}{k\mu},$$

the variables being λ μ ν, the coördinates of 8. Forming the three possible equations and bringing all the terms to the left hand member, we have in each case the factor $\lambda ps-\mu sr+\nu pk$. Hence all points of the curve that correspond to this common factor are invariant. Replacing r and k by their values we have

$$\lambda ps(\mu_7\nu-\nu_7\mu)+\mu s(\nu_7\lambda-\lambda_7\nu)+\nu p(\lambda_7\mu-\mu_7\lambda)=0$$

which is (4) above. Substituting the values of p and s in (4), we obtain

$$\lambda(\mu_7\nu-\nu_7\mu)(aG_8-bF_8)(bH_8-cG_8)g+\mu(\nu_7\lambda-\lambda_7\nu)(ag-bf)(bH_8-cG_8)G_8$$
$$+\nu(\lambda_7\mu-\mu_7\lambda)(bh-cg)(aG_8-bF_8)G_8=0. \quad (5)$$

It will be observed that at points 8 for which l_p and l_s coincide they have two intersections with all the cubics of the one parameter family through the seven fixed points and the point 8 in question. In other words the cubics of the family determined by the seven fixed points and an invariant point have a common tangent at the latter, whose equation is given equally by l_p and l_s. There is, however, one cubic of the family which has a double point at the invariant point, and for which the line l_{ps} is not in general a tangent. The envelope of this set of lines l_{ps} will be sought after a little further examination of the sextic.

In (5), note that $\mu_7\nu-\nu_7\mu$, $\nu_7\lambda-\lambda_7\nu$, $\lambda_7\mu-\mu_7\lambda$, when equated to zero are the lines 67, 47, 57 respectively; while $G_8=0$, $aG_8-bF_8=0$, $bH_8-cG_8=0$ are the conics 12345, 12356, 12346. Thus it appears that the sextic passes through the intersections of the conic 12345 with the line 67, of 12356 with 47, and of 12346 with 57. It is thus obvious from symmetry that the sextic passes through the intersections of the conic determined by any five of the seven fixed points with the line joining the other two. There are twenty-one such pairs making a total of forty-two points.

By a simple regrouping of the terms of (5) with change of signs, we obtain

$$\lambda G_8C_1+\mu(aG_8-bF_8)C_2+\nu(bH_8-cG_8)C_3=0 \quad (6)$$

where C_1 C_2 C_3 are defined as in (2). C_2 has a double point at 4, C_3 at 5, and C_1 at 6. Using (6), it is very easy to show by means of the second partial derivatives that at 4 the tangents to the sextic coincide with those of C_2, at 5 with those of C_3, and at 6 with those of C_1. It is necessary only to indicate the differentiation. To tell the terms that vanish, it suffices to remember through what points the conics pass and the double points of the cubics. Hence it appears immediately from symmetry that the sextic has a double point at each of the seven fixed points whose tangents coincide with those of the cubic that has the same double point and passes through the other six fixed points. Of the eighteen intersections of the sextic with any of the cubics of this system six are at their common double point and the remaining twelve at the other six points. The conic through any five of the fixed points has ten of its twelve intersections with the sextic at those points, and the other two are on the line determined by the two remaining points. The line joining any two of the fundamental points meets the sextic four times at those points and twice where it meets the conic through the other five. Thus the sextic fulfills seventy-seven conditions; twenty-one by reason of its seven fixed double points, fourteen by reason of the determination of the tangents at those double points, and forty-two others corresponding to the twenty-one point pairs noted in the preceding paragraph.

One other property is easily obtained. $C_1C_2C_3$ are three linearly independent cubics through the seven fixed points. Hence any of the ∞^2 cubics through them is of the form $C \equiv \xi C_1 + \eta C_2 + \zeta C_3 = 0$ where ξ, η, ζ are constants. It follows immediately from the form of the first partial derivatives of C, that if C have a double point, it must lie on the Jacobian of $C_1C_2C_3$. But from the character of the latter it follows that their Jacobian is a sextic which must have a double point at each of the seven fixed points, and must have at 6, at 4, and at 5 the same tangents as C_1, C_2, C_3 respectively. Therefore it fulfills twenty-seven conditions which determine it uniquely as coincident with the invariant sextic.

Summarizing the results of this section we have the following theorem:

Given seven points in a plane, the (forty-two) points in which the line determined by any two meets the conic determined by the other five lie on a sextic which has a double point at each of the seven given points, and whose tangents at any one of these points coincide with those of the cubic that has the same double point and passes through the other six given points.

This sextic is the locus of the double points of all nodal cubics that pass through the seven given points.

It is also the locus of points 8 *such that the single infinity of cubics determined by* 8 *and the seven given points have a common tangent at* 8.

6. *The single infinity of cubics determined by the seven fixed points and an invariant point have a common tangent at the latter; and the envelope of the system of lines thus defined is of class four and order twelve.*

These common tangents are, as has been seen in section 5, paragraph 2, the lines l_p, or l_s, which coincide at points on the sextic. Their envelope, therefore, is the envelope of l_p, or, which is the same thing, that, of l_s, when 8, whose coördinates determine their coefficients, moves on the sextic. Difficulties of elimination render the more direct methods inapplicable. The class is, however, easily obtained. For if in l_p we fix xyz, the number of lines of the system through $(x : y : z)$ will depend on the intersections of the sextic with l_p regarded as a locus in $\lambda\mu\nu$. The latter is a quartic only four of whose intersections with the sextic depend on xyz. Hence the class is four. The same is obtained from l_s. It is not feasible, however, to express the condition that two of these four intersections coincide.

But the lines whose envelope we seek join 8 and 9, which for points on the sextic are infinitely near. From (1) the equation of the line connecting 8 with 9 is

$$x(\mu_7\nu p - \nu_7\mu s)\Sigma + y(\nu_7\lambda s - \lambda_7\nu)\Sigma + z(\lambda_7\mu - \mu_7\lambda p)\Sigma = 0$$

where Σ is the left hand member of the sextic given by (4). Removing this common factor the coefficients of xyz are easily shown to be proportional to the corresponding coefficients in l_p and l_s for points on the sextic. Substituting the values of p and s in the equation of 8 9, we have

$$xgC_1+y(ag-bf)C_2+z(bh-cg)C_3=0 \tag{7}$$

which, for $\lambda\mu\nu$ variable, is a cubic through the seven fixed points, and through $(x:y:z)$ by the first of identities (β), end of section 2. It has therefore, as expected, four intersections with the sextic which depend on xyz, and at each of them it is tangent to the corresponding line of the system. Hence the four lines of the system through an arbitrary $(x':y':z')$ are the four tangents to the corresponding cubic (7) from $(x':y':z')$. If $(x':y':z')$ is an inflection, there are only three such tangents; but in that case $(x':y':z')$ is a point P on the sextic and the line of the system corresponding to P makes the fourth. For this line is the common tangent at P to all the cubics (7) determined by points $(x':y':z')$ upon it, and in particular it is the inflectional tangent to that determined by P itself. And there are no other points $(x':y':z')$ which have this property. For they must lie on the Hessian of (7), which is a sextic in $x'y'z'$ when the latter are put for $\lambda\mu\nu$, and hence must be the same sextic.

If the cubic (7) has a double point, it must be on the sextic. Thus the envelope contains the points $(x':y':z')$ whose corresponding cubics (7) have double points. For the latter count for two intersections with the sextic in each case, and there will be only three lines of the system through $(x':y':z')$. And the envelope in general contains no other points. For the cubic (7) determined by an arbitrary $x'y'z'$ is tangent at its four intersections with the sextic that depend on $x'y'z'$ to the corresponding line of the system, and in general there are no points on the sextic at which this line coincides with the sextic. The condition that (7) have a double point is the vanishing of the eliminant of the first partial derivatives of (7) and of its Hessian. This is of degree twelve in xyz.

7. *The introduction of relations among the seven original points leads to transformations of lower order, each of which has an invariant curve.*

Something should be said of the special cases produced by introducing relations among the seven original points. It was noted in section 3 that the coördinates of 9, $\lambda'\mu'\nu'$, as given by (2) contain the common factor $(\lambda'\cdot\mu_2\cdot\nu_3)$ which vanishes when 1 2 3 are collinear. This does not mean that 9 is indeterminate if three of the seven points are on a line. The difficulty is obviated by numbering them suitably. It is very easy to deal with the special cases if we recollect in connection with the elementary properties of the cubic the geometric character of the Cs and Ks in the equations of transformation (2) or (2′).

If three points are collinear they may be numbered 2 3 7. Then all the steps taken in deriving (2) may be retraced and no equation becomes illusory. But $K_1\,K_2\,K_3$ all contain a linear factor corresponding to the line 237. This linear factor is therefore common to $\lambda'\ \mu'\ \nu'$ and appears in the invariant sextic. We have, therefore, a transformation of degree seven and an invariant quintic curve, whose properties are immediately deduced by regarding it as part of the sextic.

Thus the quintic will have double points at 1 4 5 6, and will pass through 2 3 7. Since 2 3 7 are collinear, the cubic through the seven fixed points that has a double point at 2 must consist of the line 2 3 7 and the conic 1 4 5 6 2. Hence this conic and the quintic have the same tangent at 2. Corresponding relations hold at 3 and 7. Proceeding thus, the quintic is seen to fulfill fifty-three conditions. Removing the common factor, λ' μ' ν' represent curves of order seven, each of which has a triple point at 1 4 5 6 and a double point at 2 7 3. This gives forty-eight common intersections and imposes thirty-three conditions, thus satisfying the requirements for a birational transformation of degree seven.

Other cases are dealt with in similar fashion. If four of the seven points are collinear, 9 is necessarily indeterminate. If six are on a conic they may be numbered 4 5 6 1 2 7. Then C_1 C_2 C_3 all contain a corresponding quadratic factor, which appears to the first degree in the sextic, and to the second degree in λ' μ' ν'. We have thus a quartic transformation with an invariant quartic curve. This quartic has a double point at 3, and its tangents at the other six fixed points are the lines joining them respectively to 3. Hence the class is at least ten; and since 3 is known to be a double point, it is exactly ten, and the deficiency is two. The same is true if 3 should be a cusp; for from a cusp three less tangents can be drawn than from an arbitrary point. This indicates that in the general case the only singularities of the non-degenerate sextic are those at the fixed points, and that its deficiency is three. If the six points lie on a proper conic, the quartic fulfills forty-seven conditions, relatively the largest number satisfied by any of the invariant curves connected with this transformation. If the six points lie on two lines, three points on each, the quartic satisfies thirty-five conditions. The coördinates of 9 represent quartic curves which have a common triple point at 3 and single intersections common at each of the other six points—a well known type of birational transformation of degree four. Since 4 5 6 1 2 7 are on a conic, 9 must lie on 3 8. This line meets the quartic twice at 3, and 9 is the harmonic conjugate of 8 with respect to the other two intersections of 3 8 with the quartic. If 8 is at 3, 3 8 is any line through 3, and thus to 3 will correspond a curve which is in fact the polar cubic of 3 with respect to the quartic, since any line through a double point of a quartic is divided harmonically by the quartic and the polar cubic. But this polar cubic has a double point at 3 and passes through the other six fixed points, since the tangents to the quartic at those points are the lines joining them to 3. To each of these six points corresponds the line joining it with 3.

If 4 5 6 1 2 7 are on a conic and 3 is on 2 7, the transformation is of degree three, and the invariant curve is a cubic of deficiency one. For it passes through 3 1 4 5 6, and its tangents at the last four points are 3 1, 3 4, 3 5, 36. The cubic fulfills twenty-nine conditions if the fixed conic does not degenerate, and twenty if it does. λ' μ' ν' represent cubic curves having a common double point at 3 and single intersections common at 1 4 5 6. 9 is again the harmonic conjugate of 8 with respect to the other two intersections of 3 8 with the cubic. Thus to 3 corresponds its polar conic with respect to the cubic. To 1 4 5 6 correspond the lines which join them respectively to 3.

The case when 4 5 6 1 2 7 are on a conic and 3 is the intersection of 2 7 and 1 6, is particularly interesting. The transformation is quadratic, and the invariant curve is a conic which, if the fixed conic does not degenerate, satisfies sixteen conditions. For it is tangent to 3 4 at 4, and to 3 5 at 5, and passes through the intersections of 1 5 with 4 6, of 75 with 2 4, of 1 4 with 5 6, of 7 4 with 5 2, of the conic 3 2 4 5 6 with 1 7, of 3 2 1 4 5 with 6 7, of 3 4 5 1 7 with 6 2, and of 3 4 5 6 7 with 1 2. 9 is again the fourth harmonic of 8 with respect to the intersections of 3 8 with the invariant conic, that is, the intersection of 3 8 with the polar of 8. The canonical form of this transformation is $x'=yz$, $y'=xy$, $z'=zx$. Inversion with respect to the unit circle is a special case. All these curves exist apart from the transformation that revealed them. The arbitrary numbering in each case simply serves to adapt it to the original procedure. Thus in the last case we may say that if six points are taken on a proper conic, and if any vertex of the complete quadrangle of any four is taken as a seventh point, there is a conic which has the properties just described. Finally if 4 5 6 1 2 7 are on a conic, and if 2 7, 1 6, and 4 5 are concurrent at 3, the transformation is a collineation. The invariant line is the polar of 3 with respect to the fixed conic, and 9 is the harmonic conjugate of 8 with respect to 3 and the intersection of 3 8 with the invariant line. 3 is a self-corresponding point.

Whenever six of the fixed points are on a conic, 9 must lie on the line joining 8 and the remaining point, unless 8 is on the fixed conic, in which case 9 is any point on that conic. It was observed in the case of the quartic, the cubic, and the quadratic transformations that the line connecting 8 with this remaining point had two intersections with the invariant curve that depended on 8. Hence in those cases the envelope considered in section 6 is indeed of class four; but it is simply the fixed conic and the remaining point (a curve of class one) counted twice. The cubics C_1, C_2, C_3 are not linearly independent when 4 5 6 1 2 7 are on a conic, but the results just stated may be verified in certain simplified cases by use of l_p r l_s as in section 6, paragraph 1.

Transmitted May 2, 1919.

UNIVERSITY OF CALIFORNIA PUBLICATIONS
IN
MATHEMATICS

Vol. 1, No. 11, pp. 223-240 February 17, 1920

CLASSIFICATION OF INVOLUTORY CUBIC SPACE TRANSFORMATIONS

BY

FRANK RAY MORRIS

In a paper "On the Rational Transformation between two Spaces"[1] Cayley gives a general treatment and two special cases of what he calls a cubo-cubic transformation. The correspondence is between the points of two spaces which may or may not be superimposed upon one another. The term rational is used in the sense of birational, i.e., the ratios of the homogeneous coördinates of either space can be expressed as rational functions of the coördinates of the corresponding point in the other space. If these rational functions are ratios whose terms are cubic functions of the coördinates the transformation is said to be cubo-cubic, or simply cubic.

Let x, y, z, w be the coördinates of a point in the first space and x', y', z', w' those of the corresponding point in the second space. Then the transformation is

$$x : y : z : w = X' : Y' : Z' : W'$$

where X', Y', Z', W' are certain cubic functions of x', y', z', w'. It is evident that a point in the second space determines a corresponding point in the first space. However, a point in the first space determines three cubic surfaces in the second and these in general determine 27 points. Hence the cubic surfaces must be so related that only one of their common points depends upon the point in the first space. Cayley shows that this condition is satisfied if the cubics have in common a twisted sextic curve with 7 apparent double points. This curve is called Σ'. Any cubic of the form $aX'+bY'+cZ'+dW'=0$, and in particular $X'=Y'=Z'=W'=0$, passes through Σ'. With this restriction on the cubic functions it is possible to express the transformation in the form

$$x' : y' : z' : w' = X : Y : Z : W,$$

[1] *Collected Math. Papers*, vol. VII, pp. 189–240. Also *Proc. London Math. Soc.*, vol. III (1869–1873), pp. 127–180.

where X, Y, Z, W are cubic function of x, y, z, w having a similar relation to that existing between those of the second space. They determine a sextic curve Σ with the same properties as Σ'. It is assumed that the most general transformation which gives rise to the cubo-cubic transformation is given by the equations

$$x'A_1+y'A_2+z'A_3+w'A_4=0,$$
$$x'B_1+y'B_2+z'B_3+w'B_4=0,$$
$$x'C_1+y'C_2+z'C_3+w'C_4=0,$$

where the capital letters are linear functions of x, y, z, w.

The points on Σ and Σ' are exceptional. To a point on Σ corresponds a line in the second space meeting Σ' in three points. As the point moves on Σ the corresponding line generates a surface of degree 8 with Σ' as a triple curve. The equation of this surface is got by equating to zero the Jacobian of any four of the cubic surfaces through Σ'. We shall call it $J'=0$. There is a similar surface, $J=0$, in the first space corresponding to the curve Σ'.

The transformation defined above is not in general involutory and neither of the special cases which Cayley gives is. But the involutory case is especially interesting. It is easily treated from the projective point of view if the two spaces are thought of as superimposed upon one another in such a way that Σ coincides with Σ' and $J=0$ with $J'=0$. The transformation is given by three quadric surfaces and their polar planes. To any point in the space corresponds a polar plane with respect to each of the three quadrics. These three planes in general determine a single point which, by the same process, determines the original point. Thus there is a one-to-one correspondence, with certain exceptions, between the points of the same space. The exceptional points are those for which the three polar planes have a common line. Such points lie on the curve Σ and, as in the general case, to Σ corresponds the surface $J=0$. A discussion of this case is given later.

A purely synthetic treatment of the involutory case is given by D. N. Lehmer in a paper, "On Combinations of Involutions."[2] Without the use of a system of coördinates or a single equation he discovers the properties of the curve Σ and the surface $J=0$. The properties are identical with those given by Cayley for the general case except that they do not include the number of apparent double points. This point of view leads to a number of special cases of particular interest where Σ breaks up in such a way as to contain one or more lines.[3]

This paper is an analytic treatment of the involutory cubic transformation. Case 1 gives the theory of the general case in which Σ is a proper curve.[4] It also gives the notation and plan of treatment used throughout the paper. The remainder of the paper develops the transformation and properties of Σ and $J=0$ for the cases in which Σ breaks up into lower degree curves one or more of which

---

[2] *Amer. Math. Monthly*, vol. XVIII, No. 3, Mar. 1911. See also Steiner's *Gess. Werke*, vol. II, p. 651.

[3] Elizabeth J. Easton gave a synthetic treatment of six of these cases in her thesis for the degree of M. A. at the Univ. of Cal. 1917. They occur in this paper as cases 2, 3, 5, 9, 11, 12.

[4] This case is given in a number of text books on solid geometry. See Salmon's *Geometry of Three Dimensions*, Art. 238a, p. 247 (fifth edition); Snyder and Sisam's *Analytic Geometry of Space*, pp. 170–172.

are straight lines including double lines. It is hoped that these special cases may prove valuable in the study of cubic surfaces and in resolving noninvolutory transformations into the product of involutory ones.

1. *Proper sextic curve*

Capital letters are used to denote functions of the homogeneous coördinates, and small letters, except the coördinates, to denote constants. An unaccented letter and the same letter accented denote the same function of x, y, z, w and x', y', z', w' respectively. The subscripts 1, 2, 3, 4 denote the partial derivatives with respect to x, y, z, w or x', y', z', w' respectively, except for the functions A, B, C, in which case they denote half the partials.

Let the three quadric surfaces be

$$A=a_{11}x^2+a_{22}y^2+a_{33}z^2+a_{44}w^2+2a_{12}xy+2a_{13}xz+2a_{14}xw+2a_{23}yz+2a_{24}yw+2a_{34}zw=0,$$
$$B=b_{11}x^2+b_{22}y^2+b_{33}z^2+b_{44}w^2+2b_{12}xy+2b_{13}xz+2b_{14}xw+2b_{23}yz+2b_{24}yw+2b_{34}zw=0,$$
$$C=c_{11}x^2+c_{22}y^2+c_{33}z^2+c_{44}w^2+2c_{12}xy+2c_{13}xz+2c_{14}xw+2c_{23}yz+2c_{24}yw+2c_{34}zw=0.$$

The three polar planes of a point P whose coördinates are x, y, z, w are

$$x'A_1+y'A_2+z'A_3+w'A_4=0,$$
$$x'B_1+y'B_2+z'B_3+w'B_4=0,$$
$$x'C_1+y'C_2+z'C_3+w'C_4=0.$$

These equations determine a point P' which corresponds to the point P. They may be written in the form

$$xA'_1+yA'_2+zA'_3+wA'_4=0,$$
$$xB'_1+yB'_2+zB'_3+wB'_4=0,$$
$$xC'_1+yC'_2+zC'_3+wC'_4=0.$$

The last three equations represent the polar planes of a point P' and they determine the point P. Hence the transformation is involutory. The equations are linear in either set of coördinates so that the transformation is what Cayley calls a cubo-cubic. It may be expressed as

$$x' : y' : z' : w'=X : Y : Z : W,$$

where X, Y, Z, W are the determinants, with the proper sign, of the matrix

$$\begin{Vmatrix} A_1 & A_2 & A_3 & A_4 \\ B_1 & B_2 & B_3 & B_4 \\ C_1 & C_2 & C_3 & C_4 \end{Vmatrix}.$$

Each determinant is a cubic function of x, y, z, w. Similarly, we get

$$x : y : z : w=X' : Y' : Z' : W'.$$

If for a point P the four determinants of the above matrix are each equal to zero, then the three polar planes no longer determine a point P', for this is the condition that they have a line in common. Then P is on each of the four surfaces $X=Y=Z=W=0$ and its locus is the sextic curve Σ.[5] Cayley shows that to the curve Σ corresponds the Jacobian surface of the four cubic surfaces. It is

$$J=\begin{vmatrix} X_1 & X_2 & X_3 & X_4 \\ Y_1 & Y_2 & Y_3 & Y_4 \\ Z_1 & Z_2 & Z_3 & Z_4 \\ W_1 & W_2 & W_3 & W_4 \end{vmatrix}=0.$$

It is a surface of degree eight and contains Σ as a triple curve. Both the curve and the surface play an important part in the special cases which we are now ready to consider.

2. *Sextic curve made up of a line and a quintic*

The simplest form of this case is that in which two of the quadrics are unlimited, the third is a pair of planes and the three are unrelated. The equations of the first two are the same as in the general case while the third may be taken as $C=2xy=0$.

The matrix is

$$\begin{Vmatrix} A_1 & A_2 & A_3 & A_4 \\ B_1 & B_2 & B_3 & B_4 \\ y & x & 0 & 0 \end{Vmatrix}$$

from which we get the cubic surfaces

$$\begin{aligned} &X=xD=0, \quad Y=yD=0, \\ &Z=x(A_1B_4-A_4B_1)-y(A_2B_4-A_4B_2)=0, \\ &W=x(A_1B_3-A_3B_1)-y(A_2B_3-A_3B_2)=0, \end{aligned}$$

where D is the determinant $A_3B_4-A_4B_3$.

The Jacobian surface is

$$J=\begin{vmatrix} D+xD_1 & xD_2 & xD_3 & xD_4 \\ yD_1 & D+yD_2 & yD_3 & yD_4 \\ Z_1 & Z_2 & Z_3 & Z_4 \\ W_1 & W_2 & W_3 & W_4 \end{vmatrix}=0.$$

To reduce it multiply the first and second columns and divide the first and second rows by x and y respectively, subtract the second row from the first and add the first column to the second. The result is

$$J=D\begin{vmatrix} D+xD_1+yD_2 & D_3 & D_4 \\ xZ_1+yZ_2 & Z_3 & Z_4 \\ xW_1+yW_2 & W_3 & W_4 \end{vmatrix}=3D\begin{vmatrix} D & D_3 & D_4 \\ Z & Z_3 & Z_4 \\ W & W_3 & W_4 \end{vmatrix}=0.$$

[5] For a proof that these three cubics have a common sextic curve see footnote 4 and Salmon's *Lessons on Higher Algbra*, Art. 271.

The last step is obtained by adding z times the second and w times the third column to the first, and the applying Euler's theorem with regard to partials of homogeneous functions to the elements of the first column.

With these equations before us we are now in a position to examine the sextic curve and the Jacobian surface. Σ is made up of the line $x=y=0$ and the quintic curve which is the intersection of $D=0$ and $Z=0$ except the line $A_4=B_4=0$, or of $D=0$ and $W=0$ except the line $A_3=B_3=0$. Since the line of Σ is not on $D=0$, is on $Z=0$ and does not meet the line $A_4=B_4=0$, it must meet the quintic curve in two points and have three apparent double points with it. Hence the quintic must have four apparent double points in order to complete the required number seven.[6]

The Jacobian surface is made up of the quadric $D=0$, which corresponds to the line of Σ, and a sixth degree surface corresponding to the quintic of Σ. The line is a triple line on the sixth degree surface, while the quintic is a double curve on the sixth degree surface and a single curve on the quadric.

The same results could have been obtained without having the quadric $C=0$ become a pair of planes, by choosing it as a general quadric so related to the other two that the three polar planes of a point on the line $x=y=0$ meet in a line. The quadrics are so related if $c_{i3}=ha_{i3}+kb_{i3}$, where $i=1, 2, 3$, and $c_{i4}=ha_{i4}+kb_{i4}$, where $i=1, 2, 3, 4$. As a point moves on the line $x=y=0$, the polar planes with regard to the three quadrics generate three axial pencils whose axes are rulings of the same set on $D=0$. Any three planes corresponding to the same point or the line meet in a ruling of the other set of $D=0$.

If in addition to the conditions already imposed on $C=0$ we choose $h=0$, then we have three general quadrics, one independent and the other two related in such a way as to give a line on Σ. This means that two of the axes mentioned above coincide and that the polar planes of a point on the line with regard to $B=0$ and $C=0$ coincide. If the planes of reference are appropriately choosen the cubic surfaces of this case and the one of the preceding paragraph become identical with those where one quadric is a pair of planes so that Σ and the Jacobian surface are also identical.

3. *Sextic curve made up of two skew lines and a quartic*

As in the preceding case the work is much simplified by choosing two of the quadrics pairs of planes. Then the three are

$$A=0, \quad B=2zw=0, \quad C=2xy=0.$$

The matrix is

$$\begin{Vmatrix} A_1 & A_2 & A_3 & A_4 \\ 0 & 0 & w & z \\ y & x & 0 & 0 \end{Vmatrix}.$$

[6] This particular quintic is discussed by Salmon, *Geometry of Three Dimensions*, Art. 352, also in *Cam. and Dub. Math. Jour.*, vol. v, p. 39.

And if we put $zA_3-wA_4=D$, $xA_1-yA_2=E$ the cubic surfaces are

$$X=xD=0, \quad Y=yD=0, \quad Z=zE=0, \quad W=wE=0.$$

The Jacobian of these surfaces is

$$J=\begin{vmatrix} D+xD_1 & xD_2 & xD_3 & xD_4 \\ yD_1 & D+yD_2 & yD_3 & yD_4 \\ zE_1 & zE_2 & E+zE_3 & zE_4 \\ wE_1 & wE_2 & wE_3 & E+wE_4 \end{vmatrix}=0.$$

By a process similar to that used in the preceding case, it may be reduced to

$$J=DE\begin{vmatrix} D+xD_1+yD_2 & zD_3+wD_4 \\ xE_1+yE_2 & E+zE_3+wE_4 \end{vmatrix}=3DE\begin{vmatrix} D & zD_3+wD_4 \\ E & E+zE_3+wE_4 \end{vmatrix}=0.$$

Thus we see that the Jacobian is made up of the two quadric surfaces, $D=0$, $E=0$, and a quartic surface, while Σ is made up of the two skew lines, $z=y=0$, $z=w=0$, and the quartic curve $D=E=0$. The first line is a ruling on $E=0$, is a double line on the quartic surface and meets the quartic curve in two points and has two apparent double points with it. The second line bears a similar relation to $D=0$, the quartic surface and the quartic curve. The two lines have one apparent double point and the quartic two making the correct total of seven. Salmon has called this quartic one of the first family. It is a single curve on each of the quadrics and the quartic of the Jacobian. To it corresponds the quartic surface, to the line $x=y=0$ corresponds $D=0$ and to $z=w=0$ corresponds $E=0$.

The same form of the sextic and the Jacobian may be obtained by choosing a special relation between the three quadrics. This relation is defined by the equations $c_{ij}=ha_{ij}+kb_{ij}$, where the combinations of ij are 13, 23, 33, 14, 24, 34, 44, $h_{ij}=ga_{ij}$, where $ij=13$, 23, 14, 24, and $ga_{ij}+fc_{ij}=f(ha_{ij}+kb_{ij})+b_{ij}$, where $ij=11$, 12, 22. The theory with this choice of the quadrics is similar to that with the general related quadrics in the preceding case.

If in addition to the above conditions on the general quadrics we assume $h=g=0$, we then have $A=0$ and $B=0$ unrelated, but $C=0$ has a special relation to the other two. The polar planes of a point on $x=y=0$ with respect to $B=0$ and $C=0$ coincide and those of a point on $z=w=0$ with respect to $A=0$ and $C=0$ coincide. As in the preceding case, the reference planes can be so chosen that the cubic surfaces derived from the general related quadrics are the same as when two of the quadrics are pairs of planes. Hence there is no change in Σ and the Jacobian.

4. Sextic curve made up of a double line and a quartic

This form of Σ is given if two of the quadrics are tangent to each other along one of their rulings. Let $x=y=0$ be the line of tangency of $B=0$ and $C=0$. The reference planes can be so chosen that the tetrahedron of reference is a selfpolar tetrahedron of $A=0$. Then the quadrics are of the form

$$\begin{aligned}
A&=x^2+y^2+z^2+w^2=0,\\
B&=b_{11}x^2+2b_{12}xy+b_{22}y^2+2b_{13}xz+2b_{14}xw+2b_{23}yz+2b_{24}yw=0,\\
C&=c_{11}x^2+2c_{12}xy+c_{22}y^2+2b_{13}xz+2b_{14}xy+2b_{23}yz+2b_{24}yw=0.
\end{aligned}$$

The matrix may be simplified by subtracting the second row from the third to

$$\left\Vert\begin{matrix} x & y & z & w\\ B_1 & B_2 & B_3 & B_4\\ (c_{11}-b_{11})x+(c_{12}-b_{12})y & (c_{12}-b_{12})x+(c_{22}-b_{22})y & 0 & 0\end{matrix}\right\Vert.$$

Let D, H, K be defined by the following equations,

$$\begin{aligned}
D&=zB_4-wB_3,\\
H&=(c_{12}x-b_{12}x+c_{22}y-b_{22}y)B_1-(c_{12}y-b_{12}y+c_{11}x-b_{11}x)B_2,\\
K&=(c_{12}x-b_{12}x+c_{22}y-b_{22}y)x-(c_{12}y-b_{12}y+c_{11}x-b_{11}x)y.
\end{aligned}$$

Making use of these equations we have from the matrix

$$\begin{aligned}
X&=\left\{(c_{12}-b_{12})x+(c_{22}-b_{22})y\right\}D=0,\\
Y&=\left\{(c_{11}-b_{11})x+(c_{12}-b_{12})y\right\}D=0,\\
Z&=wH-B_4K=0,\\
W&=zH-B_3K=0.
\end{aligned}$$

As in case 2, the Jacobian may be reduced to

$$J=3D\begin{vmatrix} D & D_3 & D_4\\ Z & Z_3 & Z_4\\ W & W_3 & W_4\end{vmatrix}=0.$$

By the introduction of the values in this case, it becomes

$$J=3D\begin{vmatrix} D & B_4 & -B_3\\ (wH-B_4K) & wH_3 & wH_4+H\\ (zH-B_3K) & zH_3+H & zH_4\end{vmatrix}=0.$$

By expanding this determinant we obtain

$$J=D^2\left\{2K(B_4H_3+B_3H_4)-3H(zH_3+wH_4)-H^2=0.\right\}$$

It is easy to see that $x=y=0$ is a double line on $X=0$ and $Y=0$, and it can be shown that the other two cubics are tangent along this line. Hence it counts as a double line of Σ. The remainder of Σ is the intersection of $D=0$ and $Z=0$ except the lines $x=y=0$, $w=B_4=0$. Since the last two lines lie in the plane $B_4=0$ they intersect, and the quartic is again of the first family with two apparent double points. The double line meets the quartic in two points and has two apparent intersections with it, counting for four apparent double points. The double line absorbs one double point, making a total of seven. The Jacobian is made up of a double quadric and a quartic surface. The double line is on the double quadric and a double line on the quartic surface so that it has four sheets of the Jacobian through it. The quartic curve is a single curve on both parts of the Jacobian thus having three sheets through it. To the double line corresponds the double quadric, and to the quartic curve corresponds the quartic surface. This case may be derived from the preceding one by letting the two skew lines coincide, which would cause the two quadrics $D=0$ and $E=0$ to coincide.

5. *Sextic curve made up of three skew lines and a twisted cubic*

This form of Σ is obtained by choosing for the quadrics three unrelated pairs of planes. We may choose the system of reference in such a way that their equations are

$$A=2xy=0, \quad B=2xw=0, \quad C=2MN=0,$$

where $M=m_1x+m_2y+m_3z+m_4w$, $N=n_1x+n_2y+n_3z+n_4w$.

The matrix is

$$\left\| \begin{matrix} y & x & 0 & 0 \\ 0 & 0 & w & z \\ m_1N+n_1M & m_2N+n_2M & m_3N+n_3M & m_4N+n_4M \end{matrix} \right\|$$

from which we get

$$X=xD=0, \quad Y=yD=0, \quad Z=zE=0, \quad W=wE=0,$$

where

$$\begin{aligned} D&=z(m_3N+n_3M)-w(m_4N+n_4M)=M(n_3z-n_4w)+N(m_3z-m_4w),\\ E&=x(m_1N+n_1M)-y(m_2N+n_2M)=M(n_1x-n_2y)+N(m_1x-m_2y). \end{aligned}$$

The Jacobian surface is

$$J=\begin{vmatrix} D+xD_1 & xD_2 & xD_3 & xD_4 \\ yD_1 & D+yD_2 & yD_3 & yD_4 \\ zE_1 & zE_2 & E+zE_3 & zE_4 \\ wE_1 & wE_2 & wE_3 & E+wE_4 \end{vmatrix}=0.$$

This determinant can be factored and reduced, as in case 3, to the form

$$J=DE\begin{vmatrix} D+xD_1+yD_2 & zD_3+wD_4 \\ xE_1+yE_2 & E+zE_3+wE_4 \end{vmatrix}=0.$$

If this second order determinant be written in full it may be factored into

$$\begin{vmatrix} m_3z-m_4w & n_3z-n_4w \\ m_1x-m_2y & n_1x-n_2y \end{vmatrix} \cdot \begin{vmatrix} N+n_1x+n_2y & N+n_3z+n_4w \\ M+m_1x+m_2y & M+m_3z+m_4w \end{vmatrix}.$$

And the second factor can be reduced to

$$\begin{vmatrix} m_1x+m_2y & m_3z+m_4w \\ n_1x+n_2y & n_3z+n_4w \end{vmatrix}$$

Let these two factors be called F and G respectively. Then the Jacobian $J=DEFG=0$ is composed of four quadric surfaces. Σ is in this case the three single skew lines $x=y=0$, $z=w=0$, $M=N=0$ and the twisted cubic $D=E=F=0$. The line $x=y=0$ is on the three quadrics $E=0$, $F=0$, $G=0$, and to it corresponds the quadric $D=0$; the line $z=w=0$ is on $D=0$, $F=0$, $G=0$, and to it corresponds $E=0$; the line $M=N=0$ is on $D=0$, $E=0$, $G=0$, and to it corresponds $F=0$; and to the cubic curve corresponds the quadric $G=0$. The three lines each meet the cubic in two points and give one apparent double point with it. They themselves give three apparent double points and the cubic gives one making a total of seven.

6. *Sextic curve made up of a double line, a single line skew to the double line and a twisted cubic*

The analysis of this case is made simple by choosing one general quadric, the second quadric a pair of planes and the third quadric a pair of planes with the double line a ruling on the first. The second is independent of the other two. The equations are

$$A=a_{11}x^2+a_{22}y^2+2a_{12}xy+2a_{13}xz+2a_{14}xw+2a_{23}yz+2a_{24}yw=0,$$
$$B=2zw=0, \quad C=2xy=0.$$

The matrix is

$$\begin{Vmatrix} A_1 & A_2 & A_3 & A_4 \\ 0 & 0 & w & z \\ y & x & 0 & 0 \end{Vmatrix}$$

from which we get

$$X=xD=0, \quad Y=yD=0, \quad Z=zE=0, \quad W=wE=0,$$

where D and E are defined as in case 3.

Due to the fact that A_3 and A_4 are functions of x, y alone, we see that $zD_3+wD_4=D$, and the Jacobian of case 3 is

$$J=D^2E(zE_3+wE_4)=0.$$

The Jacobian is composed of the double quadric $D=0$, and the single quadrics $E=0$, $zE_3+wE_4=0$. Σ is composed of the double line $x=y=0$. the single line $z=w=0$, and the twisted cubic common to $D=0$ and $E=0$. Each line meets the cubic in two points and has one apparent intersection with it, giving three apparent double points. The two lines give two apparent double points, the cubic one and the double line absorbs one, making a total of seven. The double line is on each of the three quadrics of the Jacobian and hence has four sheets through it. It is easy to see that the remainder of Σ has three sheets through it and to determine the correspondence between the parts of Σ and the parts of the Jacobian.

This same transformation may be obtained by choosing two general quadrics tangent along a line and the third a pair of planes. A third method of obtaining it is to choose two general quadrics tangent along a line and the third related to these two in the same manner as was given in the second part of case 2. In either case the analysis is the same as given above.

7. *Sextic curve made up of three concurrent lines and a plane cubic*

To obtain this form of the transformation let one of the quadrics be unrestricted and the other two be pairs of planes with their double lines intersecting. The equations are

$$A=a_{11}x^2+a_{22}y^2+a_{33}z^2+w^2+2a_{12}xy+2a_{13}xz+2a_{13}yz=0,$$
$$B=x^2-by^2=0. \quad C=x^2-cz^2=0.$$

The matrix is

$$\left\|\begin{matrix} A_1 & A_2 & A_3 & w \\ x & -by & 0 & 0 \\ x & 0 & -cz & 0 \end{matrix}\right\|$$

from which we get

$$X=bcyzw=0, \quad Y=cxzw=0, \quad Z=bxyw=0,$$
$$W=bcyzA_1+cxzA_2+bxyA_3=0.$$

The Jacobian is

$$J=\begin{vmatrix} 0 & zw & yw & yz \\ zw & 0 & xw & xz \\ yw & xw & 0 & xy \\ W_1 & W_2 & W_3 & 0 \end{vmatrix}=0.$$

To factor and reduce the determinant multiply the columns in order by x, y, z, w and apply Euler's theorem. The result, with the exception of a constant, is

$$J=xyzw^2W=0.$$

The surface $W=0$ is a cubic cone with the vertex at $(0, 0, 0, 1)$. Thus the Jacobian surface is made up of the four planes forming the tetrahedron of reference with the $w=0$ plane as a double plane and cubic cone containing three edges of the tetrahedron. Σ is composed of the lines $x=y=0$, $x=z=0$, $z=w=0$, and the plane cubic $W=w=0$. It is easy to see that each part of Σ is a triple curve on $J=0$. Each line meets the plane cubic in one point and gives two apparent double points with it, making a total of six. Cayley remarks (op. cit. page 227, footnote 1) that he has ascertained that an actual triple point on Σ counts as an apparent double point. Since the three lines give an actual triple point, the total is raised to 7. To each of the lines, with the exception of the triple point, corresponds the plane through the other two. To the triple point corresponds the double plane and to the cubic curve the cubic cone.

The same transformation is given if one of the quadrics is a general quadric and the other two are cones with a common vertex. It is evident that one of the cones may degenerate into a pair of planes. The double plane in this case is the polar with respect to the general quadric of the common vertex. This plane cuts the cones in two conics which have a selfpolar triangle. The sides of this triangle with the vertex of the cones determine the three single planes of the Jacobian surface.

8. Sextic curve made up of a single line, a double line and a plane cubic

This case is obtained from the preceding case by letting two of the lines of Σ fall together. This happens if a plane of one of the two pairs of planes contains the double line of the other pair. Let $A=0$ and $B=0$ be unchanged, and let $C=2xz=0$.

The matrix is

$$\begin{Vmatrix} A_1 & A_2 & A_3 & w \\ x & -by & 0 & 0 \\ x & 0 & x & 0 \end{Vmatrix}$$

from which we get

$$X=bxyw=0, \quad Y=x^2w=0, \quad Z=byzw=0,$$
$$W=byxA_1+x^2A_2-byzA_3=0.$$

The Jacobian surface is

$$J=\begin{vmatrix} yw & xw & 0 & xy \\ 2xw & 0 & 0 & x \\ 0 & zw & yw & yz \\ W_1 & W_2 & W_3 & 0 \end{vmatrix}=0.$$

When the determinant is reduced by a process similar to that used in the preceding case, the result is

$$J=x^2w^2yW=0.$$

In this case Σ is made up of the single line $x=y=0$, the double line $x=z=0$, and the plane cubic $w=W=0$. The lines intersect and each meets the cubic in one point. The single line gives two apparent double points with the cubic and the double line four, and the double line itself counts for one making a total of seven. The Jacobian surface is made up of the single plane $y=0$, the two double planes $x^2=0$, $w^2=0$, and the cubic cone $W=0$. The single plane is tangent to the cone along the double line of Σ. The double plane $x^2=0$ also passes through this line, so that it has four sheets of the Jacobian through it. It is easy to analyze the remainder of Σ and the Jacobian surface and to trace the correspondence between the parts of the two.

This transformation may be obtained by taking, instead of two pairs of planes, two cones with a common vertex and a common tangent plane along one of their common elements. Also we may choose one pair of planes and a cone having the double line of the pair of planes for an element.

9. *Sextic curve made up of three concurrent lines, a conic and a line in the plane of the conic*

This case is a special form of case 7 in which the plane cubic has broken up into a line and a conic. It is obtained by the quadrics as in case 7 with the additional requirement that $A=0$ be a pair of planes also. The equations are

$$A=(n_1x+n_2y+n_3z)^2-w^2=N^2-w^2=0,$$
$$B=x^2-by^2=0, \quad C=x^2-cz^2=0.$$

The matrix is

$$\begin{Vmatrix} n_1N & n_2N & n_3N & -w \\ x & -by & 0 & 0 \\ x & 0 & -cz & 0 \end{Vmatrix}$$

from which we obtain

$$X=bcyzw=0, \quad Y=cxzw=0, \quad Z=bxyw=0,$$
$$W=(n_1bcyz+n_2cxz+n_3bxy)N=0.$$

The reduction of the Jacobian is the same as in case 7. The result is

$$J=xyzw^2N(n+bcyz+ncxz+nbxy)=0.$$

The Jacobian is composed of four planes, a double plane and a quadric cone. Σ is composed of the three concurrent lines $x=y=0$, $x=z=0$, $y=z=0$, the line $w=N=0$, and the conic determined by the double plane and the cone. It is easy to see that each part of Σ is a triple line on $J=0$. The plane $N=0$ passes through the triple point formed by the three concurrent lines and through the line of $A=0$. Each of the first three lines meets the conic and has one apparent double point with it and one apparent double point with the fourth line. With the triple point this makes a total of seven. The three lines $x=w=0$, $y=w=0$, $z=w=0$ are triple lines on $J=0$ but are not a part of Σ. The correspondence between the parts of Σ and the Jacobian is the same as in case 7 except that we must consider the parts of the plane cubic. To the line in the plane of the conic corresponds the quadric cone, and to the conic corresponds the plane determined by this line and the tr˙ple point.

10. Sextic curve made up of a single line meeting a double line, a conic and a line in the plane of the conic

Two of the three concurrent lines of the preceding case now coincide. The analysis may be derived from case[5] by choosing the plane $M=0$ through the line $x=y=0$. Without loss of generality it may be taken as $M=x+y=0$. Then from case 5 we obtain the cubics

$$\begin{aligned} X &= x(x+y)(n_3z-n_4w)=0,\\ Y &= y(x+y)(n_3z-n_4w)=0,\\ Z &= Z(x+y)(n_1x-n_2y)+(x-y)N=0,\\ W &= w(x+y)(n_1x-n_2y)+(x-y)N=0. \end{aligned}$$

And the Jacobian is

$$J=(x+y)(n_3z-n_4w)(x-y)(n_3z+n_4w)\left\{(x+y)(n_1x-n_2y)+(x-y)N\right\}=0.$$

In this case Σ is composed of the double line $x=y=0$, the single lines $z=w=0$, $x+y=N=0$, and the conic $n_3z-n_4w=(x+y)(n_1x-n_2y)+(x-y)N=0$. The last line is skew to the first two but lies in the plane of the conic. The seven apparent double points may be accounted for as in case 8, since the only change in Σ is that the plane cubic is now a line and a conic. The Jacobian is made up of two double planes, two single planes, and a quadric cone. The double line of Σ has four sheets of the Jacobian through it for the same reason as given in case 8. The correspondence between the parts of Σ and the Jacobian may be traced as in cases 8 and 9.

11. Sextic curve made up of six lines. One line has four double points on it and a skew line to it has two triple points

The double line of the pair of planes $MN=0$, used in case 5, may be so situated that it meets each of the double line of the other pairs of planes. This condition exists if $m_1=n_1$, $m_2=n_2$, $m_3=kn_3$, $m_4=kn_4$. With this restriction the cubics determining Σ are

$$\begin{aligned} X &= x(m_3z-m_4w)(N+kM)=0,\\ Y &= y(m_3z-m_4w)(N+kM)=0,\\ Z &= z(m_1x-m_2y)(N+M)=0,\\ W &= w(m_1x-m_2y)(N+M)=0. \end{aligned}$$

The Jacobian surface is

$$J=(m_1x-m_2y)(m_3z-m_4w)(m_1x+m_2y)(m_3z+m_4w)(N+M)(N+kM)=0.$$

It is easy to see that Σ is made up of the six lines $M=N=0$, $m_1x-m_2y=m_3z-m_4w=0$, $x=y=0$, $z=w=0$, $m_1x-m_2y=N+kM=0$, $m_3z-m_4w=M+N=0$. The first and second are skew lines with one apparent double point. The third and fifth meet the second in different points, and the first in the same point forming a triple point. The fourth and sixth meet the second in separate points, and the first in the same point, which is a triple point different from the first. The third and fifth are skew to the fourth and sixth, thus accounting for four apparent double points and making a total of seven triple points and apparent double points.

The six planes of the Jacobian are located as follows: four of the planes pass through $M=N=0$ and are determined by the four lines which meet this line. The two double planes pass through the line skew to $M=N=0$ and are determined by the two triple points. Although the line $M=N=0$ has four planes of the Jacobian through it and the line $M_1x-m_2y=m_3z-m_4w=0$ has two double planes through it they are only single lines on the sextic curve. Each of the other four lines are triple lines on the Jacobian.

One of the triple points is common to the four planes of $A=0$ and $C=0$. Its polar with respect to $B=O$ is the double plane of the Jacobian which does not pass through it, i.e., $m_3z-m_4w=0$. The plane $m_1x+m_2y=0$ is the plane of the double lines of $A=0$ and $C=0$ while $M+N=0$ is its conjugate with respect to $C=0$. A similar theory holds for the other triple point.

12. Sextic curve made up of six lines which are the edges of a tetrahedron

If the three quadrics have a selfpolar tetrahedron the edges of this tetrahedron are the sextic curve. If this tetrahedron is used for the tetrahedron of reference the equations of the quadrics are

$$\begin{aligned} A&=a_{11}x^2+a_{22}y^2+a_{33}z^2+a_{44}w^2=0,\\ B&=b_{11}x^2+b_{22}y^2+b_{33}z^2+b_{44}w^2=0,\\ C&=c_{11}x^2+c_{22}y^2+c_{33}z^2+c_{44}w^2=0. \end{aligned}$$

The matrix is

$$\begin{Vmatrix} a_{11}x & a_{22}y & a_{33}z & a_{44}w\\ b_{11}x & b_{22}y & b_{33}z & b_{44}w\\ c_{11}x & c_{22}y & c_{33}z & c_{44}w \end{Vmatrix}$$

By dividing out constant factors we get

$$X=yzw=0,\quad Y=xzw=0,\quad Z=xyw=0,\quad W=xyw=0.$$

The Jacobian is

$$J=\begin{vmatrix} 0 & zw & yw & yz \\ zw & 0 & xw & yz \\ yw & xw & 0 & xy \\ yz & xz & xy & 0 \end{vmatrix}=0.$$

Therefore the Jacobian is made up of the four faces, each counted twice, of the selfpolar tetrahedron, and Σ is the six edges of the tetrahedron. Each of the six lines of Σ, being the intersection of two double planes, is a fourth order line on the Jacobian. There are four triple points on Σ, viz., the four vertices of the tetrahedron. Each of the three pairs of opposite edges of the tetrahedron gives an apparent double point. This makes a total of seven triple points and apparent double points. The correspondence between the parts of Σ and the Jacobian is evident.

A special form of this case is obtained by choosing all three of the quadrics pa'rs of planes situtated so that their double lines lie in the same plane. Where they are so related they have a selfpolar tetrahedron of which the three double lines of the pairs of planes are edges. The cubic equations of the transformation are identical with those for general quadrics having a selfpolar tetrahedron.

13. Sextic curve made up of two skew double lines and two skew single lines each meeting each double line

The simplest form of the quadrics which gives this form is that in which they are all pairs of planes with the double lines of the first and second lying in separate planes of the third. Their equations are

$$A=2xy=0, \quad B=2zw=0, \quad C=2(x+y)(z+w)=0.$$

The cubics which determine Σ are

$$X=x(x+y)(z-w)=0, \quad Y=y(z+y)(z+w)=0.$$
$$Z=z(x-y)(z+w)=0, \quad W=w(x-y)(z+w)=0.$$

And the Jacobian is

$$J=(x+y)(x-y)(z+w)(z-w)=0.$$

The sextic Σ is made up of the double lines $x=y=0$, $z=w=0$, and the single lines $x+y=z-w=0$, $x-y=z+w=0$. Each of the single lines meets each of the double lines. The single lines are skew and give one apparent double point, the double lines are skew and give four apparent double points, and each of the double lines absorbs one double point making the correct total. The Jacobian surface is made up of the four double planes determined by the four lines of Σ. Thus each of the four lines has four sheets of the Jacobian through it.

This same transformation is obtained if any three quadrics, no matter whether they are general quadrics, pairs of planes, or cones, are so chosen that the third is tangent to the first and second along different rulings of the same set on the third. This is a special case of Cayley's transformation in which Σ is made up of six lines. The four skew lines coincide in pairs making the transformation involutory.

14. *Sextic curve made up of a double line and four skew lines cutting it.*

The two skew lines of Cayley's transformation in which Σ is made up of six lines will fall together if the quadrics have a line in common. Although it is much easier to study this case with three pairs of planes, it is interesting to see that three general quadrics with a common line give the same form of Σ. For this reason the latter is given. Let the equations be

$$A=a_{11}x^2+a_{22}y^2+2a_{12}xy+2a_{13}xz+2a_{14}xw+2a_{23}yz+2a_{24}yw=0,$$
$$B=b_{11}x^2+b_{22}y^2+2b_{12}xy+2b_{13}xz+2b_{14}xw+2b_{23}yz+2b_{24}yw=0,$$
$$C=2(xw+yz)=0.$$

The matrix is

$$\begin{Vmatrix} A_1 & A_2 & a_{13}x+a_{23}y & a_{14}x+a_{24}y \\ B_1 & B_2 & b_{13}x+b_{23}y & b_{14}x+b_{24}y \\ w & z & y & x \end{Vmatrix}$$

In order to simplify the cubics of the matrix use the following definitions:

$$D=(a_{13}x+a_{23}y)(b_{14}x+b_{24}y)-(a_{14}x+a_{24}y)(b_{13}x+b_{23}y),$$
$$F=A_2(b_{13}x+b_{23}y)-B_2(a_{13}x+a_{23}y),$$
$$G=A_2(b_{14}x+b_{24}y)-B_2(a_{14}x+a_{24}y),$$
$$H=A_1(b_{13}x+b_{23}y)-B_1(a_{13}x+a_{23}y),$$
$$K=A_1(b_{14}x+b_{24}y)-B_1(a_{14}x+a_{24}y),$$
$$E=A_1B_2-A_2B_1.$$

Then the cubic surfaces are

$$X=xF-yG+zD=0, \quad Y=xH-yK+wD=0,$$
$$Z=xE-zK+wG=0, \quad W=yE-zH+wF=0.$$

It is easy to see that $w=y=0$ is a double line on the first two cubic surfaces It is not so easy to see that the last two are tangent along this line, but this can be shown to be true. Therefore the line is a double line on the sextic curve. The remainder of the curve may be examined from another point of view.

Salmon gives a formula[7] for the number of points common to three surfaces absorbed by a curve common to the three surfaces. It shows that a line common to three quadric surfaces absorbs four of the eight points common to the quadrics.

[7] *Solid Geometry*, Art. 355.

He also states[8] that the sextic curve is the locus of the vertices of the cones passing through the eight points of intersection of the three quadrics.[9] Since four of these points lie on the line $x=y=0$, a quadric cone through all eight points must either have its vertex on this line or break up into a pair of planes. If the cone is a pair of planes one plane is determined by the line and one of the four points and the other plane is determined by the remaining three points. The common line of the pair of planes is a line of the sextic curve. It intersects the double line $x=y=0$. There are four such pairs of planes and four such lines meeting the double line. The four lines are in general skew to each other, and give six apparent double points. The double line absorbs one apparent double point making a total of seven.

In order to study the Jacobian surface take the simple case in which the three quadrics are pairs of planes. We need make only a slight modification of the work done in case 5 so that the quadrics are

$$A=2xy=0,\quad B=2zw=0,\quad C=2(x+z)N=2(x+z)(n_1x+n_2y+n_3z+n_4w)=0.$$

As in case 5

$$X=xD=0,\quad Y=yD=0,\quad Z=zE=0,\quad W=wE=0,$$

where now

$$D=(x+z)(n_3z-n_4w)+zN,\quad E=(x+z)(n_1x-n_2y)+xN.$$

From these equations we find that the sextic is made up of the double line $x=z=0$ and the four single lines $x=y=0$, $z=w=0$, $x+z=N=0$, $n_1x+n_3z=n_3(n_1x+n_4w)+n_1(n_2y+n_3z)=0$.

The Jacobian surfaces is $J=DEFG=0$, where F and G are now given by

$$F=z(n_1x-n_2y)-x(n_3z-n_4w),\quad G=x(n_3z+n_4w)-z(n_1x+n_2y).$$

Each of the four quadrics of the Jacobian passes through the double line of the sextic and through three of the single lines. To each single line corresponds one of the quadrics and the double line corresponds to itself.

---

[8] *Ibid.*, Art. 238a.

[9] See reference to Snyder & Sisam, footnote 4.

RECAPITULATION

1. Sextic: a proper curve with seven apparent double points.
   Jacobian: a proper eighth degree surface.

2. Sextic: a line and a quintic. The line meets the quintic in two points. The quintic has four apparent double points.
   Jacobian: a quadric and a sextic surface.

3. Sextic: two skew lines and a quartic. Each line meets the quartic in two points. The quartic has two apparent double points.
   Jacobian: two quadrics and a quartic.

4. Sextic: double line and a quartic. The line meets the quartic in two points. The quartic has two apparent double points.
   Jacobian: a double quadric and a quartic.

5. Sextic: three skew lines and a twisted cubic. Each line meets the cubic in two points. The cubic has one apparent double point.
   Jacobian: four quadrics.

6. Sextic: a single line, a double line and a twisted cubic. The two lines are skew and each meets the cubic in two points The cubic has an apparent double point.
   Jacobian: two single quadrics and a double quadric.

7. Sextic: three concurrent lines and a plane cubic. Each line meets the cubic in one point.
   Jacobian: three single planes, a double plane and a cubic cone. The common point of the planes is the vertex of the cone.

8. Sextic: a single line, a double line and a plane cubic. The two lines intersect and each meets the cubic once.
   Jacobian: a single plane, two double planes and a cubic cone.

9. Sextic: four lines and a conic. One line lies in the plane of the conic. The other three are skew to it but meet in a point. Each of the three concurrent lines meets the conic once.
   Jacobian: four single planes, a double plane and a quadric cone. The four single planes pass through the vertex of the cone.

10. Sextic: two single lines, a double line and a conic. One of the single lines lies in the plane of the conic. The other meets the double line and the conic. The double line also meets the conic.
    Jacobian: two single planes, two double planes and a quadric cone.

11. Sextic: six lines. The first and second lines are skew. The other four meet the first in four distinct points and the second in pairs in two points.
    Jacobian: four single planes and two double planes.

12. Sextic: six lines which are the edges of a tetrahedron.
    Jacobian: four double planes which are the faces of the tetrahedron of the sextic curve.

13. Sextic: two skew single lines and two skew double lines. Each single line meets each double line.
    Jacobian: two double quadrics.

14. Sextic: four skew lines meeting a double line.
    Jacobian: four quadrics having a common line.

UNIVERSITY OF CALIFORNIA PUBLICATIONS
IN
MATHEMATICS

Vol. 1, No. 12, pp. 241-248 April 12, 1920

A SET OF FIVE POSTULATES FOR BOOLEAN ALGEBRAS IN TERMS OF THE OPERATION "EXCEPTION"

BY
J. S. TAYLOR

INTRODUCTION

There are three binary operations between classes which have come into general use in Boolean algebras. These three are "logical addition," "rejection," and "exception," and are expressed respectively by the symbols "$+$," " $|$," and "$-$." Simple and elegant sets of postulates already exist for the logic of classes in terms of "logical addition,"[1] and in terms of "rejection."[2] The set of postulates by B. A. Bernstein,[3] however, to whom the third operation is due, is somewhat involved and it is therefore the purpose of this paper to present a comparatively simp'e set in terms of "exception."

I

A SET OF FIVE POSTULATES FOR BOOLEAN ALGEBRAS IN TERMS OF THE OPERATION "EXCEPTION"

Let us take as undefined ideas a *class* K of *elements* $a, b, c, \ldots\ldots$ and an operation "$-$," $a-b$ reading "a except b." Then the logic of classes may be defined as a system Σ $(K,-)$ which satisfies the following five postulates:

Postulates

I. K contains at least two distinct elements.

II. If a and b are elements of K, $a-b$ is an element of K.

III. If a, b, and the combinations indicated are elements of K,

$$a-(b-b)=a.$$

[1] E. V. HUNTINGTON, "Sets of Independent Postulates for the Algebra of Logic," *Transactions of the American Mathematical Society*, Vol. V (1904), pp. 288–309.

[2] B. A. BERNSTEIN, "A Set of Four Independent Postulates for Boolean Algebras," *Transactions of the American Mathematical Society*, vol. XVII, pp. 50–52.

[3] B. A. BERNSTEIN, "A Complete Set of Postulates for the Logic of Classes in Terms of the Operation 'Exception', and a Proof of the Independence of a Set of Postulates Due to Del Re," *Univ. Calif. Publ. Math.*, vol. I, pp. 87–96 (May 15, 1914).

IV. There exists a unique element 1 in K such that, if a, b, and the combinations indicated are elements of K,

$$a-(1-b)=b-(1-a).$$

Definition 1. $a^{\cdot}=1-a.$

V. If IV holds, and if $a, b, c,$ and the combinations indicated are elements of K,

$$a-(b-c)=[(a-b)^{\cdot}-(a-c^{\cdot})]^{\cdot}.$$

Definition 2. $a^{\cdot\cdot}=(a^{\cdot})^{\cdot}.$

Theorems

Theorem 1. $a^{\cdot\cdot}=a$

Proof. $a-(1-1)=1-(1-a)$ by IV.

but $a-(1-1)=a$ by III.

and $1-(1-a)=1-a^{\cdot}$ by Def. 1.

$=(a^{\cdot})^{\cdot}$ by Def. 1.

$=a^{\cdot\cdot}$ by Def. 2.

Theorem 2. $a^{\cdot}$ is unique (for any a in K). by IV, II.

Theorem 3. $a-b^{\cdot}=b-a^{\cdot}$ by IV, Def. 1.

Corollary 1. $a^{\cdot}-b=b^{\cdot}-a$

Corollary 2. $a-b=b^{\cdot}-a^{\cdot}$

Theorem 4. $(a-a)^{\cdot}=(b-b)^{\cdot}$

Proof. $(a-a)^{\cdot}-(b-b)=(a-a)^{\cdot}$ by III.

and $(a-a)^{\cdot}-(b-b)=(b-b)^{\cdot}-(a-a)$ by Th. 3, Cor. 1.

$=(b-b)^{\cdot}$ by III.

Theorem 5. $1=(e-e)^{\cdot}.$

Proof. First, $(e-e)^{\cdot}$ is unique. by II; Th's 2 and 4.

Secondly, $(e-e)^{\cdot}$ satisfies the equation of IV, in other words,

$$a-[(e-e)^{\cdot}-b]=b-[(e-e)^{\cdot}-a]$$

for, $a-[(e-e)^{\cdot}-b]=a-[b^{\cdot}-(e-e)]$ by Th. 3, Cor. 1.

$=a-b^{\cdot}$ by III.

and $b-[(e-e)^{\cdot}-a]=b-[a^{\cdot}-(e-e)]$ by Th. 3, Cor. 1.

$=b-a^{\cdot}$ by III.

but $a-b^{\cdot}=b-a^{\cdot}$ by Th. 3.

Theorem 6. $a-a^{\cdot}=a$

Proof. Set $b=c=a^{\cdot}$ and $a=a$ in V.

The left then becomes $a-(a^{\cdot}-a^{\cdot})=a$ by III.

The right becomes $[(a-a^{\cdot})^{\cdot}-(a-a^{\cdot\cdot})]^{\cdot}=[(a-a^{\cdot})^{\cdot}-(a-a)]^{\cdot}$ by Th. 1.

$=[(a-a^{\cdot})^{\cdot}]^{\cdot}$ by III.

$=a-a^{\cdot}$ by Th. 1.

Corollary. $a^{\cdot}-a=a^{\cdot}$

Theorem 7. $a-(b-c)=[(b^{\cdot}-a^{\cdot})^{\cdot}-(c-a^{\cdot})]^{\cdot}$ by V; Th. 3; Th. 3, Cor. 2.

Corollary. $a^{\cdot\cdot}-(b^{\cdot\cdot}-c)=[(b^{\cdot}-a^{\cdot})^{\cdot}-(c^{\cdot\cdot}-a^{\cdot})]^{\cdot}$ by Th. 1.

Theorem 8. $(b^{\cdot}-a)^{\cdot}-(b-a)=a$

Proof. Set $a=a^{\cdot}$, $c=b$ in Th. 7, which then becomes

$$a^{\cdot}-(b-b)=[(b^{\cdot}-a^{\cdot\cdot})^{\cdot}-(b-a^{\cdot\cdot})]^{\cdot}$$

but $a^{\cdot}-(b-b)=a^{\cdot}$ by III.

and $[(b^{\cdot}-a^{\cdot\cdot})^{\cdot}-(b-a^{\cdot\cdot})]^{\cdot}=[(b^{\cdot}-a)^{\cdot}-(b-a)]^{\cdot}$ by Th. 1.

hence $[(b^{\cdot}-a)^{\cdot}-(b-a)]^{\cdot}=a^{\cdot}$

hence $[(b^{\cdot}-a)^{\cdot}-(b-a)]^{\cdot\cdot}=a^{\cdot\cdot}$ by Th. 2.

and $(b^{\cdot}-a)^{\cdot}-(b-a)=a$ by Th. 1.

Corollary. $(b^{\cdot}-a)^{\cdot}-(b^{\cdot\cdot}-a)=a$

Definition 3. $a\mid b=a^{\cdot}-b$

Theorem 9. $a\mid a=a^{\cdot}$ by Th. 6, Cor. and Def. 3.

Definition 4. $a'=a\mid a$

Theorem 10. $a'=a^{\cdot}$ by Def. 4, Th. 9.

Theorem 11. $(b\mid a)\mid(b'\mid a)=a$

Proof. $(b^{\cdot}-a)^{\cdot}-(b^{\cdot\cdot}-a)=a$ by Th. 8, Cor.

but $(b^{\cdot}-a)^{\cdot}-(b^{\cdot\cdot}-a)=(b\mid a)^{\cdot}-(b^{\cdot}\mid a)$ by Def. 3.

$=(b\mid a)\mid(b'\mid a)$ by Def. 3; Th. 10.

Theorem 12. $a'\mid(b'\mid c)=[(b\mid a')\mid(c'\mid a')]'$

Proof. $a^{\cdot\cdot}-(b^{\cdot\cdot}-c)=[(b^{\cdot}-a^{\cdot})^{\cdot}-(c^{\cdot\cdot}-a^{\cdot})]^{\cdot}$ by Th. 7, Cor.

but $a^{\cdot\cdot}-(b^{\cdot\cdot}-c)=a^{\cdot\cdot}-(b^{\cdot}\mid c)$ by Def. 3.

$=a^{\cdot}\mid(b^{\cdot}\mid c)$ by Def. 3.

$=a'\mid(b'\mid c)$ by Th. 10.

and $[(b^{\cdot}-a^{\cdot})^{\cdot}-(c^{\cdot\cdot}-a^{\cdot})]^{\cdot}=[(b\mid a^{\cdot})^{\cdot}-(c^{\cdot}\mid a^{\cdot})]^{\cdot}$ by Def. 3.

$=[(b\mid a^{\cdot})\mid(c^{\cdot}\mid a^{\cdot})]^{\cdot}$ by Def. 3.

$=[(b\mid a')\mid(c'\mid a')]'$ by Th. 10.

Sufficiency

That postulates I–V are sufficient is now evident. In the light of Definition 3 postulates I and II give P_1 and P_2 of B. A. Bernstein's set of four postulates in terms of "rejection" referred to in an earlier part of this paper; while P_3 and P_4 of that set are here exhibited as theorems 11 and 12. That postulates I–V may likewise be derived from P_1-P_4 of Bernstein's set is also easily shown. Thus the two sets of postulates are equivalent.

Consistency

The consistency of the set of postulates I–V is demonstrated by the following system composed of two distinct elements e_1 and e_2 which satisfies all five postulates. As in succeeding examples, e_i-e_j will be given by means of a table; so that if, as in the present instance, $e_1-e_1=e_2$, $e_1-e_2=e_1$, $e_2-e_1=e_2$, and $e_2-e_2=e_2$, this will be stated in the form:

$-$	e_1	e_2
e_1	e_2	e_1
e_2	e_2	e_2

That this system Σ satisfies all the postulates the reader may verify without difficulty.

Independence

The independence of the five postulates is demonstrated by exhibiting five systems each satisfying all but one of the postulates, the unsatisfied postulate being I$-$V in turn. The system Σ failing to satisfy the ith postulate will be designated Σ^{-i}. $e_i-e_j=x$ means that e_i-e_j does not give an element belonging to K.

Σ^{-1}; K a class of one element e_1, with $e_1-e_1=e_1$.

Σ^{-2}; K a class of two distinct elements e_1 and e_2, with e_i-e_j defined by the accompanying table:

$-$	e_1	e_2
e_1	e_1	e_1
e_2	e_2	x

Σ^{-3}; K a class of three distinct elements, with e_i-e_j defined by the accompanying table:

$-$	e_1	e_2	e_3
e_1	e_1	e_1	e_1
e_2	e_2	e_1	e_2
e_3	e_1	e_1	e_1

III fails for $a=e_3$.

IV holds, for e_2, and e_2 only, satisfies the conditions imposed on 1.

V holds. It holds obviously for a, b, and c limited to e_1 and e_2, for that part of K is identical with the system used to demonstrate the consistency of the postulates with e_1 and e_2 simply interchanged. The other possibilities may be disposed of as follows:

(1) $a=e_1$ or e_3, $b=e_i$, $c=e_j$ $(i,j=1, 2, 3)$.

$$e_{1 \text{ or } 3}-(e_i-e_j)=e_1, \quad [(e_{1 \text{ or } 3}-e_i)^{\cdot}-(e_{1 \text{ or } 3}-e_j^{\cdot})]^{\cdot}=[e_2-e_1]^{\cdot}=e_1$$

(2) For $a=e_2$ we have the following as yet undisposed of cases:

$$e_2-(e_3-e_i)=e_2, \quad [(e_2-e_3)^{\cdot}-(e_2-e_i^{\cdot})]^{\cdot}=[e_1-e_j]^{\cdot}=e_2$$

$$e_2-(e_1-e_3)=e_2, \quad [(e_2-e_1)^{\cdot}-(e_2-e_3^{\cdot})]^{\cdot}=[e_1-e_1]^{\cdot}=e_2$$

$$e_2-(e_2-e_3)=e_1, \quad [(e_2-e_2)^{\cdot}-(e_2-e_3^{\cdot})]^{\cdot}=[e_2-e_1]^{\cdot}=e_1$$

Σ^{-4}; K a class of two distinct elements e_1 and e_2 with e_i-e_j defined by the table:

$-$	e_1	e_2
e_1	e_1	e_1
e_2	e_2	e_2

Neither e_1 nor e_2 satisfies the conditions imposed on 1 by IV.

V is satisfied vacuously.

Σ^{-5}; K a class of three distinct elements with $e_i - e_j$ defined by the table:

$-$	e_1	e_2	e_3
e_1	e_1	e_1	e_1
e_2	e_2	e_1	e_3
e_3	e_3	e_1	e_1

III is obviously satisfied.
IV is satisfied, for e_2 satisfies the conditions imposed on 1.
V fails for $a=b=c=e_3$.

II

COMPLETE EXISTENTIAL THEORY

As has already been shown, postulates I—V are independent in the ordinary sense that no one of the postulates is implied by the other four. Professor E. H. Moore,[4] however, has suggested the question in connection with sets of postulates of determining not only the implicational relations existing among the postulates as they stand, but also all the implicational relations which exist among properties defined either by the postulates themselves or by the negatives of the postulates. A set of postulates is said to be *completely independent* if, and only if, no such implicational relations exist. For example, I have shown in an earlier paper[5] that while Bernstein's set of four postulates in terms of "rejection" already referred to are independent in the ordinary sense, they are not completely independent, since the negative of the first postulate implies the third and fourth.

Any system Σ $(K, -)$ of the type prescribed earlier in this paper has with respect to the five postulates there stated one of the $2^5 = 32$ characters:

(1) $(+++++), (++++-), (+++-+), \ldots\ldots (+----), (-----)$;

the ith sign of the character being plus or minus according as Σ does or does not satisfy the ith postulate. The body of thirty-two propositions stating for the various characters represented in (1) that there exists or does not exist a system having the character in question constitutes what Professor Moore has called "the complete existential theory" of the five postulates.

For the five postulates in question the complete existential theory consists of fourteen propositions of existence and eighteen propositions of non-existence. The eighteen non-existencies arise from the fact that the negative of I implies III, IV, and V, and that also the negative of IV implies V.[6]

---

[4]E. H. Moore, "Introduction to a Form of General Analysis," *New Haven Mathematical Colloquium*, Yale University Press, p. 82.

[5]J. S. Taylor, "Complete Existential Theory of Bernstein's Set of Four Postulates for Boolean Algebras," *Annals of Mathematics, Second Series*, vol. XIX, No. 1, pp. 64–69 (September, 1917).

[6]The question naturally arises as to whether it might not be possible to modify the postulates in a way such that they would become completely independent. The writer has investigated this possibility in considerable detail but has been unable to make such a modification without a considerable loss of simplicity. The simplest change found to bring about the desired results is as follows:—

I′. *K contains at least four distinct elements.*

V′. *There exists an element ϵ in K such that, for each a, b, c choice for which there exists any element in K satisfying the condition imposed upon 1 by the equation of IV for each pair of elements in the group of elements obtained by combining a, b, and c in all possible ways, the element ϵ does so,*

Propositions of Non-Existence

The eighteen propositions of non-existence, as implied above, may be expressed by the two following propositions:[7]

(2) $\Sigma^{-1} \supset \Sigma^{345}$

(3) $\Sigma^{-4} \supset \Sigma^{5}$

The truth of these two propositions is readily perceived from the following considerations. First, the hypothesis that I is not satisfied necessitates either K^{null} (a class without any elements) or $K^{singular}$ (a class with only one element). But if K contains no elements, postulates III, IV, and V are satisfied vacuously. And if K contains only one element, then they are satisfied either evidently or vacuously, according as Σ does or does not satisfy postulate II.

Secondly, the hypothesis that IV is not satisfied obviously results in the vacuous satisfaction of V.

Propositions (2) and (3) render impossible the existence of systems with the following eighteen characters.

$(-+++-)$, $(-++-+)$, $(-+-++)$, $(-++--)$, $(-+-+-)$,
$(-+--+)$, $(-+---)$, $(--++-)$, $(--+-+)$, $(---++)$,
$(--+--)$, $(---+-)$, $(----+)$, $(-----)$, $(+++--)$,
$(++---)$, $(+-+--)$, $(+----)$.

Propositions of Existence

The fourteen propositions of existence are established by the exhibition of fourteen systems having the remaining fourteen characters; there are two examples for $K^{singular}$, seven for K^{dual}, and five for K^{triple}. In each case K contains the least number of elements possible.

Examples for K singular

Systems having the characters $(-++++)$ and $(--+++)$ respectively are the following:

System I_1. Character $(-++++)$; class composed of single element e_1, with $e_1 - e_1 = e_1$.

System I_2. Character $(--+++)$; class composed of single element e_1, with $e_{.} - e_1 \neq e_1$.

---

and such that, for such an a, b, c choice, if $e^{\cdot}$ *be defined as* $\epsilon - e$, *and if a, b, c, and the indicated combinations are elements of K,*

$$a-(b-c)=[(a-b)^{\cdot}-(a-c^{\cdot})]^{\cdot}$$

Since considerable space would be occupied by a proof of the fact that the postulates thus modified are completely independent and since the reader should meet with no very serious difficulty in establishing this fact for himself, such proof is here omitted.

[7] $\Sigma^{-i} \supset \Sigma^{jh \cdots n}$ means, "If a system Σ does not satisfy the ith postulate, then it does satisfy postulates $j, k, \ldots$, and n."

Examples for K dual

System II_1

(+++++),

$-$	e_1	e_2
e_1	e_2	e_1
e_2	e_2	e_2

;

System II_2

(+++−+),

$-$	e_1	e_2
e_1	e_1	e_1
e_2	e_2	e_2

;

System II_3

(+−+++),

$-$	e_1	e_2
e_1	e_1	e_1
e_2	e_2	x

;

System II_4

(++−+−),

$-$	e_1	e_2
e_1	e_1	e_2
e_2	e_1	e_1

;

System II_5

(++−−+),

$-$	e_1	e_2
e_1	e_1	e_1
e_2	e_1	e_1

;

System II_6

(+−+−+),

$-$	e_1	e_2
e_1	x	x
e_2	x	x

;

System II_7

(+−−−+),

$-$	e_1	e_2
e_1	x	e_1
e_2	e_1	e_1

Examples for K triple

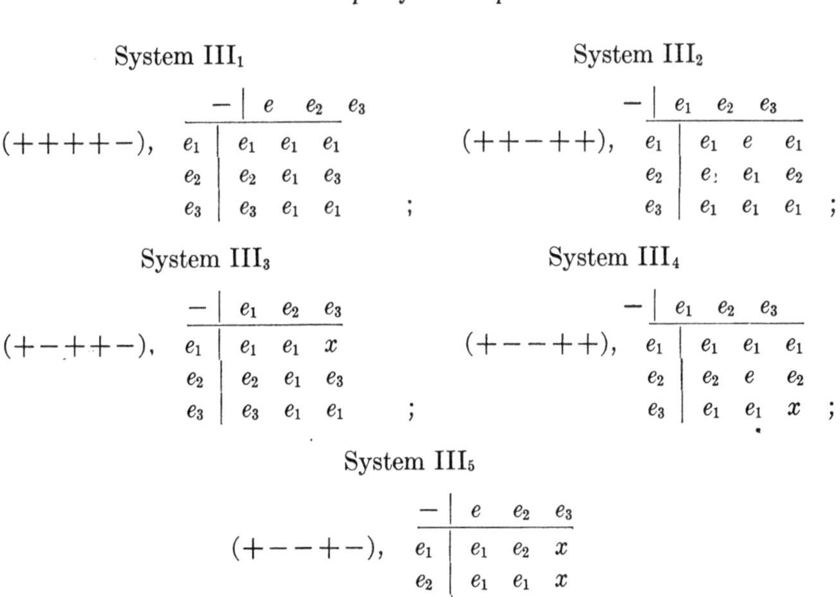

System III_1

(++++−),

$-$	e	e_2	e_3
e_1	e_1	e_1	e_1
e_2	e_2	e_1	e_3
e_3	e_3	e_1	e_1

;

System III_2

(++−++),

$-$	e_1	e_2	e_3
e_1	e_1	e	e_1
e_2	e:	e_1	e_2
e_3	e_1	e_1	e_1

;

System III_3

(+−++−),

$-$	e_1	e_2	e_3
e_1	e_1	e_1	x
e_2	e_2	e_1	e_3
e_3	e_3	e_1	e_1

;

System III_4

(+−−++),

$-$	e_1	e_2	e_3
e_1	e_1	e_1	e_1
e_2	e_2	e	e_2
e_3	e_1	e_1	x

;

System III_5

(+−−+−),

$-$	e	e_2	e_3
e_1	e_1	e_2	x
e_2	e_1	e_1	x
e_3	e_1	e_2	x

III

THE ELEMENT 1 AND NEGATION

It is interesting to note that it is impossible to express either the element 1 or negation, "not–a," directly in terms of the operation of "+" or "–," although this can be done in terms of "rejection." It has been found necessary in all sets of postulates in terms of "logical addition" or "exception," therefore, to postulate one or both of these two ideas. Curiously enough, although there are several sets in which only 1 is postulated and "not–a" then defined, there has been no set formulated in which "not–a" is postulated and the element 1 defined. This might lead one to believe that the element 1 plays a more important rôle than negation, but that this does not follow in the case of "exception," at least, is demonstrated by the following set of five postulates in which only "not–a" is postulated.

I. K contains at least two distinct elements.

II. If a and b are elements of K, $a-b$ is an element of K.

III. If a, b, and the combinations indicated are elements of K, $a-(b-b)=a$

IV. For every element e in K there exists another element $e^{\cdot}$ in K, unique for each e, such that, if the combinations indicated are elements of K,

(1) $e-e^{\cdot}=e$

and (2) $a-(b-c)=[(c-a^{\cdot})^{\cdot}-(b^{\cdot}-a^{\cdot})]^{\cdot}$,

where each dotted element is an element satisfying (1).

Definition 1. $a^{\cdot\cdot}=(a^{\cdot})^{\cdot}$

Theorem 1. $a^{\cdot\cdot}=a$

Proof. Set $b=c=a$ in IV (2)

the left becomes $a-(a-a)=a$ by III,

while the right becomes $[(a-a^{\cdot})^{\cdot}-(a^{\cdot}-a^{\cdot})]^{\cdot}=[(a-a^{\cdot})^{\cdot}]^{\cdot}$ by III,

$=[(a)^{\cdot}]^{\cdot}$ by IV (1),

$=a^{\cdot\cdot}$ by Def. 1.

Theorem 2. $a^{\cdot}-a=a^{\cdot}$ by IV (1), Th. 1.

Theorem 3. $b^{\cdot}-a^{\cdot}=a-b$

Proof. Set $a=a$, $b=b$, $c=b^{\cdot}$ in IV (2)

$a-(b-b^{\cdot})=[(b^{\cdot}-a^{\cdot})^{\cdot}-(b^{\cdot}-a^{\cdot})]^{\cdot}$

whence $a-b=[(b^{\cdot}-a^{\cdot})^{\cdot}]^{\cdot}$ by IV (1), Th. 2.

$=b^{\cdot}-a^{\cdot}$

Corollary. $a^{\cdot}-b=b^{\cdot}-a$

Theorem 4. $a-(b-c)=[(b^{\cdot}-a^{\cdot})^{\cdot}-(c-a^{\cdot})]^{\cdot}$ by Th. 3, Cor.

Theorem 5. $(b^{\cdot}-a)^{\cdot}-(b-a)=a$

Proof. Set $a=a^{\cdot}$, $c=b$ in Th. 5.

Definition 2. $1=(a-a)^{\cdot}$

The rest of the development and the proof of the sufficiency of the set of postulates follow so closely those of the set for which it has already been explained in detail that the reader is left to complete the work for himself.

(*Flow of Electricity in a Magnetic Field,* Four Lectures, by Vito Volterra, pp. 249–320, 40 text figures. Issued May 28, 1921.)

ACKNOWLEDGMENT

Page 253. The method of obtaining equations (2) by the use of the electron theory is due to Professor Corbino.

UNIVERSITY OF CALIFORNIA PUBLICATIONS
IN
MATHEMATICS

Vol. 1, No. 13, pp. 249-320, 40 text figures May 28, 1921

FLOW OF ELECTRICITY IN A MAGNETIC FIELD

FOUR LECTURES

BY
VITO VOLTERRA

FIRST LECTURE

Section 1—Theory of electrical phenomena. Section 2—Electrons. Section ꟷlectron theory of metals. Section 4—The problem of flow of electricity in a e, homogeneous, isotropic metal plate kept at a constant temperature and ect to the action of a uniform magnetic field perpendicular to it. Section 5—hematical theory in relation to the preceding problem: General formulae. ion 6—Characteristics of the problem. Section 7—Point electrodes. Section ɔther general formulae. Section 9—The principle of reciprocity.

SECTION 1

ın historical order, the first theory of electrical phenomena may be characterized ɔy the name of Coulomb. This theory assumes the real existence of particular, imponderable substances called electricity, and the instant action at a distance (activ in distans) of such substances, an action analogous to that between masses in the theory of gravitation.

A new theory of electricity corresponds to the general energetical tendencies of the second half of the last century; it is characterized by the names of Faraday, Maxwell, Hertz, Heaviside. This theory does not admit the real existence of particular substances forming electricity; it denies the possibility of immediate action at a distance. According to this theory, the actions between electrified bodies are exercised by means of ether, in which the presence of these bodies determines particular deformations and perturbations; electrical actions propagate themselves in ether with the velocity of light. In the Maxwell-Hertz theory the electric charge has a purely mathematical definition; it loses the physical concrete meaning attributed to it by the old theory. Among the advantages of Maxwell's theory, must be remembered that of having connected two parts of physics, electricity and optics, which before were always considered completely distinct.

Already in the last years of the nineteenth century the deeper study of already known phenomena and of new ones, led physicists to a new theory of electrical phenomena, i.e., the electron theory (1) which has been considerably developed in these first years of present century. This new theory takes from the older one the notion of the real existence of a particular *substratum* of electrical phenomena to which it attributes an atomistic structure, and on the other hand it holds from Maxwell's theory the principle of action by means of a medium and the general form of equations. The modifications that these undergo in the new theory, though of an important intrinsic meaning, are externally very slight.

SECTION 2

We shall recall some experimental facts that led physicists to the notion of elementary electric charges, or electric corpuscles, or electrons.

Helmholtz was the first to observe that the laws of electrolysis given by Faraday lead to the conclusion, that to the atomistic structure of matter corresponds an atomistic structure of electricity. In fact, according to these laws, the electric charge that is carried by a monovalent atom in electrolysis always retains the same absolute value; and the one carried by a bivalent atom always retains the double absolute value, etc. This law is very easily explained, if one considers that each atom of matter transfers in electrolysis a whole number, equal to the valency, of elementary electric charges.

Researches on the electrical conductivity of gases have demonstrated in a clear and suggestive manner, the existence of atoms of electricity. (2) For instance, under the action of Röentgen rays, or of radio-active substances, gases acquire, together with conductivity, the property of condensing more easily the supersaturated vapor of water; the condensation takes place in the form of electrified drops, of which it has been possible to measure the charge. For this latter either the same value of the charge carried by a monovalent atom in electrolysis has been found or a multiple of that value (between 4 and 5×10^{-10} electrostatic units C.G.S.).

Among the other experimental researches that led to the acceptance of the granular structure of electricity and the determination of the value of the elementary electric charge, we note those on the cathode rays, those on the emission of electrified particles from bodies brought to an incandescent state or exposed to ultra-violet radiations, those on the phenomena of radio-activity and on the Zeemann effect.

It will be observed, that while in a whole series of phenomena the existence of negative, free electrons (not bound to ordinary matter) has been demonstrated, the same can not be said for positive electrons.

Lastly we note that the hypothesis of bodies containing electrons of different distribution and mobility, according to the substance of which they are formed, has been able to eliminate the difficulties that arose in Maxwell's theory concerning the propagation of light in material media.

SECTION 3

To explain the different phenomena, different properties have been attributed to electrons (concerning especially their degrees of freedom) according to the nature of the substances in which they are found. It will suffice to give a short explanation of the electronic theory of metals.

To explain the conductivity of metals, it is assumed that their atoms emit with facility electrified corpuscles. According, then, to the electron theory, in metals there are not only, as in insulators, bound electrons, that follow the molecules or the atoms in their thermal oscillations and that only separate from them under particular actions, but also free electrons that move in the inter-molecular or inter-atomic spaces, with a movement analogous to that of the molecules of a gas.

Concerning the movement of these electrons, there are two different ways of viewing the question: The electron theory of metals may be developed on the assumption of Riecke, Drude, Lorentz (3) that a free electron effects a great number of collisions with molecules or with the neutral atoms before uniting again with a charge of contrary sign; —or one may assume, as J. J. Thomson proposes (4) that the electron emitted by an atom is instantly captured by the nearest atoms. We shall retain the first point of view, which is more generally adopted. We shall thus be able to assume that in a metal an equilibrium of temperature between molecules and electrons is reached and to these latter the formulae of the ordinary kinetic theory may be applied.

Let us suppose that at the absolute temperature T, the positive and negative electrons per cubic centimeter are respectively N_1 and N_2, U_1 and U_2 are the average velocities of agitation, in all directions, of the two kinds of electrons; e stands for the absolute value of the charge of a positive or negative electron. The average free path l_1 of the positive electrons will be the average of the free paths traveled between two collisions, by these electrons; the average interval of time between two collisions will be represented by $T_1=\frac{l_1}{U_1}$; l_2, and T_2 will have a similar meaning for the negative electrons.

Having once reached the thermal equilibrium, the average kinetic energy of all the electrons will be equal to that of a gas molecule at the same absolute temperature; we shall then have according to the kinetic theory $\frac{m_1U_1^2}{2}=\frac{m_2U_2^2}{2}=\alpha T$. ($\alpha$ is a universal constant; m_1, m_2 the masses of the electrons of the two kinds).

Let us now consider a conductor in which an electric field X parallel to the axis x has been created; let us suppose that X does not appreciably effect U_1, U_2, l_1, l_2. This field will communicate the accelleration $\frac{eX}{m_1}$ to a positive electron in the direction of the said axis. At the end of its free path, therefore, the electron will have acquired, in the direction x, an excess of velocity (above the velocity of agitation U_1 existing even before the creation of the electric field) $\frac{eX}{m_1}T_1=\frac{eXl_1}{m_1U}$; in the collision that ends the free path the electron loses this excess of velocity. We may, therefore, say that owing to the field X during a free path a positive electron acquires in the direction x an average excess of velocity $W_1=\frac{1}{2}\;\frac{eXl_1}{m_1U_1}=eX\frac{l_1U}{4\alpha T}$. The excess of velocity that an electron acquires in a given direction when subject to a unit force having the same direction, is called the mobility of the electron. Calling V_1 the mobility of a positive electron, we shall have $V_1=\frac{l_1}{2m_1U_1}=\frac{l_1U_1}{4\alpha T}$. In the same way, for a negative electron we shall have $W_2 = -\frac{1}{2}\frac{eX\,l_2}{m_2U_2} = -eX\frac{l_2U_2}{4aT}$; $V_2 = \frac{l_2}{2m_2U_2} = \frac{l_2U_2}{4aT}$. Under the action of the electric field, there will be added to the disorderly movement of the electrons a movement of the positive electrons in the direction of the field and of the negative electrons in the opposite direction. Because of this movement, the unit of surface perpendicular to the axis x will be crossed in the unit of time by the quantity of electricity: $eN_1W_1 - eN_2W_2 = e^2X(N_1V_1 + N_2V_2)$. Denoting by j the density of current we shall have: $j = e^2(N_1V_1 + N_2V_2)X$. This equation expresses Ohm's law. Denoting by σ the electric conductivity of the metal we obtain: $\sigma = e^2(N_1V_1 + N_2V_2 = \frac{e^2}{2}\left(\frac{N_1l_1}{m_1U_1} + \frac{N_2l_2}{m_2U_2}\right) = \frac{e^2}{4aT}(N_1l_1U_1 + N_2l_2U_2)$.

We will limit ourselves to these short notes on the general part of the electronic theory of metals, in order to consider some particular problems that, owing to this theory, it has been possible to solve.

SECTION 4

Among the phenomena which the electronic theory has been able to explain and often to foresee, there is a whole class of phenomena that take place when a conductor is traversed by an electric or thermal current and is subject to the action of a magnetic field. These lectures will treat especially of the results of researches on conductors traversed by an electric current and subject to the action of a magnetic field.

Let us consider a plane, homogeneous, isotropic, metal plate subject to the action of a uniform magnetic field H normal to it. We shall refer the points of the plate to the axes x, y, which form together with the direction H an orthogonal right-handed system. Let us denote with X, Y, the components of the electric force (parallel to the axes) of potential V; with j_x, j_y, the components of the density of current; with $\frac{d\xi_1}{dt}, \frac{dn_1}{dt}, \frac{d\xi_2}{dt}, \frac{dn_2}{dt}$, the components of the velocities of the electrons of the two kinds (these velocities are added to those of agitation that exist also without the action of the electric field and of the magnetic field). e, N_1, N_2, V_1, V_2, will keep the meaning they had before.

The components of the density of current will now be given by the formulae

$$(1)\ j_x = e\left(N_1\frac{d\xi_1}{dt} - N_2\frac{d\xi_2}{dt}\right);\ j_y = c\left(N_1\frac{dn_1}{dt} - N_2\frac{dn_2}{dt}\right).$$

A positive electron that moves in the direction l with the velocity V is equivalent to an element of current (having the same direction) if the product of i, intensity of current, by dl, length of the element, be equal to ev. If, therefore, the direction l form with the magnetic field H an angle θ, the electron is subject to an electromagnetic force, perpendicular to L and to H and having the direction indicated by the well known rule of the three fingers of the left hand; the intensity of this force is given by $evH \sin\theta$.

If the plate be kept at a constant temperature, every electron is subject only to the electric force depending on the distribution of the potential and to the electromagnetic force produced by the movement of the electron in the magnetic field.

The velocities of electrons will therefore be determined by the formulae:

$$(2)\begin{cases} \dfrac{d\xi_1}{dt}=eV_1\left(X-H\dfrac{dn_1}{dt}\right) & \dfrac{d\xi_2}{dt}=eV_2\left(X-H\dfrac{dn_2}{dt}\right) \\ \dfrac{dn_1}{dt}=eV_1\left(Y-H\dfrac{d\xi_1}{dt}\right) & \dfrac{dn_2}{dt}=eV_2\left(Y+H\dfrac{d\xi_2}{dt}\right) \end{cases}$$

In these formulae in place of X, Y, we can put $-\dfrac{dV}{dx}, -\dfrac{dV}{dy}$.

Let us solve the equations (2) with respect to $\dfrac{d\xi_1}{dt}, \dfrac{d\eta_1}{dt}, \dfrac{d\xi_2}{dt}, \dfrac{d\eta_2}{dt}$, and let us substitute the expressions thus obtained in the equations (1); then we shall have:

$$(3)\ j_x=-K\left(\frac{dV}{dx}-\lambda\frac{dV}{dy}\right),\quad j_y=-K\left(\frac{dV}{dy}+\lambda\frac{dV}{dx}\right).$$

where

$$K=e^2\left(\frac{N_1V_1}{1+e^2V_1^2H^2}+\frac{N_2V_2}{1+e^2V_2^2H^2}\right);\ \lambda=\frac{\Sigma}{K};\ \Sigma=e^3H\left(\frac{N_1V_1^2}{1+e^2H^2V_1^2}+\frac{N_2V_2^2}{1+e^2H^2V_2^2}\right).$$

The flow of current through every closed line that forms the complete boundary of a portion of the plate free from electrodes, must be zero; we therefore have

$$\frac{dj_x}{dx}+\frac{dj_y}{dy}=0.$$

We deduce this equation from (3), remembering that the potential V must (as is the case when the magnetic field is missing) satisfy the condition $\nabla^2V=0$.

If the plate be insulated, the components of the density of current in the direction of the normal to the free boundary must be zero, likewise the component of the density of current in the direction of the normal to a line of flow.

Bearing in mind (3) and indicating with t and n the tangent and the normal respectively to a line of flow, or to the insulator boundary (t, n and the direction H form a right-handed system) we deduce from the boundary conditions

$$\frac{dV}{dn}+\lambda\frac{dV}{dt}=0.$$

Therefore, if l be a direction forming with n an angle β, so that $tg\beta=\lambda$, (the positive direction l should be included in the right angle formed by the positive direction n and') we have, $\dfrac{dV}{dl}=\sigma$. (Fig. 1).

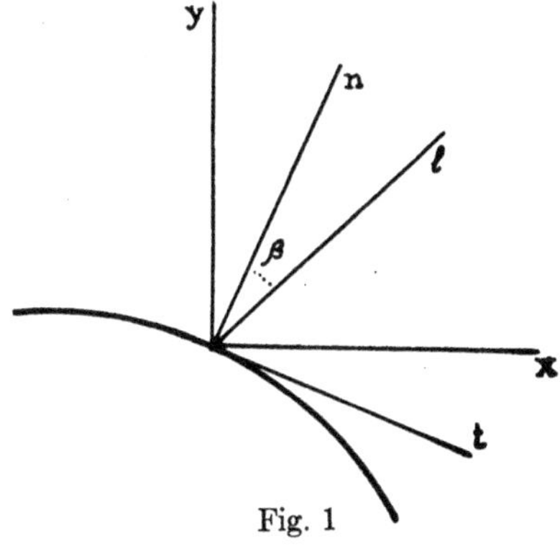

Fig. 1

Therefore, the condition to which V must answer in the interior of the plate, is the same one we should have without the magnetic field; for the condition $\frac{dV}{dn}=\sigma$, that we should have along the insulated boundary without the field, we must substitute instead the condition $\frac{dV}{dl}=\sigma$, if the field exist.

If along parts of the boundary there are electrodes of a negligible resistance, the potential V will have to assume determined constant values along them.

We may also conclude, that under the action of the magnetic field either the lines of flow, or the equipotential lines, or both the two systems of lines, are deformed so that the line of flow and the equipotential line passing through a given point of the plate, are no longer perpendicular to one another, as is the case when the magnetic field is missing, but form instead an angle $\frac{\pi}{2}-\beta$. The value of β depends on the nature of the metal and on the intensity of the magnetic field. For bismuth β can exceed 10°.

Limiting our researches to the variation of potential produced by the magnetic field, i.e., to the determination of the Hall effect, it will be sufficient to consider the function $v=V-V_o$. (V_o represents the potential when there is no magnetic field). We have already seen that by indicating with t and n the tangent and the normal to a line of flow or to the insulator boundary, we find $\frac{dV}{dn}=\lambda\frac{dV}{dt}$; as $\frac{dV_o}{dn}=\sigma$; we shall also have $\frac{dV}{dn}=-\lambda\frac{dV}{dt}$. Bearing in mind also (3) we obtain:

$$j_t=-K(1+\lambda^2)\frac{dV}{dt}.$$

From this equation and from the preceding one we deduce $\frac{dv}{dn}=\frac{\lambda}{K(1+\lambda^2)}j_t$.

Let us consider the case in which the lines of flow do not change under the action of the magnetic field; in this case, whatever the value of the field may be, the normal n to the line of flow that passes through a point of the plate, coincides with the tangent to the equipotential line, corresponding to the zero field, that passes through the same point. If s be an equipotential line, when the magnetic field does not exist and A and B be two points along it, by integrating the last equation along s, between A and B, we shall have $V_A-V_B=\frac{\lambda}{K(1+\lambda^2)}\int_A^B j_t ds=\frac{\lambda}{K(1+\lambda^2)}J$, in which J is the ratio between the intensity of the total current flowing in the plate and the thickness of this latter. From the last equation we see that between two points of the boundary that are at the same potential when the magnetic field is zero, we must observe, under the action of a magnetic field, a constant Hall effect, however the said points may be chosen. In the case we are considering, the last equation determines completely the Hall effect.

SECTION 5

Starting from the laws of which we have been speaking, I shall develop the mathematical theory of the propagation of electric currents in a plate under the action of a magnetic field. (6).

We shall now begin, and in the following lessons continue, the explanation of such a theory. The theoretical results have led to several experimental researches, of which we shall also speak.

Let us begin by establishing some fundamental formulae, which are to be used through all the following exposition. (Fig. 2).

Let s be any curve whatever, in the interior or including part or the whole of the boundary of the plate; let n be its normal so directed as to make the pair of orthogonal directions s, n, congruent in the plane to the pair x, y.

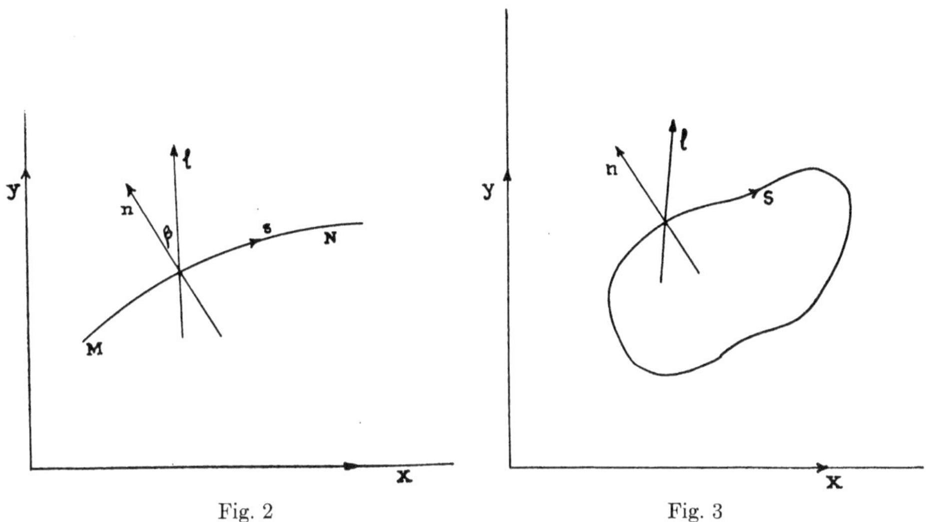

Fig. 2 Fig. 3

From the formulae (3) of Section 4, it follows that the flow of electricity through ds will be expressed by:

$$j_n ds = (j_x \cos nx + j_y \cos ny) ds = -K\left(\frac{\delta V}{\delta n} + \lambda \frac{\delta V}{\delta s}\right) ds = \frac{-K}{\cos \beta} \frac{\delta V}{\delta l} ds$$

where l is a direction forming the angle β with n.

The flow across MN, that is to say across the whole line s, will be:

$$(4) \qquad \int_s j_n \, ds = -K \int_o \frac{dV}{dn} ds - \lambda K (V_N - V_M) = \frac{-K}{\cos \beta} \int_s \frac{\delta V}{\delta l} ds,$$

We find analogously, the other formula:

$$(5) \qquad \int_s V j_n \, ds = -K \int_s V \frac{\delta V}{\delta n} ds - \frac{\lambda K}{2} (V_N^2 - V_M^2) = \frac{-K}{\cos \beta} \int_s V \frac{\delta V}{\delta l} ds.$$

For the validity of these formulae we have only to admit V to be finite all along s; and its first derivatives to be finite, or infinite of an order inferior to a number smaller than one. (Here and henceforward in defining the order of infinity of a function in a given point, we shall take as fundamental infinity the inverse of the distance to the same point.) (Fig. 3).

Let us suppose that the curve s is closed (Fig. 3). Then V being single valued, since we assume that in no region of the plate electromotive forces are acting, the formulae (4) and (5) become

$$\text{(A)} \qquad \int j_n \, ds = -K \int_s \frac{dV}{\delta n} \, ds = \frac{-K}{\cos\beta} \int_s \frac{\delta V}{\delta l} \, ds.$$

$$\text{(B)} \qquad \int_s V j_n \, ds = -K \int_s V \frac{\delta V}{\delta n} \, ds = \frac{-K}{\cos\beta} \int_s V \frac{\delta V}{\delta l} \, ds.$$

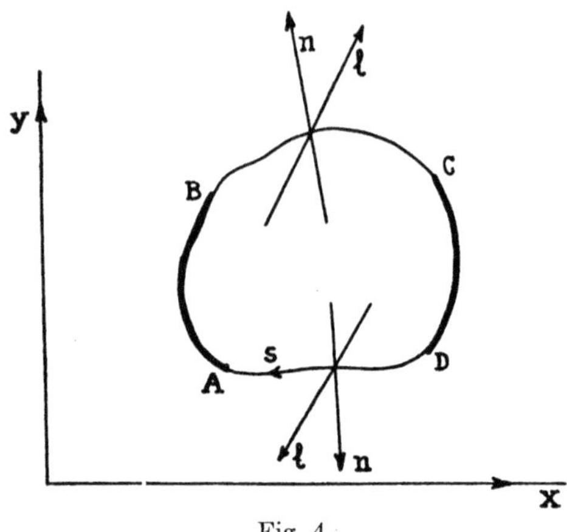

Fig. 4

SECTION 6

Let us suppose that the current enters through the portion $A\,B$ of the boundary, by means of an electrode of negligible resistance, and goes out through the portion CD of the boundary, which also forms an electrode of negligible resistance, whereas the portions BC and AD are insulated. (Fig. 4).

We shall have, then, along AB and CD respectively, $V = C_1$, $V = C_2$ (C_1, C_2 being constants); and along BC and AD $\dfrac{\delta V}{\delta l} = 0$.

For points of the plate in the interior of the curve σ, V will be regular and harmonic. Now, if we call s the whole boundary and assume as verified the above mentioned condition about the order of infinity of the derivatives of V along s, we shall have, by reason of (B)

$$\int_{AB} V j_n \, ds + \int_{CD} V j_n \, ds = -K \int_s V \frac{\delta V}{\delta n} \, ds$$

because $\dfrac{\delta V}{\delta l} = \dfrac{j_n \cos\beta}{-K}$ is null on $B\,C$ and $A\,D$. Then

$$C_1 \int_{AB} j_n ds + C_2 \int_{CD} j_n \, ds = -K \int_s V \frac{\delta V}{\delta n} \, ds.$$

Let us call J the ratio between the intensity of the total current flowing in the plate and the thickness of this latter; we have:

$$J = \int_{CD} j_n ds = \int_{AB} j_n ds$$

and also

$$(C_2 - C_1)J = -K \int_s V \frac{\delta V}{\delta n} ds$$

and according to the well known divergence theorem in harmonic functions:

$$(C_1 - C_2)J = K \int_\sigma \Delta V \delta\sigma.$$

It follows from this equation that, if $C_1 = C_2 = 0$, V must be zero at every point of σ; and, consequently, that only one solution for V may correspond to given values of C_1, C_2. We, therefore, have the theorem: *When the constant values of the potential along the electrodes AB and CD are known, the distribution of currents in the plate is fully determined.*

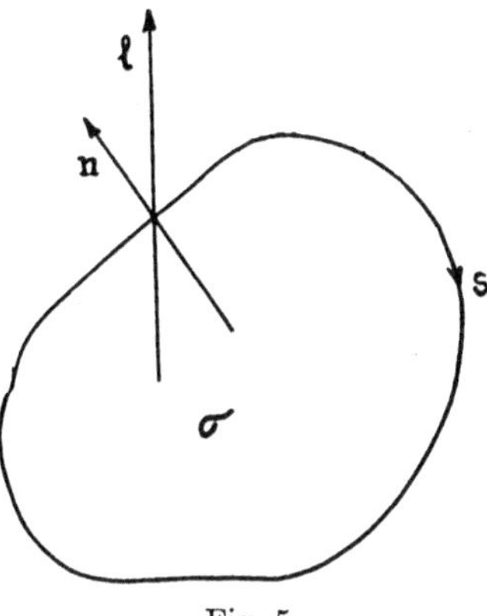

Fig. 5

In like manner we see that if $J = 0$, V must be constant; and therefore the distribution of current is determined also if the intensity of the total current flowing in the plate be known.

The preceding theorem may easily be extended. Let σ be an area (Fig. 5) in the interior of which the harmonic function V is regular, and let s be its boundary. In consequence of (B) we shall have (if on the boundary V be finite and its derivatives be finite also or infinite of orders inferior to a number smaller than 1):

$$\int_s V \frac{\delta V}{\delta l} ds = \cos \beta \int_s V \frac{\delta V}{\delta n} ds = \cos \beta \int_s \Delta V d\sigma.$$

Hence, if V be zero in some portions of the boundary and $\frac{\delta V}{\delta l}$ zero in (Fig. 5) the other portions of it, V must be zero in σ; and therefore if V be known in certain portions of the boundary and $\frac{\delta V}{\delta l}$ be known in the other portions of it, V must be determined in σ. If only $\frac{\delta V}{\delta l}$ be known along the whole boundary, V is determined except for an additive constant. (A) gives us the condition with which $\frac{\delta V}{\delta l}$ must comply: (A^1) $\int_s \frac{\delta V}{\delta l} ds = 0.$

SECTION 7

Let us now suppose that the current enters into the plate and comes out of it through two point electrodes A and B (Fig. 6). Let us then go back to the formula (A) and suppose at first that the curve s is the curve sa which surrounds the point A (Fig. 6). If we call J the ratio between the intensity of the current and the thickness of the plate, and suppose n to be directed externally to the area which sa encloses, we shall have:

$$J = \int_{sa} j_n ds_a = -K \int_{sa} \frac{\delta V}{\delta n} ds_a$$

whereas, if we call sb a curve surrounding the point B and n the normal external to the area which sb encloses, we shall have:

$$J = \int_{sa} j_n ds_a = -K \int_{sa} \frac{\delta V}{\delta n} ds_a$$

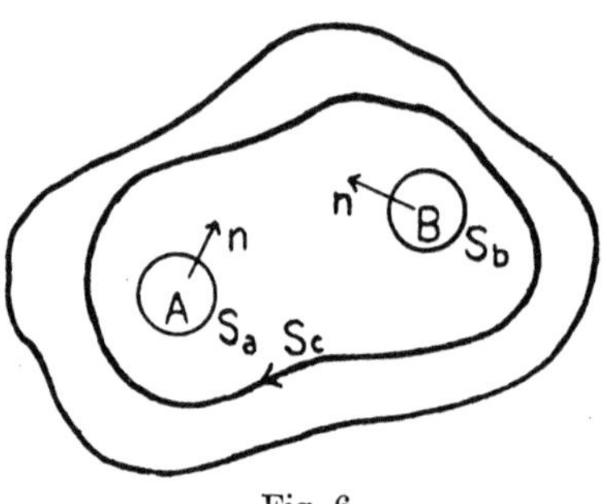

Fig. 6

whereas, if we call sb a curve surrounding the point B and n the normal external to the area which sb encloses, we shall have:

$$J = \int_{sb} j_n ds_b = -K \int_{sb} \frac{\delta V}{\delta n} \delta s_b.$$

If, however, we take a line sc including both A and B or excluding both, we shall have:

$$0 = \int_{sc} j_n ds_c = -K \int_{sc} \frac{\delta V}{\delta n} ds_c.$$

In the above-mentioned hypothesis, according to what is done in the ordinary case when there is no magnetic field, we shall assume:

$$V = \frac{J}{2\pi K}(\log r_B - \log r_A) + W \tag{6}$$

W being harmonic and regular, and r_A, r_B, being the distances between the generic point xy and A and B respectively. On the boundary of the area of the plate the condition:

$$\frac{\delta V}{\delta l} = 0 \tag{7}$$

must be fulfilled; which is the same as the condition:

$$\frac{\delta W}{\delta l}=\frac{\delta}{\delta l}\left(\frac{J}{2\pi K}(\log r_B-\log r_A)\right). \tag{7^1}$$

Thus the problem of the distribution of currents is brought back to the determination of the harmonic regular function W, the values $\frac{\delta W}{\delta l}$ of which are known on the boundary, being given by (7^1). It is easy to see that the condition (A^1) of Section 6, $\int_s \frac{\delta V}{\delta l}\,ds=0$, is fulfilled. W will therefore be determined but for an arbitrary additive constant, which had evidently no influence on the law of the distribution of currents.

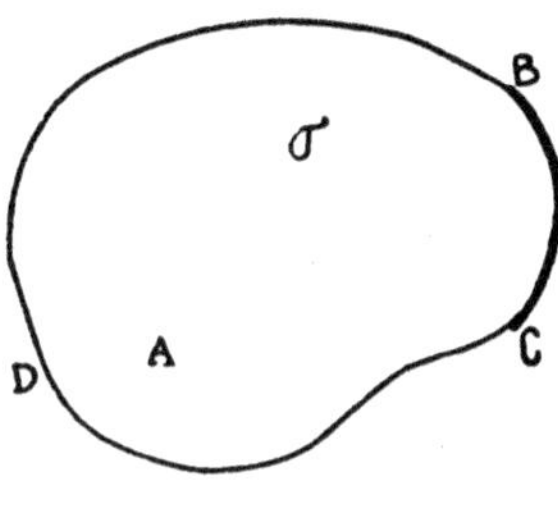

Fig. 7

Let us now suppose that the current enters through a point electrode A and comes out through an electrode BC, placed on the boundary and of negligible resistance; and let us give to J the usual meaning. In this case (Fig. 7) according to what we have already seen in Section 6, we have:

$$V=\frac{-J}{2\pi K}\log r_A+W$$

where W is a function harmonic and regular in the interior of σ. Along the free and insulated portion BCD of the boundary we must have:

$$\frac{\delta W}{\delta l}=\frac{\delta}{\delta l}\left[\frac{J}{2\pi K}\log r_A\right]$$

and along BC,

$$W=C+\frac{J}{2\pi K}\log r_A$$

where C is the constant value that V is to assume along the electrode BC. Hence, if the function W be finite and its derivatives along s be finite or infinite of an order inferior to a number smaller than l, this function will be determined within σ (see Section 6) and therefore the law of the distribution of currents will be determined also.

We shall assume as is generally done, that if a current of intensity I comes out through an electrode, the fact may also be described by saying that a current of intensity $-I$ enters through it. Let J be the ratio between the intensity of the current (positive or negative) that penetrates into the plate through the point electrode and the thickness of the plate. In the neighborhood of the electrode, the potential will be:

$$V = \frac{-J}{2\pi K} \log r + W$$

where r is the distance from the generic point xy to the electrode and W is a function harmonic and regular in the neighborhood of the same point.

The portion of potential:

$$\frac{-J}{2\pi K} \log r = \frac{J}{2\pi K} \log \frac{1}{r}$$

will be called potential of the electrode and will be the logarithmic potential of a point, the mass of which is $\frac{J}{2\pi K}$.

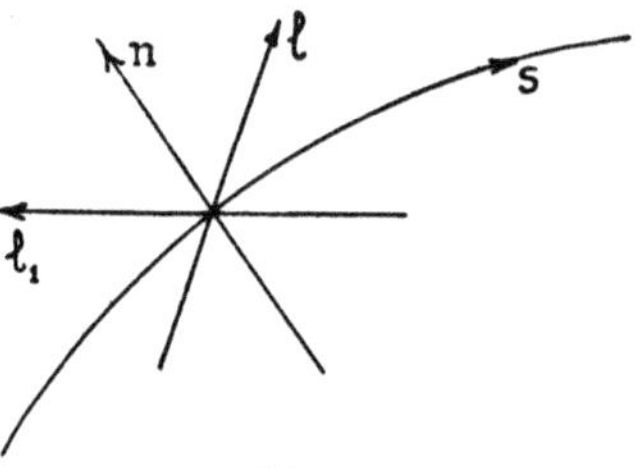

Fig. 8

SECTION 8

Let us now fix some other fundamental formulae, to be added to those which we found in Section 5.

Let us see what happens when we invert the sense of the magnetic field, that is to say change H into $-H$. In order to distinguish one case from the other, we shall say that in the former the magnetic field is *direct*, in the latter the magnetic field is *inverted*. From the expressions given in Section 4 for K and λ, it follows that, when the field is inverted, K remains unchanged, whereas the sign of λ is changed, which is the same as saying that the sign of the angle β is changed; consequently, instead of the direction l we must consider the direction symmetric to it with (Fig. 8) respect to the normal n, which new direction we shall call l_1. If we give an index 1 to all the elements corresponding to this case, we shall have the formulae:

$$j_{1x} = -K\left(\frac{\delta V_1}{\delta x} + \lambda \frac{\delta V_1}{\delta y}\right) \quad j_{1y} = -K\left(\frac{\delta V_1}{\delta y} - \lambda \frac{\delta V_1}{\delta x}\right)$$

and also

$$\frac{\delta V_1}{\delta l_1} = 0$$

along the portions of the boundary that are free and insulated.

Let now s be any closed line whatever, in the interior or reaching the boundary of the plate; let n be the normal to s, directed as we explained in Section 5. We shall have:

$$V_1 j_n - V j_{1n} = V_1(j_x \cos nx + j_y \cos ny) - V(j_{1x} \cos nx + j_{1y} \cos ny) =$$
$$= -K\left[V_1\left(\frac{\delta V}{\delta n} + \lambda\frac{\delta V}{\delta s}\right) - V\left(\frac{\delta V_1}{\delta n} - \lambda\frac{\delta V_1}{\delta s}\right)\right] = -K\left(V_1\frac{\delta V}{\delta n} - V\frac{\delta V_1}{\delta n}\right) - \lambda K\frac{\delta W}{\delta s} =$$
$$= \frac{-K}{\cos\beta}\left(V_1\frac{\delta V}{\delta l} - V\frac{\delta V_1}{\delta l_1}\right)$$

and integrating along the whole line s:

$$\text{(C)} \qquad \int_s (V_1 j_n - V j_{1n}) ds = -K\int_s \left(V_1\frac{\delta V}{\delta n} - V\frac{\delta V_1}{\delta n}\right) ds = \frac{-K}{\cos\beta}\int_s \left(V_1\frac{\delta V}{\delta l} - V\frac{\delta V_1}{\delta l_1}\right)\delta s$$

which is one of the formulae that we wanted to establish. This formula may be extended. In fact, let us denote by S not merely one closed curve but a number of closed curves bounding a field σ' internal to the area σ of the plate. As the above formula is true for every closed curve, it will be true also if we substitute the curves S for the single curve s; hence if V and V_1 be regular in the interior of the area σ', we shall have, by taking n external to the field σ'

$$\text{(D)} \qquad \int_s (V_1 j_n - V j_{1n})\, \delta\!\int = \frac{-K}{\cos\beta}\int_s \left(V_1\frac{\delta V}{\delta l} - V\frac{\delta V_1}{\delta l_1}\right) ds = 0$$

since according to Green's theorem, we have

$$\int_s \left(V_1\frac{\delta V}{\delta n} - V\frac{\delta V_1}{\delta n}\right) ds = 0.$$

Let us suppose S to be formed by the two systems of closed curves S' and S''; then the formula (D) may be written also as:

$$\text{(D')} \qquad \frac{1}{\cos\beta}\int_{S'} \left(V_1\frac{\delta V}{\delta l} - V\frac{\delta V_1}{\delta l_1}\right) dS' + \int_{S''} \left(V_1\frac{\delta V}{\delta n} - V\frac{\delta V_1}{\delta n}\right) dS'' = 0.$$

The formulae (D) and (D′), which we have thus found, are extensions of Green's theorem.

Let us now proceed to some applications of the preceding formulae. Let us take V as given by (6) with the condition (7); let us suppose S' to be formed by the boundary s of σ, and S'' to consist of two circles sa and sb having their centres in A and B. Let us suppose (Fig. 9) V regular in the interior of σ. The area σ' will be obtained by subtracting from σ the areas enclosed within sa and sb; σ' will therefore be limited by s, sa, sb. In V and V_1, being regular, we may apply (D′); and remembering that, on s, $\frac{\delta V}{\delta l} = 0$, we shall have:

$$\frac{-1}{\cos\beta}\int_s V\frac{\delta V_1}{\delta l_1}\, ds + \int_{sa} \left(V_1\frac{\delta V}{\delta n} - V\frac{\delta V_1}{\delta n}\right) ds_a + \int_{sb} \left(V_1\frac{\delta V}{\delta n} - V\frac{\delta V_1}{\delta n}\right) ds_b = 0.$$

But if we make the circles sa and sb grow indefinitely smaller, we see that

$$\lim \int_{sa}\left(V_1\frac{\delta V}{\delta n}-V\frac{\delta V_1}{\delta n}\right)ds_a=\frac{J}{K}V_{1A};\quad \lim \int_{sb}\left(V_1\frac{\delta V}{\delta n}-V\frac{\delta V_1}{\delta n}\right)ds_b=\frac{-J}{K}V_{1B}$$

where V_A and V_B denote the values of V at the points A and B; hence

$$V_{1A}-V_{1B}=\frac{K}{J\cos\beta}\int_s V\frac{\delta V_1}{\delta l_1}ds.$$

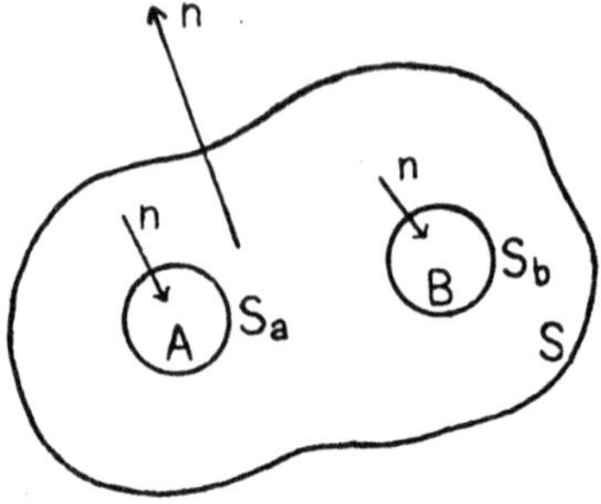

Fig. 9

From this formula the following proposition is deduced: *If the distribution of currents in a plate be known, when the current enters through A and goes out through B and the magnetic field is direct, the difference of the values of a harmonic regular function in the points A and B may be determined provided its derivatives be known at the boundary in the direction l.*

It is evident that also another proposition subsists with it, i.e.: *If the distribution of currents in a plate be known, when the current enters through A and goes out through B, and the magnetic field is inverted the difference of the values of a harmonic regular function in the points A and B may be determined provided its derivative be known at the boundary in the direction l.*

In other words, the potential V, corresponding to the *direct magnetic field* and to the pair of point electrodes A and B, is a *function analogous to Green's* in the case when *the values of the derivative of a harmonic regular function* are known on the boundary in the direction l_1; whereas the potential corresponding to the *inverted magnetic field* and to the pair of point electrodes A and B, is a *function analogous to Green's* in the case when the *values of the derivative of a harmonic regular function* are known on the boundary in the direction l.

Let us now consider the case when the current enters through a point electrode A and goes out through an electrode BC placed on the boundary and of negligible resistance (Section 7, Fig. 7); let us suppose the electrode BC to be at potential zero. Let V_1 be a function harmonic and regular within σ. Let us take the boundary s of the plate as S', and a circle sa having its centre in A as S''.

Applying the formula (D′) we shall obtain:

$$\frac{1}{\cos\beta}\int_{BCD}V\frac{\delta V_1}{\delta l_1}\,ds-\frac{1}{\cos\beta}\int_{BC}V_1\frac{\delta V}{\delta l}\,ds-\int_{sa}\left(V_1\frac{\delta V}{\delta n}-V\frac{\delta V_1}{\delta n}\right)ds_a=0$$

and making the circle sa grow indefinitely smaller

$$V_{1A}=\frac{K}{J\cos\beta}\int_{BCD}V\frac{\delta V_1}{\delta l_1}\,ds-\frac{K}{J\cos\beta}\int_{BC}V_1\frac{\delta V}{\delta l}\,ds.$$

Hence, *if the distribution of currents be known when the current enters through A and goes out through the electrode BC and the magnetic field is direct, the value of a harmonic regular function in A may be determined provided its value be known along BC and the value of its derivative be known along ADC in the direction l.*

The analogous proposition subsists when the magnetic field is inverted.

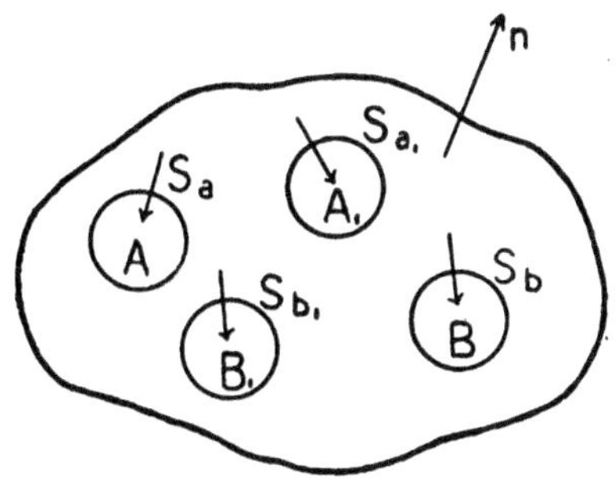

Fig. 10

Section 9

Let us now take V as given by (6) with the condition (7), and V as given by

$$V_1 = \frac{J_1}{2\pi k}(\log r_{B1} - \log r_{A1}) + W_1$$

where W is a function harmonic and regular within σ; let A_1 and B_1 be two new points chosen in this field (Fig. 10); along the boundary s of σ let us have $\frac{\delta V_1}{\delta l_1} = 0$.

Let us suppose S' to be the boundary s and S'' the system of four circles sa, sb, sa_1, sb_1, having their respective centres in A, B, A_1 and B_1. The area σ' will be obtained by subtracting from σ the areas enclosed within the four circles (Fig. 10). In σ', V and V_1 are regular; then applying the (D′) and remembering that on s $\frac{\delta V}{\delta l} = \frac{\delta V_1}{\delta l_1} = 0$, we shall obtain:

$$\int_{sa}\left(V_1\frac{\delta V}{\delta n} - V\frac{\delta V_1}{\delta n}\right)ds_a + \int_{sa}\left(V_1\frac{\delta V}{\delta n} - V\frac{\delta V_1}{\delta n}\right)ds_b + \int_{sa1}\left(V_1\frac{\delta V}{\delta n} - V\frac{\delta V_1}{\delta n}\right)ds_{a1} +$$

$$+ \int_{sb1}\left(V_1\frac{\delta V}{\delta n} - V\frac{\delta V_1}{\delta n}\right)ds_{b1} = 0.$$

Now, making the four circles grow indefinitely smaller, we easily have

$$\lim \int_{sa}\left(V_1\frac{\delta V}{\delta n} - V\frac{\delta V_1}{\delta n}\right)ds_a = \frac{J}{K}V_{1A}; \quad \lim \int_{sb}\left(V_1\frac{\delta V}{\delta n} - V\frac{\delta V_1}{\delta n}ds_b = \frac{-J}{K}V_{1B}\right.$$

$$\lim \int_{sa1}\left(V_1\frac{\delta V}{\delta n} - V\frac{\delta V_1}{\delta n}\right)ds_{a1} = \frac{-J_1}{K}V_{A1}; \quad \lim \int_{sb1}\left(V_1\frac{\delta V}{\delta n} - V\frac{\delta V_1}{\delta n}\right)ds_{b1} = \frac{J_1}{K}V_{B1}$$

hence

$$J(V_{1B} - V_{1A}) = J_1(V_{B1} - V_{A1});$$

and, if J and J_1 be equal:

$$V_{1B} - V_{1A} = V_{B1} - V_{A1}.$$

From which the following law of reciprocity is deduced: *If under the action of a certain magnetic field a current of given intensity passes through a conducting plate, entering through the point A and going out through the point B, and a difference of potential be obtained in two points A_1, B_1, the same difference will be obtained between the potentials of the points A and B when a current of the same intensity is caused to enter through A_1 and to go out through B_1 and the magnetic field is inverted.*

We shall demonstrate in the third lecture that, by the same principle of reciprocity now established, the condition of homogeneity may be omitted.

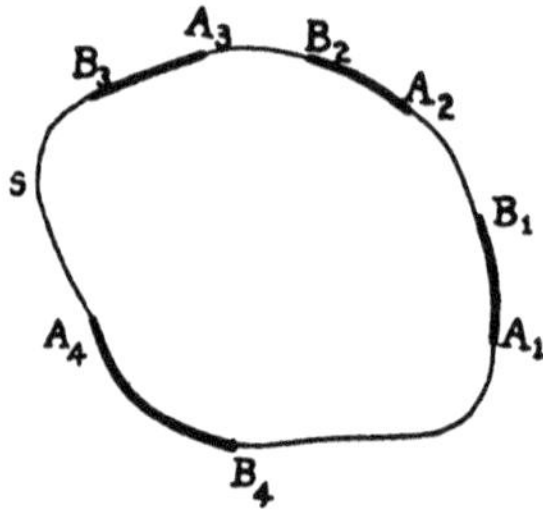

Fig. 11

Let us suppose that on the boundary of the plate, there are the electrodes A_1B_1, A_2B_2, A_3B_3, A_4B_4, ... of negligible resistance and that the other portions of the boundary are free and insulated. Let us suppose moreover that no electrode exists in the interior. Let $J^{(1)}$, $J^{(2)}$, $J^{(3)}$, $J^{(4)}$...... be the ratios between (Fig. 11) the intensities of the currents entering through these electrodes and the thickness of the plate $C^{(1)}$, $C^{(2)}$, $C^{(3)}$, $C^{(4)}$, ... the values of the potential V along the electrodes when the magnetic field is direct. Let $J_1^{(1)}$, $J_1^{(2)}$, $J_1^{(3)}$, $J_1^{(4)}$, ..., $C_1^{(1)}$, $C_1^{(2)}$, $C_1^{(3)}$, $C_1^{(4)}$, V_1 be the corresponding quantities when the magnetic field is inverted. It is obvious that we shall have:

$$\Sigma_k J^{(k)} = \Sigma_k J_1^{(h)} = 0$$

a result which we might also have obtained from the formula (A).

Let us now apply the formula (D), taking the boundary s of the plate as S. We shall have

$$0 = \Sigma_h \int_{A_hB_h} (V_1 j_n - V j_{1n}) ds = \Sigma n (C^{(h)} J_1^{(h)} - C_1^{(h)} J^{(h)})$$

hence

$$\Sigma_h C_1^{(h)} J^{(h)} = \Sigma_h C^{(h)} J_1^{(h)}.$$

If the J^h's are all zero, except $J^{(1)}$ and $J^{(2)}$; and the $J_1^{(h)}$ be also all zero except $J_1^{(3)}$ and $J_1^{(4)}$, we shall have:

$$J^{(1)} = J^{(2)} = I \qquad J_1^{(3)} = -J_1^{(4)} = I_1$$

hence:

$$(C_1^{(1)} - C_1^{(2)}) J = (C^{(3)} - C^{(4)}) J_1$$

and if $J = J_1$,

$$C_1^{(1)} - C_1^{(2)} = C^{(3)} - C^{(4)}.$$

These formulae express new theorems of reciprocity, the interpretation of which is easy.

Let us finally suppose that, the magnetic field being direct, there are besides the electrodes A_1, B_1, A_2, B_2, ... the boundary, some internal point electrodes $M^{(1)}$, $M^{(2)}$, $M^{(3)}$......; let us call $J^{(1)}$, $J^{(2)}$, $J^{(3)}$, the ratios between the intensities of the currents which these electrodes admit into the plate and the thickness of this latter. And let us give an analogous meaning to $M_1{}^{(1)}$, $M_1{}^{(2)}$, $M_1{}^{(3)}$......, $J_1{}^{(1)}$, $J_1{}^{(2)}$, $J_1{}^{(3)}$, in the case of the inverted magnetic field. We shall have then:

$$\Sigma_h J^{(h)} + \Sigma_k I^{(k)} = \Sigma_h J_1{}^{(h)} + \Sigma_{k1} I_1{}^{(k1)} = 0$$

and

$$\Sigma_h C_1{}^{(h)} J^{(h)} + \Sigma_k V_1{}^{(k)} I^{(k)} = \Sigma_h C^{(h)} J_1{}^{(h)} + \Sigma_{k1} V^{(k1)} I_1{}^{(k1)}$$

where $V_1{}^{(k)}$ is the value of the potential V_1 in the point $M^{(k)}$ and $V^{(k1)}$ the value of the potential V in the point $M_1{}^{(k1)}$.

From the last formula theorems of reciprocity analogous to that already stated may be likewise deduced.

SECOND LECTURE

Section 1—The fundamental function. Section 2—Circular plate: Two principles of images. Section 3—The principle of point electrodes at the boundary. Section 4—Cyclic plates. Section 5—The problem of linear electrodes of a negligible resistance, situated at the boundary.

Section 1

Let us denote by V' the conjugate function of V: then we have

$$\frac{\delta V}{\delta x} = \frac{\delta V'}{\delta y}; \qquad \frac{\delta V}{\delta y} = \frac{\delta V'}{\delta x}.$$

The equations (3) of Section 4 of the preceding lecture will then become

$$(3)' \qquad j_x = -K\frac{\delta(V+\lambda V')}{\delta x}; \qquad j_y = -K\frac{\delta(V+\lambda V')}{\delta y}.$$

If we put $V+\lambda V' = U$, we shall have

$$(3)'' \qquad j_x = -K\frac{\delta U}{\delta x}; \qquad j_y = -K\frac{\delta U}{\delta y}.$$

The function U will satisfy first, the condition $\Delta^2 U = 0$ at every point of the plate and second, the condition $\frac{\delta U}{\delta n} = 0$ at every point of the insulated boundary.

Therefore the distribution of current in the plate happens as if there were no magnetic field, but as if the potential were U instead of V, and the conductivity were K. As K for a given value of the field is constant the lines of flow are independent of K.

We shall call U the *fundamental function of the distribution of current in the plate* or, *more simply*, the fundamental function. U coincides with the *potential* only if the magnetic field be null. When we know the potential V we can have U by the operation

$$U = V + \lambda V' = \frac{V \cos\beta + V' \sin\beta}{\cos\beta}. \tag{E}$$

Now let us resolve the problem of the *calculation of the potential when the fundamental function is known.*

Let U' be the conjugate function of U; we shall have

$$U' = V' - \lambda V. \tag{8}$$

Then, from the formula (E), we have

$$V = \frac{U - \lambda V}{1 + \lambda^2} = (U \cos\beta - U' \sin\beta)\cos\beta. \tag{E)'$$

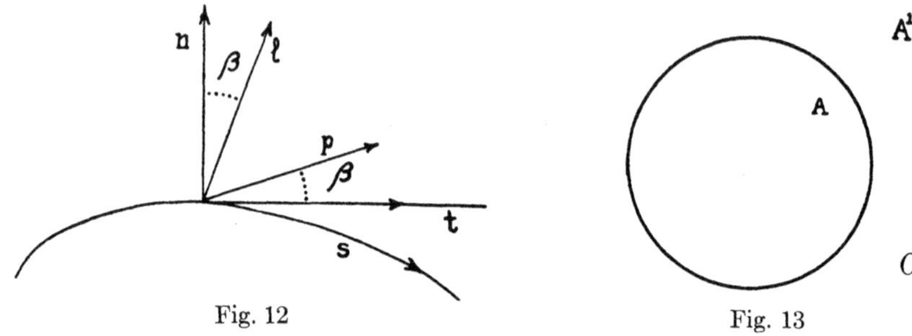

Fig. 12

Fig. 13

The proposed problem is thus solved.

The preceding formulae easily provide us with *the condition which the fundamental function must fulfill all along the edge of the electrodes of negligible resistance.* Let s be a portion of one of these edges. At every point of s we shall have

$$V = (U \cos\beta - U' \sin\beta)\cos\beta = constant$$

$$0 = \frac{\delta V}{\delta s} = \left(\frac{\delta U}{\delta s} \cos\beta - \frac{\delta U^1}{\delta s} \sin\beta\right)\cos\beta = \left(\frac{\delta U}{\delta s} \cos\beta + \frac{\delta U}{\delta n} \sin\beta\right)\cos\beta.$$

Assuming then $\beta < \frac{\pi}{2}$, the condition will be

$$\frac{\delta U}{\delta s} \cos\beta + \frac{\delta U}{\delta n} \sin\beta = \frac{\delta U}{\delta \beta} = 0;$$

β is a line which forms with the tangent t and with the normal n to the arc s the same angles which l forms respectively with n and with t. (Fig. 12).

From the preceding consideration it follows that in the case of the homogeneous plate placed in a uniform magnetic field the distribution of currents depends on four harmonic functions V, V', U, U' connected by the relations previously established. The lines $V = const.$ give us the *equipotential lines,* the lines $U' = const.$ give us the lines of flow; therefore the function $\frac{U'}{K}$ may be considered as the *function of the currents.*

We can easily relate among themselves the functions U, U', V, V' by means of functions of a complex variable.

Indeed, if we put $x+iy=z$, $U+iU'=f(z)$, $V+iV'=\varphi(z)$ we shall have

$$f(z)=\frac{e^{-i\beta}}{\cos\beta}\varphi(z). \tag{E)''$$

The conjugate function of $\log r_A$ is θ_A, where we denote by θ_A the angle that the radius vector which has its origin in A, forms with a fixed direction; $\frac{-m}{2\pi}\theta_A$ is the potential of a vortex situated in A and having the moment m. (We suppose that the axes x, y, are situated as in figure 2 of Section 5 of the first lecture and that the angle θ_A is measured in the sense of the rotation by which the positive direction x may reach the positive direction y; also we assume that a vortex having a positive moment rotates in the sense opposite to that just mentioned). Then, if we have in A a point electrode through which the current I enters, putting $I=\nu J$ (ν thickness of the plate), the corresponding potential will be $\frac{-J}{2\pi K}\log r_A$ and the corresponding fundamental function will be $\frac{-J}{2\pi K}\left(\log r_A+\lambda\theta_A\right)=\frac{J}{2\pi K}\log\frac{1}{r_A}-\frac{\lambda J}{2\pi K}\theta_A$.

Therefore we can say that the *fundamental function corresponding to a point electrode A, through which the current of intensity I enters, is the logarithmic potential of a mass* $\frac{J}{2\pi K}$ *and of a vortex having the moment* $\frac{\lambda J}{K}$, *both situated in the point A.*

This result may be stated by saying that the action of the magnetic field on the distribution of currents consists in modifying the conductivity, which becomes equal to K, and in modifying the fundamental function by adding to each point electrode, through which the current of intensity I enters, a vortex having the moment $\frac{\lambda J}{K}$; the normal derivative of the fundamental function remains zero at every point of the insulated edge of the plate. (The addition of the vortex by the action of the magnetic field appears quite natural if we think of the deflection of the movement of electrons, in the neighbourhood of the electrode, produced by the same field).

The function θ_A is many valued and therefore the fundamental function U is also many valued; U has the electrodes as branch points. Thus, the potential is a single valued function becoming logarithmically infinite at the point electrodes, whereas the fundamental function has the same points as points of logarithmic infinity and as branch points.

Section 2

Now, let us suppose that the plate is circular and that the current enters and goes out through internal point electrodes. We shall calculate the effect of the boundary C (that consists in annulling the normal derivative of the fundamental function), by adding masses and vortices situated in the inverse points (images) of the electrodes, with respect to the circumference C. (Fig. 13).

Let A_1 be the inverse point of the electrode A. If we put in A_1 the mass $\frac{J}{2\pi K}$ the logarithmic potential of the two masses $\frac{J}{2\pi K}$ situated in A and A_1 will have along C the normal derivative equal to $\frac{-J}{2\pi KR}$, R being the radius of C; and, if we put in A_1 a vortex having the moment $\frac{-\lambda J}{K}$ the potential of the two vortices $\frac{-\lambda J}{K}$ and $\frac{\lambda J}{K}$ situated in A_1 and A will have the normal derivative zero at the boundary.

Bearing in mind that the algebraic sum of the intensities of the currents which enter through the various internal electrodes is zero, we shall obtain for the fundamental function the following expression:

$$U=-\Sigma\frac{J}{2\pi K}\ \{\log r_A+\log r_{A1}+\lambda\theta_A-\lambda\theta_{A1}\{\ ; \tag{9}$$

the sum being extended to all the internal electrodes.

The same expression can be written:

$$U=-\Sigma\frac{J}{2\pi K}(\log r_A+\lambda\theta_A)+\varphi$$

where

$$\varphi=-\Sigma\frac{J}{2\pi K}(\log r_{A1}-\lambda\theta_{A1}):$$

evidently φ is regular within the area occupied by the plate, because all the points where it becomes infinite and the branch points are external of this area.

The result now obtained can be thus expressed: The distribution of the currents, which enter and go out through point electrodes in a circular plate subject to a magnetic field, is identical with the distribution that we should have if the plate were indefinitely extended, if the magnetic field were suppressed, and if the following conditions were fulfilled: the addition of a vortex having the moment $\frac{\lambda J}{K}$ to each electrode through which the current $I=\nu J$ enters (ν thickness of the plate); the addition in the inverse point of each internal electrode, of an image electrode traversed by the same current I and of a vortex, the image of the internal vortex and having the moment $\frac{-\lambda J}{K}$.

The formula (9) gives us the fundamental function of the distribution of currents; and from this, bearing in mind the rule (E′), we can obtain the expression of the potential.

The conjugate function of U is

$$U'=-\Sigma\frac{J}{2\pi K}(\theta_A+\theta_{A1}-\lambda\log r_A+\lambda\log r_{A1});$$

therefore we shall have

$$V=\frac{U-\lambda U'}{1+\lambda^2}=-\Sigma\frac{J}{2\pi K}\left(\log r_A+\frac{1-\lambda^2}{1+\lambda^2}\log r_{A1}-\frac{2\lambda}{1+\lambda^2}\vartheta_{A1}\right). \tag{10}$$

Now, as $\lambda = \log \beta$ we have

$$\frac{1-\lambda^2}{1+\lambda^2} = \cos 2\beta_1 \frac{2\lambda}{1+\lambda^2} = \sin 2\beta$$

hence

$$V = -\Sigma \frac{J}{2\pi K}(\log r_A + \cos 2\beta \log r_{A1} - \sin 2\beta \theta_{A1}). \tag{10'}$$

This formula can be written in the following manner:

$$V = -\Sigma \frac{J}{2\pi K} \log r_A + \varphi \tag{10''}$$

where

$$\varphi = -\Sigma \frac{J}{2\pi K}(\cos 2\beta . \log r_{A1} - \sin 2\beta . \theta_{A1}). \tag{10_a}$$

The first term of (10″) is the potential of the electrodes; the second term φ is a function regular within the area occupied by the plate, because the infinities and branch points are external.

The result now obtained can be stated in this way: *If in a circular plate, subject to a magnetic field, the currents enter and go out through point electrodes, we shall obtain the potential by adding to that of each electrode where the intensity of current is $I = \nu J$ the potential of an image electrode, situated in the inverse point and traversed by the current I cos 2β, and the potential of a vortex situated in the same inverse point and having the moment* $\frac{-J \sin 2\beta}{K}$.

Thus, in the case in which the plate is subject to the magnetic field, we have *two different principles of images*; one of these concerns the distribution of currents and therefore the *fundamental function*, the other concerns the *electric potential*.

The same results can be easily obtained by using the functions of complex variables introduced in section 1. Let us suppose that a and a' are the values of the complex variable $z = x + iy$ in two points inverse to each other with respect to a circle; that M and m are real constants and that $\Sigma M = 0$; then along the circumference that limits the said circle the real part of the function

$$e^{im} \Sigma M \log (z-a) - e^{-im} \Sigma M \log (z-a') \tag{11}$$

and the imaginary part of the function

$$e^{im} \Sigma M \log (z-a) + e^{-im} \Sigma M \log (z-a') \tag{11'}$$

are constant.

Now, if we denote with a the values of the complex variable z corresponding to the point electrodes, the potential of these latter is the real part of the function

$$\varphi = -\Sigma \frac{J}{2\pi K} \log (z-a);$$

according to the formula (E″) the corresponding fundamental function will be the real part of $f = \frac{-e^{-i\beta}}{\cos \beta} \Sigma \frac{J}{2\pi K} \log (z-a)$.

According to what we have said before, since along the circular boundary of the plate U' must be constant (see section 1), the fundamental function U will be the real part of

$$F=-\Sigma\frac{J}{2\pi K\cos\beta}[e^{-i\beta}\log(z-a)+e^{-i\beta}\log(z-a')];$$

and, owing to the formula (E″), the electric potential will be the real part of

$$\Phi=-\Sigma\frac{J}{2\pi K}[\log(z-a)+e^{ri\beta}\log(z-a')].$$

From the two last formulae we can easily deduce the formulae (9) and (10). Hitherto we have supposed the electrodes to be internal, now let us suppose that they are at the boundary. It will be sufficient to suppose, in the formulae (9) and (10), that the points A and A_1 coincide, that is it will be sufficient to put

$$\log r_A=\log r_{A1}{}^1\theta_A=\theta_{A1}.$$

Therefore we shall have

$$U=-\Sigma\frac{J}{\pi K}\log r_A;\qquad V=-\Sigma\frac{J}{\pi K}(\cos\beta\log r_A-\sin\beta\ \theta_A)\cos\beta.$$

These formulae show us that, if the point electrodes be at the boundary, the magnetic field does not alter the distribution of currents, but alters the electric potential.

In the case of two electrodes A and B the formulae (9) and (10) become

$$U=\frac{J}{2\pi K}\left[\log\left(\frac{r_Br_{B1}}{r_Ar_{A1}}\right)+\lambda(\Omega_{AB}-\Omega_{A1B1})\right]$$

$$V=\frac{J}{2\pi K}-\log\frac{r_Br_{B1}}{r_Ar_{A1}}\ \frac{-J}{\pi K}\sin\beta\left[\log\left(\frac{r_{B1}}{r_{A1}}\right)\sin\beta+\Omega_{A1B1}\cos\beta\right]$$

where Ω_1, Ω_2,.are the angles under which the points A, B, and A_1, B_1 are seen from a generical point $x\ y$.

If the two electrodes are on the boundary, we have

$$U=\frac{J}{\pi K}\log\frac{r_B}{r_A},\qquad V=\frac{J}{\pi K}\left(\cos\beta.\log\frac{r_B}{r_A}-\sin\beta\Omega_{AB}\right)\cos\beta.$$

We have thus completely solved the problem of the flow of electricity under the action of the magnetic field, in the case of a circular plate; therefore we are able to solve the same problem in all the cases in which the area occupied by the plate can be conformally represented in a circle.

From this result we can immediately deduce a consequence; in fact, if we know the law of the distribution of currents in an acyclic plate, provided only with point electrodes and not subject to the magnetic field, we are able to make the conformal representation of the area occupied by the plate in a circle, and therefore we are also able to determine the distribution of currents and the potential when the currents enter and go out through any point electrodes and the plate is subject to the magnetic field.

SECTION 3

Let us consider the case in which the point electrodes are on the boundary, then, provided the plate be acyclic whatever form it may have, the distribution of currents is not altered by the action of the magnetic field.

This result, obtained in the case of the circle (Section 2) and extended to the case of a generical acyclic field by means of the conformal representation, can be also deduced from the conditions that U must fulfill. (Section 1) In fact, if the plate be acyclic and the point electrodes are on the boundary, U is a single valued function, at the boundary $\frac{\delta U}{\delta n}=0$, in the neighborhood of an electrode U must be the sum of a logarithmic term and of a regular part, and lastly if, by means of an arc of any curve, we separate from the plate the region occupied by the electrode, we must have $\int_\Sigma \frac{\delta U}{\delta n} d_\Sigma = -\frac{J}{K}$ (where J is the ratio between the intensity of the current which flows through the electrode and the thickness of the plate). Then, if the intensity of the total current be not altered, U can differ from the electric potential corresponding to the case in which the magnetic field is zero, only by a constant factor of proportionality, equal to the ratio between the conductivity of the plate before the action of the magnetic field, and K

We will call this proposition *the principle of the point electrodes at the boundary.* Evidently, if the plate be not acyclic, we cannot represent its area conformally in a circle and we cannot say that U must be a single valued function; therefore the demonstration which we have given does not hold good in this case, and in fact, as we shall see in the following paragraph, the preceding proposition generally is not true, when the plate is not acyclic. From what we have previously said, we can deduce that, *if we know the law of the distribution of currents in a generical acyclic plate, provided with an arbitrary number of point electrodes situated at the boundary when the magnetic field does not exist, we have immediately U and, by applying the formula* (E′) *of Section* 1, *we can calculate the electric potential V, when the magnetic field exists.*

For instance, if a rectangular plate be not subject to a magnetic field, we can express (according to Betti's calculations) (1) the distribution of currents when the electrodes are in the middle points of two opposite sides, by means of the elliptic function Δam; then, by applying the preceding results we are able to solve the analogous problem, when the plate is subject to a magnetic field.

Let us consider an acyclic plate traversed by currents of given intensities entering and going out through point electrodes situated at the boundary, let us suppose that these intensities remain constant before and after the creation of the magnetic field. Let us denote respectively with u and with U the fundamental function when the magnetic field does not exist and when it exists; then we have $U=\varsigma u$, where ς is the ratio between the conductivity of the plate, (without the magnetic field) and K. Let us denote with u' the conjugate function of the harmonic function u; by applying the formula (E′) we shall have: $V=\varsigma\,\frac{u-\lambda u'}{1+\lambda^2}$.

Without the magnetic field, let MN and QP be lines of flow, MQ and NP.

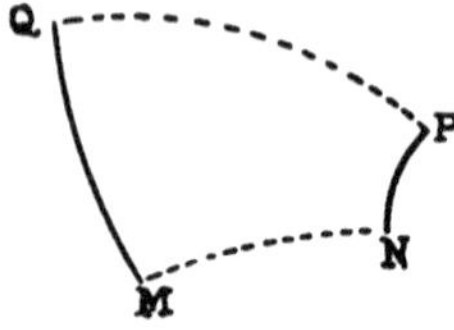

Fig. 14

equipotential lines. We shall have

$$U_M = U_2, \qquad U_N = U_{P1} \qquad U_M' = U_2' = U_P'$$

and hence

$$V_M = V_N + V_P - V_2 = 0$$

we can then state the following theorem: *When the point electrodes are on the boundary, if we consider a quadrangle formed by lines of flow and equipotential lines corresponding to the value zero of the magnetic field, and if we determine the values of the electric potential in the four vertices when the magnetic field exists, the difference of the values in two adjacent vertices is equal to the difference in the other two vertices.*

This proposition, which we will call *the theorem of the four vertices*, can be easily verified experimentally.

Let us consider the particular case of a circular plate in which the currents enter and go out through the point electrodes A and B situated at the boundary. Let

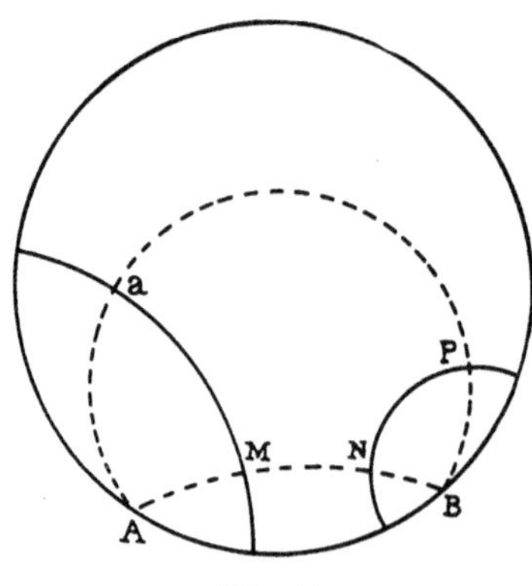

Fig. 15

MN and QP be circles passing through A and B; MQ and NP circles orthogonal to the preceding. Whatever the intensity of the magnetic field may be for the quadrangle formed by these circles we have

$$V_M - V_N + V_P - V_2 = 0.$$

Section 4

Various results among those that we have considered, hold good in both the cases of an acyclic and of a cyclic plate; others hold good only in the case of an acyclic plate (we have explicitly declared this, when we stated those results).

The formulae and the theorems of reciprocity, stated in the last paragraph of the first lecture, evidently hold good also in the case of a cyclic plate; but in this case they may assume a different aspect, because the portions of the boundary where the potential is constant may be constituted by some of the whole closed curves which form the total boundary of the cyclic area occupied by the plate. Let us suppose the plate to be bounded by the closed curves s_1', s_2', ..., s_n' and s_1'', s_2'', ..., s_m''; let us assume the first contours to be insulated, the second kept at a constant potential. We shall call S' the system of all the contours s_r', S'' the system of all the contours, s_t''; we shall exclude, for the sake of simplicity, the existence of point electrodes.

With the direct magnetic field let $C^{(1)}$, $C^{(2)}$, ..., $C^{(m)}$ be the values of the potential V along s_1'', s_2'', ..., s_m''; with the inverted magnetic field let $C_1^{(1)}$, $C_1^{(2)}$, ..., $C_1^{(m)}$ be the values of the potential V_1, along the same curves. Let us denote by $J^{(1)}$, $J^{(2)}$, ..., $J^{(m)}$ the ratios between the intensities of the currents which enter, through the lines $s_1^{('')}$, s_2'', ..., s_m'', in the plate and the thickness of this latter; let $J_1^{(1)}$, $J_1^{(2)}$, ..., $J_1^{(m)}$ have the same meaning when the magnetic field is inverted. By applying the results of Section 9 of the first lecture, we shall have:

$$\Sigma_k J^{(k)} C_1^{(k)} = \Sigma_k J_1^{(k)} C^{(k)}.$$

If we consider s_1'', s_2'', ..., s_m'' as edges of electrodes of negligible resistance, we can easily deduce, from the preceding formula, the principles of reciprocity (analogous to those stated in the first lecture) in the case in which the currents enter into the plate through internal electrodes having a finite area and a negligible resistance.

Now let us consider the case of the direct magnetic field and let us apply the formula (E) (Section 1) in order to pass from the potential V to the fundamental function U. We shall have:

(12) $$U = V + \lambda V'.$$

If we draw a closed contour s in the area occupied by the plate and denote by s_α'', s_β'', ..., s_s'' the curves s'' within s, we have

$$\int_s \frac{\delta V'}{\delta s}\, ds = -\int_s \frac{\delta V}{\delta n}\, ds = \frac{1}{K}(J^{(\alpha)} + J^{(\beta)} + \ldots\ldots + J^{(s)});$$

therefore we can say that V' is a many valued function, that the closed curves enclosing curves s'' are the cycles, and that the numbers $\frac{J^{(\alpha)}}{K}$ are the cyclic constants.

As V is a single valued function, (see first lecture, Section 5) U will have the same sort of multiplicity as V', except that the cyclic constants will be changed in the ratio λ. By applying the formulae (A) and (B) obtained in the Section 5 of the first lecture, we deduce (see first lecture, Section 6)

$$\Sigma_k J^{(h)} = 0 \qquad \Sigma_k J^{(k)} C^{(k)} = K \int_\sigma \Delta V d\sigma.$$

From the last equation we deduce that if $J^{(k)}$ be zero along some of the lines s''_h and $C_1^{(h)}$ be zero along the remaining lines s''_k, V must be zero; and if all the J's be zero, V must be constant; We can therefore state that the knowledge of

$J^{(h)}$ determines V except for a constant and the knowledge of some values of $J^{(h)}$ and of the remaining values of $C^{(h)}$ determines V completely.

Let us suppose that in the plate there are no insulated edges s'_h and that all values of $C^{(h)}$ are known; then V will be determined and independent of K. Whereas if all values of $J^{(h)}$ be known, V will vary inversely as K. Lastly if $C^{(h)}$ be known for some of the lines s''_h and $J^{(k)}$ be known for the other lines s''_h, V will depend on K. From what we have said we can deduce *that if each of the various curves which form the boundary has a constant and unalterable potential, the electric potential V will not depend on the magnetic field, but this latter will alter the fundamental function U and therefore also the distribution of the currents.*

It follows that, in this case, if we know the distribution of currents without the magnetic field, and therefore the potential V (with or without the field), we shall be able to calculate, by means of the formula (E), the alteration of the distribution of the currents caused by the magnetic field.

Likewise we shall have similar results by supposing known and inalterable the J^h.

These results agree perfectly with those obtained by Professor Corbino in the case of a plate bounded by two closed curves kept at constant potentials.(2) Let us consider the particular case in which the two closed curves are concentric circumferences having the radii R_1 and R_2. Then we shall have

$$V=\frac{(C^{(1)}-C^{(2)})\log r+C^{(2)}\log R_1-C^{(1)}\log R_2)}{\log R_1-\log R_2}, \qquad V'=\frac{(C^{(1)}-C^{(2)})}{\log R_1-\log R_2}\theta,$$

$$U=\frac{(C^{(1)}-C^{(2)}(\log r+\lambda\theta)}{\log R_1-\log R_2}+C$$

where r and θ are the polar coördinates of the points of the circular ring which constitutes the plate (having taken as origin the centre of the ring) and C is a constant.

If the current enters through the external circle of radius R_1 and its intensity be $I=\nu J$ (ν thickness of the plate) we shall have

$$V=\frac{J}{2\pi K}\log r, \qquad V^1=\frac{J}{2\pi K}\theta, \qquad U=\frac{J}{2\pi K}(\log r+\lambda\theta).$$

The results obtained at the beginning of this paragraph can be extended. Let us suppose we have a cyclic plate bounded by the closed curves s_1, s_2 ......, s_n. Without making any hypothesis about the values of the potential corresponding to these curves, or the way in which point or linear electrodes are situated along them, we shall indicate with J_1, J_2, ...J_n the ratios between the intensities of the currents which across the said curves penetrate into the plate and the thickness of this latter. Let M_1, M_2, ...M_m be internal point electrodes; let I_1, I_2, ..., I_m be the ratios between the intensities of the currents which enter through these electrodes and the thickness of the plate. We shall have $\sum_1^n{}_k J_h+\sum_1^m{}_K I_K=0$, and, if a closed line s within the plate enclose the lines s_α, s_β, ..., s_τ and the points M_A, M_B, ..., M_{t1}

$$\int\frac{\delta V'}{\delta s}\,ds=-\int\frac{\delta V}{\delta n}\,ds=\frac{1}{K}(J_\alpha+J_\beta+\ldots\ldots+J_\tau+I_a+I_b+\ldots\ldots+I_t).$$

This proves that V', and therefore U, is a many-valued function, unless J_1, J_2, ..., J_n, I_1, I_2, ..., I_m be all zero.

We can then enunciate the theorem: *The necessary and sufficient condition that the fundamental function be uniform is that no internal electrodes exist and that the total quantity of electricity which enters through the electrodes distributed along each closed curve making part of the boundary be zero.*

Whereas, in the case in which the area occupied by the plate is acyclic, *it is sufficient, for the uniformity of the fundamental function, that no internal point electrodes exist; while owing to the presence of these electrodes the fundamental function becomes many-valued.*

Let us make an application of the results now obtained. Let us suppose that the area is cyclic and that the electrodes are punctiform and distributed along the curves s_1, s_2, ..., s_n which form the boundary. Let $i_p^{(1)}$, $i_p^{(2)}$, ..., $i_p^{(hs)}$ be the ratios between the intensities of the currents which enter through the electrodes distributed along the curve s_p and the thickness of the plate; then the necessary and sufficient condition, in order that the fundamental function be single-valued, is that $\sum_1^p {}_t i_p^{(t)} = \theta$, $(p = 1_1, 2_1 \ldots . n)$. Now if the fundamental function be single-valued, by repeating a *reasoning* made at the beginning of the preceding paragraph, we shall be able to demonstrate that the magnetic field does not alter the distribution of currents, while, if the same function be many valued, we shall see that the action of the magnetic field will have to change the distribution of currents, because without the field the fundamental function coincides with the potential and therefore is single-valued.

We shall have thus the following theorem: *If the plate be cyclic and all the electrodes be point electrodes and situated on the boundary, the action of the magnetic field changes the distribution of currents unless the sum of the intensities of the currents, which penetrate through the electrodes placed on each closed curve that forms part of the boundary, is zero.*

When the preceding condition is fulfilled and one knows the law of the distribution of currents without the magnetic field, one can, by means of the formula (E′), calculate the potential corresponding to a given magnetic field and therefore one can completely solve the problem.

It remains to consider the case in which the above mentioned condition is not fulfilled.

Let us denote by s_1 the closed curve which forms the outer boundary of the cyclic plate, with s_2, s_3, ..., s_n the closed curves that form the inner boundaries. Let us trace a curve l which joins s_1 with s_h without meeting other parts of the boundary and let us imagine that along l there is an electromotive force having the value l and directed in the sense in which the arc s_h increases. Supposing that the plate is not subject to the magnetic field, the electric potential φ_h will be a regular and harmonic function within the area occupied by the plate and will have a discontinuity l along l. Moreover we shall have

$$\frac{\delta \varphi_h}{\delta n} = \frac{\delta \varphi_h}{\delta n_2} = \ldots\ldots\ldots\ldots = \theta, \quad \int_{sh} \frac{\delta \varphi_h}{\delta s_h} ds_h = -1, \quad \int_{sg} \frac{\delta \varphi_h}{\delta s_g} ds_g = \theta, \quad 1 \angle g \not\angle h.$$

We can also consider φ_h as a finite, continuous and many-valued function, in the area σ when there is no cut l; the cycles of this function enclose s_h and its cyclic constant is -1. The functions $\varphi_2, \varphi_3, \ldots\ldots$ will only depend on the geometrical form of the field σ, and we will call them the *elementary cyclic potentials* of σ. We will indicate by $\varphi_2', \varphi_3', \ldots\ldots$ the conjugate functions of φ_2, φ_3. (3).

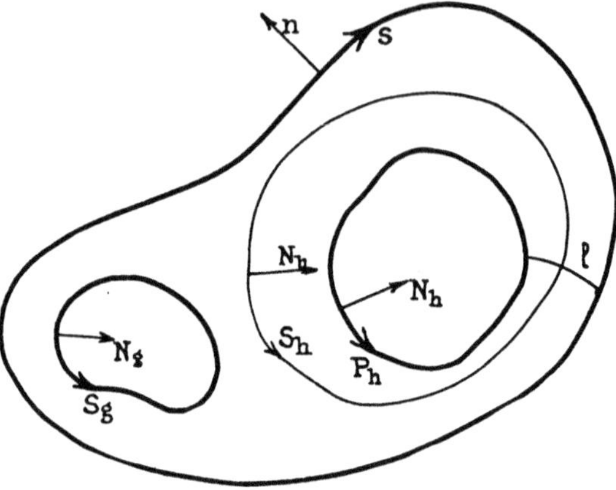

Fig. 16

Now let us suppose that every electromotive force in the interior of the plate is taken away; let us call W the electric potential without the magnetic field (but reduced in the ratio between the conductivity of the plate before the action of the same field and K), when the currents enter and go out through point electrodes disposed along the various closed contours which form the boundary; let us call V the electric potential with the magnetic field, U the fundamental function.

Let I_h be the algebraic sum of the intensities of the currents which enter and go out through the electrodes distributed along s_h; let us put $I_h = \nu J_h$ (ν thickness of the plate). If S_h be a line which encloses only s_h we shall have

$$J_h = K\int_{Sh}\frac{\delta V}{\delta N_h}\,dS_h = K\int_{Sh}\frac{\delta W}{\delta N_h}\,dS_h = -K\int_{Sh}\frac{\delta V'}{\delta S_h}\,dS_h = -K\int_{Sh}\frac{\delta W'}{\delta S_h}\,dS_h$$

where V' and W' are the conjugate functions of V and W. Hence follows (see formula (E))

$$\int_{Sh}\frac{\delta U}{\delta S_h}\,dS_h = \int_{Sh}\frac{\delta(V+\lambda V')}{\delta S_h}\,dS_h = \frac{-\lambda J_h}{K}.$$

We can therefore take

$$U = W + \frac{\lambda}{K}\sum_2^n{}_h J_h\varphi_h$$

(3) The elementary cyclic potentials are considered in classical hydrodynamics in order to obtain the irrotational motions of a fluid in a cyclic space bounded by rigid walls. They correspond in the theory of elasticity to the *distortions*. Let $\Psi_2, \Psi_3, \ldots, \Psi_n$ be regular harmonic functions; let each of these functions Ψ_h be zero along all the lines $s_1, s_2, \ldots, s_n$ excepted, s_h where Ψ'_k has the value 1. If we know $\Psi_2, \Psi_3, \ldots, \Psi_{4n}$ we shall be able to obtain (by combining them linearly with constant coefficients) the function φ'_2 and then we shall be able to calculate the functions φ_h. The knowledge of the φ_h or of the Ψ_h is therefore analytically equivalent.

because the second member of this equation satisfies all the conditions which U must fulfill.

In order to obtain W we have only to apply the rule (E′); we shall obtain thereby

$$V=\frac{W-\frac{\lambda^2}{K_1}\sum_h^n J_h\varphi'_h-\lambda\left(W'-\frac{1}{K_2}\sum_h^n J_h\varphi_h\right)}{1+\lambda^2}.$$

Therefore *if one know the elementary cyclic potentials of the cyclic area occupied by the plate, one can determine the perturbation produced by the magnetic field on the electric currents whatever they may be, provided they enter and go out through point electrodes situated on the boundary.* The two last formulae express *the principle of point electrodes on the boundary* modified for the case of the cyclic plates (see the preceding paragraph).

Let us consider the particular case in which the plate is a ring bounded by two concentric circles. In this case the elementary cyclic potential is $\frac{-\theta}{2\pi}$, where θ

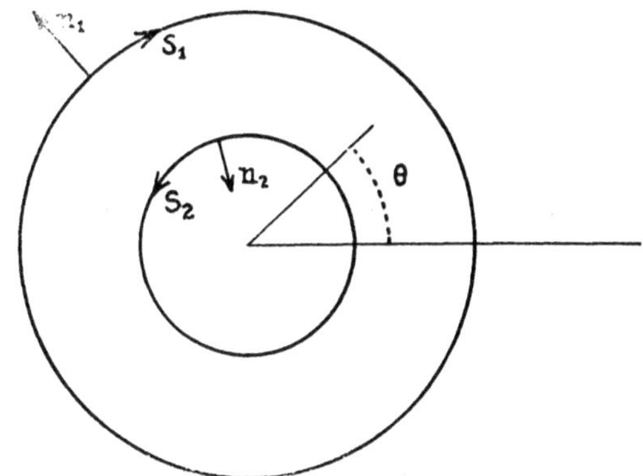

Fig. 17

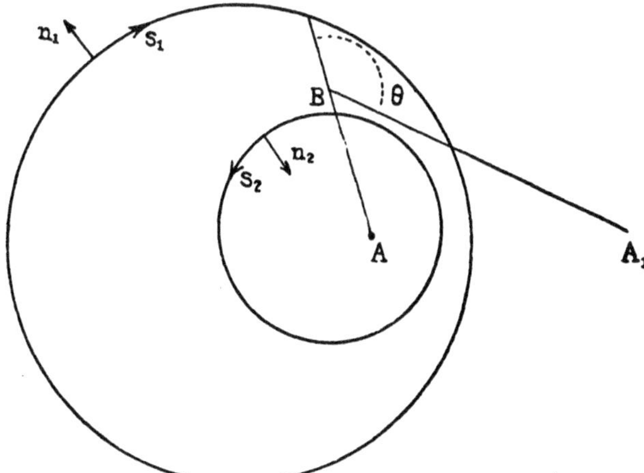

Fig. 18

is the angle which the vector radius, having its origin in the center of the plate, forms with a fixed direction; the conjugate function of this elementary cyclic potential is $\frac{\log r}{2\pi}$.

Therefore the two last formulae become

$$U = W - \frac{\lambda}{2\pi K} J_2\theta, \qquad V = \frac{W - \frac{\lambda^2}{2\pi K} J_2 \log r - \lambda\left(W' + \frac{1}{2\pi K}\,\theta\right)}{1+\lambda^2}.$$

If the two circles be not concentric we take the points A and A_1, *images* of each other with respect to the two circles contemporaneously, and the elementary cyclic potential will be $\frac{-\theta}{2\pi}$ where θ is the supplement of the angle under which from the generical point B of the plate one sees the segment $A\,A_1$; the conjugate function will be $\frac{1}{2\pi}\log.\frac{AB}{A_1B}$.

SECTION 5

We shall now pass to the solution of the problem in the case in which the electrodes, supposed to be of negligible resistance, constitute portions of the boundary. Therefore let us go back to the conditions examined in Section 6 of the first lecture.

Let us suppose what we have been able to make the conformal representation of the acyclic area σ in a parallelogram *abcd* of the plane $\xi\gamma$ in such a way that $\hat{bad} = \frac{\pi}{2} - \beta$. Let us suppose also that the sides *ab* and *cd* are parallel to the axis γ and that *db* and AB, *bc* and BC, *cd* and CD, *da* and DA correspond respectively to each other.

Let us take the function

$$V = M\xi + N$$

where M and N are two constants, and let us consider V as function of ξ and γ. V will be constant along the sides *ab* and *cd* and will satisfy the condition $\frac{\delta V}{\delta\gamma} = 0$

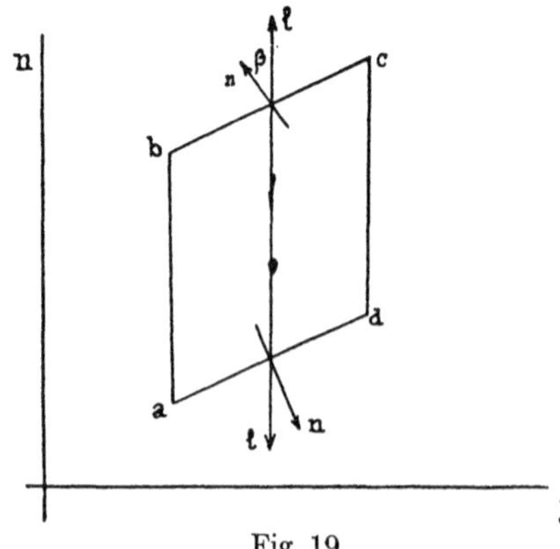

Fig. 19

along bc and ad. It is easy to see that along bc and ad the directions γ and $-\gamma$ form respectively the angle β with the external normals n to the same sides.

Let us now consider ξ as function of x and y and let us refer the function V to points of the area σ in the plane xy. V will result harmonic and regular, will be constant along the portions AB and CD of the boundary and along the portions BC and AD will satisfy the condition $\frac{\delta V}{\delta l}=0$.

Making use of the arbitrariness of the constants M and N, we shall be able to reduce the values of V along AB and CD to the given values, and there'ore V will be the required potential. It is easy to recognize the order of *infinity* of the derivatives of ξ (with respect to x and y) in the angular points of the boundary.

Let us now suppose σ to be a square. Let us begin by taking on the real axis of the complex plane z two points a and $-a$ and let us put

(13) $$Z=\int_0^z(a^2-z^2)^{\nu-1}dz$$ (13') $$Z_1=\int_0^z(a^2-z^2)^{\mu-1}dz$$

While z moves in the half plane corresponding to the positive coefficient of the imaginary part of z, Z and Z_1 move respectively in the two isosceles triangles ABC, $A_1B_1C_1$ the angles at the base of which have respectively the amplitudes $\nu\pi$ and $\mu\pi$. Therefore, applying the principle of symmetry, we see that, while

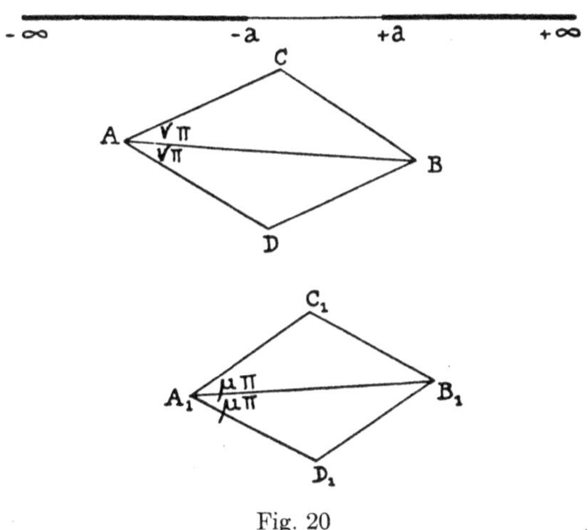

Fig. 20

z moves in the whole plane, sectioned by the two *cuts* $-a-\infty$ and $+a+\infty$, Z and Z_1 move respectively in the two rhombs $ABCD$ and $A_1B_1C_1D_1$. Taking the first rhomb becomes a square, and taking $\mu=\frac{1}{\Psi}-\frac{\beta}{2\pi}$ the second rhomb becomes a parallelogram having an angle equal to $\frac{\pi}{2}-\beta$. By means of the rotation $Z_1'=Z_1e^{i\left(\frac{1}{2}-\mu\right)\pi}$ one transforms the second rhomb so that it has a coup'e of sides parallel to an axis, that is one reduces this rhomb to be in the condition of the parallelogram represented in the figure 19. The conformal representation of the square in the parallelogram which we have thus obtained solves the problem of *determining the potential and the distribution of currents in a square plate subject to a magnetic field, when two opposite sides of the plate are electrodes of negligible resistance through which the current enters and goes out, and the other two sides are free and insulated.*

It is obvious that, having taken $\nu = \frac{1}{\Psi}$, the integral (13) is elliptic, and in fact, if we put $a^2 z^2 = X^4$ and we take $a = 1$, we have

$$Z = \int_o \frac{dz}{(a^2 - r^2)^{3/4}} = 2 \int_1^x \frac{dx}{\sqrt{1 - x^4}}.$$

In the first paragraph of the following lecture we shall study the case of linear electrodes of negligible resistance, situated at the boundary of a circular plate.

THIRD LECTURE

Section 1—The problem of the flow of electricity, under the action of magnetic field, in a circular plate, provided with a linear electrode of negligible resistance, situated on the boundary and with point electrodes. Section 2—The problem of the flow of electricity in a curved and non-homogeneous plate subject to the action of a non-uniform magnetic field; general formulae and results. Section 3—Comparison between the problem relative to the plane homogeneous plate and uniform magnetic field and the problem relative to the curved and non-homogeneous plate and the non-uniform field. Section 4.—The theorem of reciprocity in the case of the curved and non-homogeneous plate subject to the action of a non-uniform magnetic field.

SECTION 1

Let us consider a circular plate in which the current enters through an internal point electrode and goes out through a linear electrode, of negligible resistance, situated along the arc AB of the boundary. (See Fig. 21).

In this case, for studying the flow of electricity under the action of the magnetic field, it will be sufficient to determine a regular, harmonic function W, its value along the arc BC and the value of $\frac{\delta W}{\delta l}$ along the arc CDB being known. (See Section 7 of the first lecture). Let us represent conformally the circle in an angle of the plane $\xi\gamma$ having the amplitude $\frac{\pi}{2} - \beta$ (Fig. 21), in such a way that the side bc corresponds to the arc BDC, the vertex b corresponds to B and the point at infinity C^∞ to the point C. Let us transfer to the angular figure the values of W; then $\frac{\delta W}{\delta l}$ will be known on the two sides of the angle and, as it is harmonic, we

---

[1]Betti, *Opere*, vol. II, p. 267.

[2]Corbino, *Nuovo Cimento*, 1911, vol. I, p. 397.

shall be able to determine its values within the angle and therefore we shall be also able to calculate W.

Evidently, we should have the same result if a side of the angle were parallel to and corresponded to the arc CC and the other inclined side corresponded to the arc BDC, and if we considered $\frac{\delta W}{\delta \xi}$ instead of $\frac{\delta W}{\delta \gamma}$.

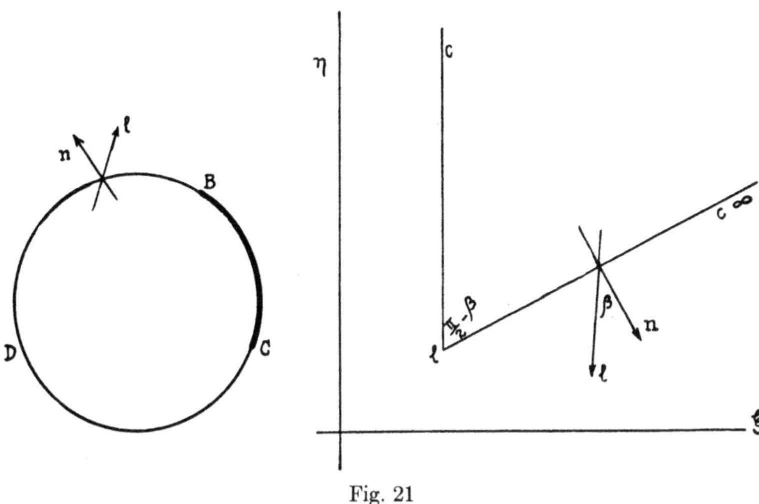

Fig. 21

But, in the case that we are considering, it is useful to employ the functions of a complex variable which we introduced in Section 1 of the preceding lecture.

The case in which $\beta=\frac{n-2}{n}\pi$ is particularly interesting, because in this case, by a convenient conformal representation and by the application of the two principles of images (see Section 2 of the preceding lecture) we can obtain the solution more simply.

By means of the function

$$Z=\frac{ze^{-\frac{i\theta}{2}}-e^{\frac{i\theta}{2}}}{Z-1}$$

we can represent the area within a circle situated in the plane z, having the centre in the origin and the radius equal to 1, (Fig. 22), in the half-plane $Z=X+iY$ corresponding to the positive values of Y. If we consider the points of the circumference of the circle (for which $z=e^{i\omega}$), we find for Z

$$Z=\frac{\sin\frac{\omega-\theta}{2}}{\sin\frac{\omega}{2}};$$

therefore these points correspond to the points of the real axis in the plane Z, and more precisely, the portion of the circumference for which $0<\omega<\theta$ corresponds to the negative real semiaxis and the portion of the circumference for which $\omega<\theta<2\pi$ corresponds to the positive real semiaxis.

Let us suppose $\theta<\mu<1$ and let us put $S=Z^{\mu}$; thus we obtain the conformal representation of the semiplane Z, corresponding to the positive values of Y, in an angle having the amplitude $\pi\mu$, of the plane S (Fig. 24), in such a way that the

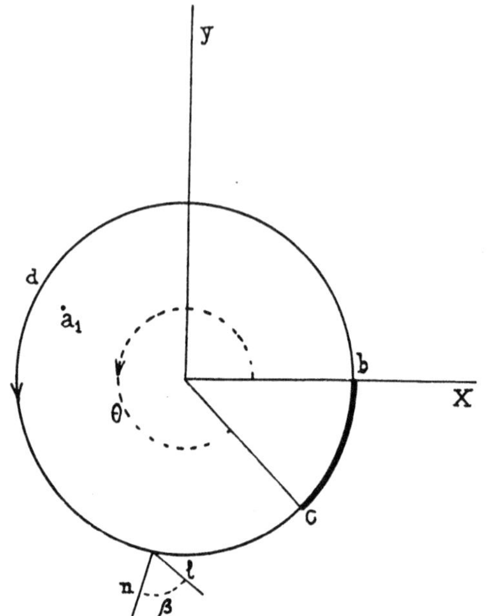

Fig. 22

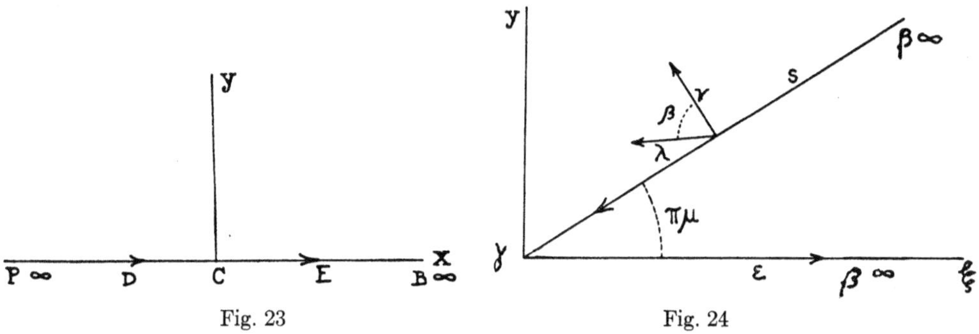

Fig. 23

Fig. 24

positive real semiaxis of Z corresponds to the positive real semiaxis of S and the negative real semiaxis of Z corresponds to the side $j\beta\infty$ (Fig. 24) of the angle $\pi\mu$; then, by putting

$$(14) \qquad S=\left(\frac{Ze^{-\frac{i\theta}{2}}-e^{\frac{i\theta}{2}}}{z-1}\right)^{\mu}$$

we shall represent the circle (Fig. 22) in the angle (Fig. 24). By putting

$$z=re^{iw}, \quad S=\xi+i\gamma=se^{i\varphi}$$

for $0<w<\theta$, $r=1$, we shall have

$$S=\left(\frac{\sin\frac{w-\theta}{2}}{\sin\frac{w}{2}}\right)^{\mu} e^{i\mu\pi} \qquad \text{that is } \xi=\left(\frac{\sin\frac{w-\theta}{2}}{\sin\frac{w}{2}}\right)^{\mu}, \quad \varphi=\mu\pi$$

and for $\theta<w<2\pi$, $r=1$ we shall have

$$S=\left(\frac{\sin\frac{w-\theta}{2}}{\sin\frac{w}{2}}\right)^{\mu} \qquad \text{that is } \xi=\left(\frac{\sin\frac{w-\theta}{2}}{\sin\frac{w}{2}}\right)^{\mu}, \quad Y=0.$$

Differentiating the two equations

$$S=\left(\frac{\sin\frac{\theta-w}{2}}{\sin\frac{w}{2}}\right)^{\mu} \qquad\qquad \xi=\left(\frac{\sin\frac{w-\theta}{z-\theta}}{\sin\frac{w}{2}}\right)^{\mu}$$

with respect to S and ξ respectively, we find

$$1=\frac{d}{dw}\left(\frac{\sin\frac{\theta-w}{2}}{\sin\frac{w}{2}}\right)^{\mu}\frac{dw}{dS}=-\frac{\mu}{2}\frac{\sin\frac{\theta}{2}}{\left(\sin\frac{\theta-w}{2}\right)^{1-\mu}\left(\sin\frac{w}{2}\right)^{1+\mu}}\frac{dw}{dS},$$

$$1=\frac{d}{dw}\left(\frac{\sin\frac{w-\theta}{2}}{\sin\frac{w}{2}}\right)^{\mu}\frac{dw}{d\xi}=\frac{\mu}{2}\frac{\sin\frac{\theta}{2}}{\left(\sin\frac{w-\theta}{2}\right)^{-1\mu}\left(\sin\frac{w}{2}\right)^{1+\mu}}\frac{dw}{d\xi};$$

that is

$$\text{(15)} \qquad \frac{dw}{dS}=\frac{-2}{\mu}\frac{\left(\sin\frac{\theta-w}{2}\right)^{1-\mu}\left(\sin\frac{w}{2}\right)^{1+\mu}}{\mathrm{Sin}\frac{\theta}{2}}, \quad 0<w<\theta$$

$$\text{(15)}' \qquad \frac{dw}{d\xi}=\frac{2}{\mu}\frac{\left(\sin\frac{w-\theta}{2}\right)^{1\mu}\left(\sin\frac{w}{2}\right)^{1+\mu}}{\sin\frac{\theta}{2}}, \quad 0<w<2\pi.$$

Now, let $w(z)=u+iv$ be a function regular within the circle and along its boundary. By inverting (14) we can have z as function of S, and by substituting the expression thus obtained in the preceding function, we shall obtain

$$w(z(S))=w_1(S)$$

which will be regular in the interior and on the boundary of the angle j (Fig. 24).

Let us calculate

$$\text{(16)} \qquad u_2+iv_2=w_2(S)=\frac{dw_1}{dS}=\frac{du}{d\xi}+i\frac{dv}{d\xi}$$

and for this purpose let us suppose that u is known along the arc *ceb* as function of w. Let us denote by $\angle(w)$ this function and by $\angle'(w)$ its derivative with respect to w. Further let l be a direction which forms the angle $\beta=\frac{\pi}{2}-\pi\mu$ with the external normal n to the circle and let $\mu(w)$ be the derivative $\frac{du}{dl}$. Then, owing to (15′) along the positive semiaxis ξ, that is along the side $j\Sigma\beta\infty$ of the angle $\pi\mu$ (Fig. 24) we shall have

$$u_2=\frac{\delta u}{\delta\xi}=\frac{\delta u}{\delta w}\frac{dw}{d\xi}=\angle'(w)\frac{2}{\mu}\frac{\sin\left(\frac{w-\theta}{2}\right)^{1-\mu}\left(\sin\frac{w}{2}\right)^{1+\mu}}{\sin\frac{\theta}{2}}$$

and along the side $j\delta\beta\infty$ of the same angle, denoting with λ a direction which forms the angle β with the normal ν, we shall have

$$u_2=\frac{\delta u}{\delta\xi}=-\frac{\delta u}{\delta\lambda}=-\frac{\delta u}{\delta l}\left(-\frac{dw}{dS}\right)=-\mu(w)\frac{2}{\mu}\frac{\left(\sin\frac{\theta-w}{2}\right)^{1-\mu}\left(\sin\frac{w}{2}\right)^{1+\mu}}{\sin\frac{\theta}{2}}$$

If in (16) we substitute for S the value (14), we obtain

$$w_2(S(z))=u_2(x_1y)+iv_2(x_1y);$$

u_2 along the arc bdc will be equal to

$$-\frac{2}{\mu}\mu(w)\frac{\left(\sin\frac{\theta-w}{2}\right)^{1-\mu}\left(\sin\frac{w}{2}\right)^{1+\mu}}{\sin\frac{\theta}{2}}$$

and along the arc ceb will be equal to

$$\frac{2}{\mu}\angle'(w)\frac{\sin\left(\frac{w-\theta}{2}\right)^{1-\mu}\left(\sin\frac{w}{2}\right)^{1+\mu}}{\sin\frac{\theta}{2}}.$$

Thus we know at the boundary of the circle the values of the real part u_2 of the function $w_2(S(z))$ and therefore, by applying a well known formula we can easily calculate w_2 in the circle. The formula which we employ is

$$W_2(S(z))=\frac{1}{2\pi}\int_0^{2\pi}u_2(w)\frac{e^{iw}+z}{e^{iw}-z}\,dw+iC$$

(w) the values of u at the boundary of the circle which we denote with u_2 and with C a real constant.

Then, since for $z=1$ w_2 must be zero, we shall have

$$w_2(S(z))=-\frac{1}{\pi\mu}\int_0^\theta M_{(w)}\frac{\left(\sin\frac{\theta-w}{2}\right)^{1-\mu}\left(\sin\frac{w}{2}\right)^{1+\mu}}{\sin\frac{\theta}{2}}\left(\frac{e^{iw}+z}{e^{iw}-z}-\frac{e^{iw}+1}{e^{iw}-1}\right)dw$$

$$+\frac{1}{\pi\mu}\int_0^{2\pi}\angle'(w)\frac{\left(\sin\frac{w-\theta}{2}\right)^{1-\mu}\left(\sin\frac{w}{2}\right)^{1+\mu}}{\sin\frac{\theta}{2}}\left(\frac{e^{iw}+z}{e^{iw}-z}-\frac{e^{iw}+1}{e^{iw}-1}\right)dw.$$

But from formula (14) we deduce

$$\frac{dS}{dz}=2\mu i e^{\frac{i\theta}{2}(1-\mu)}\frac{\sin\frac{\theta}{2}}{(z-e^{i\theta})^{1-\mu})(z-1)^{1+\mu}}$$

and therefore with single operations we are able to calculate

$$\frac{dw}{dz}=\frac{dw}{dS}\frac{dS}{dz}=w_2(S(z))\frac{dS}{dz};$$

we obtain

$$(17)\qquad \frac{dw}{dZ}=\frac{2}{\pi}\frac{e^{i\frac{\theta}{2}(1-\mu)}}{(z-e^{i\theta})^{1-\mu}(z-1)^{\mu}}\left\{\int_o^\theta M(w)\frac{\left(\sin\frac{\theta-w}{2}\right)^{1-\mu}\left(\sin\frac{w}{2}\right)^{\mu}e^{\frac{iw}{2}}}{z-e^{iw}}dw\right.$$

$$-\int_o^{2\pi}\angle'(w)\frac{\left(\sin\frac{w-\theta}{2}\right)^{1-\mu}\left(\sin\frac{w}{2}\right)^{\mu}e^{\frac{iw}{2}}}{z-e^{iw}}dw$$

Now, by an integration, we shall calculate w except for a constant, and, by separating the real part from the imaginary part, we shall have the harmonic function u, leaving a constant that will be easily determined.

Therefore, if along the arc bdc we know $\frac{\delta u}{\delta l}$, and along the arc ceb we know u, we are able to calculate the harmonic function u within the circle. According to what we have stated in Section 7 of the first lecture, we can say that the problem of the *distribution of currents in a circular plate, when the current enters through one or even through several, internal point electrodes and goes out through a linear electrode, of negligible resistance, situated along an arc of the boundary,* can be completely solved by the preceding formulae.

The integration with respect to z that we have to make for obtaining from (17), w, can be calculated by elementary function when $\mu=\frac{1}{n}$, n being a whole number. But in this case the method of the images permits us to obtain the solution more easily, as we shall now see.

Let a_1 be a point in the interior of the angle $dj\Sigma$ (Fig. 24) having the amplitude $\pi\mu=\frac{\pi}{n}$ (n, whole number).

Let us complete the division of the plane in angles all equal to $\frac{\pi}{n}$ and let us consider the successive images $a_2, a_3 \ldots\ldots a_n$ of the point a_1 with respect to the various sides of the angles (see Fig. 25); in this figure we have taken $n=6$, $\beta=60°$. The smallest value of n is 2, in which case we have $\beta=0$).

Let us consider the two functions of the complex variable S

$$(18)\qquad \Phi=\sum_1^n{}_h e^{\frac{-2(h-1)\pi i}{n}}\left[\log(S-a_{2h-1})-e^{\frac{-2\pi i}{n}}\log(S-a_{2h})\right]$$

$$(18')\qquad F=\sum_1^n{}_h e^{\frac{-2(h-1)\pi i}{n}}\left[{}_e^{\pi}\log(S-a_{2h-1})-e^{\frac{-\pi i}{n}}\log(S-a_{2h})\right]$$

which are connected by the relation

$$(18'') \qquad F = e^{\frac{\pi i}{n}} \phi$$

and let us study their properties. We can easily recognize that each term of these functions does not vary by changing h in $h+n$. In fact, since after $2n$ reflections the images reproduce themselves, we have

$$a_g = a_g + 2_{n1}$$

and also

$$e^{\frac{-2(h+n-1)\pi i}{n}} = e^{\frac{-2(h-1)\pi i}{n}}.$$

We can therefore substitute in any term h' for h, without altering its value, provided we have $h' = h$ (mod. n).

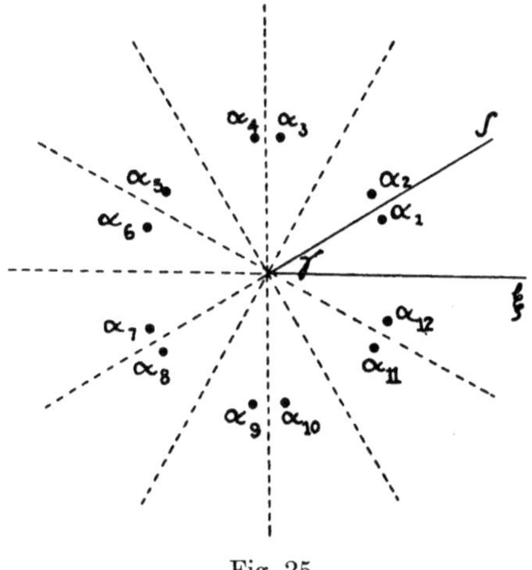

Fig. 25

Now let us observe that a_{2h-1} and a_{2k} are images each of the other with respect to the side $j\Sigma$ when $2h-1+2k=2n+1$ and that a_{2h-1} and $a_{2k'}$ are images each of the other with respect to the side jd when $2h-1+2k'=2n+3$. From the two last equations we obtain

$$k = n-h+1, \qquad k' = n-h+2.$$

Let us consider now the two following terms of the sum (18);

$$e^{\frac{-2(h-1)\pi i}{n}} \log(z - a_{2h-1})$$

and

$$-e^{\frac{-2(k-1)\pi i}{n}} e^{\frac{-2\pi i}{n}} \log(z - a_{2k}) = -e^{\frac{-2(h-1)\pi i}{n}} \log(z - a_{2k})$$

their sum will be

$$e^{\frac{-2(h-1)\pi i}{n}} \log(z - a_{2h-1}) - e^{\frac{-2(h-1)\pi i}{n}} \log(z - a_{2k})$$

and will have its real part constant along the side $j\Sigma$ (see Section 2 of the second lecture).

Thus, the terms of $\varnothing$ can be coupled in such a way that the real part of the sum of each couple is constant along $j\Sigma$; therefore along this side the real part of $\varnothing$ is also constant.

Let us consider the two terms of the sum (18′)

$$e^{\frac{-2(h-1)\pi i}{n}} e^{\frac{\pi i}{n}} \log(S-a_{2h-1})$$

and

$$e^{\frac{-2(k'-1)\pi i}{n}} e^{\frac{-\pi i}{n}} \log(S-a_{2k'} = e^{\frac{-2(k-1)\pi i}{n}} e^{\frac{-\pi i}{n}} \log(S-a_{2k'})$$

their sum will be

$$e^{\frac{-(2h-3)\pi i}{n}} \log(S-a_{2h-1}) - e^{\frac{-(2h-3)\pi i}{n}} \log(S-a_{2k'})$$

and will have this real part constant along the side $j\delta$. We can thus couple the terms of F in such a way that the real part of the sum of each couple is constant along $j\delta$; therefore also F will have the real part constant along this side.

Now let us take

$$\varphi=\frac{-J}{2\pi k}\Phi \tag{19}$$

$$f=\frac{-J}{2\pi k_i \sin\frac{\pi}{n}}F. \tag{19′}$$

Evidently φ will have the real part constant along the side $j\Sigma$ and f will have the imaginary part constant along $j\delta$, that is the normal derivative of the real part of f will be zero along $j\delta$.

But we can easily verify, owing to (18″) that we have $f=\frac{e^{-i\beta}}{\cos\beta}\varphi$ when we take $\beta=\left(\frac{n-2}{2n}\right)\pi$. Therefore, if we assume the real part of φ as electric potential, the real part of f will be the corresponding fundamental function (Section 1 of the second lecture). Let us separate in f and φ the real from the imaginary part, by writing

$$\varphi=V+iV', \qquad f=U+i^1U';$$

then V and W are respectively the electric potential and the fundamental function corresponding to the distribution of the current in an indefinite plate bounded by the two radii $j\Sigma$ and $j\delta$ and subject to the action of a magnetic field, when the current of intensity $I=\nu J$ (ν thickness of the plate) enters through the point a_1 and goes out through an indefinite electrode of negligible resistance placed along the side $j\Sigma$, when the side bdc is insulated and the angle β is equal to $\left(\frac{n-2}{2n}\right)\pi$.

Now let us put in (19) and (19′) (see formula (14))

$$S=\left(\frac{ze^{-\frac{i\theta}{2}}-e^{\frac{i\theta}{2}}}{z-1}\right)^{\frac{1}{n}}$$

we shall thus obtain the conformal representation of the angle $\Sigma j \delta$ (Fig. 25) in the circle (Fig. 22). To the point a_1 will correspond a point α_1 in the interior of the circle; a_1 and α_1 will be connected by the relation

$$\left(\frac{a_1 e^{-\frac{i\theta}{2}} - e^{\frac{i\theta}{2}}}{a_1 - 1}\right)^{\frac{1}{n}} = \alpha_1$$

and therefore the complex numbers $\alpha_1, \alpha_3, \alpha_5, \ldots\ldots, \alpha_{2n-1}$ will be the n values of

$$\sqrt[n]{\frac{a_1 e^{-\frac{i\theta}{2}} - e^{\frac{i\theta}{2}}}{a-1}}$$

while $\alpha_2, \alpha_4, \ldots., \alpha_{2n}$ will be the number conjugates of the preceding, that is the n values of

$$\sqrt[n]{\frac{a_1' e^{\frac{i\theta}{2}} - e^{-\frac{i\theta}{2}}}{a_1' - 1}}$$

having denoted with a_1' the complex number conjugate to a_1. Therefore: *If we have*

$$\beta = \frac{n-2}{2M}\pi \qquad S = \left(\frac{z e^{-\frac{i\theta}{2}} - e^{\frac{i\theta}{2}}}{z-1}\right)^{\frac{1}{n}}$$

and if the φ_{2h-1} *and the* α_{2h} *denote respectively the n values of*

$$\sqrt[n]{\frac{a_1 e^{-\frac{i\theta}{2}} - e^{\frac{i\theta}{2}}}{a_1 - 1}} \quad \text{and} \quad \sqrt[n]{\frac{a_1' e^{\frac{i\theta}{2}} - e^{-\frac{i\theta}{2}}}{a_1' - 1}}$$

(a' being the conjugate of a), *the real parts of*

$$\varphi = \frac{-J}{2\pi K}\sum_1^n{}_h e^{\frac{-2(h-1)\pi i}{n}}\left[\log(S-\alpha_{2h-1}) - e^{\frac{-2\pi i}{n}}\log(S-\alpha_{2h})\right] \quad \text{and}$$

$$f = \frac{-J}{2\pi K i \sin\frac{\pi}{n}}\sum_1^n{}_h e^{\frac{-2(h-1)\pi i}{n}}\left[e^{\frac{\pi i}{n}}\log(S-\alpha_{2h-1}) - e^{\frac{-\pi i}{n}}\log(S-\alpha_{2h})\right]$$

are respectively the electric potential and the fundamental function of the distribution of currents under the action of the magnetic field in a circle having the radius equal to 1 (Fig. 22), *when the current enters through an internal point electrode a and goes out through a linear electrode, of negligible resistance, placed along the arc bec of the boundary, while the part bdc of the same boundary is insulated.*

Now let us suppose that the point electrode a_1 *coincides with a point of the arc bdc of the boundary*; then we shall have $\alpha_{2h-1} = \alpha_{2h}$ and if we put $a_1 = e^{iw}$ with $w < \theta$, these numbers α will be the n values of

$$\sqrt[n]{\frac{e^{i\left(a-\frac{\theta}{2}\right)} - e^{i\frac{\theta}{2}}}{e^{iw} - 1}}$$

that is the n values of

$$\left(\frac{\sin\frac{\theta - w}{2}}{\sin\frac{w}{2}}\right)^{\frac{1}{n}} e^{\frac{(2h-1)\pi i}{n}}$$

for $h=1, 2, \ldots\ldots, n$, therefore we shall have

$$f=\frac{-J}{\pi K}\sum_{1}^{n}{}_h\, e^{\frac{-2(h-1)\pi i}{n}}\log S-\sqrt[n]{\frac{\sin\frac{\theta-w}{2}}{\sin\frac{w}{2}}}\,e^{\frac{(2h-1)\pi i}{n}}$$

$$\varphi=-i\,\frac{J\sin\frac{\pi}{n}}{\pi K}\sum_{1}^{n}{}_h\, e^{\frac{-(2h-1)\pi i}{n}}\log S-\sqrt[n]{\frac{\sin\frac{\theta-w}{2}}{\sin\frac{w}{2}}}\,e^{\frac{(2h-1)\pi i}{n}}$$

where we have to take for $\sqrt[n]{\frac{\sin\frac{\theta-w}{2}}{\sin\frac{w}{2}}}$ the positive value and we must always suppose the expression $\left(\frac{ze^{-\frac{i\theta}{2}}-e^{\frac{i\theta}{2}}}{z-1}\right)^{\frac{1}{n}}$ substituted for S.

Section 2

We will now study the case in which the plate is not plane and the magnetic field is not uniform.

Let us subdivide the plate into infinitesimal elements. We can consider each of these elements as plane and subject to a uniform field, and therefore we can apply to it the fundamental formulae found in the case of a plane plate subject to the action of a uniform magnetic field; we may conveniently develop these formulae by using curvilinear coördinates.

Then, by passing through an element to the contiguous element, we shall be able to state the relations which are to be verified on the whole surface and from these relations we shall deduce the general differential equations and the boundary conditions.

Let us consider an infinitesimal, plane element of the plate, adjacent to a point A, and let us remember the equations (3)

$$j_x=-K\left(\frac{\delta V}{\delta x}-\lambda\frac{\delta V}{\delta y}\right);\qquad j_y=-K\left(\frac{\delta V}{\delta y}+\lambda\frac{\delta V}{\delta x}\right).$$

Let ds be a linear element passing through A and dn the corresponding element of normal; let ds and dn be situated as we have said in Section 5 of the first lecture; then we shall have:

$$\frac{dx}{ds}=\frac{dy}{dn},\qquad \frac{dy}{ds}=-\frac{dx}{dn}$$

$$j_n=j_x\frac{dx}{dn}+j_y\frac{dy}{dn}=-K\left(\frac{\delta V}{\delta n}+\lambda\frac{\delta V}{\delta s}\right)$$

where j_n denotes the density of the current normal to ds. By employing a system of curvilinear coördinates u and v, we shall have

$$(20) \quad j_n = -K\left[\frac{\delta V}{\delta u}\frac{\delta u}{\delta n}+\frac{\delta V}{\delta v}\frac{\delta v}{\delta n}+\lambda\left(\frac{\delta V}{\delta n}\frac{\delta u}{\delta s}+\frac{\delta V}{\delta v}\frac{\delta v}{\delta s}\right)\right].$$

(Let us regard positive directions of the curves $V=const.$ and $u=const.$, as the directions in which u and v increase respectively. We shall make the convention that the rotation of the positive direction of the line $v=const.$ towards the positive direction of the line $u=const.$, through the angle smaller than β, has the same sense as the rotation of 90° by which the positive direction of the line s can reach the positive direction of the line n; moreover we shall here and in what follows, take for $\sqrt{EG-F^2}$ the positive value).

Now, if the square of the linear element be

$$ds^2 = Edu^2+2Fdudv+Gdv^2$$

we have

$$(a)\begin{cases}\dfrac{\delta u}{\delta n} = -\dfrac{1}{\sqrt{EG-F^2}}\left(F\dfrac{\delta u}{\delta s}+G\dfrac{\delta v}{\delta s}\right)\\[2ex] \dfrac{\delta v}{\delta u} = \dfrac{1}{\sqrt{EG-F^2}}\left(E\dfrac{\delta u}{\delta s}+F\dfrac{\delta v}{\delta s}\right)\end{cases} \qquad (a_1)\begin{cases}\dfrac{\delta u}{\delta s} = \dfrac{1}{\sqrt{EG-F^2}}\left(F\dfrac{\delta u}{\delta n}+G\dfrac{\delta v}{\delta n}\right)\\[2ex] \dfrac{\delta v}{\delta s} = -\dfrac{1}{\sqrt{EG-F^2}}E\left(\dfrac{\delta u}{\delta n}+F\dfrac{\delta v}{\delta n}\right);\end{cases}$$

and the equation (20) becomes

$$(20') \qquad j_n = -K\left[\frac{E\dfrac{\delta V}{\delta v}-F\dfrac{\delta V}{\delta u}}{\sqrt{EG-F^2}}+\lambda\frac{\delta V}{\delta u}\right]\frac{\delta u}{\delta s}+K\left[\frac{G\dfrac{\delta V}{\delta u}-F\dfrac{\delta V}{\delta v}}{\sqrt{EG-F^2}}-\lambda\frac{\delta V}{\delta v}\right]\frac{\delta v}{\delta s}.$$

Let us suppose the line s to coincide with the line $u=const.$; then the preceding equation will become

$$(21) \qquad j_{nu} = K\left[\frac{G\dfrac{\delta V}{\delta u}-F\dfrac{\delta V}{\delta v}}{\sqrt{EG-F^2}}-\lambda\frac{\delta V}{\delta u}\right]\frac{1}{\sqrt{G}}$$

where j_{nu} is the density of the current normal to the element of the line $u=const.$ Analogously we shall obtain

$$(21_1) \qquad j_{nv} = -K\left[\frac{E\dfrac{\delta V}{\delta v}-F\dfrac{\delta V}{\delta u}}{\sqrt{EG-F^2}}+\lambda\frac{\delta V}{\delta \nu}\right]\frac{1}{\sqrt{E}}$$

($\sqrt{G}$ and $\sqrt{E}$ have to be taken with the positive sign.)

All the preceding formulae hold good for an infinitesimal plane element: now if *the plate be curved and be placed in an uniform or non-uniform magnetic field*, the preceding formulae hold good for each infinitesimal element of the surface, we have only to suppose that K and $\lambda = tg\beta$ change from element to element; in other words, the preceding formulae will hold good in the more general case of a metallic plate, of any form, situated in any magnetic field, provided K and λ be considered as known functions of u and v. In order to obtain the values of these functions in a point of the plate it is sufficient to substitute in the formulae (see Section 4 of the first lecture):

$$K=e^2\left(\frac{N_1v_1}{1+e^2v^2{}_1H^2}+\frac{N_2v_2}{1+e^2v^2{}_2H^2}\right),\quad \lambda=\frac{\Sigma}{K},\quad \Sigma=e^3H\left(\frac{N_1v_1{}^2}{1+e^2v^2{}_1H^2}+\frac{N_2v_2{}^2}{1+e^2v^2{}_2H^2}\right)$$

for H the component of the magnetic field in the direction of the normal to the plate; this component will vary from point to point of the plate because the angle that the normal forms which the direction of the magnetic field and the intensity of this latter vary.

(We do not take into account, here, the secondary actions by which a plate loses the character of isotropy, with respect to its electrical conductivity, when it is subject to the action of a magnetic field, the direction of which does not coincide, in every point, with the normal to the plate [see fourth lecture, Section 1.] We observe only that the perturbations caused by these secondary actions do not take place when the magnetic field is perpendicular to the plate at every point of this latter).

From the formula (20′) which holds good on all the surface, we obtain

$$(22)\quad \int_s j_n ds=\int_s\left\{-K\left[\frac{E\frac{\delta V}{\delta v}-F\frac{\delta V}{\delta u}}{\sqrt{EG-F^2}}+\lambda\frac{\delta V}{\delta u}\right]\frac{\delta u}{\delta s}+K\left[\frac{G\frac{\delta V}{\delta u}-F\frac{\delta V}{\delta v}}{\sqrt{EG-F^2}}-\lambda\frac{\delta V}{\delta v}\right]\frac{\delta v}{\delta s}\right\}ds.$$

If s be a closed line which encloses no electrode, the first member of the preceding equation is zero and therefore

$$(23)\quad -K\left[\frac{E\frac{\delta V}{\delta v}-F\frac{\delta V}{\delta n}}{\sqrt{EG-F^2}}+\lambda\frac{\delta V}{\delta u}\right]du+K\left[\frac{G\frac{\delta V}{\delta u}-F\frac{\delta V}{\delta v}}{\sqrt{EG-F^2}}-\lambda\frac{\delta V}{\delta v}\right]dv$$

must be an exact differential; calling it dW, we shall have

$$(24)\quad \begin{cases} -\sqrt{G}\,j_{nu}=-K\left[\frac{G\frac{\delta V}{\delta u}-F\frac{\delta V}{\delta v}}{\sqrt{EG-F^2}}-\lambda\frac{\delta V}{\delta v}\right]=-\frac{\delta W}{\delta v} \\ \sqrt{E}\,j_{nu}=-K\left[\frac{E\frac{\delta V}{\delta v}-F\frac{\delta V}{\delta u}}{\sqrt{EG-F^2}}+\lambda\frac{\delta V}{\delta u}\right]=\frac{\delta W}{\delta u}. \end{cases}$$

Therefore V must fulfil the differential equation

$$(F)\quad \frac{\delta}{\delta u}\left\{K\left[-\frac{G\frac{\delta V}{\delta u}-F\frac{\delta V}{\delta v}}{\sqrt{EG-F^2}}-\lambda\frac{\delta V}{\delta v}\right]\right\}+\frac{\delta}{\delta v}\left\{K\left[\frac{E\frac{\delta V}{\delta v}-F\frac{\delta V}{\delta u}}{\sqrt{EG-F^2}}+\lambda\frac{\delta V}{\delta u}\right]\right\}=0.$$

By solving the equations (24) with respect to $\frac{\delta V}{\delta u}$ and $\frac{\delta V}{dv}$ we obtain

$$-\frac{1}{K(1+\lambda^2)}\left[\frac{G\frac{\delta W}{\delta u}-F\frac{\delta W}{\delta v}}{\sqrt{EG-F^2}}+\lambda\frac{\delta W}{\delta v}\right]=\frac{\delta V}{\delta v},\quad \frac{-1}{K(1+\lambda^2)}\left[\frac{E\frac{\delta W}{\delta v}-F\frac{\delta W}{\delta u}}{\sqrt{EG-F^2}}-\lambda\frac{\delta W}{\delta u}\right]=-\frac{\delta V}{\delta u}$$

therefore W fulfills the differential equation

$$(G)\frac{\delta}{\delta u}\left\{\frac{1}{K(1+\lambda^2)}\left[\frac{G\frac{\delta W}{\delta u}-F\frac{\delta W}{\delta v}}{\sqrt{EG-F^2}}+\lambda\frac{\delta W}{\delta v}\right]\right\}+\frac{\delta}{\delta v}\left\{\frac{1}{K(1+\lambda^2)}\left[\frac{E\frac{\delta W}{\delta v}-F\frac{\delta W}{\delta u}}{\sqrt{EG-F^2}}-\lambda\frac{\delta W}{\delta u}\right]\right\}=0.$$

Now let us suppose that in the formula (22) the integration is extended to an open line s. Owing to (24), we shall have

$$\int_s j_n ds=\int_s dW=W_2-W_1$$

where W_1 and W_2 are respectively the values of W at the origin and at the extreme of the arc s.

Therefore *along the insulated portions of the boundary* (*as along each line of flow*), *W will be constant, while along all the electrodes of negligible resistance V will be constant.*

If we suppose that in the interior of the plate there are no electromotive forces, *V will be a single-valued function, while W will become many-valued if we go along any closed curve which encloses some electrodes through which a total quantity of electricity different from zero enters into the plate.* (We shall say that some electrodes are enclosed by the curve if they be on the same side of the said curve).

If the plate be acyclic and all the electrodes be at the boundary, W will be evidently single-valued.

We shall call W *the function of currents.* In the case in which K and λ are constant, for a given value of H, we have $W=KU'$, where U' is the conjugate function of the *fundamental function U.* (see Section 1 of the second lecture).

The expression for j_n (formula (20′)) can be transformed in various ways. In fact, by using (24), we can write

$$(20'') \qquad j_n=\frac{\delta W}{\delta s}$$

whereas, by applying (a'), we have

$$(20''') \qquad j_n=-K\left\{\left(\frac{\delta V}{\delta n}-\frac{\lambda\left(E\frac{\delta V}{\delta v}-F\frac{\delta V}{\delta u}\right)}{\sqrt{EG-F^2}}\right)\frac{\delta u}{\delta n}+\left(\frac{\delta V}{\delta v}+\frac{\lambda\left(G\frac{\delta V}{\delta u}-F\frac{\delta V}{dv}\right)}{\sqrt{EG-F^2}}\right)\frac{\delta v}{\delta n}\right\}.$$

Lastly, bearing in mind the formula (20″), that is, $j_n=\frac{\delta W}{\delta u}\frac{\delta u}{\delta s}+\frac{\delta W}{\delta v}\frac{\delta v}{\delta s}$ and (a'), we find

$$(20'''') \qquad j_n=\frac{1}{\sqrt{EG-F^2}}\left\{\left(F\frac{\delta W}{\delta u}-E\frac{\delta W}{\delta v}\right)\frac{\delta u}{\delta n}+\left(G\frac{\delta W}{\delta u}-F\frac{\delta W}{\delta v}\right)\frac{\delta v}{\delta n}\right\}.$$

Bearing in mind (24) the formulae (21) and (21′) can also be written in this way

$$(21') \qquad j_{nu}=\frac{1}{\sqrt{G}}\frac{\delta W}{\delta v}, \qquad j_{nu}=\frac{1}{\sqrt{E}}\frac{\delta W}{\delta u}.$$

Let us call j_n and j_v the orthogonal projections of the density of the current in the directions of the lines $u=const.$ and $v=const.$; we may obtain their values by supposing that n coincides successively with the directions of the lines $u=const.$ and $v=const.$; we shall have

$$j_u=\frac{-K}{\sqrt{G}}\left(\frac{\delta V}{\delta v}+\lambda\frac{G\frac{\delta V}{\delta n}-F\frac{\delta V}{\delta v}}{\sqrt{EG-F^2}}\right);\quad j_v=\frac{-K}{\sqrt{E}}\left(\frac{\delta V}{\delta u}-\lambda\frac{E\frac{\delta V}{\delta v}-F\frac{\delta V}{\delta u}}{\sqrt{EG-F^2}}\right) \tag{25}$$

By employing the function W, the preceding formulae become

$$j_u=\frac{1}{\sqrt{G}}\left(\frac{G\frac{\delta V}{\delta n}-F\frac{\delta W}{\delta v}}{\sqrt{EG-F^2}}\right);\qquad j_v=\frac{1}{\sqrt{E}}\left(\frac{F\frac{\delta W}{\delta u}-E\frac{\delta W}{\delta v}}{\sqrt{EG-F^2}}\right). \tag{25'}$$

Lastly, let us c nsider the components of density of the current parallel to the directions of the lines $u=const.$ and $v=const.$; we obtain formulae

$$\left\{\begin{aligned} i_u&=-\frac{K\sqrt{G}}{EG-F^2}\left(E\frac{\delta V}{\delta v}-F\frac{\delta V}{\delta u}+\lambda\sqrt{EG-F^2}\,\frac{\delta V}{\delta n}\right);\\ i_v&=\frac{-K\sqrt{E}}{EG-F^2}\left(G\frac{\delta V}{\delta n}-F\frac{\delta V}{\delta v}-\lambda\sqrt{EG-F^2}\,\frac{\delta V}{\delta v}\right)\end{aligned}\right. \tag{26}$$

which, by employing the function W, become

$$i_u=\frac{\sqrt{G}}{\sqrt{EG-F^2}}\,\frac{\delta W}{\delta u},\qquad i_v=\frac{-\sqrt{E}}{\sqrt{EG-F^2}}\,\frac{\delta W}{\delta v}. \tag{26'}$$

From any one of these groups of formulae we obtain the square of the density of current, that is

$$\left\{\begin{aligned} j^2&=K^2(1+\lambda^2)\left[\frac{E\left(\frac{\delta V}{\delta v}\right)^2-2F\frac{\delta V}{\delta u}\,\frac{\delta V}{\delta v}+G\left(\frac{\delta V}{\delta u}\right)^2}{EG-F^2}\right]\\ j^2&=\frac{E\left(\frac{\delta W}{\delta v}\right)^2-2F\frac{\delta W}{\delta u}\,\frac{\delta W}{\delta v}+G\left(\frac{\delta W}{\delta u}\right)^2}{EG-F^2}\end{aligned}\right. \tag{27}$$

These formulae, by using the symbol of the differential parameter of the first order, become

$$\left\{\begin{aligned} j^2&=K^2(1+\lambda^2)\triangle_1 V\\ j^2&=\triangle_1 W\end{aligned}\right. \tag{27'}$$

From the expressions that we have obtained for j_n we can deduce other forms for the condition $j_n=0$ which has to be fulfilled along the insulated portions of the boundary. For instance, by using the formula (20‴) we can write the said condition in this way:

$$\left(\frac{\delta V}{\delta u}-\lambda\frac{E\frac{\delta V}{\delta v}-F\frac{\delta V}{\delta u}}{\sqrt{EG-F^2}}\right)\frac{\delta u}{\delta n}+\left(\frac{\delta V}{\delta v}+\lambda\frac{G\frac{\delta V}{\delta u}-F\frac{\delta V}{\delta v}}{\sqrt{EG-F^2}}\right)\frac{\delta v}{\delta n}=0.$$

SECTION 3

When the plate is homogeneous and the magnetic field H is zero, we have $K=const.$ and $\lambda=0$; and the equations (F) and (G) of the preceding paragraph become

$$(\beta)=\left\{\frac{\delta}{\delta u}\left\{\frac{G\frac{\delta V}{\delta u}-F\frac{\delta V}{\delta v}}{\sqrt{EG-F^2}}\right\}+\frac{\delta}{\delta v}\left\{\frac{E\frac{\delta V}{\delta v}-F\frac{\delta V}{\delta u}}{\sqrt{EG-F^2}}\right\}=0,\right.$$

$$\frac{\delta}{\delta u}\left\{\frac{G\frac{\delta W}{\delta u}-F\frac{\delta W}{\delta v}}{\sqrt{EG-F^2}}\right\}+\frac{\delta}{\delta v}\left\{\frac{E\frac{\delta W}{\delta v}-F\frac{\delta W}{\delta u}}{\sqrt{EG-F^2}}\right\}=0$$

and can also be written, by using the symbol of the differential parameter of the second order, in this way

$$(\beta')\quad \triangle_2 V=0,\quad \triangle_2 W=0.$$

Therefore V and W, in this case, are two harmonic functions on the surface, as we could also have foreseen. Moreover the equations (24) become

$$(j)\quad \frac{G\frac{\delta V}{\delta u}-F\frac{\delta V}{\delta v}}{\sqrt{EG-F^2}}=\frac{\delta\frac{W}{K}}{\delta v},\qquad \frac{E\frac{\delta V}{\delta v}-F\frac{\delta V}{\delta u}}{\sqrt{EG-F^2}}=-\frac{\delta\frac{W}{K}}{\delta u}.$$

The conditions (β') are the necessary and sufficient conditions in order that V and W be harmonic on the surface; the conditions (j) are the necessary and sufficient conditions in order that $V+i\frac{W}{K}$ be a complex variable on the surface (1).

Now, when $K\lambda$ is not constant, the equations (F) and (G) are essentially different from (β) and therefore V and W are not harmonic on the surface. In the case of the plane and homogeneous plate subject to the action of a uniform magnetic field, we saw in the preceding lectures that the potential V was harmonic, as it was without the magnetic field; for only the condition which V must fulfill along the insulated portions of the boundary changes. Whereas, in the case of the curved and non-homogeneous plate, subject to the action of a non-uniform magnetic field, the action of the magnetic field alters (as we have just now seen) not only the condition which the potential must fulfill along the free and insulated boundary, but also the nature of the potential in all the area occupied by the plate.

We can say, summarizing the results obtained, that when we pass from the case of the plane plate and of the uniform field to the case of the curved homogeneous or non-homogeneous plate of the uniform or non-uniform magnetic field, we pass, (from the analytical point of view from Laplace's equation to new equations of different character (the equations (F) and (G)). Only two of the four functions considered in the first case (*potential, fundamental function* and their *conjugate functions*) remain also in the second case: the *potential* and the function that we have called *function of the currents.* But the one and the other lose the character of harmonic functions on the surface occupied by the plate, and this is the reason for which the other two functions cease to subsist; in fact, as the potential V and the function of the currents W are not harmonic (that is have not zero the second differential parameter) but instead fulfill the equations (F) and (G), the conjugate functions of V and W, in the sense of the theory of the functions of complex variables on a surface, cannot exist.

From all we have now said, we can deduce that the principle of the point electrodes at the boundary exists no more in the general case of the curved non-homogeneous plate, subject to the action of a non-uniform magnetic field; the same can be said for the principle that the potential remains unaltered, under the action of the field, in the case of a cyclic plate, when the lines which form the boundary are electrodes of negligible resistance, kept at a constant potential. On the contrary, we shall see, in the following paragraph, that the *principle of reciprocity* (see the Section 9 of the first lecture) holds good also in the general case just said.

Section 4

From the formula (20′) of Section 2, denoting by V_1 an arbitrary function and by S the boundary (formed by one or several lines) of a part of the plate, and supposing V and V_1 to be regular, we deduce

$$(28)\quad \left\{\begin{aligned} \int_s V_1 j_n dS = &\int_S \Bigg\{ -KV_1\left[\frac{E\frac{\delta V}{\delta v}-F\frac{\delta V}{\delta u}}{\sqrt{EG-F^2}}+\lambda\frac{\delta V}{\delta u}\right]\frac{\delta u}{\delta s}+ \\ &+KV_1\left[\frac{G\frac{\delta V}{\delta u}-F\frac{\delta V}{\delta v}}{\sqrt{EG-F^2}}-\lambda\frac{\delta V}{\delta v}\right]\frac{\delta v}{\delta s}\Bigg\}dS= \\ =\int_\sigma \frac{K}{\sqrt{EG-F^2}}&\Bigg\{\frac{G\frac{\delta V}{\delta u}\frac{\delta V_1}{\delta u}-F\left(\frac{\delta V_1}{\delta u}\frac{\delta V}{\delta v}+\frac{\delta V}{\delta u}\frac{\delta V_1}{\delta v}\right)+E\frac{\delta V}{\delta v}\frac{\delta V_1}{\delta v}}{}+ \\ &+\lambda\left(\frac{\delta V}{\delta u}\frac{\delta V_1}{\delta u}-\frac{\delta V_1}{\delta u}\frac{\delta V}{\delta v}\right)\Bigg\}=d\sigma. \end{aligned}\right.$$

(We suppose the normal n drawn towards the interior of σ and thus we fix implicitly the direction s).

Let us denote with $\Delta_1 VV_1$ the *intermediate* or mixed *differential parameter* of the functions V and V_1, that is let us put

$$\Delta_1 VV_1=\frac{G\frac{\delta V}{\delta u}\frac{\delta V_1}{\delta u}-F\left(\frac{\delta V_1}{\delta u}\frac{\delta V}{\delta v}+\frac{\delta V}{\delta u}\frac{\delta V_1}{\delta v}\right)+E\frac{\delta V}{\delta v}\frac{\delta V_1}{\delta v}}{EG-F^2}$$

and let us consider the determinant

$$\begin{vmatrix} \frac{\delta V}{\delta u}, & \frac{\delta V}{\delta v} \\ \frac{\delta V_1}{\delta u}, & \frac{\delta V_1}{\delta v} \end{vmatrix}=\frac{d(V_1V_1)}{d(u_1v)}.$$

We can immediately recognize that $\triangle_1 VV_1$ is symmetrical with respect to V and V_1 and that the determinant $\frac{\delta(VV_1)}{d(uv)}$ changes sign by exchanging V with V_1.

The equation (28) can be written thus

$$(L) \qquad \int_S V_1 j_n dS = \int_\sigma K\left(\triangle_1 VV_1 + \frac{\lambda}{\sqrt{EG-F^2}}\,\frac{d(V_1V_1)}{d(uv)}\right)d\sigma.$$

Now let us suppose V_1 to be the electric potential when the magnetic field is inverted; let us denote with j_{1n} the corresponding density of current normal to S: since on reversal of the field K remains unaltered and λ changes only in sign, together with the formula (L) we shall have the other
Hence

$$(M) \qquad \int_s (Vj_{1n} - V_1 j_n)dS = 0.$$

If we compare the formula (D) (see Section 8 of the first lecture) with the formula (M), we recognize that from this latter we can deduce the same consequences which we have deduced from the former, and in particular the *theorems of reciprocity* (see Section 9 of the first lecture). For instance, let us suppose that with the direct magnetic field the current of intensity $I=\nu J$ (ν thickness of the plate) enters through the point electrode A and goes out through the point electrode B, and with the inverted field the current $I_1=\nu J_1$ enters and goes out respectively through the two point electrodes A_1 and B_1. Let us suppose S to be formed by boundary s of the plate and by the four geodetical circumferences s_a, s_b, s_{a1}, s_{b1}, having the center s in A, B, A_1 and B_1. We shall suppose that, at least when these circumferences are sufficiently small, if j_n be drawn from the inside towards the outside of the same circumferences, j_n is positive along s_a, negative along s_b, and j_{1n} is positive along s_a, and negative along s_b; moreover we shall suppose that, along s_a, s_b, s_{a1}, and s_{b1}, V and V_1 are finite or infinite of an order smaller than 1. Since on s j_n and j_{1n} are void, we shall have

$$\int_{sa} j_{1n} V ds_a + \int_{sb} j_{1n} V ds_b + \int_{sa1} j_{1n} V ds_{a1} + \int_{sb1} j_{1n} V d_{sb1} -$$
$$- \int_{sa} j_n V_1 ds_a - \int_{sb} j_n V_1 ds_b - \int_{sa1} j_n V_1 ds_a - \int_{sb1} j_n V_1 d_{sb1} = 0.$$

If we pass the limit, making the four geodetical circles grow indefinitely smaller, we obtain

$$J_1(V_{A1} - V_{B1}) = J(V_{1A} - V_{1B})$$

and hence, if $J_1 = J$,

$$V_{A1} - V_{B1} = V_{1A} - V_{1B}.$$

In an analogous manner, *all the theorems of reciprocity* (relative to internal electrodes of finite areas and negligible resistances or to electrodes situated at the boundary and also of negligible resistances) *can be extended from the case of the plane and homogeneous plate situated in a uniform field, to the case of the curved homogeneous or non-homogeneous plate, situated in a uniform or non-uniform field.*

The equation (M) can also be written in this way

$$\int_S \left(V\frac{\delta W_1}{\delta s} - V_1\frac{\delta W}{\delta s}\right)dS = 0$$

therefore *$VdW_1 - V_1dW$ will be an exact differential*: this will be another manner of expressing the theorem of reciprocity.

From the formula (20′), denoting with S the boundary (formed by one or several curves) of a part σ of the plate, we can deduce the other formula

$$\text{(H)} \qquad \int_s Vj_n dS = \int_\sigma K \triangle_1 V d\sigma$$

which can also be written

$$\text{(H}'\text{)} \qquad \int_S V\frac{\delta W}{\delta s} dS = \int_\sigma K \triangle_1 V d\sigma$$

or also

$$\text{(H}''\text{)} \qquad \int_S Vj_n dS = \int_\sigma \frac{1}{K(1+\lambda^2)} j^2 d\sigma$$

or also

$$\text{(H}'''\text{)} \qquad \int_S Vj_n dS = \int_\sigma \frac{1}{K(1+\lambda^2)} \triangle_1 W d\sigma.$$

The first members of the preceding equations are proportional to the quantity of energy which, in the unity of time, penetrates in the area σ through its boundary S. (The factor of proportionality is the thickness ν of the plate). If σ be infinitely small, $\frac{\nu}{K(1+\lambda^2)} j^2$ is the quantity of heat that, for Joules' effect, is developed by the current in each element of surface of the plate.

From the formula (H′) we can also deduce that, if V be zero along a certain part of the boundary S and W be constant along the other parts, V is zero within the area σ. Therefore, if some portions of the boundary be electrodes of negligible resistance, where either the value of the potential or the intensity of the current which penetrates in the plate is known, and if the other portions of the boundary be free and insulated, the distribution of the currents will be determined. (See first lecture, Section 6).

FOURTH LECTURE

Section 1—The problem of the flow of electricity in a non-homogeneous anisotropic conductor, kept at constant temperature and subject to the action of a non-uniform magnetic field. Section 2—Generalization of the principle of reciprocity. Section 3—On the change of resistance of a conductor subject to the action of a magnetic field. Section 4—The electromagnetic phenomena that can be observed when a disc traversed by a radial current is placed in a uniform magnetic field. Section 5—An reversible generator for steady currents founded on the action of magnetic field on the electrons of conductors placed in the same field. Section 6—Experimental verifications of the theorem of reciprocity in the different cases. Section 7—A consequence of the theorem of reciprocity. Section 8—An applica-

[1]Beltrami, *Dele variabili complesse sopra una superficie qualunque.* Opere, vol. I, p. 318.

tion of the principle of the point electrodes at the boundary. Section 9—Experimental tests of the theory of flow of electricity in a circular plate perpendicular to the lines of force of a uniform magnetic field.

SECTION 1

We shall now speak of the generalization to the case of a three-dimensional conductor of the problem previously considered, of the flow of electricity in a plate under the action of a magnetic field. This generalization was made, in 1916, by Miss Freda. (1).

The experimental results that have been obtained by studying the properties of a conductor placed in a magnetic field have led physicists to make the hypothesis that the field determines a transitory alteration of the specific properties of the substances subject to its action. According to this hypothesis the substance also if homogeneous and isotropic without the field, under the action of the latter acquires electrical properties different in different points, if the field is not uniform, and different in the different directions that pass through a point according to the angle that the same directions form with that of the field.

If we do not exclude this hypothesis, in order to study analytically the flow of electric currents in a three-dimensional medium subject to any magnetic field, it is necessary to consider media that are heterogeneous and anisotropic.

(The said hypothesis on the other hand does not introduce difficulties in the case of a plate coincident with an equipotential surface of the magnetic field, because then, for reasons of symmetry, the plate will remain homogeneous and isotropic under the action of the field if it was such before the creation of this latter; we have only to consider, in the formulae which we have already seen, N_1, N_2, v_1, v_2, as functions of the absolute value of H.)

Let x, y, z, be three axes that form an orthogonal right-handed system. Let $\frac{d\xi_1}{dt}, \frac{d\gamma_1}{dt}, \frac{dS_1}{dt}, \frac{d\xi_2}{dt}, \frac{d\gamma_2}{dt}, \frac{dS_2}{dt}$ be the components of the velocities of a positive electron and of a negative electron the charges of which have both the absolute value e. Let N_1, N_2, be the numbers of positive electrons and of negative electrons per cubic centimeter of the conductor. The components j_x, j_y, j_z, of the density of current are expressed in terms of the components of the velocities of the electrons by the equations:

$$(29) \quad \Bigg\{ j_x = e\left(N_1\frac{d\xi_1}{dt} - N_2\frac{d\xi_2}{dt}\right); \quad j_y = e\left(N_1\frac{d\gamma_1}{dt} - N_2\frac{d\gamma_2}{dt}\right); \quad j_z = e\left(N_1\frac{dS_1}{dt} - N_2\frac{dS_2}{dt}\right).$$

Let $eE_{1x}, eE_{1y}, eE_{1z}, eE_{2x}, eE_{2y}, eE_{2z}$ be the components of the total electromotive forces acting respectively on a positive electron and on a negative electron; let H_x, H_y, H_z, be the components of the magnetic field. Let $X = -\frac{\delta V}{\delta_x}$; $Y = -\frac{\delta V}{\delta y}$; $Z = -\frac{\delta V}{\sigma_z}$ be the components of the electric force which admits the potential V.

According to Ohm's law for anisotropic media, the components of the velocities of the electrons of the two kinds are expressed, by linear equations, in terms of the components of the total electromotive forces which act on the same electrons. We have therefore

$$(30)\quad \begin{cases} \dfrac{d\xi_1}{dt}=e(a_{11}E_{1x}+a_{12}E_{1y}+a_{13}E_{1z}) \\ \dfrac{d\gamma_1}{dt}=e(a_{21}E_{1x}+a_{22}E_{1y}+a_{23}E_{1z}) \\ \dfrac{dS_1}{dt}=e(a_{31}E_{1x}+a_{32}E_{1y}+a_{33}\mathrm{E}_{1z}) \end{cases}
\qquad (31)\quad \begin{cases} \dfrac{d\xi_2}{dt}=-e(b_{11}E_{2x}+b_{12}E_{2y}+b_{13}E_{2z}) \\ \dfrac{d\gamma_2}{dt}=-e(b_{21}E_{2x}+b_{22}E_{2y}+b_{23}E_{2z}) \\ \dfrac{dS_2}{dt}=-e(b_{31}E_{2x}+b_{32}E_{2y}+b_{33}E_{2z}) \end{cases}$$

The experimental results that induce us to infer an alteration of the specific properties of a conductor caused by the magnetic field induce us also to infer that this alteration does not depend on the sense of the field. We shall therefore assume that the coefficients N_1, N_2, a_{mn}, b_{mn} (which characterize the specific properties of the metal) can eventually depend, not only on x, y, z, but also on H_x, H_y, H_z, but do not vary when we change the sign of all three components of the field.

As the conductor is kept at a constant temperature, the total electromotive force that acts upon an electron is the resultant of the electric force depending on the distribution of the potential and of the electromotive force caused by the movement of the electron in the field. We therefore have

$$(32)\quad \begin{cases} E_{1x}=X+H_y\dfrac{dS_1}{dt}-H_z\dfrac{d\gamma_1}{dt} \\ E_{1y}=Y+H_z\dfrac{d\xi_1}{dt}-H_x\dfrac{dS_1}{dt} \\ E_1Z=Z+H_x\dfrac{d\gamma_1}{dt}-H_y\dfrac{d\xi_1}{dt} \end{cases}
\qquad (33)\quad \begin{cases} E_{2x}=X+H_y\dfrac{dS_2}{dt}-H_z\dfrac{d\gamma_2}{dt} \\ E_{2y}=Y+HZ\dfrac{d\xi_2}{dt}-H_x\dfrac{dS_2}{dt} \\ E_{2z}=Z+H_x\dfrac{d\gamma_2}{dt}-H_y\dfrac{d\xi_2}{dt} \end{cases}$$

By means of the equations (30) (31) (32) (33) $\dfrac{d\xi_1}{dt}, \dfrac{d\gamma_1}{dt}, \dfrac{dS_1}{dt}, \dfrac{d\xi_2}{dt}, \dfrac{d\gamma_2}{dt}, \dfrac{dS_2}{dt}$, can be expressed as linear functions of X, Y, Z. We obtain thus the equations:

$$(34)\quad \begin{cases} \dfrac{d\xi_1}{dt}=e(\alpha_{11}X+\alpha_{12}Y+\alpha_{13}Z) \\ \dfrac{d\gamma_1}{dt}=e(\alpha_{21}X+\alpha_{22}Y+\alpha_{23}Z) \\ \dfrac{dS_1}{dt}=e(\alpha_{31}X+\alpha_{32}Y+\alpha_{33}Z) \end{cases}
\qquad (35)\quad \begin{cases} \dfrac{d\xi_2}{dt}=-e(\beta_{11}X+\beta_{12}Y+\beta_{13}Z) \\ \dfrac{d\gamma_2}{dt}=-e(\beta_{21}X+\beta_{22}Y+\beta_{23}Z) \\ \dfrac{dS_2}{dt}=-e(\beta_{31}X+\beta_{32}Y+\beta_{33}Z) \end{cases}$$

The coefficients α_{rs}, β_{rs} are functions of the components of the magnetic field and of the coefficients a_{rs} or of the coefficients b_{rs} respectively; for the sake of brevity we do not transcribe their expressions.

By means of the equations (23) (34) (35), j_x, j_y, j_z, can be expressed as functions of the derivatives of the potential V. We have

$$
(36)\quad \begin{cases}
j_x = -e^2\left[(N_1\alpha_{11}+N_2\beta_{11})\dfrac{\delta V}{\delta x}+(N_1\alpha_{12}+N_2\beta_{12})\dfrac{\delta V}{\delta y}+(N_1\alpha_{13}+N\beta_{13})\dfrac{\delta V}{\delta z}\right] \\
j_y = -e^2\left[(N_1\alpha_{21}+N_2\beta_{21})\dfrac{\delta V}{\delta x}+(N_1\alpha_{22}+N_2\beta_{22})\dfrac{\delta V}{\delta y}+(N_1\alpha_{23}+N_2\beta_{23})\dfrac{\delta V}{\delta z}\right] \\
j_y = -e^2\left[(N_1\alpha_{31}+N_2\beta_{31})\dfrac{\delta V}{\delta x}+(N_1\alpha_{32}+N_2\beta_{32})\dfrac{\delta V}{\delta y}+(N_1\alpha_{33}+N_2\beta_{33})\dfrac{\delta V}{\delta z}\right]
\end{cases}
$$

The condition

$$
(37)\qquad \frac{\delta j_x}{\delta x}+\frac{\delta j_y}{\delta y}+\frac{\delta j_z}{\delta z}=0
$$

gives us the differential equation that V must fulfill. If n is the normal at any point of the free surface of the conductor, from the condition $j_n=0$ we deduce a boundary condition for V. If along the surface of the conductor there be electrodes of negligible resistance kept at constant potentials V_1, V_2, we have also for V the conditions $V=V_1$ along a part of the boundary $V=V_2$ along another part of the boundary.

Now let us suppose that the conductor is homogeneous and isotropic when the field does not exist; that the field is uniform and has the lines of force parallel to the axis z. Then, if the magnetic field does not alter the specific properties of the conductor, the equations (36) become

$$
(36')\qquad j_x=-K\left(\frac{\delta V}{\delta x}-\lambda\frac{\delta V}{\delta y}\right);\quad j_y=-K\left(\lambda\frac{\delta V}{\delta x}+\frac{\delta V}{\delta y}\right);\quad jz=-\sigma\frac{\delta V}{\delta z}
$$

(K and λ have the same values they had in the formulae (3) of the first lecture; σ is given by the formula $\sigma=e^2(N_1v_1+N_2v_2)$.

SECTION 2

From the formulae (36) and (37) of the preceding paragraph, Miss Freda has deduced that, if the coefficients a_{rs} b_{rs} of the equations (30) and (31) satisfy the conditions

$$
(38)\qquad a_{rs}=a_{sr} \qquad\qquad b_{rs}=b_{sr}
$$

we have (39) $\int_s(V_1j_n-Vj_{1n})dS=0$.
(We denote with V, j, V_1, j_1, the potential and the density of current when the field is direct and when the field is inverted; with S the complete boundary of the conductor or of a part of it).

The formula (39) is perfectly analogous to the formula (D) of the first lecture. From the formula (39), by specifying conveniently the boundary S, Miss Freda has stated (analogously to what we have already seen in the case of a plate) that my theorem of reciprocity holds good also in the case of a non-uniform magnetic field and of a non-homogeneous anisotropic conductor for which the conditions (38) are satisfied.

The four electrodes of the conductor can be point electrodes (in the interior or at the boundary) or else laminar, situated at the boundary and of negligible resistance, or else three-dimensional, situated in the interior and of negligible resistance.

If the magnetic field be zero, the necessary and sufficient condition in order that the equation (39) may be fulfilled is

$$N_1 a_{rs} + N_2 b_{rs} = N_1 a_{sr} + N_2 b_{sr} \tag{38'}$$

If we consider the physical meaning of the coefficients a_{rs} b_{rs} of the formulae (30) and (31) we see that the formulae (38) express some elementary laws of reciprocity. For instance, the property expressed by the formula $a_{12} = a_{21}$ can be thus translated in words:

The component, parallel to the axis y, of the velocity that a positive electron acquires when subject to an electromotive force equal to 1 and parallel to the axis x, is equal to the component, parallel to the axis x, of the velocity that the same electron acquires when subject to an electromotive force equal to 1 and parallel to the axis y.

My theorem of reciprocity in the case of anisotropic conductors is, then, a consequence of these elementary laws of reciprocity.

Maxwell, speaking of the flow of electricity in an anisotropic medium (not subject to the action of a magnetic field) says that in every case of aelotropy (with the possible exception of magnets) we can grant that the determinant of the coefficients, in the equations that express the components of the electric current linearly in terms of the components of the electromotive force, is a symmetrical determinant (2).

If we refer to the electronic theory, this hypothesis of Maxwell's is equivalent to the conditions (38′).

If not only the conditions (38′) but also (38) be satisfied in every case of aeolotropy, the same generality holds good for my theorem of reciprocity. We can however easily determine a case in which the conditions (38) are certainly satisfied. Among others there is the case, the most interesting here, of a conductor that is isotropic without the magnetic field and that acquires a temporary aeolotropy only under the action of the field. To state this it is sufficient to bear in mind that for evident reasons of symmetry, all the directions, passing through a point P of the conductor and tangent to the equipotential surface that passes through P, are equivalent from the electromagnetic point of view.

Section 3

As is known, the conductors show, under the action of a magnetic field, a change of resistance. The experimental researches on this phenomenon are very numerous; but the results obtained and the empirical formulae proposed by the various experimenters to represent the phenomenon are often discordant and even incompatible among themselves. This depends in part, as will be seen from what we shall say in this paragraph, on not having clearly defined what we mean by change of resistance produced by the field.

One result with reference to which the said experimental researches may be said to be concordant is the following: the ferromagnetic substances (iron, nickel, cobalt) show an increase of resistance parallel to the magnetic lines of force and a decrease in the perpendicular direction; the diamagnetic substances, among which bismuth has the first place, show an increase of resistance in all directions (different in the different directions); for all the other substances the change of resistance is very slight.

To explain these experimental results the hypothesis has been made that the field causes a real and true temporary alteration of the specific properties of the conducting substances subject to its action. The theoretical researches that have been made in order to give an explanation of the said results, taking as basis not the hypothesis above alluded to, but the laws of the movement of electricity in a conductor subject to the action of a magnetic field, are few compared with the experimental researches.

Miss Freda after having briefly examined these theoretical researches proposed to herself the problem of stating what the electronic theory, to which in these lectures we have referred, predicts about the experimental results that can be interpreted as a change of electric resistance caused by the magnetic field; whether there are or not phenomena that the theory does not foresee and to explain which it will be necessary to introduce the hypothesis of an alteration of specific properties produced by the field in the conductors subject to its action. (3) We shall briefly resume the results of this research.

Let us again consider the formulae (36)′ of the first paragraph, formulae that hold good in the case of a conductor which remains homogeneous and isotropic also under the action of a uniform magnetic field H. Let P be a point of the conductor; when this conductor is not subject to the action of a magnetic field, let l_0 be the line of flow passing through P, let j_0 be the density of current, V_0 the potential in P, c_0 the specific conductivity of the substance (c_o and σ have the same value). We shall have $j_o = -c_0 \dfrac{\delta V_0}{\delta l_0}$. Let now l be the line of flow passing through P when the conductor is subject to the action of the magnetic field; let j and V be the values of the density of current and of the potential at the point P. From the equations (36)′ we can easily deduce

$$(40)\qquad j = \left[K(1+\lambda^2) + [\sigma - K(1+\lambda^2)] \times \frac{K(1+\lambda^2)\cos^2 lz}{K(1+\lambda^2)\cos^2 lz + \sigma(1-\cos^2 lz)}\right\} \frac{\delta V}{\delta l} = -c\,\frac{\delta V}{\delta l}.$$

It will be plainly seen, that unless $\cos^2 lz = o$, in which case we have $c = c_0$, we always have $c < c_0$.

Therefore *the present theory predicts for a homogeneous isotropic conductor kept at a constant temperature and subject to the action of a uniform magnetic field, an apparent increase of specific resistance in all directions, that of the magnetic lines of force excepted; in this direction the theory does not foresee an apparent alteration of the specific resistance.*

K, σ, λ^2 do not vary when we change H to $-H$; then the apparent specific conductivity c does not vary by inverting the magnetic field, if $\cos^2 lz$ does not vary. (This condition is satisfied, for instance, in the case of a plane plate situated in a transverse magnetic field and in the case of a wire situated any way in the field.) c, for a conductor of arbitrary form, will be different generally in the different points, because it depends on $\cos^2 lz$. The dependence of c on $\cos^2 lz$ shows us that the theory foresees an apparent aelotropy of the conductor. (That is shown moreover also by the equations (36)′).

Let us consider a conductor in which, *with or without the action of the field, the lines of flow remain the same, and at every point of these lines the value of the density of current remains the same if we leave unaltered the total current flowing in the conductor.*

From what we have previously seen we can deduce that, *if the conditions now stated be fulfilled, the difference of potential between two points of a same line of flow l increases under the action of the field, if l be not parallel to the axis z; only in this latter case the difference of potential remains constant. The increase of the said difference of potential can be interpreted as an increase of the total resistance of the conductor.*

The said conditions (invariability of the lines of flow and of the density of current) are satisfied, as we can easily prove in the case of a conductor in which the lines of flow are straight lines parallel to the axis z, when $H=0$. Then the theory does not foresee an apparent increase of the total resistance of such a conductor.

The same conditions are also verified in the case of a plate provided, along the boundary, with point electrodes and subject to the action of a uniform magnetic field normal to it. (Second lecture, Section 3). Besides in this case the apparent specific conductivity c, corresponding to a given value of H, is constant at every point of any line of flow (because at every point we have $\cos lz = o$), and does not depend on the sense of the field. In this case the ratio of the values that the difference of potential between two points, of a same line of flow and not coincident with the electrodes assumes before and after the creation of the magnetic field, is equal to the ratio of the specific resistance $\frac{1}{c_o}$ that the plate has without the field and of the apparent specific resistance $\frac{1}{c}$ determined with the field. Therefore, for a plate placed in a transverse magnetic field and provided with point electrodes at the boundary, the theory predicts an apparent increase $\Delta\sigma$ of the total resistance; the ratio between $\Delta\sigma$ and the resistance σ_o without the field depends only on the intensity of this latter and on the specific properties of the conducting substance; this ratio does not vary by changing H to $-H$. The same can be said for a rectilinear metallic wire, inclined with respect to the lines of force of the magnetic field. (Also in this case $\cos lz$ is constant at every point of the conductor).

In a plate placed transversally in the field but not provided with point electrodes situated at the boundary (for instance in a rectangular plate provided along two opposite sides with electrodes of negligible resistance) under the action of the field not only the distribution of the potential changes, but also that of the current.

If we consider, in this case, the ratio of the value that the difference of potential, between two points of the same line of flow or between two electrodes of negligible resistance, assumes for $(H)=0$, and for $(H)>0$ we find that the ratio depends not only on the apparent increase of specific resistance of which we have already spoken, but also on the altered distribution of currents; in this case the ratio $\frac{\Delta\sigma}{\sigma_0}$ will depend generally not only on the specific properties of the conducting substance and on the intensity of the field, but also on the form and on the dimensions of the plate and on the position of the electrodes.

The preceding observations may partly explain the divergences of the results obtained by the physicists who have measured in different conditions the change of resistance of a conductor in the field.

Another fact, to explain which has been introduced the hypothesis of an alteration of specific properties caused by the field in the conductors subject to its action, is the law of dependence of the Corbino effect (see the next paragraph) on the intensity of the field. But Miss Freda has remarked that the dependence of the Corbino effect on H is measured by the factor $\lambda = tg\beta$; that λ increases more slowly than H, and that, therefore, theory and experiments (as far as the metals that conduct themselves like bismuth are concerned) agree, at least qualitatively.

From all that we have seen in this paragraph it follows that by confining ourselves to the consideration of the substances only which conduct themselves like bismuth, we can assert that *the theory of flow of electricity perpendicularly to the magnetic lines of force agrees, at least qualitatively, with the phenomena interpretable as a change of resistance that the experiments have verified.*

To be able to say whether there is or is not the said agreement also from the quantitative point of view and therefore whether the hypothesis that we have already several times mentioned in this paragraph is necessary or not would require having the exact values of v_1, v_2, N_1, N_2, but for the present we can not say that we have them.

Whereas, if we consider the case in which the flow of electricity takes place parallel to the lines of force of the field, we must admit that the theory does not foresee the increase of resistance that the experiments have verified. It is not easy to decide whether that depends on a real alteration of the properties of the conductors determined by the field, or on the fact that the conditions supposed in the theory are verified only approximately in the experiments, or on the superposition of other phenomena to those that the theory considers.

We shall, lastly, observe that all these results hold good only on the hypothesis that the mobility of the positive electrons is not zero.

Section 4

We shall now speak of various experiments which are applications or verifications of some points of the theory previously stated.

We shall begin by remembering in this paragraph some results (obtained already in 1911 by Professor Corbino) relative to the flow of electricity in a metallic plate bounded by two concentric circles, provided along these circles with two electrodes of negligible resistance and subject to the action of a uniform magnetic field (4). The same results could also be deduced from what we have said in the second lecture on the cyclic plates.

Professor Corbino found that, if the plate is perpendicular to the direction of the magnetic field, under the action of the latter the equipotential lines remain circles having their centres in the centre of the plate; whereas the lines of flow, which are radial without the magnetic field, become, under the action of this latter, logarithmic spirals. We can therefore consider the disc as traversed by radial currents and by circular currents. These circular currents enable the disc to induce a current in a concentric circular coil, in the instant in which we open or close the circuit of which the disc itself forms a part. The inductive action is evidently proportional to the intensity of the circular current; this intensity will be easily determined if we bear in mind the theory explained in the preceding lecture. As the lines of flow form the angle β with the concentric circles which coincide with the equipotential lines, the intensity of the circular current will be $I tg\beta = \frac{\Sigma}{K} I$, if I be the intensity of the total current flowing in the plate. Then the inductive

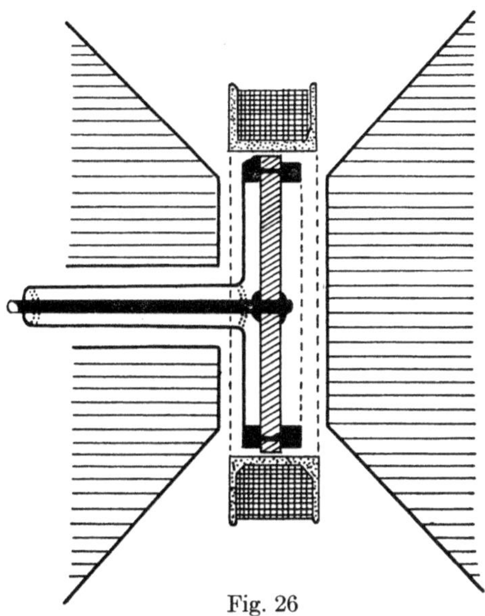

Fig. 26

action (Corbino effect) is proportional to $\frac{\Sigma}{K} I$ (the factor of proportionality depends on the dimensions of the plate). This action has been effectively proved and measured by Professor Corbino for bismuth, and by Adams and Chapman (5) for many other metals. It we assume that the current is carried only by negative electrons, we have $\frac{\Sigma}{K} = ev_2 H = -\frac{e\tau_2}{2m_2} H$ (v_2 is the mobility, m_2 the mass of the negative electrons; τ_2 average interval of time between two collisions). In this hypothesis from the measure of the Corbino effect, the value of τ_2 can be deduced. That has been done for many metals by Adams and Chapman and by other physicists.

A metallic disc traversed by a radial current, as we have seen, under the action of the magnetic field is transformed into a particular magnetic shell perpendicular to the lines of force of the field and produces thus a slight alteration of this latter; reciprocally the creation of the field should create a radial electromotive force in the disc.

Let us suppose, lastly that the disc traversed by a radial current is suspended in such way that its plane forms with the magnetic lines of force an angle α. In such conditions an electromagnetic couple tends to turn the disc; it can easily be verified that the moment of this couple is proportional to $I tg\beta \sin 2\alpha$.

These two last theoretical predictions were also experimentally verified in 1911, by Professor Corbino.

Section 5

With the phenomena mentioned in the preceding paragraph are connected some other experiments made in 1915 by Professors Corbino and Trabacchi. (6) Let us consider a rectangular prismatic surface (Fig. 27) the faces of which are thin metallic plates; α is of bismuth, β, γ, and δ of copper.

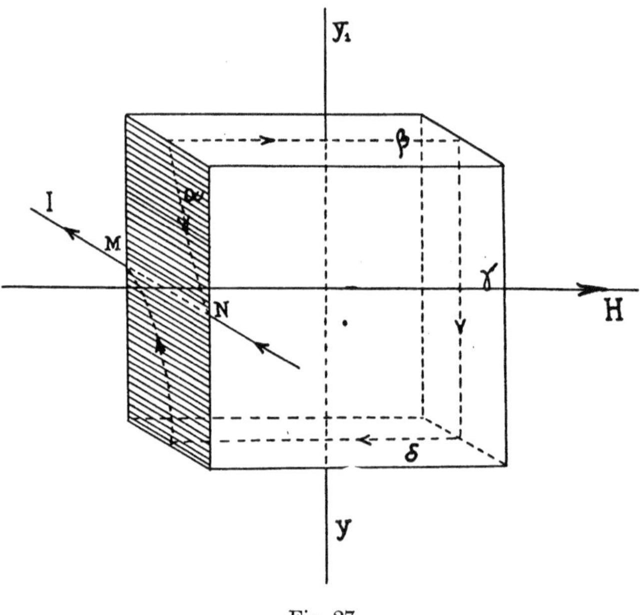

Fig. 27

Two copper wires are soldered in the middle points M and N of the bismuth plate; through these wires a current I can be sent in the plate. Let us suppose that the prismatic surface, which can rotate about the axis $Y\ Y'$, is placed in a uniform magnetic field perpendicular to the bismuth plate. Then the current I will be partially distorted and a part of it will flow in the prismatic surface $\alpha\beta j\delta$ transforming this surface into a magnetic shell which will tend to turn in the field about the axis YY'. If θ be the angle that the axis of the prism bounded by $\alpha\beta j\delta$ forms with H in a phase of the rotation, the component of the field perpendicular to the bismuth plate will be $H \sin \theta$; the intensity i of the distorted current can be said to be proportional approximately to $IH \sin \theta$. The moment of the couple acting on the prismatic surface will be proportional to $iH \sin \theta$; its approximate value will be then (except for a constant factor) $IH^2 \sin^2 \theta$.

If another identical prismatic surface can be rotated about the same axis YY' and is turned 90° from the first prismatic surface, this second surface will be acted on by a couple the moment of which will be approximately proportional to $IH^2 \cos^2\theta$. Then the rigid system of the two prismatic surfaces will be acted on by a couple the moment of which will be proportional to IH for any value of 0.

The figure 28 represents the intersection of the two prismatic surfaces with a plane which cuts them in half and is perpendicular to the axis. YY',NM,$N'M'$,

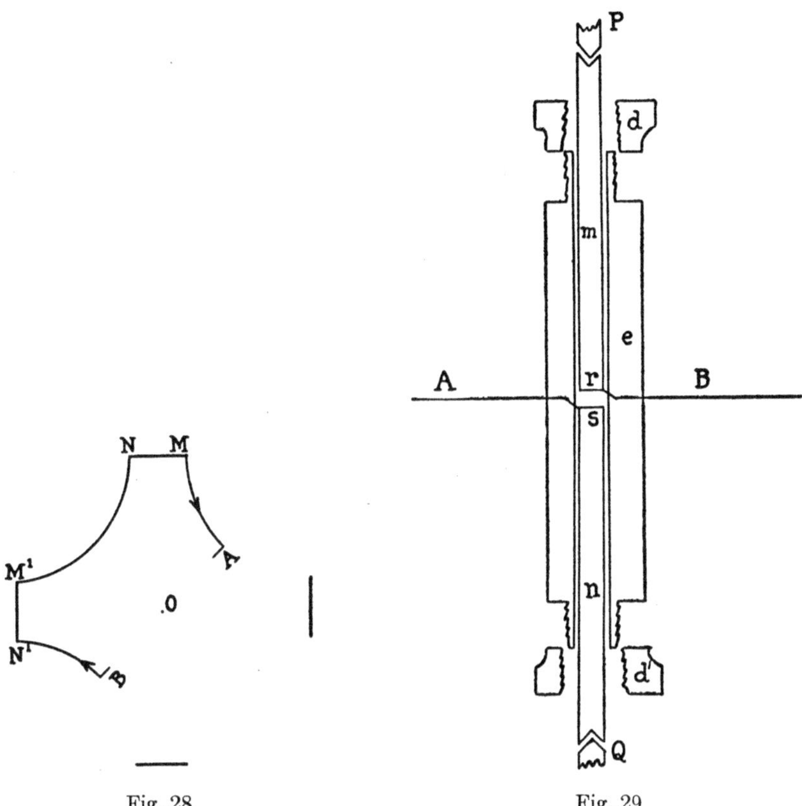

Fig. 28 Fig. 29

are the intersections with the two bismuth plates. The electrical connections are made by the wires AM, NM', $M'B$; the ends A and B are soldered respectively to the upper half m and to the lower half n of the axis of the apparatus (Fig. 29); the said wires are situated in a plane perpendicular to the axis of rotation YY'. The two halves mn of this axis are metallic, joined together mechanically but insulated electrically from each other; they are connected through the points of support PQ to the source of electricity.

According to what we have said, the system of the two prismatic surfaces represents a model of an armature which turns uniformly (with a constant couple proportional to the square of the magnetic field) when we send in it a steady current through two "fixed" contacts. (P and Q). Vice-versa *by turning the armature in the field with a constant velocity we can produce a steady constant electromotive force between the points P and Q without sliding contacts.* Lastly, *by holding fast the armature in a Ferraris rotary field we obtain between the same two fixed points P and Q a steady and constant electromotive force. The apparatus acts, therefore, as an armature provided with Pacinotti collector and with brushes situated in such a way that the straight line passing through them is perpendicular to the exterior field however this field or the armature may turn.*

For the details of the experiments which have made it possible to verify these properties we refer to the memoir cited at the beginning of this paragraph. The apparatus described cannot have any practical utilization because of the slight magnitude of the described phenomena, but is interesting on account of its properties which we have noted.

Section 6

We shall speak now in these last paragraphs of the experiments which have been made with the actual object of testing some points of the theory developed in

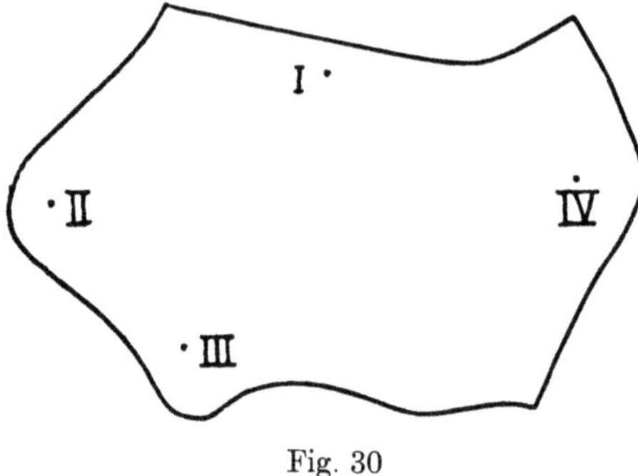

Fig. 30

the preceding lectures. We shall begin with the experimental verifications of my theorem of reciprocity.

Dr. Tasca Bordonaro, in 1915, verified this theorem first in the case of a bismuth plane plate situated in a uniform magnetic field normal to it (7) and afterwards in the case of a bismuth plane plate coincident with an equipotential surface of a non-uniform magnetic field. (8) We shall briefly relate the conditions in which these experiments were made and the results obtained.

The magnetic field was produced by the large Weiss electromagnet of the Physical Institute of Rome. In order to obtain a uniform magnetic field, the electromagnet was provided with cylindrical pole pieces having a diameter of 10 centimeters. (When the pole faces were 1.4 centimeters apart, by sending in the electromagnet a current of 4 amperes a field of about 6700 gausses was obtained.

The plate had the form indicated in Fig. 30; it was provided, in the points indicated in the figure, with point electrodes obtained by soldering at the said points four thin copper wires. (These wires were protected by rubber tubes to prevent their touching the pole faces of the electromagnet). To avoid the thermoelectric forces, the plate was enveloped in cotton. By means of two conveneient interrupters it was possible to connect two electrodes A and B of the plate with the poles of a battery of accumulators and two other electrodes *C* and *D* with a Hartman and Braun galvanometer provided with movable coils, and vice-versa. In the circuit of the battery and of the plate was inserted a high resistance in order to maintain constant the current when the electrodes *A* and *B* were exchanged with the electrodes *C* and *D*. As the current in the plate was interrupted soon after the reading of the impulsive deflection of the galvanometer, it was possible to avoid the influence of the thermoelectric phenomena which accompany the Hall effect. The zero of the galvanometer was observed after the creation of the magnetic field. The following are the results obtained by Dr. Tasca with a field of 6000 gausses and a current in the plate of 0.2 amperes:

	Direct field	Inverted field
$V_{II} - V_{IV}$	420	104
$V_I - V_{III}$	104	420.5

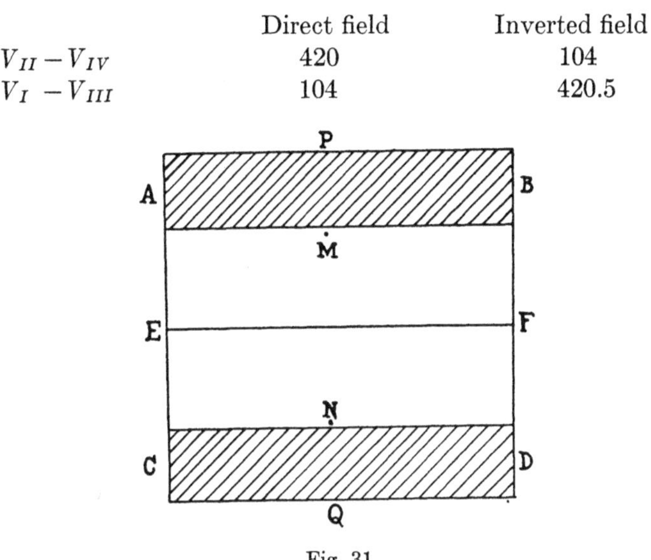

Fig. 31

(The differences of potential were measured by means of the corresponding deflections of the galvanometer expressed in millimeters. It is understood that when the difference of potential was measured between *II* and *IV* the current entered and went out through *I* and *III*, and vice-versa).

These results show clearly the law of reciprocity to be verified, as theory requires, only when, by exchanging the electrodes the field is also inverted; the asymmetry of the effect without the inversion of the field is very remarkable.

Dr. Tasca afterwards substituted for the plate represented in the Fig. 30 a rectangular bismuth plate (Fig. 31) provided with copper electrodes of negligible resistance soldered along the two longest opposite sides and with two thin copper wires soldered in the middle points of the two free sides. The results obtained are the following

	Direct field	Inverted field
$V_E - V_F$	51.5	54.5
$V_P - V_Q$	54	51

Also in this case the theorem of reciprocity is completely confirmed.

Dr. Tasca has made the following observation on the mechanism by which the law of reciprocity is maintained in the case of wide copper electrodes: the copper electrodes P and Q, when the current enters and goes out through them, cause the recombination of the charges which owing to Hall effect, reach the free edges; P and Q thus diminish this effect. Whereas, when the current enters and goes out through the point electrodes E and F, the copper electrodes P and Q absorb a good part of the current; the current which flows in bismuth is therefore only a part of the total current sent into the plate and thus the Hall effect is diminished also in this case. Theory demonstrates and experiments confirm that the diminution is identical in the two cases; we could not have foreseen intuitively this result.

Dr. Tasca has lastly tested my theorem of reciprocity in the case of a plane bismuth plate situated along an equipotential surface of a non-uniform magnetic field. (See third lecture, Section 4); this latter was obtained by providing the

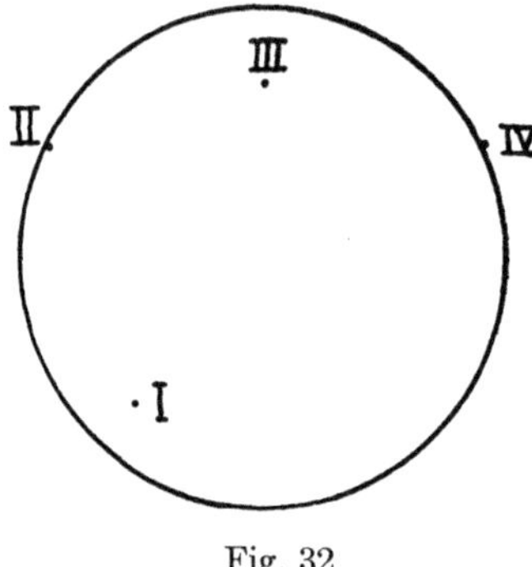

Fig. 32

Weiss electromagnet with pole pieces having the form of truncated cones and terminating with circles of 5 millimeters diameter; in those conditions the intensity of the magnetic field assumed values extremely different in the different points, from some hundred of units to about 15000 units.

The plate, being circular in form and provided with point electrodes asymmetrically soldered (fig. 32), was situated along the equatorial plane of the electromagnet; for evident reasons of symmetry, this plane is an equipotential surface of the magnetic field.

In this case the following experimental results were obtained:

	Direct field	Inverted field
$V_I - V_{III}$	256	21
$V_{II} - V_{IV}$	20	256

The action of the field is very remarkable; in fact the value of the difference of potential by exchanging only the electrodes passes from 256 to 20, but by inverting the field we have again the original value.

In the study of the problem of which we have spoken in the preceding lectures I had not taken into account the secondary actions for which a conductor loses the character of isotropy, with respect to its electrical conductivity, when it is subject to the action of a magnetic field; but I remarked that the perturbations caused by these secondary actions do not take place when the magnetic field is perpendicular to the plate at every point of this latter. (Third lecture Section 2).

It is for this reason that Dr. Tasca, in his experiments, tested my theorem of reciprocity only in the case of a plate placed along an equipotential surface of the field.

In 1916, Miss Freda, starting from the theoretical results obtained by her (see Section 1 and 2 of this lecture) was able to remove this restriction, and wished to verify experimentally that my theorem of reciprocity holds good also when the plate is situated in any way in a uniform or non-uniform magnetic field. For this verification (as for those of which we shall speak shortly) the same electromagnet was used that had already served for Dr. Tasca, with the same pole pieces, having the form of cylinders or of truncated cones, capable of producing respectively a uniform or a non-uniform field.

The experimental arrangement adopted was the same of which we have already spoken with reference to the experiments of Dr. Tasca.

For the said verification a bismuth disc was used, of diameter cm. 4.5 and of thickness 2 mm, provided with four point electrodes *ABCD* asymmetrically situated.

Let us denote with d the least distance between the pole pieces, with I the intensity of the current in the electromagnet, with i the intensity of the total current in the conductor.

With a uniform field, for $d=5$ cm., $I=5$ amperes, $i=0.2$ amperes, having placed the disc in such a way that its plane forms an angle of about 45° with the lines of force of the field, the following results were found:

	Direct field	Inverted field	Without field
V_A-V_B	26	53	38
V_C-V_D	53.5	26	38.5

With a non-uniform magnetic field, for $d=2.6$ cm., $I=6.3$ amperes, $i=.2$ amperes, having placed the disc in such manner that its plane forms an acute angle with the equatorial plane of the electromagnet, these results have been obtained

	Direct field	Inverted field	Without field
V_A-V_B	18	71.5	38
V_C-V_D	71.5	18.5	38.5

According to the theoretical results summarized in the Section 2 of this lecture, my theorem of reciprocity can be extended to the case of a three-dimensional conductor, isotropic without the magnetic field. Miss Freda wished to verify experimentally this result obtained by her. She employed for this verification a bismuth prism having as base a square, with the side of 1.5 cm. and having the height of 6 cm. The prism was provided with electrodes, situated in any way along its surface (fig. 33) or with point electrodes and laminar copper electrodes of negligible resisttance(fig. 34). The following results were obtained by using the four point electrodes $ABCD$ (fig. 33) with a uniform magnetic field, for $d=3.4$ cm., $I=7.8$ amperes $i=0.43$ amperes, the prism having been placed with its lateral faces inclined with respect to the lines of force of the field:

	Direct field	Inverted field	Without field
V_A-V_B	411	421	395.5
V_C-V_D	421	411.5	395.5

With a non-uniform magnetic field, for $d=1.8$ cm., $I=7.8$ amperes, $i=0.58$ amperes, by using the four electrodes $H'B'C'D'$ (fig. 33) the differences of potential had these values:

	Direct field	Inverted field	Without field
$V_{A'}-V_{B'}$	20	13.5	15
$V_{C'}-V_{D'}$	13.5	20.5	15.5

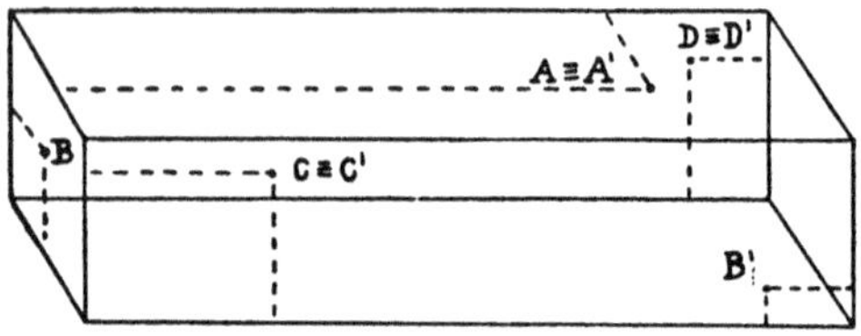

Fig. 33

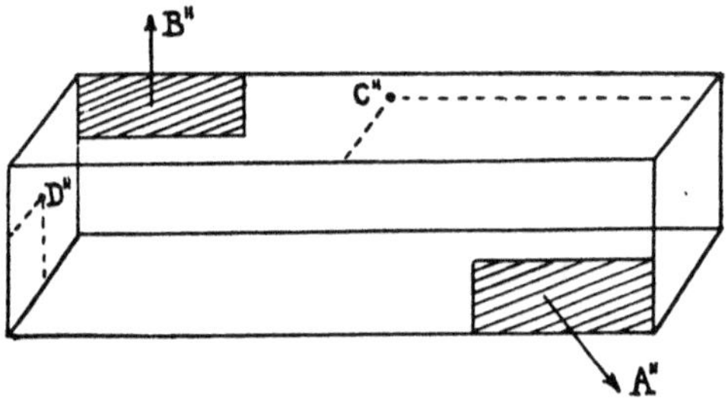

Fig. 34

(In this experiment the face provided with the electrodes $B'C'$ and the opposite face provided with the electrode D' were parallel to the equatorial plane of the electromagnet).

After having soldered to the prism the two laminar electrodes A'' B'', the theorem of reciprocity was tested by using these electrodes and the point electrodes $C''D''$. (fig. 34) Then with a non-uniform magnetic field, for $d=1.8$ cm., $I=7.7$ amperes, $i=0.59$ amperes, having placed the prism in such a manner that the faces provided with the laminar electrodes were inclined with respect to the equatorial plane of the electromagnet, the following results were obtained:

	Direct field	Inverted field	Without field
$V_{A''}-V_{B''}$	383	387	367.5
$V_{C''}-V_{D''}$	387	383	367

As we see, in the case of the prism the action of the field is less remarkable than in the case of the disc, but the theorem of reciprocity is always verified.

Miss Freda wished then to show if this theorem held good for a piece of bismuth crystal (obtained artificially) of a quite irregular form and provided with four point electrodes asymmetrically soldered along its surface. (It is to be remarked that if a negative result hadbeen obtained with or without the action of the magnetic field the existence would have been proved of cases of aeolotropy for which the conditions (38) or (38′) of the Section 2 are not fulfilled). The following are the values obtained for the differences of potential with the cylindrical pole pieces, for $d=4.8$ cm., $I=7$ amperes, $i=0.31$ ampere and for any position of the crystal:

	Direct field	Inverted field	Without field
V_A-V_B	104	157	119
V_C-V_D	156.5	104	119

The values obtained with the non-uniform field, for $d=2$ cm., $I=7$ amperes, $i=0.37$ ampere, the crystal being placed in an arbitrary position in the field are the following:

	Direct field	Inverted field	Without field
V_A-V_B	115	215	136
V_C-V_D	215	114.5	136.5

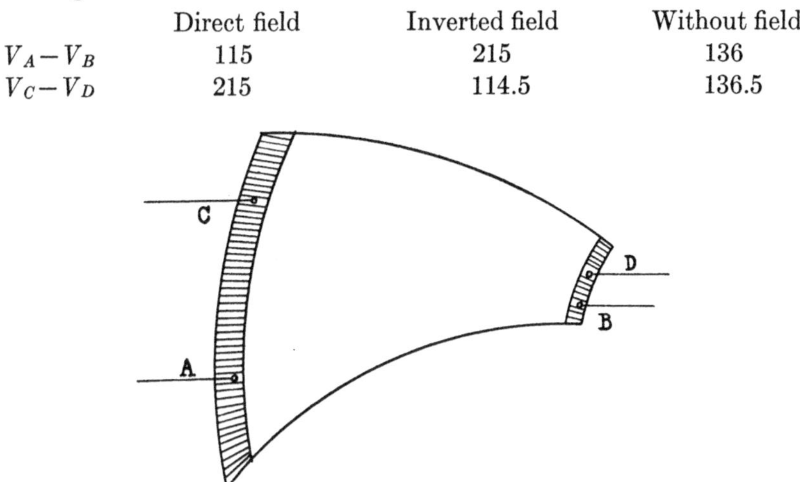

Fig. 35

As we see, also in the case of the crystal the theorem of reciprocity is completely verified.

In all these tests of the theorem of reciprocity not only the experiments of which we have shown the results have been made, but many other experiments by either exchanging the electrodes with each other, or by changing the position of the conductor in the field. The action of this latter was more or less remarkable in the various cases, but the agreement with the theory was always equally good.

Section 7

We shall now speak of a consequence of my theorem of reciprocity stated and verified by Professors Corbino and Trabacchi (9).

Let us consider a disc in which the current enters and goes out through two concentric circular electrodes of negligible resistance; the lines of flow will be radial without the magnetic field, but will become logarithmic spirals under the action of this latter. Let us supose the disc cut along two of these spirals; then the distribution of the currents is not altered and therefore we can limit ourselves to consider only the plate that we have detached (fig. 35) and that has its free boundary formed by spirals. If the magnetic field be inverted the lines of flow will undergo a marked deformation, whereas the specific resistance of the metal will maintain the same value. It would seem that the remarkable alteration of the lines of flow which takes place on reversal of the field had to be accompanied by a change of the total resistance of the plate, that is by a change of the ratio between the difference of potential of the two electrodes and the intensity of the current flowing in the plate. When the experiment was made, however, a negative result was obtained.

This result could neither have been foreseen nor explained, intuitively; whereas we can easily demonstrate that it is a consequence of my theorem of reciprocity. In fact, let $ABCD$ be four point electrodes placed on the wide copper electrodes of the plate (fig. 35). The theorem of reciprocity, which holds good also for large electrodes of negligible resistance, can be applied to the points $ABCD$; therefore the difference of potential, that we have between C and D, when the field is direct and the current enters and goes out through A and B, must be equal to the difference of potential that we have between A and B, when the field is inverted and the same current enters and goes out through C and D. But the said differences of potential (between C and D, between A and B) both coincide with the difference of potential that we have, respectively in the two cases, between the large electrodes of negligible resistance; therefore *the inversion of the field* (however the distribution of the lines of flow may be altered by this inversion) *can not change the total resistance of a plate in which the current enters and goes out through electrodes of negligible resistance.*

Miss Freda has extended this result to the case of a three dimensional conductor provided with electrodes of negligible resistance and has verified it in the case of the bismuth prism provided with laminar electrodes, represented in fig. 34 (1).

SECTION 8

Professor Corbino in his memoir cited in the first lecture, has observed that, if a rectangular plate be provided along two opposite sides AB and CD with electrodes of negligible resistance P and Q (fig. 31), owing to the presence of these electrodes, the free sides AC and BD of the plate are short-circuited. Then the Hall effect, owing to the presence of electrodes of negligible resistance, undergoes a remarkable diminution which rapidly increases as one goes from the center of the plate (points E and F) towards the electrodes.

Dr. Tasca (7) measured this diminution using the same rectangular bismuth plate (fig. 31) with which we had tested the theorem of reciprocity (Section 6). The plate had the side AB 52 mm. long and the side AC 22 mm. long. With a uniform field of 6000 units, sending the current through P and Q, Dr. Tasca obtained $V_E - V_F = 52.5$; taking off P and Q and sending the current through the point electrodes M and N, he obtained $V_E - V_F = 318$. As we see the depression of the Hall effect caused by the electrodes of negligible resistance was in this case very remarkable. With a square bismuth plate Dr. Tasca observed a smaller, but always observable diminution.

In Section 4 of the first lecture we have stated that, if the lines of flow do not vary under the action of the field, by considering two points A and B at the boundary and at the same potential when the magnetic field is zero, we have $V_A - V_B = = \frac{\lambda}{K(1+\lambda^2)}$; then the Hall effect in this case depends only on the intensity of the field, on the intensity of the current flowing in the plate, and on the nature of the metal by which this latter is formed. Therefore, bearing in mind *the principle of the point electrodes at the boundary, we can* state that *if the current enters and goes out through point electrodes situated at the boundary of the plate, the Hall effect* (difference of potential between two points of the boundary at the same potential when $H=0$) *depends neither on the form nor on the dimensions of the plate, nor on the position of the electrodes, nor of the points between which the effect is measured, but only on the intensity of the field, on the intensity of the current and on the nature of the metal.*

On the other hand, for a plate provided with electrodes of negligible resistance the Hall effect depends on the form and on the dimensions of the plate, on the nature and on the position of the electrodes and on the position of the points between which the effect is measured. This observation can partly explain the divergences of the results obtained by the various physicists in the measure of the Hall effect.

On this basis Dr. Tasca has proposed to designate by *normal Hall effect* the effect measured using point electrodes at the boundary. He has observed that from the historical point of view it is to be remarked that the condition of the point electrodes on the boundary corresponds to the primitive arrangement of Hall. This arrangement was afterwards abandoned by physicists because they thought the conditions were theoretically simpler with the rectangle provided with large electrodes.

Section 9

In Section 2 of the second lecture we have studied the flow of electricity in a circular plate subject to the action of a uniform magnetic field normal to it. According to the results then stated, we can say that in the case in which the plate is provided with two point electrodes A and B the potential V is given by the formula

$$V = \frac{-J}{2\pi K}\left(\log\frac{r_A}{r_B} + \cos 2\beta \log\frac{r_{A1}}{r_{B1}} - \sin 2\beta\Omega\right);$$

A_1 and B_1 are the inverse points of A and B with respect to the circumference that forms the boundary of the plate; Ω is the angle under which the points A_1 and B_1 are seen from a generical point.

In 1915 Miss Alimenti considered the problem of constructing the equipotential lines corresponding to the preceding formula, in the case in which the electrodes are situated in the middle points of two perpendicular radii of the plate. She observed, therefore, that the electric field corresponding to the potential V, can be considered as resulting from the superposition of three electric fields, the potentials V_1, V_2, V_3, of which are given (neglecting a constant factor common to all three) by the formulae:

$$V_1 = \log\frac{r_A}{r_B}; \quad V_2 = \cos 2\beta \log\frac{r_{A1}}{r_{B1}}; \quad V_3 = \sin 2\beta\Omega.$$

By combining the curves $V_1 = const.$ (circles orthogonal to that passing through A and B) with the curves $V = const.$ (circles orthogonal to that passing through A_1 and B_1), Miss Alimenti obtained the curves $V_1 + V_2 = const.$ represented in the figure 36; by combining these latter with the curves $V_3 = const.$ (circles passing through A and B) she obtained the curves $V_1 + V_2 + V_3 = const.$, that is the curves $V = const.$ represented in the figure 37. As other phenomena (Ettingshausen, Nernst, etc.) superpose themselves on the phenomena which the theory considers, it would have been difficult to verify experimentally the form of the curves $V = const.$ Therefore Miss Alimenti verified otherwise some points of the theory.

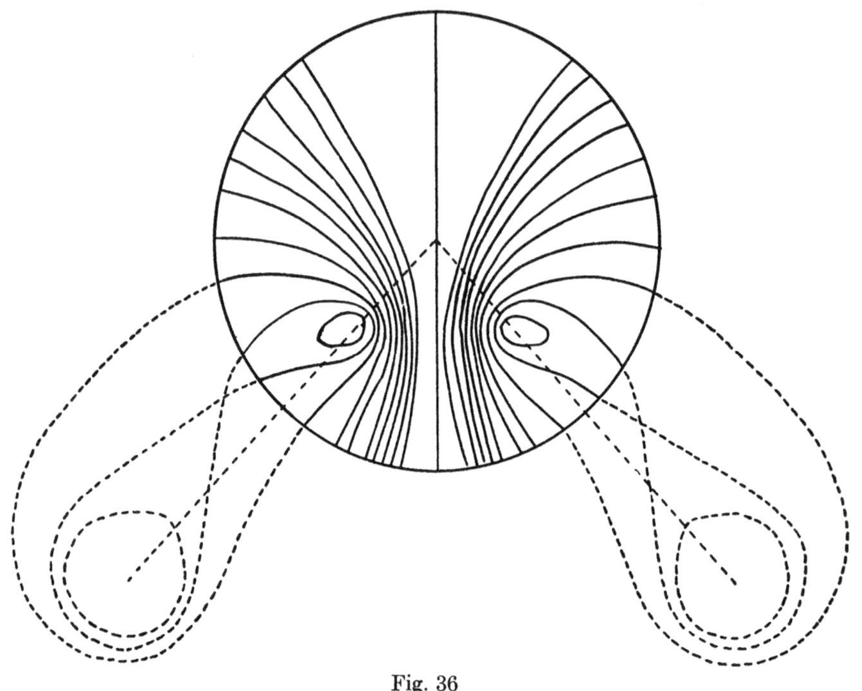

Fig. 36

The difference of the potential between two points I and II of a circular plate, perpendicular to the lines of force of a uniform magnetic field H and provided with two point electrodes A and B, is given by the formula:

$$\delta = V_1 - V_{11} = \frac{-J}{2\pi K}\left\{\left(\log\frac{r_{A1}}{r_{B1}} - \log\frac{r_{A11}}{r_{B11}}\right) + \cos 2\beta\left(\log\frac{r_{A_1}1}{r_{B_1}1} - \log\frac{r_{A_1}11}{r_{B_1}11}\right) - \sin 2\beta(\Omega_1 - \Omega_{11})\right\}$$

Let us denote with V' the potential corresponding to the inverted field; let us put $\delta = V' - V'_{11}$; since on reversal of the field only the sign of changes, we shall have $\frac{\delta - \delta'}{2} = \frac{-J}{2\pi K}\sin 2\beta(\Omega_1 - \Omega_{11})$.

(The values of v_1, v_2, N_1, N_2, in the expressions of β and K may depend on the absolute value of H but not on its sign; therefore these values do not change on reversal of the field (see Section 1 of this lecture).

Miss Alimenti has remarked that $\frac{\delta - \delta'}{2}$ must have the following properties:

ς—It must have a maximum corresponding to the two points at the extremes of the diameter perpendicular to the straight line which passes through A and B; for two points, however situated on this diameter, δ and δ' must have equal absolute values but contrary signs.

σ—It must be zero for two points however situated on the same arc of a circle passing through A_1 and B_1; i.e. the inversion of the field must leave unaltered the difference of potential between two points situated in the said manner.

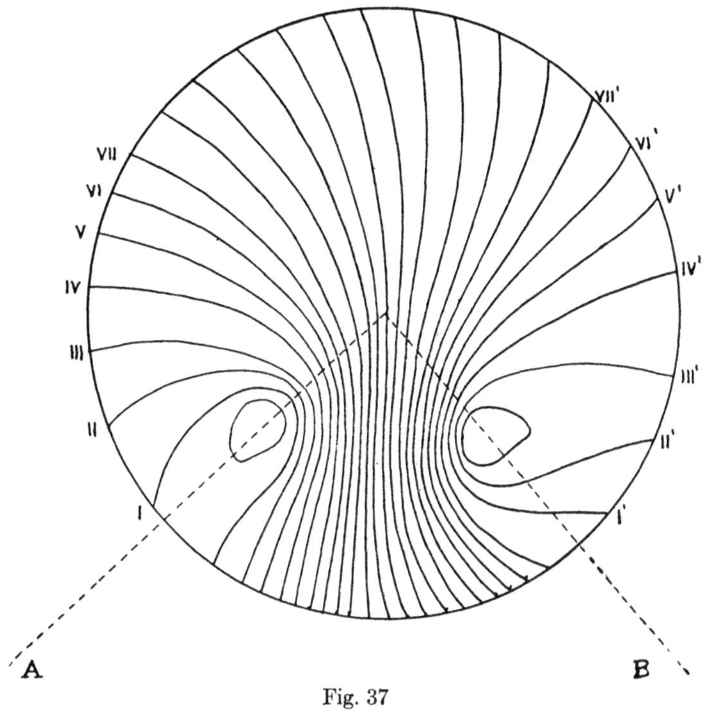

Fig. 37

θ—It must be constant for two points however chosen on two circles both passing through A and B. Miss Alimenti verified experimentally the two last properties (σ and θ) and *then the theorem of the four vertices* (see second lecture Section 3) in the particular case of a circular plate.

For making the experimental tests, she used a bismuth disc situated perpendicularly to the lines of force of a uniform magnetic field, obtained by means of an electromagnet. The difference of potential between two points was measured by means of the corresponding deflexion (expressed in mm.) of a galvanometer connected with the same points.

In the first experiment six thin copper wires were soldered to the plate in the points A,B,C,D,E,F, (fig. 38); the points A and B, through which the current entered in the plate, were situated in the middle points of two perpendicular radii; the points C,D, and E,F, coincided respectively with two generical points of two circles passing through the inverse points of A and B. In these conditions the following results were obtained:

Points connected with the galvanometer	δ	δ'	$\frac{\delta - \delta'}{2}$
C D	532	532	0
D E	325	280	22.5 }
C F	420	375	22.5 }

The values of the first line confirm the property σ; that of the second and of the third line confirm the property θ.

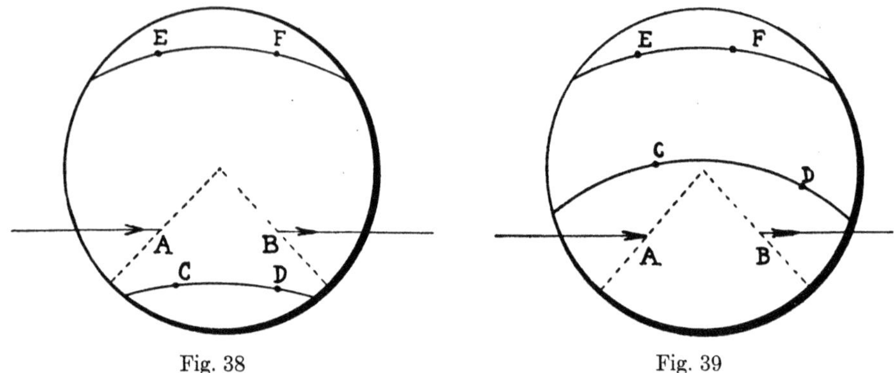

Fig. 38 Fig. 39

In a second experiment the points C and D were placed on another circular arc passing through the inverse points of A and B (Fig. 39) then for $\delta\delta'$ and $\frac{\delta - \delta'}{2}$ these values were obtained:

Points connected with the galvanometer	δ	δ'	$\frac{\delta - \delta'}{2}$
C D	458	458	0
E F	253	253	0
D E	342	246	48 }
C F	435	339	48 }
C E	227	108	59.5 }
D F	116	−3	59.5 }

The values of the two first lines confirm also in this case the property σ; the values of the third and of the fourth line, of the fifth and of the sixth line, confirm the property θ.

Let us observe that, when the point electrodes A and B are at the boundary, the points A_1 and B_1 coincide with A and B; the circles passing through A and B coincide with the lines of flow (with or without the magnetic field.

Now let us consider the results of the third experiment made by Miss Alimenti. She placed the points A, B, C, D, E, F as is indicated in figure 40, that is, the points A and B were at the boundary, the points C, D, E, F coincided with the vertices of a quadrilateral formed by two circles passing through A and B (lines of flow) and by two other circles orthogonal to the first (equipotential lines corresponding to

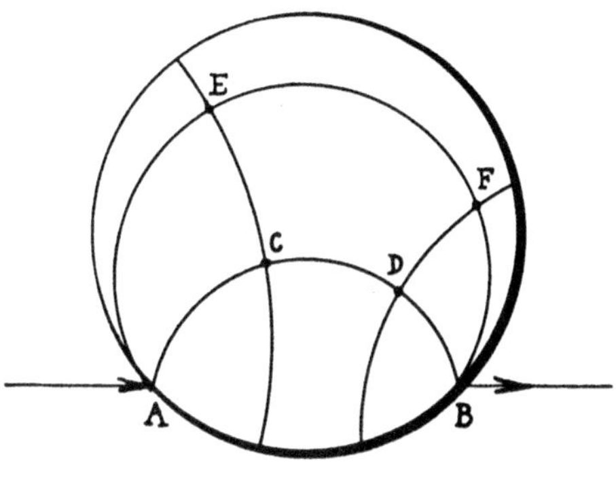

Fig. 40

the value zero of the magnetic field); thus she could also test the *theorem of the four vertices.* (See second lecture, Section 3), in the case of a circular plate. These are the results obtained:

Points connected with the galvanometer	Difference of potential without the field	δ	δ'	$\bar{\delta}$ corrected	$\bar{\delta}'$ values	$\frac{\delta-\delta'}{2} = \frac{\bar{\delta}-\bar{\delta}'}{2}$
C E	-7	41	55	$41+7=48$	$-55+7=-48$	48 }
D F	2	50	46	$50-2=48$	$-46-2=-48$	48 }

The slight differences of potential existing between C and E, and between D and F without the magnetic field, show that the equipotential lines were not constructed with a perfect exactness; therefore the values of δ and δ' have been corrected, taking into account the said differences of potential and the sense of the deflections of the galvanometer. The corrected values of δ and δ' confirm the *theorem of the four vertices*; the values of $\frac{\delta-\delta'}{2}$ confirm the property θ; we have the confirmation of this latter also without correcting the values of δ and δ', and it must be so because this property is independent of the position of the points $C, D,$ and $E, F,$ situated respectively on two circles passing through A and B.

All these experimental results, together with that of which we have spoken in the preceding paragraphs, constitute a completely satisfying confirmation of the theory developed in these lectures.

BIBLIOGRAPHY

(1) Elena, Freda: Rendiconto Accademia dei Lincei, 2° sem. 1916, p. 28, p. 60.

(2) Maxwell: A Treatise on Electricity and Magnetism, sec. 297, p. 346; sec. 303, p. 350. Oxford, Clarendon Press, 1873.

(3) Elena Freda: Rend. Acc. dei Lincei, 20 sem. 1916, p. 104, p. 142; Nuovo Cimento, 1916, XII, p. 177.

(4) Corbino: Nuovo Cimento, 1911, t. I, p. 397.

(5) Adams and Chapman: Phil. Mag., 1914, Vol. XXVIII, p. 692. Keith K. Smith, Physical Review, 1916, Vol. VIII, p. 402.

(6) Corbino and Trabacchi: Nuovo Cimento, 1915, t. IX, p. 80.

(7) Tasca Bordonaro: Rend. Acc. dei Lincei, 1° sem. 1915, p. 336.

(8) Tasca Bordonaro: Rend. Acc. dei Lincei, 1° sem. 1915, p. 709.

(9) Corbino and Trabacchi: Nuovo Cimento, 1915, t. IX, p. 118.

(10) Adalgisa Alimenti: Nuovo Cimento, 1915, t. IX, p. 109; 1916, t. XI, p. 217.

UNIVERSITY OF CALIFORNIA PUBLICATIONS
IN
MATHEMATICS
Vol. 1, No. 14, pp. 321-343 November 11, 1920

THE HOMOGENEOUS VECTOR FUNCTION

AND

DETERMINANTS OF THE P–TH CLASS

BY

JOHN D. BARTER

CONTENTS

I. THE HOMOGENEOUS VECTOR FUNCTION

1. *General Outline*

The object of the present paper is to present the properties of a vector function, which may be considered as the generalization of the linear vector function. At the same time we shall attempt to develop the various analytical representations, and to a certain extent indicate the applications.

Our homogeneous vector function of the mth degree is defined as satisfying the relation:

$$k^m \cdot \varphi(\boldsymbol{r}) = \varphi(k\boldsymbol{r}),$$

where $\boldsymbol{r}$ is any vector, and k is a constant.

Rather more general would be a vector function satisfying the relation:

$$f(k) \cdot \varphi(\boldsymbol{r}) = \varphi(k\boldsymbol{r}).$$

We shall not in this part specifically occupy ourselves with the latter, beyond indicating that it may in general be expressed in terms of the former. Moreover we may, and shall, confine ourselves to the case when m is a positive integer. The case when m is a negative integer involves the investigation of the inverse function, and will be treated under that head.

In this present part we further confine ourselves to a method of representation analogous to that of matrices or dyadics, employed in the case of the linear vector function. It will appear that there is a marked contrast between the properties of the latter and those of the general vector function. However in this case also there present themselves singular vectors which play much the same rôle as the axes of the linear vector function, and are in consequence assigned the same term. On the other hand those characteristic numbers which are termed the latent roots of the linear vector function, will readily be seen to have no analogues in the general case. We shall by this method, partly for their own sake, partly for the sake of later comparison, when a different mode of representation, that by means of quarternions, is studied, develop certain of the most fundamental properties, which we believe to have an interest of their own.

In the second part will be given the algebraic theory leading to the determination of the axes of these vector functions. The method will be found closely analogous to that employed in the case of the algebraic theory of eliminants, and its results are, we believe, of some value in the theory of differential equations.

The third part is to be confined to a survey of the geometrical applications, the vector function being regarded as determining a transformation of space.

The fourth part is to be devoted to a brief development of the theory by way of quaternions, and to a comparison of results.

2. *Survey of Applications*

The concept of the vector function enters into three branches of mathematics. In mathematical physics are encountered two distinct types. The first is that determined by two localized vector sheafs, functionally related. As examples of this type may be quoted the relation between the stress and strain vectors in the theory of elasticity, and which, it may be stated in passing, is completely and adequately represented by the self-conjugate linear vector function. Further, there present themselves such relations as that existing between the vectors of electric force and displacement, or between magnetic force and induction. In these latter cases a more general form of representation is possibly called for than that provided by the linear vector function.

The second type is that of the general vector field in which the characteristic vector is to be regarded as a function of the vector of position. This form can probably be most satisfactorily treated by means of quaternionic functions, as we hope to show at a later point.

In geometry the general vector function represents, as already stated, a very general transformation of space. In consequence of the property it may be employed to present many geometric relations. The transformation is a collineation only in the case of the linear vector function. In the general case only lines through the origin, which remains fixed, are transformed into straight lines which likewise pass through the origin. It will furthermore appear that in the case of vector functions of odd degree the surface defined is closed, while in the case of even degree the surface is entirely confined to the positive semispace.

In purely analytic applications this function can be caused to play an auxiliary rôle, in virtue of the relations between quantics which are uncovered by an investigation of its properties. In the second part this aspect will be given some consideration.

3. *Notation and Definitions*

In the present paper the symbolism plays such an important part that it may be considered as an example of the applications of the method, capable, we hope, of much wider extension. We shall accordingly proceed with a precision of the terms employed.

Let there be given a system of n^{m+1} distinct terms, and let each be represented by a symbol of the form

$$a_{p_1 p_2 \dots p_{m+1}}$$

in which each of the $m + 1$ suffixes may assume all values from 0 to n, there being obviously the requisite number of such distinct symbols (n^{m+1}). The aggregate of these symbols will be denoted:

$$(a_{\bar{n} \dots (m+1) \dots \bar{n}}) = (a_{\bar{n}}{}^{(m+1)}), \tag{1}$$

and will be termed a matrix of the nth order and class $m + 1$.

Then the product of a matrix of class $m + 1$ and one of class 1, will be defined thus:

$$(a_{\bar{n} \dots (m+1) \dots \bar{n}})(b_{1\bar{n}}) = (\Sigma a_{\bar{n} \dots (m) . \bar{n} \bar{p}_1} b_{1 p_1}),$$

being evidently a matrix of class m and order n. We may accordingly again multiply this matrix by a matrix of class 1, obtaining a matrix of class $m - 1$:

$$(a_{\bar{n} \dots (m+1) \dots \bar{n}})(b_{1\bar{n}})(b_{2\bar{n}}) = \left(\sum_{p_1=1}^{w} a_{\bar{n} \dots (m) \dots \bar{n} p_1} b_{1 p_1} \right)(b_{2\bar{n}}) = \left(\sum_{p_1 p_2=1}^{w} a_{\bar{n} \dots \bar{n} p_1 p_2} b_{1 p_1} b_{2 p_2} \right) \tag{2}$$

By proceeding in this manner we may successively reduce the class of the matrix at each multiplication by unity, obtaining after m such multiplications a matrix of class 1:

$$(a_{\bar{n}\ldots(m+1)\ldots\bar{n}})(b_{1\bar{n}})(b_{2\bar{n}})\cdots(b_{m\bar{n}}) = \left(\sum_{p..=1}^{w} a_{\bar{n}p_1\ldots pm}b_{1p_1}b_{2p_2}\cdots b_{m-1pm-1}b_{mpm}\right) \quad (3)$$

where the summation is to be extended as before, each subscript separately being ascribed all the values from 1 to n. We may, and in the sequel frequently shall, denote a vector by a matrix of class 1, in which the constituents are simply the components of the vector. If required, the fundamental vectors of reference may be introduced by means of another matrix of class 1, whose constituents are these unit vectors. Then any vector may be expressed as the product of two matrices of class 1, one of which has numerical constituents, and the other unit vector constituents, thus:

$$\boldsymbol{\beta} = (\mathbf{a}_{\bar{n}})(a_{\bar{n}}).$$

In general we shall consider the vector matrix as understood, and denote a vector simply by the corresponding numerical matrix.

Now it is evident that by the process above indicated, we are able to define a matrix of class $m - n$ as a function of n given vectors merely by multiplying a matrix of class m by these n vectors, which may or may not be distinct. In general of course the order in which the multiplication is considered as executed is essential. In the special case when $n = m - 1$, we have thus defined a vector as a function of n given vectors. This is the case specifically treated in this paper, but it may be pointed out, at this stage, that under certain circumstances it may be advisable to consider the more general standpoint. Thus for example we may consider n points in space as defining at each point a set of n vectors, which in turn may by multiplication with a matrix of class m, in the manner indicated, be considered as defining a matrix or vector operator of class $m - n$. It is not difficult to see that in some problems such a device may be of real service.

4. *Homogeneous Vector Function of the* m*th Degree*

We are now in a position, by means of the symbolic resources developed in last section, to set up the expression for the homogeneous vector function of the mth degree, as this term was defined in the first section, as follows:

$$\begin{aligned} \varphi_m(\boldsymbol{\beta}) &= \varphi_m\cdot\{(b_{\bar{n}})(\mathbf{a}_{\bar{n}})\} \equiv \varphi_m(b_{\bar{n}}), \\ &= (a_{\bar{n}}{}^{(m+1)})(b_{\bar{n}}\overset{m}{)(}\mathbf{a}_{\bar{n}})^m \equiv (a_{\bar{n}}{}^{(m+1)})(b_{\bar{n}})^m. \end{aligned} \quad (4)$$

This expression is evidently a matrix of class 1, and accordingly, as already pointed out, may be regarded as a vector, and functionally related to the given vector as follows:

$$\varphi_m(kb_{\bar{n}}) = k^m\cdot\varphi_m(b_{\bar{n}}),$$

which, by definition, is characteristic of the homogeneous vector function of the mth degree. We shall accordingly employ this notation to develop its properties.

We shall in the first place develop the addition theorems, that is to say, set up the expressions for the sum of the vector functions of a set of vectors, and also the expression for the vector function of a sum of vectors. Each of these relations will be developed first for the special case of $m = 2$, not because the general case presents any additional difficulty, but because in the latter certain features, which are prominent in the former, become more subsidiary, and less interesting.

We shall also determine an alternative form for this function, analogous to, and to be regarded as a generalization of the m-adic of J. W. Gibbs.

5. *First Addition Theorem*

We shall give the proof of this theorem in its most general form, since from this standpoint its essential properties are more sharply defined than would be the case if we made a more special assumption regarding m.

We have then:

$$\begin{aligned}\varphi_m(b_{1\bar{n}} + b_{2\bar{n}}) &= (a_{\bar{n}}{}^{(m+1)})(b_{1\bar{n}} + b_{2\bar{n}})^m, \\ &= \sum_{r_1 \ldots r_m = 1}^{n} a_{\bar{n} r_1 \ldots r_m}(b_1 + b_2)_{r_1}(b_1 + b_2)_{r_2} \cdots (b_1 + b_2)_{r_m}\end{aligned} \tag{5}$$

where the subscripts outside the brackets are the second, those inside the first subscripts to be ascribed the b's.

Let $m = q_1{}^{a_1} q_2{}^{a_2} \cdots q_\sigma{}^{a_\sigma}$, where each of the q's is a prime. If, as customary, we define the special mth roots of unity to be those mth roots of unity which are not also nth roots of unity, where $n < m$, then the following propositions are well known, or may be readily deduced:

I. There are $m\left(1 - \dfrac{1}{q_1}\right)\left(1 - \dfrac{1}{q_2}\right) \cdots \left(1 - \dfrac{1}{q_\sigma}\right)$ special mth roots of unity.

II. If ω is any special mth root of unity, then ω^2 when τ is prime to m, is likewise a special mth root and all the special mth roots are of this form.

III. If ω is a special mth root, then all the roots are given by the series:

$$\omega^0 \omega^1 \omega^2 \omega^3 \cdots \omega^{m-1}.$$

IV. The sum of all the mth roots is zero, whence it follows immediately that if ω is a special mth root:

$$\sum_{\mu=0}^{m-1} \omega^\mu = 0,$$

V. More generally, if ϵ is any mth root of unity then:

$$\sum_{\kappa=1}^{m} \epsilon^\kappa = 0.$$

VI. It therefore follows immediately that, if ω is any mth special root, then:

$$\sum_{\kappa=1}^{m} (\omega^q)^\kappa = 0,$$

for all values of $q < m$.

VI. The sum of all the special mth roots is 0.

VII. If ω_μ be any special μth root, then:

$$\sum_{\kappa=1}^{\mu\rho} (\omega_\mu{}^q)^\kappa = 0, \tag{6}$$

for all values of q, prime to μ.

We shall now apply these properties of the special mth roots to the problem in hand. Let ω be such a special mth root of unity. Consider the product:

$$(a_{\bar{n}}{}^{m+1})(b_{1\bar{n}} + b_{2\bar{n}}\omega^\kappa)^{1n} = \sum_{r_1 \ldots r_m=1}^{n} a_{\bar{n}r_1\ldots r_m}(b_1 + \omega^\kappa b_2)_{r_1} \cdots (b_1 + \omega^\kappa b_2)_{r_m}. \tag{7}$$

In this product any term which involves b_2 q times, and consequently b_1 $(m - q)$ times, has the factor $\omega^{\kappa q}$, in addition to a vectorial factor which we may reserve for future consideration. If then we form all the distinct products obtained by ascribing to k all values from 1 to m, and take their sum, then in the latter such a term as above indicated will be multiplied by the factor:

$$\sum_{\kappa=1}^{m} (\omega^{\kappa q})$$

which in virtue of the properties of the special mth roots of unity, as above developed, is zero. Accordingly this sum simply reduces to the sum of the vector functions of $b_{1\bar{n}}$ and $b_{2\bar{n}}$, multiplied by m. That is to say:

$$\begin{aligned}\sum_{\kappa=1}^{m} (a_n{}^{(m+1)})(b_{1\bar{n}} + \omega^\kappa b_{2n}) &= m\{(a_{\bar{n}}{}^{(m+1)})[(b_{1\bar{n}}) + (b_{2\bar{n}})]^m\},\\ &= m\{\varphi_m(b_{1\bar{n}}) + \varphi_m(b_{2\bar{n}})\},\\ &= \sum_{\kappa=1}^{m} (a_{\bar{n}}{}^{(m+1)})(\omega^\kappa b_{1\bar{n}} + b_{2\bar{n}}),\end{aligned} \tag{8}$$

where the latter relation follows from considerations of symmetry, and requires evidently:

$$\varphi_m(\omega^\kappa b_{1\bar{n}} + b_{2\bar{n}}) = \varphi_m(b_{1\bar{n}} + \omega^\kappa b_{2\bar{n}}).$$

This latter condition is satisfied, for:

$$\begin{aligned}\varphi_m\{\omega^{m-q}(b_{1\bar{n}} + \omega^q b_{2\bar{n}})\} = \varphi_m(\omega_1{}^{m-q} b_{1\bar{n}} + b_{2\bar{n}}) &= \omega^{m(m-q)} \cdot \varphi_m(b_{1\bar{n}} + \omega_1{}^q b_{2\bar{n}})\\ &= \varphi_m(b_{1\bar{n}} + b_{2\bar{n}}\omega^q).\end{aligned}$$

By a repetition of the procedure we obtain:

$$\begin{aligned}\sum_{\lambda_2=1}^{m}\sum_{\lambda_1=1}^{m} \varphi_m(b_{1\bar{n}} + \omega^{\lambda_1}b_{2\bar{n}} + \omega^{\lambda_2}b_{3\bar{n}}) &= m \cdot \sum_{\lambda_1=1}^{m} (b_{1\bar{n}} + \omega^{\lambda_1}b_{2\bar{n}}) + m^2\varphi_m(b_{3\bar{n}}),\\ &= m^2\{\varphi_m(b_{1\bar{n}}) + \varphi_m(b_{2\bar{n}}) + \varphi_m(b_{3\bar{n}})\}.\end{aligned} \tag{9}$$

It is evident that the general expression is obtained by successive repetitions of the above process, in the form

$$\sum_{\lambda_p=1}^{m} \cdots \sum_{\lambda_2=1}^{m} \varphi_m\left(b_{1\bar{n}} + \sum_{s=1}^{p} \omega^{\lambda_2} b_{s\bar{n}} \right) = m^{p-1} \cdot \sum_{s=1}^{p} \varphi_m(b_{s\bar{n}}),$$

representing an identical relation between two sums of vector functions of the same order and degree.

It may be pointed out that these are not the only identities which may be obtained by an analogous process.

6. *First Addition Theorem in Case $m = 2$*

As a special case of the above theorem, when $m = 2$, we have since then

$$\omega = -1, \qquad \omega^2 = 1,$$

the following:

$$\varphi_2(b_{1\bar{n}} + b_{2\bar{n}}) + \varphi_2(b_{1\bar{n}} - b_{2\bar{n}}) = 2 \cdot \{\varphi_2(b_{1\bar{n}}) + \varphi_2(b_{2\bar{n}})\}.$$

Now on each side of this equation occurs a sum of two vectors. Consequently if this is regarded as giving an identical expression for the sum of the vector functions of the vectors $(b_{1\bar{n}})$, $(b_{2\bar{n}})$, on reading from right to left, then the application of this same theorem to the sum of the vector functions of $(b_{1\bar{n}} + b_{2\bar{n}})$, $(b_{1\bar{n}} - b_{2\bar{n}})$, on the left must give an expression reducible ultimately to that occurring on the right of the above equation. That in this simple instance this is as a matter of fact the case, is obvious. For we have:

$$(b_{1\bar{n}} + b_{2\bar{n}}) + (b_{1\bar{n}} - b_{2\bar{n}}) = 2b_{1\bar{n}},$$

$$(b_{1\bar{n}} + b_{2\bar{n}}) - (b_{1\bar{n}} - b_{2\bar{n}}) = 2b_{2\bar{n}},$$

and therefore:

$$\begin{aligned} \varphi_2(b_{1\bar{n}} + b_{2\bar{n}}) + \varphi_2(b_{1\bar{n}} - b_{2\bar{n}}) &= \tfrac{1}{2}\{\varphi_2(2b_{1\bar{n}}) + \varphi_2(2b_{2\bar{n}})\}, \\ &= 2\{\varphi_2(b_{1\bar{n}}) + \varphi_2(b_{2\bar{n}})\}, \end{aligned}$$

which is the same result as above obtained.

Consider similarly the more general case of this theorem when $m = 2$:

$$\begin{aligned} \varphi_2(b_{1\bar{n}} + b_{2\bar{n}} + b_{3\bar{n}}) + \varphi_2(b_{1\bar{n}} + b_{2\bar{n}} - b_{1\bar{n}}) + \varphi_2(b_{1\bar{n}} - b_{2\bar{n}} + b_{3\bar{n}}) \\ + \varphi_2(b_{1\bar{n}} - b_{2\bar{n}} - b_{3\bar{n}}) = 2^2 \cdot \{\varphi_2(b_{1\bar{n}}) + \varphi_2(b_{2\bar{n}}) + \varphi_2(b_{3\bar{n}})\}. \end{aligned} \tag{10}$$

If we consider this as giving an identity for the vector function sum on the right, then conversely the application of this addition theorem to the sum on the left into which enter the vectors:

$$(b_{1\bar{n}} + b_{2\bar{n}} + b_{3\bar{n}}), \quad (b_{1\bar{n}} + b_{2\bar{n}} - b_{3\bar{n}}), \quad (b_{1\bar{n}} - b_{2\bar{n}} + b_{3\bar{n}}), \quad (b_{1\bar{n}} - b_{2\bar{n}} - b_{3\bar{n}})$$

an expression must be obtained which ultimately reduces to the expression on the right. The reader may readily convince himself that such is the case, though the formal proof is too tedious to be given here.

Finally consider the most general case of the theorem for $m = 2$:

$$\Sigma_2 \varphi_2(\Sigma_1 b_{s\bar{n}}) = 2^{p-1} \cdot \sum_{s=1}^{p} \varphi_2(b_{s\bar{n}}), \tag{11}$$

where the first summation Σ_1, is to be extended over the b's, ascribing b_1 the positive sign and to the remaining b's either the $+$ or $-$ sign, and then extending the summation Σ_2 over all such expressions without repetition, obtained by taking all possible combinations of sign. There are evidently 2^{p-1} such terms.

Now those relations which have been indicated in the preceding simple cases must also hold here. That is to say, the application of the general law obtained by reading the equation from left to right or right to left must ultimately yield the same expression. That such is the case is not self-evident, in fact it gives rise to a theorem, which however does not appear to be of sufficient interest to state explicitly.

7. *Transformation of Matrix Expression*

In the preceding we have based the definition of the vector function operator φ_m on a matrix expression $(a_{\bar{n}}{}^{(m+1)})$. We shall now transform latter into an expression which is the generalization of the dyadic expression of Gibbs.

Evidently

$$\varphi_m(b_{\bar{n}}) = (a_{\bar{n}}{}^{(m+1)})(b_{1\bar{n}})^m(\boldsymbol{\varepsilon}_{\bar{n}}) = \sum_{p_1=1}^{\omega} \boldsymbol{\varepsilon}_{p_1} \mathbf{a}_{p_1} \cdot \boldsymbol{\beta},$$

where we assume that the matrix $(a_n{}^{(m+1)})$ represents the vector function φ_m with respect to the set of independent vectors $(\epsilon_{\bar{n}})$, and the vector $(\epsilon_{\bar{n}})(b_{\bar{n}})$ is written $\boldsymbol{\beta}$. Further in this equation we have written:

$$\begin{aligned} \mathbf{a}_{p_1} &= (a_{p_1\bar{n}\ldots(m)\ldots\bar{n}})\overset{m-1}{(b_{\bar{n}})}(\boldsymbol{\varepsilon}_{\bar{n}}) = \left(\sum_{p_2=1}^{\omega} \boldsymbol{\varepsilon}_{p_2} a_{p_1p_2\bar{n}(m-1)\cdots\bar{n}}\right)(b_{\bar{n}})^{m-1}, \\ &= \left(\sum_{p_2=1}^{\omega} \boldsymbol{\varepsilon}\mathbf{a}_{p_1p_2}\right) \cdot \boldsymbol{\beta}, \end{aligned}$$

and

$$\begin{aligned} \mathbf{a}_{p_1p_2} &= (a_{p_1p_2\bar{n}(m-1)\ldots\bar{n}})\overset{m-2}{(b_{\bar{n}})}(\boldsymbol{\varepsilon}_{\bar{n}}), \\ &= \left(\sum_{p_3=1}^{\omega} \boldsymbol{\varepsilon}\mathbf{a}_{p_1p_2p_3}\right) \cdot \boldsymbol{\beta}, \end{aligned} \tag{12}$$

and

$$\begin{aligned} \mathbf{a}_{p_1p_2p_3} &= (a_{p_1p_2p_3\bar{n}(m-2)\ldots\bar{n}})\overset{m-3}{(b_{\bar{n}})}(\boldsymbol{\varepsilon}_{\bar{n}}), \\ &= \left(\sum_{p_4=1}^{\omega} \boldsymbol{\varepsilon}_{p_4}\mathbf{a}_{p_1\ldots p_4}\right) \cdot \boldsymbol{\beta}, \end{aligned}$$

etc., etc.

Finally:

$$\begin{aligned} \mathbf{a}_{v_1\ldots p_{m-1}} &= (a_{p_1\ldots p_{m-1}\bar{n}\bar{n}})(b_{\bar{n}})(\epsilon_{\bar{n}}), \\ &= \left(\sum_{p_m=1}^{\omega} \epsilon_{p_m}\mathbf{a}_{p_1\ldots p_m}\right) \cdot \boldsymbol{\beta}, \end{aligned} \tag{13}$$

and

$$\mathbf{a}_{p_1 \dots p_m} = \left(\sum_{p_{m+1}=1}^{\omega} \boldsymbol{\varepsilon} a_{p_1 \dots p_{m+1}} \right).$$

Consequently we obtain:

$$(a_{\bar{n}}{}^{(m+1)})(b_{\bar{n}})^{m}(\boldsymbol{\varepsilon}_{\bar{n}}) = \sum_{p_1=1}^{\omega} \boldsymbol{\varepsilon}_{p_1} \underset{1}{\Bigg\{} \sum_{p_2=1}^{\omega} \boldsymbol{\varepsilon}_{p_2} \underset{2}{\Bigg\{} \sum_{p_3=1}^{\omega} \boldsymbol{\varepsilon}_{p_3}$$
$$\times \underset{3}{\Big\{} \cdots \{ \boldsymbol{\varepsilon}_{p_{m+1}} a_{p_1 \dots p_{m+1}} \}_m \cdot \boldsymbol{\beta} \}_{m-1} \cdot \boldsymbol{\beta} \cdots \Big\} \cdot \boldsymbol{\beta} \Bigg\} \qquad (14)$$

Consider now the special case that each of the forms

$$(a_{r_1 \bar{n} \dots (m) \dots \bar{n}})(b_{\bar{n}})^m, \qquad p_1 = 1 \cdots n,$$

is resolvable into linear factors, so that we have:

$$(a_{p_1 \bar{n} \dots (m) \dots \bar{n}})(b_{\bar{n}})^m = (c_{p_1 1 \bar{n}})(b_{\bar{n}})(c_{p_1 2 \bar{n}})(b_{\bar{n}}) \cdots (c_{p_1 m \bar{n}})(b_{\bar{n}}),$$
$$= (\boldsymbol{\gamma}_{p_1 1} \cdot \boldsymbol{\beta})(\boldsymbol{\gamma}_{p_1 2} \cdot \boldsymbol{\beta}) \cdots (\boldsymbol{\gamma}_{p_1 m} \cdot \boldsymbol{\beta}), \qquad (15)$$

where

$$\boldsymbol{\gamma}_{p_1 r} = (c_{p_1 r \bar{n}})(\boldsymbol{\varepsilon}_{\bar{n}}).$$

Then the above expression reduces to the simpler form:

$$\varphi_m(\boldsymbol{\beta}) = (\boldsymbol{\varepsilon}_1 \boldsymbol{\gamma}_{11} \boldsymbol{\gamma}_{12} \boldsymbol{\gamma}_{13} \cdots \boldsymbol{\gamma}_{1m} + \cdots + \boldsymbol{\varepsilon}_n \boldsymbol{\gamma}_{n_1} \boldsymbol{\gamma}_{n_2} \boldsymbol{\gamma}_{n_3} \cdots \boldsymbol{\gamma}_{mn}) \cdot (\boldsymbol{\beta}). \qquad (16)$$

This is the direct generalization of Gibbs' Dyadic, and will be termed the m-adic.

8. *Derivation of Alternative Expressions for Vector Function Operator*

In the above we showed that a vector function which was represented by a matrix of the $(m + 1)$th class was as a matter of fact a homogeneous vector function of the mth degree. In this section we shall show conversely that every homogeneous vector function may be expressed as such a matrix.

Let Φ be a vector function operator operating distributively, but not necessarily commutatively, so that we have:

$$\Phi \cdot (\mathbf{a}_1 + \mathbf{a}_2 + \cdots + \mathbf{a}_n) = \Phi(\mathbf{a}_1) + \Phi(\mathbf{a}_2) + \cdots + \Phi(\mathbf{a}_n),$$

each summand being again a vector operator. Whence:

$$\Phi \cdot (\mathbf{a}_1 + \mathbf{a}_2 + \mathbf{a}_3 + \cdots + \mathbf{a}_n)^m = \Phi \cdot [\mathbf{a}_1 + \mathbf{a}_2 + \cdots + \mathbf{a}_n]^m,$$

symbolically, where we assume the multinomial on the right to be expanded, the order of the $\mathbf{a}$'s being assumed as retained in every case, and then each term in that form operated on by Φ. By making use of the notion of a gap expression, due to Grassmann, we may without loss of generality, replace the operator Φ by an operator Ψ, which is to be assumed as operating commutatively.

Consider an expression into which enter p symbols

$$l_1 l_2 l_3 \cdots l_p,$$

each in a perfectly distinct manner, and which we shall denote $L \cdot (l_1 l_2 l_3 \cdots l_p)$.

Thus of course any algebraic expression into which enter both stated quantities or numerical coefficients and general symbols, is such a gap expression. The number of gaps in such an expression is that number of general symbols which it is necessary to define before the expression for the purpose in hand, is determined. Thus the number may vary, as in the following case of a polynomial in a set of variables. If the form of the polynomial alone be in question, the gaps are the constant coefficients, the variables being left undefined. If however the value of the polynomial be considered, then in addition it is necessary to define the values of the variables.

Then we define as the product of L. and a set of p values (or at least domains of values), that value or expression which is obtained by replcaing all the gaps $l_1 l_2 l_3 \cdots l_p$, by the given values in all possible distinct combinations without repetition, and dividing the sum of the results by their number. It is then evident that the order of the operands is immaterial, that is to say, the operation is commutative.

Replace now $\Sigma\Phi\cdot(\beta_{r_1}^{a_1'}\beta_{r_2}^{a_2'}\cdots\beta_{r_s}^{a_s'})$ (17), where the indices $a_1'a_2'a_3' \cdots a_s'$, are to be considered as constant for each sum, and the summation is to be extended over all possible sequences of the β's, by the gap product operator Ψ. Operating on the symbolic product $(\beta^{a_1}\beta^{a_2}\cdots\beta_n^{a_n})$ in which now the factors may be considered as commutative, multiplied of course by the multinomial coefficient

$$\frac{m!}{a!\,a!\cdots a!}$$

Then we may write for above expression:

$$\Psi\cdot(\beta_1+\beta_2+\beta_3\cdots+\beta_n)^m = \Psi\cdot[\beta_1+\beta_2+\beta_3\cdots+\beta_n]^m,$$

where the expression on the right is to be assumed as expanded by the multinomial theorem, and then each term operated on by Ψ. It then follows from the property characteristic of the homogeneous vector operator that

$$\Psi\cdot(c_1\mathbf{a}_1+c_2\mathbf{a}_2+c_3\mathbf{a}_3+\cdots+c_n\mathbf{a}_n)^m = \Psi\cdot[c_1\mathbf{a}_1+c_2\mathbf{a}_2+\cdots+c_n\mathbf{a}_n]^m, \quad (18)$$
$$= (\boldsymbol{A}_1\boldsymbol{A}_2\boldsymbol{A}_3\cdots\boldsymbol{A}_\omega \between c_{\bar{n}})^m,$$

in the ordinary notation for an n-ary m-ic. Here the coefficients $\boldsymbol{A}_{\omega_1\omega_2\cdots\omega_n}$, (18$a$) etc., are the products $\Psi\cdot(\beta_1^{\omega_1}\beta_2^{\omega_2}\cdots\beta_n^{\omega_n})$ (19) as above defined. Now when we write for the above form the matrix representation,

$$(\boldsymbol{A}_{\bar{n}}^{(m)})(c_{\bar{n}})^m,$$

in which the coefficients of the first matrix are vectors (or, as we also consider them, simply extensive magnitudes), and which is to be considered as symmetric, we have obtained an expression precisely equivalent to that which we have hitherto made the basis of our presentation. Thus if we assume, as may be done without loss of generality, that the matrix expression referred to the set of vectors

$(\mathfrak{a}_n)$ be $(a_{\bar{n}}{}^{(m+1)})$, and symmetric with respect to all the subscripts except the first, then:

$$(a_{\bar{n}(\omega_1)\cdots(\omega_n)})(\mathfrak{a}_{\bar{n}}) = \boldsymbol{A}_{\omega_1\cdots\omega_n}, \text{ etc.,} \tag{20}$$

where $a_{\bar{n}(\omega_1)\cdots(\omega_n)}$ signifies a term in which ω_1 of the subscripts are equal to 1, ω_2 equal to 2, etc., all terms of this common form being by assumption equal.

9. *Products of Extensive Magnitudes*

We may also obtain the above results by considering products of extensive magnitudes, as follows. Write

$$\text{(21)} \quad \begin{aligned} \boldsymbol{A} &= (a_{\bar{n}}{}^{(m+1)}) = \sum_{p=1}^{\omega} a_{p_1 p_2 \cdots p_{m+1}} \boldsymbol{\varepsilon}_{1 p_1} \boldsymbol{\varepsilon}_{2 p_2} \cdots \boldsymbol{\varepsilon}_{m+1 p_{m+1}} \\ \boldsymbol{B} &= (b_{\bar{n}}) = \sum_{p_1=1}^{\omega} b_{p_1} \Big| \boldsymbol{\varepsilon}_{1 p_1} = \sum_{p_2=1}^{\omega} b_{p_2} \Big| \boldsymbol{\varepsilon}_{2 p_2} = \cdots \end{aligned}$$

where $|\boldsymbol{\varepsilon}_p$ denotes as usual the complement of $\boldsymbol{\varepsilon}_p$ and accordingly $\boldsymbol{\varepsilon}_r | \boldsymbol{\varepsilon}_s = 1$, or $= 0$, according as $r \neq s$, or $r = s$. Then evidently the product is of the same form as the products of the corresponding matrices, as above defined.

Suppose that we multiply $\boldsymbol{A}$ by $\boldsymbol{B}$ μ times, where the class of $\boldsymbol{A}$ (which is equal, by definition, to the class of the corresponding matrix) is $\geq \mu$. Then we have for

$$\boldsymbol{A} \cdot \boldsymbol{B}^{\mu}$$

an algebraic expression, or homogeneous form, of the μth degree in n variables, whose coefficients are extensive magnitudes of class $m - \mu + 1$. We may treat this form as equivalent to a system of forms with extensive coefficients of lower class, in various ways, according to the requirements of the problem. The spirit of the method consists then in this, that a theorem which holds for any form, or system of forms, with coefficients which are extensive magnitudes of any class, may be readily, with certain limitations, which will become obvious in the sequel, generalized for all such forms. By this means the invariant relations are exhibited most naturally. Before investigating this side of the subject it is however necessary to develop to some extent the theory of determinants of the pth class.

10. *Transformation of Matrix Expression*

Suppose that with reference to a set of n independent vectors $(\mathfrak{a}_{\bar{n}})$, the matrix of the vector function operator be $(a_{\bar{n}}{}^{(m+1)})$. It is required to determine the transformation which this matrix undergoes when we pass to a second set of independent vectors $(\boldsymbol{\beta}_{\bar{n}})$, related to the former by the relation:

$$(\mathfrak{a}_{\bar{n}}) = (d_{\bar{n}\bar{n}})^{-1}(\boldsymbol{\beta}_{\bar{n}}). \tag{22}$$

Evidently

$$\text{(23)} \quad \begin{aligned} \Phi \cdot (\boldsymbol{K}) &= (\mathfrak{a}_{\bar{n}})(a_{\bar{n}}{}^{(m+1)})(k_{\bar{n}})^m, \\ &= (\boldsymbol{\beta}_{\bar{n}})(d_{\bar{n}\bar{n}})^{-1} : (a_{\bar{n}}{}^{(m+1)}) : (d_{\bar{n}\bar{n}})(k_{\bar{n}}')^m, \\ &= (\boldsymbol{\beta}_{\bar{n}})(b_{\bar{n}}{}^{(m+1)})(k_{\bar{n}}')^m. \end{aligned}$$

Consequently

$$(b_{\bar{n}}^{(m+1)}) = (d_{\bar{n}\bar{n}}')^{-1}(a_{\bar{n}}^{(m+1)})(d_{\bar{n}\bar{n}})^m,$$

$$(24) \qquad = (D_{\bar{n}\bar{n}}')(a_{\bar{n}}^{(m+1)})(d_{\bar{n}\bar{n}})^m,$$

$$= \left(\sum_{p=1}^{\omega} a_{p_{m+1}\dots p_1} d_{p_1\bar{n}} d_{p_2\bar{n}} \cdots d_{p_m\bar{n}} D_{p_{m+1}\bar{n}}\right)$$

The result is evidently again a matrix of class $(m + 1)$, in which the second suffixes of the d's and of D are the determining suffixes.

II. DETERMINANTS OF THE P–TH CLASS

1. *Introductory*

Let $\boldsymbol{e}_{1\bar{n}}\boldsymbol{e}_{2\bar{n}}\boldsymbol{e}_{3\bar{n}} \cdots \boldsymbol{e}_{p-1\bar{n}}$, be $p - 1$ unrelated sets of independent extensive magnitudes, or alternate numbers, which accordingly, for all values of the first subscripts satisfy the relations:

$$[\boldsymbol{e}_{r\lambda_1}\boldsymbol{e}_{r\lambda_2}\boldsymbol{e}_{r\lambda_3} \cdots \boldsymbol{e}_{r\lambda_4}] = 0, \text{ when an element is repeated,}$$

$$= \pm 1, \text{ when no element is repeated,}$$

the $+$ or $-$ sign being ascribed according as the number of inversions of the second subscripts is even or odd.

Then an extensive magnitude of the $(p - 1)$th class and nth order is defined as the product:

$$\boldsymbol{A} = (a_{\bar{n}}^{(m+1)})(\boldsymbol{e}_{1\bar{n}})(\boldsymbol{e}_{2\bar{n}}) \cdots (\boldsymbol{e}_{p-1\bar{n}}).$$

The product of n such magnitudes is evidently a scalar, and will be termed the determinant of the matrix $(\boldsymbol{A}_{\bar{n}}) = (a_{\bar{n}\bar{n}\dots(p-1)\dots\bar{n}})$, determined by these magnitudes, with respect to the first suffix. Thus

$$(25) \quad |a_{\bar{n}\dots(p)\dots\bar{n}}| = [\boldsymbol{A}_1\boldsymbol{A}_2 \cdots \boldsymbol{A}_n] = \prod_{q=1}^{\omega}\left(\sum_{\kappa=1}^{n} a_{q\kappa_1\dots\kappa_{p-1}}\boldsymbol{e}_{1\kappa_1}\boldsymbol{e}_{2\kappa_2} \cdots \boldsymbol{e}_{p-1\kappa_{p-1}}\right).$$

In a similar manner may be defined the determinants of the matrix with respect to the remaining suffixes. Each of these determinants will consist of the algebraic sum of products of the form:

$$\pm\, a_{\alpha_1\dots\alpha_p} a_{\beta_1\dots\beta_p} \cdots a_{\nu_1\dots\nu_p},$$

there being in each product n factors, supposed ordered according to the suffix of reference, in this case the first, and every value from 1 to n being represented in case of each suffix, once and once only. The $+$ or $-$ sign is to be taken in the case of any given product, according as the number of inversions of the suffixes when, as stated, the suffixes of reference are ordered naturally, is even or odd. The summation is then to be extended over all possible products of this type, without repetition.

It is then evident that the products which enter into the expressions of any two distinct determinants are, except possibly for sign, the same in each case. We now propose to investigate the change of sign which any particular term undergoes when we pass from a determinant with respect to any given suffix to that with respect to any other.

Let the first determinant be that with respect to the σth suffix, the second that with respect to the τth. The sign of any particular term in the former case will be determined by the number of inversions by which each of the remaining suffixes is affected, when the σth suffixes are ordered naturally; while in the latter case it will be determined by the corresponding sum of inversions when the τth suffixes are ordered naturally.

Denote former sum thus:

$$I_\sigma = q_1 + q_2 + q_3 + \cdots + q_{\sigma-1} + q_{\sigma+1} + \cdots + q_\tau + \cdots + q_p.$$

Now if $q_1'q_2'$, etc., be the corresponding number of inversions (increased or decreased by an even number, which does not affect the present question, which is simply to determine whether I is even or odd) in for the same term, in the case of the second determinant, i.e. when the suffixes are ordered naturally, then:

$$q_r' = q_r + q_\tau, \text{ etc.},$$

$$q_\sigma' = q_\tau.$$

Consequently $I_\tau = I_\sigma + (p - 2)q_\tau$, and therefore the sign of this term in the two suffixes differs by

$$(-1)^{pq_\tau} = (-1)^{pq_\sigma}.$$

Whence:

Theorem.—Determinants of even class are independent of the suffix of reference. In the case of determinants of odd class, the determinants are all distinct, the only term which has the same sign for all such determinants being the principal term $a_{1\ldots1}a_{2\ldots2}a_{3\ldots3}\cdots a_{n\ldots n}$.

From these facts alone, it will be inferred that the determinants of the matrices of forms of even degree, which of course are determinants of even class, will play a much more important rôle than the determinants of forms of odd degree. We shall see indeed that they are invariants of these forms. On the other hand, for systems of forms, whose matrix will have one of the suffixes distinguished from the remainder, determinants of odd class will be significant, and will indeed also be found to have the invariant property. The theory of forms of even degree will, from this standpoint, be seen to differ essentially from that of forms of odd degree. This is in harmony with the results of other methods of investigation. However it will be seen later that the theory of forms of odd degree can also be treated by this method. It will however first be necessary to develop the theory of determinants of even class, which we now proceed to do.

We shall first determine certain algebraic properties of these determinants, following immediately from their definition in terms of alternate numbers, as already given. Then we shall proceed to the consideration of the multiplication of determinants, and the relation which exists between the determinant of the product of matrices and the product of the determinants of the separate factors. From this will flow readily the invariant property of determinants of even forms.

2. *Fundamental Properties of Determinants of Even Class*

We shall set up the following:

Definition.—The matrix of class $(p - 1)$ obtained by ascribing to the rth suffixa certain value, π_1 say, and ascribing to the remaining suffixes, as before the values 1 to n, will be termed a first face of the matrix, and denoted $(r_1|^{\pi_1})$.

Similarly a first face of this matrix will be termed a second face of the original matrix, and so forth.

Theorem.—If in the case of matrices of even class, the significance of two faces with respect to the same suffix be interchanged, then the absolute value of the determinant is unchanged, but it changes sign.

By this is simply meant, that if we assume the determinant as expressed in terms of the general symbols $a_{\lambda_1 \ldots \lambda_p}$, then we shall obtain two quantities when we replace these symbols by definite values, first so that the symbols of a pair of faces with respect to the same suffix assume a given set of values, and secondly when these faces assume these same values, but in the reverse order. Then the theorem states that the absolute values of these determinants will be the same in each case, but that they will have opposite signs. The theorem is now evident from this explicit statement. For such an interchange is equivalent to an interchange of two alternate numbers in each product, which effects a change of sign. Thus interchange of the values to be ascribed to the corresponding letters of the faces $(r|^{\pi_1})$ $(r|^{\pi_2})$, is equivalent to the interchange of the alternate numbers $\mathbf{e}_{r\pi_1}$ and $\mathbf{e}^{r\pi_2}$ throughout, which evidently changes the sign of every term, provided the quantities thus attributed to the letters are extensive magnitudes of even class (under this term of course being included those of zero class, that is, ordinary numbers.) The permutation of such numbers does not affect the sign of the product.

If the face be one with respect to the suffix of reference, then the interchange of the values of two such faces, in the case of determinants of even class evidently also effects a change of sign, as may be inferred either from the equivalence of the different determinants, or from the fact that interchange of two such faces is equivalent to the interchange of $p - 1$ sets of alternate numbers, which effects a change in sign.

When p is odd, however, the latter argument shows that in this case no change in sign is effected. Consequently we have:

Theorem.—In the case of determinants of odd class, interchange of two faces which are not faces of reference, effects a change in sign, interchange of two faces of reference leaves sign unchanged.

3. *Note on Determinants of Odd Class*

It has already been pointed out that the determinants of matrices of odd class are in general all distinct. We propose to investigate the nature of the sum of all these determinants, confining ourselves in the first place to cubic determinants.

Adopt the following symbolism: For the sum of the terms, which are such that when the first suffix is taken as suffix of reference, have an even number of inversions of the second suffix, and likewise an even number of inversions of the third suffix, write $T(ree)$, and similarly for the other terms $T(reo)$, $T(roe)$, $T(roo)$. Then we have:

$$D = T(ree) - T(reo) - T(roe) + T(roo),$$

$$= T(eee) - T(eeo) - T(eoe) + T(eoo),$$

as it may also be written more symmetrically. Further,

$$D_1 + D_2 = 2T(eee) - 2T(eeo),$$

$$D_1 + D_3 = 2T(eee) - 2T(eoe),$$

and

$$D_1 + D_2 + D_3 = 3T(eee) - T(eeo) - T(eoe) - T(oee).$$

This result may be readily generalized. These sums are then completely determined by the matrix in question, without ambiguity, but it may be easily seen that they are not invariants of the corresponding forms.

4. *Transformation of Forms*

Consider the n-ary p-ic $(a_{\bar{n}}{}^{(p)})(x_{\bar{n}})$ whose matrix is $(a_{\bar{n}\ldots(p)\ldots\bar{n}})$. If for the variables $x_{\bar{n}}$ be substituted the variables $y_{\bar{n}}$, related to the former by the equation $x_{\bar{n}} = y_{\bar{n}} \cdot \tau_{\bar{n}\bar{n}}'$, then the quantic is transformed into the n-ary p-ic in these y's, given by:

$$(a_{\bar{n}\ldots(p)\ldots\bar{n}}) : (\tau_{\bar{n}\bar{n}})(y_{\bar{n}})^p,$$

whose matrix is

$$(a_{\bar{n}\ldots(p)\ldots\bar{n}}) : (\tau_{\bar{n}\bar{n}})^p = \left(\sum_{q=1}^{n} a_{q_1 q_2 \ldots q_p} \tau_{q_1 \bar{n}} \cdots \tau_{q_p \bar{n}} \right), \tag{27}$$

where the determining suffixes are now the second suffixes of the τ's.

The same relations hold of course in the case of a system of m n-ary forms, whose matrix is $(a_{\bar{m}\bar{n}\ldots(p)\ldots\bar{n}})$ and whose transformed matrix is $(a_{\bar{m}\bar{n}\ldots(p)\ldots\bar{n}}) : (\tau_{\bar{n}\bar{n}})^p$.

More generally, suppose in the above quantic that the variables $x_{\bar{n}}$ be replaced by quantics not necessarily linear, in the variables $y_{\bar{n}}$. That is, we write:

$$(x_{\bar{n}}) = (\tau_{\bar{n}}{}^{(\sigma+1)})(y_{\bar{n}})^\sigma.$$

Then the original matrix is transformed into an n-ary $(p \cdot \sigma)$ic. Its matrix is given by

$$(a_{\bar{n}\ldots(p)\ldots\bar{n}}) : (\tau_{\bar{n}\ldots(\sigma)\ldots\bar{n}})^p = \left(\sum_{q_1 \cdots q_p = 1}^{n} a_{q_1 q_2 \ldots q_p} \tau_{q_1 \bar{n}\ldots\bar{n}} \cdots \tau_{q_p \bar{n}\ldots\bar{n}} \right), \tag{28}$$

in which the last σ suffixes of the τ's are now the determining suffixes.

As before, the same relations hold if, instead of considering a single quantic, we consider a system of such quantics, the resulting expression being obtained by merely adding another subscript to the a's.

Further it may be convenient to generalize the relation between the original variables, and the substituted variables in a different manner. Thus instead of substituting for the former a single definite system, we may consider them as substituted by a set of such systems, thus:

$$(x_{\bar{n}}) = (\tau_{\bar{n}}{}^{(\sigma+\mu)})(y_{\bar{n}}),$$

which, of course is merely equivalent to considering, and grouping in a single expression, the various equations determined by a set of substitutions. Nevertheless such a treatment is on occasion valuable. The set of forms, thus obtained, is given by the expression:

$$(a_{\bar{n}\ldots\bar{n}}) : (\tau_{\bar{n}\ldots(\sigma)\bar{n}\bar{n}\ldots(\mu)\ldots\bar{n}})^{p}. \tag{30}$$

5. *Products of Determinants of Higher Class*

(1) First expression for product. Let

$$|a_{\bar{n}\ldots\bar{n}}|_1 = \prod_{r=1}^{n} \{(a_{r\bar{n}\ldots\bar{n}})(\mathbf{n}_{1\bar{n}})(\mathbf{n}_{2\bar{n}}) \cdots (\boldsymbol{\varepsilon}_{p\bar{n}}),$$

$$|b_{\bar{n}\ldots\bar{n}}|_1 = \prod_{r=1}^{n} \{(b_{r\bar{n}\ldots\bar{n}})(\mathbf{n}_{1\bar{n}})(\mathbf{n}_{2\bar{n}}) \cdots (\mathbf{n}_{\pi\bar{n}}),$$

each being the determinant of the corresponding matrix with respect to the first suffix. Then the expression:

$$|a_{\bar{n}\ldots\bar{n}}|_1 \cdot |b_{\bar{n}\ldots\bar{n}}|_1 = \sum_{r=1}^{n} \{(a_{r\bar{n}\ldots\bar{n}}b_{r\bar{n}\ldots\bar{n}})(\boldsymbol{\varepsilon}_{1\bar{n}}) \cdots (\boldsymbol{\varepsilon}_{p\bar{n}})(\mathbf{n}_{1\bar{n}}) \cdots (\mathbf{n}_{\pi\bar{n}})\} \tag{31}$$

for the product is obtained by combining corresponding pairs of factors in the two determinants into a single factor of the resulting determinant; the special selection of pairs in this case being those which are characterized by the same values of the first suffix. Any other combination of factors might also have been adopted. The different expressions, so obtained, have of course all the same value, except possibly for sign. The class of the product, expressed in this manner as a determinant, is $p + \pi + 1$. The process is evidently applicable to determinants of both odd and even class, in which respect it differs from the second expression for the product, next to be considered, which is applicable only in the case of determinants of even class in its full generality.

(2) Second expression for product.

This is the expression which will be of most value in the sequel.

Let $\mathbf{E}_n$ be a system of extensive magnitudes of odd class so that:

$$\mathbf{E}_r \cdot \mathbf{E}_s = - \mathbf{E}_s \cdot \mathbf{E}_r.$$

Further if $(\mathbf{E}_{\bar{n}}) = (b_{\bar{n}\ldots\bar{n}})(\mathbf{n}_{1\bar{n}}) \cdots (\eta_{\pi\bar{n}})$, where the $\mathbf{n}$'s are assumed to be simple extensive magnitudes, then the product of the $\mathbf{E}$'s given by

$$[\mathbf{E}_1\mathbf{E}_2 \cdots \mathbf{E}_n] = |b_{\bar{n}\ldots\bar{n}}|,$$

is a determinant of even class.

More generally, on one hand:

$$\prod_{r=1}^{\omega}\{(a_{\bar{n}\ldots\bar{n}})(\mathbf{E}_{1\bar{n}})(\boldsymbol{\varepsilon}_{2\bar{n}}) \cdots (\boldsymbol{\varepsilon}_{p\bar{n}})\} = \prod_{r=1}^{\omega}\{(a_{r\bar{n}\ldots\bar{n}}) : (b_{\bar{n}\ldots\bar{n}})(\mathbf{n}_{1n}) \cdots (\boldsymbol{\varepsilon}_{1n}) \cdots\}$$

$$= \prod_{r=1}^{\omega}\{(a_{r\bar{n}\ldots q}b_{q\bar{n}\ldots\bar{n}})(\mathbf{n}_{1\bar{n}}) \cdots (\mathbf{n}_{\pi\bar{n}})(\boldsymbol{\varepsilon}_{1\bar{n}}) \cdots (\boldsymbol{\varepsilon}_{p\bar{n}})\}, \quad (32)$$

and on other hand:

$$= \prod_{r=1}^{\omega}\{(a_{r\bar{n}\ldots\bar{n}})(\boldsymbol{\varepsilon}_{1\bar{n}})(\boldsymbol{\varepsilon}_{2\bar{n}}) \cdots (\boldsymbol{\varepsilon}_{p\bar{n}})\}[\mathbf{E}_1\mathbf{E}_2 \cdots \mathbf{E}_n],$$

$$= |a_{\bar{n}\ldots\bar{n}}| \cdot |b_{\bar{n}\ldots\bar{n}}|.$$

Consequently we are able to express the product of two determinants of classes $p + 1$ and $\pi + 1$, where one at least of the determinants is of even class, as a determinant of class $p + \pi$.

Consider now the determinant of the matrix of the transformed quantic:

$$(a_{\bar{n}\bar{n}\ldots(p)\ldots\bar{n}}) : (b_{\bar{n}\ldots(\sigma)\ldots\bar{n}})^p = \left(\sum_{q=1}^{\omega} a_{q_1 \cdot q_2 \ldots q_p} b_{q_1\bar{n}\ldots\bar{n}} b_{q^2\bar{n}\ldots\bar{n}} \cdots b_{qp\bar{n}\ldots\bar{n}}\right) \quad (34)$$

(where we shall assume the matrix in the b's to be even) taken, for simplicity, with respect to the first suffix. Then by preceding, the determinant

$$\prod_{r=1}^{\omega}\left\{\left(\sum_{r=1}^{\omega} a_{r\bar{n}\ldots\bar{n}}b_{r\,q\ldots\bar{n}}\right) : (b_{\bar{n}\ldots\bar{n}})^{p-1}\cdot(\boldsymbol{\varepsilon}_{1\bar{n}}) \cdots (\boldsymbol{\varepsilon}_{p-1\bar{n}})(\mathbf{n}_{1\bar{n}}) \cdots (\mathbf{n}_{\sigma-1\bar{n}})(\quad) \cdots\right\} \quad (35)$$

(where it may be recalled that the first suffixes of the b's are to be considered as connected with the remaining suffixes of the a's in succession) is equal to:

$$\prod_{q=1}^{\omega}\left\{\left(\sum_{r=1}^{\omega} a_{r\bar{n}\ldots\bar{n}}b_{rq\bar{n}\ldots\bar{n}}\right)(\boldsymbol{\varepsilon}_{1\bar{n}}) \cdots (\boldsymbol{\varepsilon}_{p-1\bar{n}})(\mathbf{n}_{1\bar{n}}) \cdots (\mathbf{n}_{\sigma-1\bar{n}})\right\}\cdot|b_{\bar{n}\ldots\bar{n}}|^{p-1}. \quad (36)$$

If now the first of the above determinants be even, which implies that the determinant in the a's is even, then it will be equal to the determinant of its matrix with respect to any other suffix. On this supposition above determinant will be equal to:

$$\prod_{r=1}^{\omega}\{(a_{\bar{n}r\bar{n}\ldots(p-2)\ldots\bar{n}}) : (b_{\bar{n}\ldots\bar{n}})(\boldsymbol{\varepsilon}_{2\bar{n}}) \cdots (\boldsymbol{\varepsilon}_{p-1\bar{n}})(\mathbf{n}_{1\bar{n}}) \cdots (\mathbf{n}_{\sigma-1\bar{n}})\}$$

$$= \prod_{r=1}^{\omega}\{(a_{\bar{n}r\bar{n}\ldots\bar{n}})(\boldsymbol{\varepsilon}_{1\bar{n}})(\boldsymbol{\varepsilon}_{2\bar{n}}) \cdots (\boldsymbol{\varepsilon}_{p_1\bar{n}})\}\cdot|b_{\bar{n}\ldots\bar{n}}|. \quad (37$$

Consequently the determinant of the transformed matrix, as above defined, is, under the assumption that both the original matrix and the matrix of transformation are of even class:

$$|a_{\bar{n}\ldots\bar{n}}|\cdot|b_{\bar{n}\ldots\bar{n}}|^{p}.$$

Therefore:

Theorem.—The determinant of any even p-ic is an invariant with respect to any transformation whose matrix is of even class, that is to say, whereby the original variables are substituted by quantics of odd degree.

A particular case is that of linear transformation, which is that generally employed in geometry. The advantage of the present method is that the more general transformation indicated presents substantially no additional difficulties, while the scope of the theory is greatly increased, as will, for instance, become evident when Tschirnhausen's transformation is considered. It is further immediately evident that the weight of these invariant determinants is equal to the degree of the quantic.

6. *Example of Binary Quartic*

Consider the binary quartic

$$(abcde\between x_{\bar{2}})^4 \equiv (a_{\bar{2}\bar{2}\bar{2}\bar{2}})(x_{\bar{2}})^4,$$

the former of these two expressions being the familiar one classical since Cayley, while the latter alternative is in harmony with the symbolism employed in this paper.

The value of its determinant is found to be

$$\begin{aligned}|a_{2222}| &= a_{1111}a_{2222} - a_{1112}a_{2221} + a_{1221}a_{2112}\\ &\qquad - a_{1121}a_{2212} + a_{1122}a_{2211}\\ &\qquad - a_{1211}a_{2122} + a_{1212}a_{2121}\\ &\qquad - a_{1222}a_{2111}\\ &= ae - 4bd + 3c^2.\end{aligned}$$

This, by our general theorem, is an invariant of the quartic with respect to any transformation of even class. In the particular case of linear transformation this fact, under somewhat different form, has been known for a long time, this being in fact the invariant generally designated by the symbol I. It furthermore follows from our general considerations that its weight is two.

From the above binary quartic we derive the ternary sextic by multiplying through by ζ^2. In the determinant of this sextic, all the terms which do not involve exactly 2 suffixes of value 3, are zero. Consequently in the expansion of this determinant, any summand in which all three factors are not of this description, will be zero. Now, if we denote the number of suffixes equal respectively to 1 and 2, in the first and second factors of such a summand by n_1', n_2''; n_1', n_2''; n_1', n_2''; then it is evident that the following equations must be satisfied by positive integral values of the n's:

$$n_1' + n_2' = 4, \qquad n_1' + n_1'' + n_1''' = 6, \qquad n_1'n_2''$$
$$n_1'' + n_2'' = 4, \qquad\qquad\qquad\qquad\qquad n_1'n_2'' \Big\} = 4.$$
$$n_1''' + n_2''' = 4; \qquad n_2' + n_2'' + n_2''' = 6, \qquad \underbrace{n_1'n_2''}_{6''}$$

These equations are evidently consistent, and if, for example, the first three, and one of the last two are satisfied, the remaining equation is likewise satisfied. The diagram on the right merely summarizes the relations, indicating that the sum of the n's in rows is 4, in columns 6. The following are all the possible solutions:

40	40	31	31	22
22	13	22	31	22
04	13	13	04	22.

Their interpretation in terms of the coefficients is evidently:

a	a	b	b	c
c	d	c	b	c
e	d	d	e	c

each multiplied possibly by a numerical factor, whose value is, as a matter of fact, seen from the following to have in each case the value 1/15:

$$\begin{aligned}
\left(\frac{4}{4}\right) a &= \left(\frac{6}{24}\right)(40) \quad \text{or} \quad (40) = 1/15\, a \\
\left(\frac{4}{3}\right) b &= \left(\frac{6}{23}\right)(31) \quad \text{or} \quad (31) = 1/15\, b \\
\left(\frac{4}{2}\right) c &= \left(\frac{6}{22}\right)(22) \quad \text{or} \quad (22) = 1/15\, c \\
\left(\frac{4}{1}\right) d &= \left(\frac{6}{21}\right)(13) \quad \text{or} \quad (13) = 1/15\, d \\
\left(\frac{4}{0}\right) e &= \left(\frac{6}{20}\right)(04) \quad \text{or} \quad (04) = 1/15\, e
\end{aligned} \tag{38}$$

Now it is evident that if this ternary sextic be transformed by any substitution of the variables which leaves the third variable ζ unchanged, then it will pass into a sextic of exactly the same form, bearing the same relation to the corresponding transformed quartic, as the original sextic bore to the original quartic. The modulus of transformation of the sextic is equal to that of the quartic.

Further, the determinant of the sextic being an invariant of weight six this syncopated determinant is an invariant of weight six of the quartic, provided of course that it does not vanish identically. The latter however is readily seen not to be the case, even without determining the actual numerical coefficients of the several terms in this determinant, since it involves terms, of the same form, in odd number. By the preceding it will then consist of a sum of numerical multiples of ace, ad^2, bcd, b^2e, c^3, and the complete solution then requires the determination of these numerical coefficients in the above determinant, which is to be assumed as symmetric. In the next section this problem will be attacked, in a more special case. In the meantime we shall anticipate the result sufficiently to state that the invariant so obtained is

$$ace + 2bcd - ad^2 - b^2e - c^3.$$

It is in fact the invariant which has been designated by the symbol J.

These two invariants, I, J, are the only two independent invariants of the quartic, as is known from other considerations, a fact which forms an illustration of a principle to be developed at a later stage. If we attempt to derive another determinant, say by multiplying the quartic through by $\zeta\eta$, obtaining a quarternary sextic, there results the following. In the first place each summand in this determinant will have four factors, and will be zero unless each factor does not contain one and only one subscript of value 3 and 4. Using same symbolism and diagram as before we shall have accordingly:

$$\underbrace{\left.\begin{matrix} n_1' + n_2' \\ n_1'' + n_2'' \\ n_1''' + n_2''' \\ n_1^{iv} + n_2^{iv} \end{matrix}\right\} = 4}_{6}$$

which would require $4 \cdot 4 = 2 \cdot 6$, and consequently the equations are inconsistent, and there is no summand in the determinant of this form. This is a special case of the following:

Theorem.—In order that the κ-ary π-ic, derived from the q-ary p-ic by the process above indicated of addition of auxiliary variables by multiplication, have a determinant in which all the summands are not zero, it is necessary that the relation

$$\kappa \cdot p = \pi \cdot q$$

be satisfied.

The truth of this theorem is evident from the above special cases, since their generalization involves no new principles. In the case under discussion, we have, since $p = 4$, $q = 2$, for any such derived κ-ary π-ic:

$$2 \cdot \kappa = \pi.$$

This condition is satisfied by the ternary sextic already investigated in detail, and furthermore this sextic or ternary form is the only one that satisfies it. The next quantic which also satisfies this condition is a quaternary octavic, which may be derived either by multiplication by $\zeta^3\eta$ or $\zeta^2\eta^2$.

7. *Coefficient of a Given Term in a Symmetric Determinant*

Consider any term in the expansion of a symmetric determinant of the second class and order n. It will consist of the product of n letters each with 2 subscripts. It is required to find the coefficient of such a term in the expression for the determinant.

Let 1 occur in such a factor as first suffix in combination with p_1 as second suffix. Then p_1 as first suffix in combination with p_2 as second suffix, and so on. Write down the resulting sequence as a cycle of a substitution:

$$1p_1p_2p_3 \cdots p_\sigma.$$

If this cycle does not exhaust all the numbers from 1 to n, repeat the operation as often as required to do so.

Then we say:

Theorem.—The coefficient of this term is 2^κ, where κ is the number of cycles of order ≥ 2, with either $-$ or $+$ sign, according as the number of inversions is odd or even.

(The cycle qpq is termed of order 2.)

If two terms are equal, without being identical, they can differ only in the order in which one or more of their cycles is to be taken; that is to say according as we consider the operation of substitution to be from first to second suffix or inversely. Since each cycle may be reversed independently of the remainder, the number of such equal terms is obtained by considering the number of all possible such pairs of cycles. Furthermore each of these terms has the same sign in the expansion of the determinant. The theorem consequently follows.

In the case of determinants of higher class this method is evidently inapplicable.

8. *Differential Invariants and Covariants*

(1) Hessian.—Let

$$d^2U = \left(\frac{\partial^2 U}{\partial x_{\bar{n}}\partial x_{\bar{n}}}\right)(dx_{\bar{n}})^2$$

and let the variables be transformed as follows:

$$dx_{\bar{n}} = (\tau_{\bar{n}\bar{n}})(dy_{\bar{n}}),$$

so that

$$d^2U = \left(\frac{\partial^2 U}{\partial x_{\bar{n}}\partial x_{\bar{n}}}\right) : (\tau_{\bar{n}\bar{n}})^2(dy_{\bar{n}})^2$$

$$= \left(\frac{\partial^2 U}{\partial y_{\bar{n}}\partial y_{\bar{n}}}\right)(dy_{\bar{n}})^2,$$

where

$$\left(\frac{\partial^2 U}{\partial y_{\bar{n}} \partial y_{\bar{n}}}\right) = \left(\frac{\partial^2 U}{\partial x_{\bar{n}} \partial x_{\bar{n}}}\right)(\tau_{\bar{n}\bar{n}})^2,$$

and consequently

$$\left|\frac{\partial^2 U}{\partial y_{\bar{n}} \partial y_{\bar{n}}}\right| = \left|\frac{\partial^2 U}{\partial x_{\bar{n}} \partial x_{\bar{n}}}\right| |\tau_{\bar{n}\bar{n}}|^2.$$

Therefore:

Theorem.—The Hessian of a function of n independent variables is a covariant of weight 2.

(2) Generalized Hessian.—Let

$$d^m U = \left(\frac{\partial^m U}{\partial x_{\bar{n}} \cdots \partial x_{\bar{n}}}\right)(dx_{\bar{n}}),$$

and

$$dx_{\bar{n}} \doteq (\tau_{\bar{n}\bar{n}})(dy_{\bar{n}}),$$

so that

$$d^m U = \left(\frac{\partial^m U}{\partial x_{\bar{n}} \cdots \partial x_{\bar{n}}}\right) : (\tau_{\bar{n}\bar{n}})^m (dy_{\bar{n}})^m$$

$$= \left(\frac{\partial^m U}{\partial y_{\bar{n}} \cdots \partial y_{\bar{n}}}\right) : (dy_{\bar{n}})^m,$$

and consequently:

$$\left|\frac{\partial^m U}{\partial y_{\bar{n}} \cdots \partial y_{\bar{n}}}\right| = \left|\frac{\partial^m U}{\partial x_{\bar{n}} \cdots \partial x_{\bar{n}}}\right| |\tau_{\bar{n}\bar{n}}|^m.$$

Whence:

Theorem.—General Hessian $\left|\dfrac{\partial^m U}{\partial x_{\bar{n}} \cdots \partial x_{\bar{n}}}\right|$ of a function U of n independent variables, is a covariant of weight m.

(3) Jacobian.—Let

$$(df_{\bar{n}}) = \left(\frac{\partial f_{\bar{n}}}{\partial x_{\bar{n}\downarrow}}\right)(dx_{\bar{n}}),$$

where the arrow signifies order in which subscripts are to be taken, and

$$(dx_{\bar{n}}) = (\tau_{\bar{n}\bar{n}})(dy_{\bar{n}}),$$

so that

$$(df_{\bar{n}}) = \left(\frac{\partial f_{\bar{n}}}{\partial x_{\bar{n}\downarrow}}\right)(\tau_{\bar{n}\bar{n}})(dy_{\bar{n}}) = \left(\frac{\partial f_{\bar{n}}}{\partial y_{\bar{n}\downarrow}}\right)(dy_{\bar{n}}),$$

and consequently

$$\left|\frac{\partial f_{\bar{n}}}{\partial y_{\bar{n}}}\right| = \left|\frac{\partial f_{\bar{n}}}{\partial x_{\bar{n}}}\right| \cdot |\tau_{\bar{n}\bar{n}}|.$$

Whence:

Theorem.—Jacobian $\left|\dfrac{\partial x_{\bar{n}}}{\partial f_{\bar{n}}}\right|$ of n independent functions of n independent variables, is a covariant of weight 1.

(4) Generalized Jacobian.—Type I.

Let

$$(d^{m-1}f_{\bar{n}}) = \left(\frac{\partial^{m-1}f_{\bar{n}}}{\partial x_{\bar{n}} \cdots \partial x_{\bar{n}}}\right)_{\downarrow} : (dx_{\bar{n}}),$$

$$= \left(\frac{\partial^{m-1}f}{\partial x_{\bar{n}} \cdots \partial x_{\bar{n}}}\right)_{\downarrow} : (\tau_{\bar{n}\bar{n}} \langle dy_{\bar{n}})^{m-2},$$

$$= \left(\frac{\partial^{m-1}f_{\bar{n}}}{\partial y_{\bar{n}} \cdots \partial y_{\bar{n}}}\right)_{\downarrow} : (dy)^{m-1}.$$

Accordingly

$$\left|\frac{\partial^{m-1}f_{\bar{n}}}{\partial y_{\bar{n}} \cdots \partial y_{\bar{n}}}\right| = |\tau_{\bar{n}\bar{n}}|^{m-1} \left|\frac{\partial^{m-1}f_{\bar{n}}}{\partial x_{\bar{n}} \cdots \partial x_{\bar{n}}}\right|,$$

and consequently:

Theorem.—Generalized Jacobian $\left|\frac{\partial^{m-1}f_{\bar{n}}}{\partial x_{\bar{n}} \cdots \partial x_{\bar{n}}}\right|$ is a covariant of weight $m - 1$.

(5) Type II. In a similar manner it may readily be shown that the generalized Jacobian

$$\left|\frac{\partial^{m-q}f_{\bar{n}}}{\partial x_{\bar{n}} \cdots \partial x_{\bar{n}}}\right|$$

is a covariant of weight $m - q$.

Note.—For a different treatment and for extensions along certain lines, the reader may consult E. R. Hedrick, *Annals of Mathematics*, Vol. I, p. 49. The above results were obtained independently, and I am indebted to Professor E. B Wilson for this reference.

The significance of the above results in connection with the theory which we are now investigating, that of the general vector function, lies in the fact that any vector field, or rather the corresponding vector function of the position vector $\boldsymbol{r}$, may be expressed as a series of homogeneous vector functions of $\boldsymbol{r}$, whose determinants are of the form of these differential covariants and invariants, being consequently independent of the axes of reference. We shall have occasion to treat this point at length at a later stage.

UNIVERSITY OF CALIFORNIA PUBLICATIONS
IN
MATHEMATICS
Vol. 1, No. 15, pp. 345-358 November 8, 1923

INVOLUTORY QUARTIC TRANSFORMATIONS IN SPACE OF FOUR DIMENSIONS

BY
NINA ALDERTON

§1. Cayley in his paper "On the Rational Transformation between two Spaces"[1] gives a general discussion of the quadric transformation between two planes and the cubo-cubic transformation between two spaces. The cubic transformation in space was further studied by F. R. Morris[2] who gives an analytic treatment of the cases in which the Jacobian, the sextic curve whose points go into lines by means of this transformation, breaks up into curves of lower degree, one or more of which are straight lines. A synthetic treatment of the general case of the cubic space transformation is given by D. N. Lehmer[3] in his paper "On Combinations of Involutions," and of six of the special cases by Elizabeth J. Easton.[4]

§2. A discussion of the general involutory quartic transformation in space of four dimensions has been given by P. H. Schoute[5] in an article, "La Surface de Jacobi d'un systeme lineaire d'hyperquadriques $Q^3{}_2$ dans l'espace E^4 a quatre dimensions." The writer was not acquainted with this article when work upon the present paper was begun and consequently began with a general consideration of the involutory transformation by means of four hyperquadrics. Schoute makes his transformation with respect to a pencil of hyperquadrics and defines the transform P' of the point P as being the intersection of the polar spaces of P with respect to the triple infinitude of hyperquadrics of the pencil. This is equivalent, however, to using four hyperquadrics, for four independent hyperquadrics determine the pencil. The present paper gives the discussion of the general involutory quartic transformation with respect to four hyperquadrics, as originally planned, before going on to a consideration of the cases in which the Jacobian, now a surface, breaks up into

[1] Proc. London Math. Soc., vol. 3 (1869-1873), pp. 127-180.

[2] "Classification of Involutory Cubic Space Transformations," Univ. Calif. Publ. Math., vol. 1, pp. 223-240.

[3] Am. Math. Monthly, vol. 18, no. 3 (March 1911). Also Steiner's Ges. Werke, vol. 2, p. 651.

[4] Ms., Master's Thesis, "Certain Special Cubo-cubic Space Transformations," 1917, in Univ. Calif. Library, Dept. of Mathematics.

[5] Archives du Musée Teyler, série 2, 7, 1900-01.

surfaces of lower degree including at least one plane. Although it has seemed advisable to examine the subject analytically as well as synthetically to a considerable extent, the synthetic treatment only will be presented in most cases in this paper.

§3. NOTATION

The exponent of the symbol of a surface will be used to denote the infinitude of points on the surface and the subscript to denote the degree of the surface; thus S_2^3 designates a quadric hypersurface.

§4. DEFINITIONS

1. All of the 3-spaces through a line form what we shall call an axial pencil of 3-spaces. All of the 3-spaces through a plane form a plane pencil of 3-spaces. There are ∞^2 3-spaces in an axial pencil and ∞ in a plane pencil.

2. Harmonic 3-spaces of a plane pencil are four 3-spaces which are cut by any line in four harmonic points.

3. The polar 3-space of a point P with respect to a hyperquadric is the locus of a point which is the fourth harmonic to P and the two points in which any line through P cuts the hyperquadric.

4. The simplex of reference in 4-space is a figure bounded by five 3-spaces. The five 3-spaces intersect two at a time in ten planes, three at a time in ten lines, and four at a time in five points which are the vertices of the simplex.

§5. PRELIMINARY THEOREMS

1. In a plane pencil (§4, 1) of 3-spaces, if four planes are cut by one line in four harmonic points they are cut by every line in four harmonic points.

Proof.—Upon any line p take four harmonic points. These points together with the plane of the plane-pencil determine four harmonic 3-spaces; for, cut across the plane-pencil by any 3-space through p. We then have an axial pencil of planes and we know that planes corresponding to four harmonic points of the line are four harmonic planes which are cut by any line in four harmonic points. Since this is true in any 3-space through p, we see that any line cuts the four 3-spaces corresponding to four harmonic points of the line in four harmonic points. Hence as a point P moves along p, the polar 3-spaces of P with respect to four hyperquadrics will form four projective plane-pencils. Similarly, as P moves over a plane, the polar 3-spaces of P will form four projective axial pencils.

2. There are ∞^2 lines cutting four planes in 4-space. Proof: Call the planes α_1, α_2, α_3, α_4. If we pass a 3-space through α_1, it will cut α_2, α_3, and α_4 in three lines. There will be ∞ lines in the 3-space cutting these three lines, and these lines will also cut α_1, since they lie in the same 3-space with α_1. Hence in every 3-space through α_1, there will be ∞ lines cutting α_1, α_2, α_3 and α_4. But there are ∞ 3-spaces about α_1 (§4, 1) and consequently ∞^2 lines cutting α_1, α_2, α_3 and α_4.

Case I

GENERAL TRANSFORMATION BY MEANS OF FOUR HYPERQUADRICS

§6. We may set up an involutory one-to-one correspondence between the points of 4-space by means of four arbitrarily chosen hyperquadrics. To a point P corresponds a point P', the intersection of the four polar 3-spaces of P with respect to the four hyperquadrics. Since the polar 3-spaces of P' must all pass through P and since four 3-spaces can intersect, in general, in only one point, the point P also corresponds to the point P' and we have an involutory, one-to-one correspondence.

§7. Thus, in general, to a point P will correspond a point P', but there are certain points to which correspond a whole line of points. The locus of such a point is the Jacobian. We shall show that

Theorem I. The locus of all points whose transform is a line is a surface of the tenth degree in 4-space, $J^2{}_{10}$. (2, §3). Proof: Let the four hyperquadrics be

$$\begin{aligned} A &= \Sigma a_{ij} x_i x_j = 0 \\ B &= \Sigma b_{ij} x_i x_j = 0 \\ C &= \Sigma c_{ij} x_i x_j = 0 \\ D &= \Sigma d_{ij} x_i x_j = 0 \end{aligned} \qquad \begin{bmatrix} i = 1 \ldots 5 \\ j = 1 \ldots 5 \end{bmatrix}$$

The polar 3-spaces of a point P with respect to A, B, C and D are then

$$\begin{aligned} x'_1A_1 + x'_2A_2 + x'_3A_3 + x'_4A_4 + x'_5A_5 = 0 \\ x'_1B_1 + x'_2B_2 + x'_3B_3 + x'_4B_4 + x'_5B_5 = 0 \\ x'_1C_1 + x'_2C_2 + x'_3C_3 + x'_4C_4 + x'_5C_5 = 0 \\ x'_1D_1 + x'_2D_2 + x'_3D_3 + x'_4D_4 + x'_5D_5 = 0 \end{aligned}$$

Ordinarily these four 3-spaces will intersect in a point. If, however, the equations are linearly dependent they will intersect in a line. The condition for this is that the matrix of the coefficients be of rank three; i.e. that all of the four-rowed determinants of the matrix vanish. Hence from the matrix

$$\left\| \begin{matrix} A_1 & A_2 & A_3 & A_4 & A_5 \\ B_1 & B_2 & B_3 & B_4 & B_5 \\ C_1 & C_2 & C_3 & C_4 & C_5 \\ D_1 & D_2 & D_3 & D_4 & D_5 \end{matrix} \right\|$$

we shall have five four-rowed determinants equal to zero. Each of these equations represents a quartic hypersurface. The Jacobian is the locus of points lying on all five of these hyperquartics. The hyperquartic we get by omitting the fourth column and the one we get by omitting the fifth column will intersect in a surface of degree sixteen, $S^2{}_{16}$. But they have in common the matrix formed of the first three columns which represents a surface of the sixth degree through which the other hyperquartics do not pass. Hence the Jacobian, the surface through which all five hyperquartics pass, is a $J^2{}_{10}$.

§8. *Theorem II.* The lines which are the transforms of the points of J^2_{10} form a ruled hypersurface, j^3_{15}.

Proof: If we denote by $X=0$, $Y=0$, $Z=0$, $W=0$, and $V=0$ the hyperquartics of the matrix of §7 which we get by omitting the first column, then the second, etc., we know that the equation of the Jacobian hypersurface, which is made up of the lines which are the transforms of points of J^2_{10}, may be found by equating to zero the determinant of the partial derivatives of X, Y, Z, W and V with respect to x_1, x_2, x_3, x_4 and x_5; thus

$$j=\begin{vmatrix} X_1 & X_2 & X_3 & X_4 & X_5 \\ Y_1 & Y_2 & Y_3 & Y_4 & Y_5 \\ Z_1 & Z_2 & Z_3 & Z_4 & Z_5 \\ W_1 & W_2 & W_3 & W_4 & W_5 \\ V_1 & V_2 & V_3 & V_4 & V_5 \end{vmatrix}=0$$

Each element of this determinant is of degree three in x_1, x_2, x_3, x_4 and x_5 and hence the equation represents a hypersurface of degree fifteen which we shall designate as j^3_{15}.

§9. Thus we see that the points of 4-space go by this transformation into other points with the exception of points on a J^2_{10} whose points transform into the lines of a ruled j^3_{15}.

§10. It seems at first thought as though there might be points in 4-space whose polar 3-spaces meet in planes, but this is not true in general. The condition for this would be that the matrix of the coefficients of the four polar 3-spaces of §7 be of rank two. The forty cubic hypersurfaces obtained by setting each three-rowed determinant equal to zero would all have to pass through the points which transform into planes. Taking the first three rows of the matrix,

$$\left\|\begin{matrix} A_1 & A_2 & A_3 & A_4 & A_5 \\ B_1 & B_2 & B_3 & B_4 & B_5 \\ C_1 & C_2 & C_3 & C_4 & C_5 \end{matrix}\right\|$$

the three cubic hypersurfaces represented by columns 1, 2, 3; 1, 2, 4; and 1, 2, 5 intersect in a cubic surface and a curve. The other seven cubic hypersurfaces represented by this matrix will not pass through the cubic surface but will pass through the curve. Similarly, taking the matrix

$$\left\|\begin{matrix} B_1 & B_2 & B_3 & B_4 & B_5 \\ C_1 & C_2 & C_3 & C_4 & C_5 \\ D_1 & D_2 & D_3 & D_4 & D_5 \end{matrix}\right\|$$

we find that the ten cubic hypersurfaces whose equations are the different three-rowed determinants of the matrix set equal to zero pass through another curve. Two curves do not, in general, intersect in 4-space and therefore the twenty cubic hypersurfaces so far examined have no points in common and hence the forty have none. Consequently there are no points whose transforms are planes when the hyperquadrics are unrelated.

§11. *Theorem III.* If a point P moves along a line p, its corresponding point P' moves along a quartic curve in 4-space.

Proof: As the point P moves along the line p, its polar 3-spaces with respect to the four hyperquadrics revolves about four planes forming what we shall call plane pencils of 3-spaces. These four plane pencils are projective pencils for there is a one-to-one correspondence between the points of p and the 3-spaces of the four plane pencils, and to four harmonic points of the line correspond four harmonic 3-spaces of the pencils. (2, §4, and Theorem I, §5.) The locus of the intersection of corresponding 3-spaces of the four projective plane pencils will be the transform of the line p. If we cut across the plane pencils by a 3-space, we have four projective axial pencils of planes in a 3-space. Four and only four sets of corresponding planes of these pencils meet in points, (Reye, Geometrie der Lage, vol. 2, XII, p. 93), so there are four points of the locus in every 3-space and hence the transform of the line p is a quartic curve in 4-space. The single infinitude of points of the curve corresponds to the single infinitude of points of the line p.

§12. *Theorem IV.* If the point P moves over a 3-space, the corresponding point P' moves over a quartic hypersurface.

Proof: If the point P moves over a 3-space, the polar 3-spaces of P with respect to the four hyperquadrics revolve about four points forming what we shall call four points of 3-spaces. There will be a triple infinitude of points P' corresponding to the triple infinitude of points P and hence points P' lie on a hyper-surface. In order to find the degree of this hypersurface, cut across it by a line. Transforming, the line goes into a quartic curve, as we have just seen, and the hypersurface back into the 3-space, giving off also the $j^3{}_{15}$. The hypersurface must give off the $j^3{}_{15}$ when transformed, for it contains $j^2{}_{10}$, since the 3-space of which it is the transform cuts all of the lines of $j^3{}_{15}$. The points of the hypersurface which do not transform into lines but into points will go back into the points of the 3-space on account of the involutory relation between the points under this transformation. But the quartic curve cuts the 3-space in four points. Therefore the line cuts the hypersurface in four points and it is a quartic hypersurface.

§13. *Theorem V.* If P moves over a plane, its corresponding point P' moves over an $S_6{}^2$.

Proof: The plane may be considered as the intersection of two 3-spaces, R_1 and R_2. When we transform, these two 3-spaces go into two quartic hypersurfaces which intersect in a surface $S^2{}_{16}$. But we have seen (§12) that every quartic hypersurface which is the transform of a 3-space must pass through $J^2{}_{10}$. Hence $J^2{}_{10}$ is a part of $S^2{}_{16}$ and the remaining part is an $S_6{}^2$ which is then the transform of the plane.

§14. *Theorem VI.* The multiplicity of lines of $j^3{}_{15}$ through points of $J^2{}_{10}$ and the multiplicity of points of $J^2{}_{10}$ on lines of $j^3{}_{15}$ is four.

Proof: If p' is a line which is the transform of a point P of $J^2{}_{10}$, then the polar 3-spaces of all points of p' will pass through P. Ordinarily the four polar 3-spaces of points of p' intersect in only one point and this must be the point P. Since P is always a common point of four polar 3-spaces of points of p', in order for p' to transform into a quartic curve it must go into four lines passing through P. But P is any point of J^2_{10} and hence the Jacobian must be a surface of j^3_{15} of multiplicity four. The points of p' all go into the point P except four points whose transforms are the four lines through P. Hence there are four points of J^2_{10} on every line of j^3_{15}, or the lines of j^3_{15} cut J^2_{10} in four points.

§15. Summary.—The principal facts which have been established concerning the general involutory quartic transformation in 4-space are: that lines go into quartic curves, planes into surfaces of the sixth degree, and 3-spaces into quartic hypersurfaces; also, that the locus of points whose transforms are lines is a surface of the tenth degree, and the lines which are the transforms of these points form a hypersurface of the fifteenth degree.

CASE II

ONE FUNDAMENTAL HYPERQUADRIC IS A SPACE-PAIR

§16. (a) In Case I the four hyperquadrics of the transformation were perfectly general. We shall now consider the case in which one of the hyperquadrics A is a space-pair whose 3-spaces intersect in a plane α_1. It is evident that we may still set up a one-to-one correspondence between the points of 4-space, for ordinarily to a point P will correspond a point P', the intersection of polar 3-spaces of P with respect to B, C and D and the 3-space conjugate with respect to the 3-spaces of A to that determined by P and the plane α_1.

§17. The polar 3-space with respect to A of a point P on α_1 is indeterminate and hence to such a point corresponds the line of intersection of the polar 3-spaces with respect to B, C and D. Hence α_1 is a part of J^2_{10} and

Theorem VII. The Jacobian is a plane and an S^2_9 when one of the fundamental hyperquadrics is a space-pair.

§18. *Theorem VIII*. The Jacobian hypersurface is an S^3_3 and an S^3_{12} when one of the fundamental hyperquadrics is a space-pair.

Proof: The transforms of points of α_1 are lines of j^3_{15}. As P moves over α_1, the polar 3-spaces of P revolve about three lines forming three projective axial pencils. (1, §5). Now three projective axial pencils of 3-spaces intersect in a hypersurface of the third degree, for if we cut across them by any 3-space we have three points of planes intersecting in a cubic surface. Hence j^3_{15} breaks down into a hypersurface of the third degree and one of the twelfth degree.

§19. (b) If the two 3-spaces of A coincide; i.e. if A is composed of two coincident 3-spaces we cannot set up an involutory relation between points of 4-space, for the polar 3-space of any point will be A itself and the points of 4-space will transform into those of a 3-space.

Case III

TWO FUNDAMENTAL HYPERQUADRICS ARE SPACE-PAIRS

§20. (a) Suppose A and B are the two space-pairs and α_1 and α_2, the planes of intersection of the pairs of 3-spaces of A and B respectively, have only a point in common. It is evident that we may still set up an involutory one-to-one correspondence between the points of 4-space. The planes α_1 and α_2 are part of $J^2{}_{10}$ and we have the theorem

Theorem IX. The Jacobian is composed of two planes and an $S_8{}^2$ when two of the fundamental hyperquadrics are space-pairs.

§21. The planes α_1 and α_2 both go into cubic hypersurfaces of $j^3{}_{15}$. Hence

Theorem X. The Jacobian hypersurface is composed of two $S^3{}_3$'s and an $S^3{}_9$ when two of the fundamental hyperquadrics are space-pairs.

§22. There is a plane lying on both of the cubic hypersurfaces and hence forming part of their intersection; namely the transform of the point of intersection of the two planes α_1 and α_2. This point transforms into a plane since its polar 3-spaces with respect to A and B are indeterminate, and its transform is then the intersection of its polar 3-spaces with respect to C and D.

§23. A line l cutting either α_1 or α_2 will transform into a line and a cubic curve. Suppose it cuts α_1. The line l_1 is the transform of the point where the given line l cuts α_1. To get the transform of the rest of l, allow the point P to move along l. The point P and the plane α_1 always determine the same 3-space and hence all points of l have the same polar 3-space with respect to A. The polar 3-spaces of points of l with respect to B, C and D form the three projective plane pencils of 3-spaces and these intersect in lines of a ruled $S^2{}_3$, for if we cut across them by a 3-space we have three projective axial pencils whose corresponding planes intersect in points of a cubic curve. The polar 3-space of points of l with respect to A will cut this $S_3{}^2$ in a twisted cubic. Hence l transforms into a line and a twisted cubic lying in the polar 3-space with respect to A of points of l. Similarly, a line cutting α_2 will transform into a line and a twisted cubic lying in the polar 3-space with respect to B of points of the line.

§24. A line l cutting both α_1 and α_2 will transform into two lines l_1 and l_2 and a conic C_2 lying in the plane of intersection of the two polar 3-spaces of points of l with respect to A and B.

§25. Other lines which do not transform into quartic curves are those lying on α_1 (or α_2). A line lying on α_1 (or α_2) and not passing through P_1, the intersection of α_1 and α_2, becomes an $S_3{}^2$, the locus of intersections of corresponding 3-spaces of three projective plane pencils of 3-spaces. If the line passes through this point P_1, also, the cubic surface breaks up into a plane, the transform of P_1, and an $S_2{}^2$ lying in the polar 3-space with respect to B (or A) of points of the line.

§26. The surface S_6^2 which is the transform of a plane has two lines lying on it, the transforms of the points of intersection of the plane with α_1 and α_2. If these two points coincide at P_1, the S_6^2 breaks down into a plane which is the transform of P_1 and an $S^2{}_5$. Further degeneration of the S_6^2 occurs when the plane intersects α_1 (or α_2) and both α_1 and α_2 in lines.

§27. A 3-space containing α_1 (or α_2) will transform into a ruled cubic hypersurface which is the transform of α_1 (or α_2) and the 3-space which is the polar of points of the 3-space with respect to A (or B).

(b) Now suppose A and B are so related that α_1 and α_2 intersect in a line L_3; i.e. that all of the 3-spaces of A and B pass through L_3. Then we have the theorem

Theorem XI. The Jacobian is composed of three planes and a $S^2{}_7$ when space-pairs A and B have a line in common.

Proof: The points of L_3 transform into the planes of a planed hyperquadric which are the intersections of two plane pencils of polar 3-spaces of points of L_3 with respect to C and D. Hence the plane α_1 (or α_2) will transform into the planes of a planed hyperquadric and ∞^2 lines of the 3-space which is the polar of points of α_1 (or α_2) with respect to B (or A). Hence α_1 and α_2 are still parts of $J^2{}_{10}$ and have a line lying on them whose points transform into planes. Now α_1 and α_2 determine a 3-space since they have a line in common. Points of this 3-space will have a single polar 3-space with respect to A and a single polar 3-space with respect to B and these two 3-spaces will intersect in a plane β_3 passing through L_3. The polar 3-space of points of β_3 with respect to both A and B will be the 3-space determined by α_1 and α_2, and this 3-space will cut the polar 3-spaces with respect to C and D in lines. Hence β_3 is also a part of $J^2{}_{10}$, as are the planes α_1 and α_2.

§28. *Theorem XII.* The Jacobian hypersurface is composed of a planed hyperquadric counted twice, three 3-spaces, and an S_8^3 when the four 3-spaces of A and B have a line in common.

Proof: The plane α_1 (or α_2) goes into a hyperquadric which is the transform of L_3 and the 3-space which is the polar 3-space with respect to B (or A) of points of α_1 (or α_2). The plane β_3 goes into the 3-space determined by α_1 and α_2. Hence three 3-spaces and a hyperquadric counted twice will be part of the $j^3{}_{15}$. It should be noted that a line on $J^2{}_{10}$ whose transform is counted twice is a triple line on $J^2{}_{10}$; in this case the three planes α_1, α_2, and β_3 pass through L_3.

§29. Again, lines which have a special position with respect to α_1 and α_2 will not transform as usual into quartic curves. Any line cutting L_3 will transform into a plane and a conic lying in the plane of intersection of the two polar 3-spaces of points of the line with respect to A and B. The transform of a plane cutting L_3 or passing through L_3 will be a degenerate S_6^2. The 3-space determined by α_1 and α_2 will go into the planed hyperquadric and the two 3-spaces which are the remainder of the transforms of α_1 and α_2. All points of the 3-space not on α_1 or α_2 go into points of β_3.

§30. (c) Suppose one of the 3-spaces of B (or A) passes through the plane of intersection of the 3-spaces of A (or B). If A and B are so related that one of the

3-spaces, R_3 of B, passes through α_1, then α_1 and α_2 determine R_3 and must intersect in a line L_3 as in (b). The plane β_3 will now coincide with α_1, the intersection of the conjugate 3-space of R_3 with respect to A and R_3 itself, since it is self-conjugate with respect to B. Hence

Theorem XIII. The Jacobian is composed of a single plane, a double plane, and an S_7^2 when one of the 3-spaces of B (or A) passes through α_1 (or α_2).

§31. *Theorem XIV.* The Jacobian hypersurface is composed of a planed hyperquadric counted twice, a single 3-space, a double 3-space and an S_8^3. The single 3-space is the conjugate with respect to A of R_3 and the double one is R_3 itself.

§32. (d) If α_1 and α_2 coincide in α_1, the points of 4-space go into points of α_1 and we no longer have a one-to-one correspondence.

Case IV

THREE FUNDAMENTAL HYPERQUADRICS ARE SPACE-PAIRS

§33. (a) Suppose A, B and C are space-pairs and their planes, α_1, α_2, and α_3 respectively, have only points in common.

Theorem XV. The Jacobian is composed of three planes and an S_7^2 when three of the fundamental hyperquadrics are space-pairs. The planes α_1, α_2 and α_3 which form part of the $J^2{}_{10}$ intersect two at a time in three points P_1, P_2 and P_3. These points transform into planes. There are two such points on each of the three planes.

§34. *Theorem XVI.* The Jacobian hypersurface is composed of three hypercubics and an S_6^3 when three of the fundamental hyperquadrics are space-pairs.

The three hypercubics are transforms of the planes α_1, α_2 and α_3.

§35. If α_1 and α_2 intersect in P_3, α_2 and α_3 in P_1, and α_3 and α_1 in P_2, then a line such as the one joining P_1 with any point of α_1 transforms into a plane and two lines. A plane through two of the points P_1, P_2, P_3, say P_1 and P_2, will go into two planes which are the transforms of P_1 and P_2, another plane which is the transform of the other points of the line P_1 P_2, and a cubic surface lying in the 3-space conjugate to that determined by the plane to be transformed and α_3. The plane P_1, P_2, P_3, goes into six planes.

§36. (b) Suppose two of the three planes α_1, α_2 and α_3, say α_1 and α_2, intersect in a line L_3.

Theorem XVII. The Jacobian is composed of four planes and an $S^2{}_6$ when two of the three fundamental space-pairs A, B and C have a line in common.

The four planes are the planes α_1, α_2 and α_3 and the plane β_3 which is the intersection of the two polar 3-spaces of points of the 3-space determined by α_1 and α_2 with respect to A and B.

§37. *Theorem XVIII.* The Jacobian hypersurface is composed of three 3-spaces, a planed hyperquadric counted twice, a hypercubic and an S_5^3 when two of the three fundamental space-pairs A, B, C have a line in common.

§38. (c) Suppose the three fundamental space-pairs A, B, C are so related that a 3-space R_3 of B passes through α_1. This implies that α_1 and α_2 have a line L_3 in common.

Theorem XIX. The Jacobian is composed of two planes, a double plane, and an S_6^2 when three of the fundamental hyperquadrics are space-pairs and a 3-space of one of them passes through the plane of another.

This is true since two of the four planes of Theorem XVII now coincide in α_1. (Compare Theorem XIII.)

§39. *Theorem XX.* The Jacobian hypersurface is composed of one single 3-space, one double 3-space, a planed hyperquadric counted twice, a hypercubic and an S_5^3 when three of the fundamental hyperquadrics are space-pairs and a 3-space of one of them passes through a plane of another.

§40. (d) The three fundamental space-pairs may be so related that A and B have a line L_3 in common and B and C have a line L_1 in common. In this case

Theorem XXI. The Jacobian is composed of six planes and an S_4^2 when A and B have a line in common and B and C have a line in common.

The six planes are α_1, α_2, α_3, the plane β_3 which is the intersection of the two polar 3-spaces of points of the 3-space R' determined by α_1 and α_2 with respect to A and B, the plane β_1 which is the intersection of the two polar 3-spaces of points of the 3-space R'' determined by α_2 and α_3 with respect to B and C, and the plane β_2 which is the intersection of the polar 3-space of points of R' with respect to A and the polar 3-space of points of R'' with respect to C.

§41. The lines L_3 and L_1 intersect since they both lie in α_2 and this point P which lies on each of the three planes α_1, α_2 and α_3 transforms into a 3-space. It should be noted that the planes β_1, β_2 and β_3 also pass through P and hence a point on $J^2{}_{10}$ whose transform is a 3-space of $j^3{}_{15}$ counted three times has six sheets of the surface passing through it. The 3-space into which all points of α_1 except those lying on L_3 transform is the 3-space determined by α_2 and α_3, and similarly for α_3. This may be shown by taking A, B and C as space-pairs through three planes of the simplex of reference. (4, §4.) The 3-spaces determined by α_1 and α_2 and by α_2 and α_3 are then two of the five faces of the simplex. Hence in this case

Theorem XXII. The Jacobian hypersurface is composed of one triple 3-space, four double 3-spaces and an S_4^3 when the three fundamental space-pairs A, B and C are so related that A and B have a line in common and B and C have a line in common. Two of the double 3-spaces are of course the 3-spaces determined by α_1 and α_2 and by α_2 and α_3.

§42. Any line lying in α_2 and not passing through P transforms into three planes since it cuts both L_1 and L_3. If it passes through P it transforms into a 3-space and a plane.

§43. (e) Let one of the 3-spaces R_3 of B pass through α_1 while α_2 and α_3 still have a line L_1 in common.

Theorem XXIII. The Jacobian is composed of four single planes, a double plane and an S_4^2 when one of the 3-spaces of B passes through α_1 and B and C have a line in common.

This is due to the coincidence of two of the planes of Theorem XXI, the double plane being the plane α_1.

§44. *Theorem XXIV.* The Jacobian hypersurface is composed of two triple 3-spaces, two double 3-spaces, a single 3-space and an S_4^3 when one of the 3-spaces of B passes through α_1 and B and C have a line in common.

The 3-space R_3 is one of the triple 3-spaces and the plane determined by α_2 and α_3 is the single 3-space.

§45. (f) Suppose one of the 3-spaces R_3 of B passes through α_1 and one of the 3-spaces R_5 of C passes through α_2. Then

Theorem XXV. The Jacobian is composed of two single planes, two double planes, and an S_4^2 when one of the 3-spaces of B passes through α_1 and one of the 3-spaces of C passes through α_2. The two double planes are α_1 and α_2 while α_3 is one of the single planes.

§46. *Theorem XXVI.* The Jacobian hypersurface is composed of a triple 3-space, four double 3-spaces, and an S_4^3 when one of the 3-spaces of B passes through α_1, and one of the 3-spaces of C passes through α_2.

The 3-spaces R_3 and R_5 are double 3-spaces.

§47. (g) Suppose one of the 3-spaces, R_3, of B passes through α_1 and the other, R_4, passes through α_3.

Theorem XXVII. The Jacobian is composed of two single planes, two double planes and an S_4^2 when one 3-space of B passes through α_1 and the other through α_3.

The planes α_1 and α_3 are double planes while α_2 is now a single plane.

§48. *Theorem XXVIII.* The Jacobian hypersurface is composed of a triple 3-space, four double 3-spaces, and an S_4^3 when one 3-space of B passes through α_1 and the other through α_3. R_3 and R_4 are double 3-spaces.

§49. (h) Let the planes α_1, α_2, and α_3 intersect in lines two at a time.

Theorem XXIX. The Jacobian is composed of six planes and an S_4^2 when the planes α_1, α_2 and α_3 intersect two at a time in lines.

Call the intersection of α_1 and α_2 L_3, the intersection of α_2 and α_3 L_1, and the intersection of α_3 and α_1 L_2. The planes α_1, α_2 and α_3 now lie in a 3-space R and intersect in a point P so the lines L_1, L_2 and L_3 are concurrent. The 3-space R will have conjugate 3-spaces with respect to A, B and C which will intersect two at a time in planes β_3, β_1 and β_2 which together with planes α_1, α_2 and α_3 are the six planes of the J^2_{10}.

§50. *Theorem XXX.* The Jacobian hypersurface is composed of one triple 3-space, four double 3-spaces and an S_4^3 when the planes α_1, α_2 and α_3 intersect two at a time in lines. This is true since the plane α_1 transforms into lines of β_1 and the three 3-spaces which are the transforms of L_2 and L_3, α_2 into lines of β_2 and the three 3-spaces which are the transforms of L_1 and L_3, and α_3 into lines of β_3 and the three 3-spaces which are the transforms of L_1 and L_2.

§51. The transform of points of R with respect to A, B and C will be three 3-spaces intersecting in a line L_4. Hence the planes β_1, β_2 and β_3 all transform into R, the transform of points of L_4, which is then one of the double 3-spaces of j^3_{15}. (§28.) Points of this line L_4 will transform into planes of R. Hence there are four lines, L_1, L_2, L_3 and L_4, whose transforms are 3-spaces of planes.

§52. Any line in α_1, α_2 or α_3 will transform into three planes if it does not pass through P, for it will cut two of the lines L_1, L_2 and L_3. If it does pass through P it will transform into a 3-space and a plane through one of the lines L_1, L_2 or L_3. The planes β_1, β_2 and β_3 all pass through L_4 and hence any line in β_1, β_2 or β_3 will transform into two planes since it cuts L_4 and either L_1, L_2 or L_3.

§53. (i) Suppose one of the 3-spaces R_3 of B in (h) passes through α_1. The planes α_1, α_2 and α_3 now all lie in R_3 and hence R_3 passes through α_3. Hence part of J^2_{10} is α_1 taken twice, α_3 taken twice, α_2 and β_2 or

Theorem XXXI. The Jacobian is composed of two double planes, two single planes, and an S_4^2 when a 3-space R_3 of B passes through α_1 and α_3.

§54. *Theorem XXXII.* The Jacobian hypersurface is composed of one triple 3-space, four double 3-spaces, and an S_4^3 when a 3-space R_3 of B passes through α_1 and α_2.

The 3-space R_3 is one of the double 3-spaces.

§55. (j) If L_2 coincides with L_1, then L_3 also coincides with L_1. In this case all of the points of 4-space go into points of L_1 and we no longer have a one-to-one correspondence.

CASE V

THE FOUR FUNDAMENTAL HYPERQUADRICS ARE SPACE-PAIRS

§56. (a) Let A, B, C and D be four space-pairs any two of which have only a point in common. The planes α_1, α_2, α_3, and α_4 of A, B, C and D respectively are planes of the J^2_{10}. Hence

Theorem XXXIII. The Jacobian is composed of four planes and an S_6^2 when the four fundamental hyperquadrics are space-pairs intersecting in pairs in points.

§57. *Theorem XXXIV.* The Jacobian hypersurface is composed of four hypercubics and an S_3^3 when the four fundamental hyperquadrics are space-pairs intersecting in pairs in points.

§58. There are ∞^2 lines cutting α_1, α_2, α_3 and α_4. (2, §5.) Any one of these lines will transform into four lines which are rulings of the $j^3{}_{15}$. It cuts $J^2{}_{10}$ four times so the point into which the remainder of the line transforms is the point through which the four lines pass. Hence the four lines are concurrent. Lines cutting only three of the planes α_1, α_2, α_3 and α_4 also transform into four lines but these lines are not concurrent.

§59. (b) Suppose the 3-spaces of A and B have a line L_3 in common.

Theorem XXXV. The Jacobian is composed of five planes and an $S_5{}^2$ when two of the four fundamental space-pairs have a line in common.

The planes are α_1, α_2, α_3 and α_4 and the plane β_3 which transforms into lines of the 3-space determined by α_1 and α_2.

§60. *Theorem XXXVI.* The Jacobian hypersurface is composed of three 3-spaces, a planed hyperquadric counted twice, two cubic hypersurfaces and an $S_2{}^3$.

§61. (c) Suppose a 3-space R_3 of B passes through α_1.

Theorem XXXVII. The Jacobian is composed of three single planes, a double plane, and an $S_5{}^2$ when a 3-space of B passes through α_1.

The double plane is the plane α_1.

§62. *Theorem XXXVIII.* The Jacobian hypersurface is composed of a single 3-space, a double 3-space, a planed hyperquadric counted twice, and an $S_2{}^3$, when a 3-space of B passes through α_1. The double 3-space is R_3.

§63. (d) Suppose A and B have a line L_3 in common and C and D a line L_2 in common.

Theorem XXXIX. The Jacobian is composed of six planes and an $S_4{}^2$ when A and B have a line in common and C and D have a line in common.

The planes are α_1, α_2, α_3, α_4 and also β_3 and β_2 which transform into the 3-spaces determined by α_1 and α_2 and by α_3 and α_4 respectively.

§64. *Theorem XL.* The Jacobian hypersurface is composed of two hyperquadrics counted twice and seven 3-spaces when A and B have a line in common and C and D have a line in common.

§65. (e) If the lines L_2 and L_3 of (d) intersect in a point P, then every point in 4-space transforms into the point P and we no longer have a one-to-one correspondence.

§66. (f) Suppose the space-pairs A, B, and C are so related that A and B have a line L_3 in common and B and C have a line L_1 in common.

Theorem XLI. The Jacobian is composed of seven planes and an $S_4{}^2$ when A and B have a line in common and B and C have a line in common.

The planes are α_1, α_2, α_3 and α_4 and the planes β_1, β_2 and β_3 of Theorem XXI.

§67. *Theorem XLII.* The Jacobian hypersurface is composed of a triple 3-space, four double 3-spaces, a single 3-space, and a hypercubic.

§68. (g) Suppose the space-pairs A and B have a line L_3 in common, B and C a line L_1 in common, and C and D a line L_2 in common.

Theorem XLIII. The Jacobian breaks down into the ten planes of the simplex of reference (4, §4) when the four fundamental hyperquadrics are space-pairs and three pairs of them intersect in lines. This is more easily seen analytically. Take as the fundamental space-pairs

$$A = x^2{}_1 - bx^2{}_2 = 0$$
$$B = x^2{}_1 - cx^2{}_3 = 0$$
$$C = x^2{}_3 - dx^2{}_4 = 0$$
$$D = x^2{}_4 - ex^2{}_5 = 0$$

The Jacobian turns out to be, in addition to the planes α_1, α_2, α_3 and α_4, the planes

$$\begin{cases} x_2 = 0 \\ x_3 = 0 \end{cases} \quad \begin{cases} x_1 = 0 \\ x_4 = 0 \end{cases} \quad \begin{cases} x_3 = 0 \\ x_5 = 0 \end{cases} \quad \begin{cases} x_1 = 0 \\ x_5 = 0 \end{cases} \quad \begin{cases} x_2 = 0 \\ x_4 = 0 \end{cases} \quad \text{and} \begin{cases} x_2 = 0 \\ x_4 = 0 \end{cases}$$

§69. *Theorem XLIV.* The Jacobian hypersurface breaks down into the five faces of the simplex of reference counted three times when the four fundamental hyperquadrics are space-pairs and the three pairs of them intersect in lines.

The equation of the Jacobian hypersurface turns out to be

$$x^3{}_1 \cdot x^3{}_2 \cdot x^3{}_3 \cdot x^3{}_4 \cdot x^3{}_5 = 0$$

UNIVERSITY OF CALIFORNIA PUBLICATIONS
IN
MATHEMATICS
Vol. 1, No. 16, pp. 359-369 December 19, 1923

ON THE INDETERMINATE CUBIC EQUATION $x^3+Dy^3+D^2z^3-3Dxyz=1$

BY

CLYDE WOLFE

1. The object of the following paper is to develop and to discuss some processes for solving the equation $x^3+Dy^3+D^2z^3-3Dxyz=1$.

2. In a letter to M. Liouville, Dirichlet[1] sets forth this beautiful theorem: If $f(s)\equiv s^n+as^{n-1}+\ .\ .\ .\ +gs+h\equiv(s-\alpha)\ (s-\beta)\ .\ .\ .\ (s-\omega)=0$ has at least one real root, then the equation $F(xy\ .\ .\ .\ z)=\Phi(\alpha)\Phi(\beta)\ .\ .\ .\ \Phi(\omega)=m$ has an infinite number of integral solutions, where $\Phi(\alpha)=x+\alpha y+\ .\ .\ .\ +\alpha^{n-1}z$, and similarly for $\beta\ .\ .\ .\ \omega$. To establish this theorem, Dirichlet shows that there is at least one value of m for which the equation $F(xy\ .\ .\ .\ z)=m$ has an infinite number of solutions; in particular, two solutions, $F(xy\ .\ .\ .\ z)=m$ and $F(x'y'\ .\ .\ .\ z')=m$, such that $x\equiv x'$, $y\equiv y'$, . . . $z\equiv z'$, (mod m), whence multiplying the fraction $\frac{x'+\alpha y'+\ .\ .\ .\ +\alpha^{n-1}z'}{x+\alpha y+\ .\ .\ .\ +\alpha^{n-1}z}$ by $\Phi(\beta)\ .\ .\ .\ \Phi(\omega)$ one obtains $X+\alpha Y+\ .\ .\ .\ +\alpha^{n-1}Z$, where $X\equiv Y\equiv\ .\ .\ .\ \equiv Z\equiv 0$, (mod m), whence $\frac{X}{m}, \frac{Y}{m}\ .\ .\ .\ \frac{Z}{m}$ is an integral solution of the equation $F\ (xy\ .\ .\ .\ z)=1$. In the case of the binary quadratic, Euler[2] had shown that if X, Y is a particular solution of the equation $f(xy)=ax^2+2bxy+cy^2=m$ of discriminant $D=b^2-ac$, and if T, U is a particular solution of the equation $F(tu)=t^2-Du^2=1$, then an infinite set of solutions of $f(xy)=m$ is obtained from the equation $ax+(b+\sqrt{D})y=[aX+(b+\sqrt{D})Y](T+U\ \ D)^n$ by giving to n different positive and negative integral values, and equating the rational and irrational parts on the two sides of the equation. Lagrange[3] showed that this process gives all the solutions of the equation $f(xy)=m$ corresponding to the particular solution X, Y when, and only when, T, U is the fundamental solution of $F(tu)=1$, that is, the numerically least solution, excepting of course the solution, 1, 0.

3. Dirichlet[4] points out that this relation divides the totality of solutions of $f(xy)=m$ into sets, all the solutions of any set being obtained from any solution of the set by Euler's method with Lagrange's restriction, whence the field to be investigated for particular solutions is limited to $\sigma<ax+(b+\sqrt{D})y<\sigma(T+U\sqrt{D})$,

where σ is any positive number whatsoever. Generalizing to the third degree, as a special case of the general theorem set forth to Liouville, Dirichlet[4] shows that if $f(s)=s^3+as^2+bs+c\equiv(s-\alpha)(s-\beta)(s-\gamma)=0$, α, β, and γ, being irrational, and only α real, and if X, Y, Z is a particular solution of $F(xyz)\equiv(x+\alpha y+\alpha^2 z)(x+\beta y+\beta^2 z)(x+\gamma y+\gamma^2 z)=m$, and finally if T, U, V is one of the two fundamental solutions of $f(tuv)=1$, then the equation $x+\alpha y+\alpha^2 z=(X+\alpha Y+\alpha^2 Z)(T+\alpha U+\alpha^2 V)^n$ gives all the solutions of $F(xyz)=m$ corresponding to the particular solution X, Y, Z by giving to n different positive and negative integral values and equating the coefficients of like powers of α on the two sides of the equation. Here again Dirichlet[4] shows that this relation divides the totality of solutions of $F(xyz)=m$ into sets, all the solutions of any set being obtained from any solution of the set by Euler's method with Lagrange's restriction, whence the field to be investigated for particular solutions is limited to $\sigma<x+\alpha y+\alpha^2 z<\sigma(T+\alpha U+\alpha^2 V)$, σ being any number whatever.

4. Indeed, in the case of the binary quadratic form, Fermat[5] had already announced that the equation $t^2-Du^2=1$ has an infinite set of solutions all of which may be derived as powers of the fundamental solution, that is, if T_1U_1 is the fundamental solution, the others are obtained as Dirichlet has pointed out in the general case from the equation $T_n+U_n\sqrt{D}=(T_1+U_1\sqrt{D})^n$ by giving to n different positive and negative integral values, and equating the coefficients of the rational and irrational parts on the two sides of the equation. In the case of the ternary cubic, as Mathews[6] points out, there are two reciprocally related fundamental solutions of $F(tuv)=1$, namely $T_1U_1V_1$ and $T_{-1}U_{-1}V_{-1}$ such that $T_{-1}+\alpha U_{-1}+\alpha^2 V_{-1}=(T_1+\alpha U_1+\alpha^2 V)^{-1}$. The solution $T_{-1}U_{-1}V_{-1}$ will be called the reciprocal of the fundamental solution, while the fundamental solution will hereafter signify that one for which $T_1U_1V_1$ is the set of least *positive* integers satisfying the equation $F(tuv)=1$. $T_0U_0V_0$ will signify the solution, 1,0,0, and $T_nU_nV_n$ the nth derived solution obtained by putting $T_n+\alpha U_n+\alpha^2 V_n=(T_1+\alpha U_1=\alpha^2 V_1)^n$ and equating like powers of α on the two sides of the equation.

5. The generalization of the binary quadratic form $F_D(xy)\equiv(x+ty)(x-ty)$ $\equiv x^2-Dy^2\equiv\begin{vmatrix} x & y \\ Dy & x\end{vmatrix}$, where t and $-t$ are the square roots of D, becomes in the case of the ternary cubic form, $F_D(xyz)\equiv(x+ty+t^2z)(x+t\omega y+t^2\omega^2 z)(x+t\omega^2 y+t^2\omega z)$ $\equiv x^3+Dy^3+D^2z^3-3Dxyz\equiv\begin{vmatrix} x & y & z \\ Dz & x & y \\ Dy & Dz & x\end{vmatrix}$, where t is the real cube root of D, and ω and ω^2 are the imaginary cube roots of unity. The solution of the quadratic equation depends upon the expansion of the square root of D into a simple continued fraction, which Lagrange[7] first proved must become periodic, and this periodic continued fraction expansion is closely connected with the reduction of the general binary quadratic form. Hermite[8] raised the question as to whether these relations held for forms of higher order, and Jacobi[9], in 1868, devised an extension of the continued fraction process for determining approximations to the ratios of three numbers involving a cubic irrationality, each approximation being expressed

linearly in terms of the three preceding ones, the expansion in some cases becoming periodic, and giving a solution of $F_D(xyz)=1$ just as in the case of the ordinary continued fraction. Jacobi's fractions may be called proper, regular, ternary continued fractions—"proper" because in his expansion the multiplier of the last set of convergents at any point is never less than the multiplier of the next to the last set of convergents; "regular" because the multiplier of the second from the last set of convergents is always $+1$. In semi-regular fractions, the third multiplier may be $+1$ or -1, which always keeps the determinant of three successive sets of convergents equal to $+1$ or -1. In irregular fractions, the third multiplier may be any positive or negative integer whatsoever, which corresponds to the general binary continued fraction where the numerators are not unity, and where variety of form is obtained at the expense of uniqueness.

6. Bachmann[10] in 1872 showed that periodicity does not occur in Jacobi's expansion unless the two inequalities, $\frac{x_i}{z_i}-\sqrt[3]{D^2}<\frac{k}{z_i^{3/2}}$, and $\frac{y_i}{z_i}-\sqrt[3]{D}<\frac{k}{z_i^{3/2}}$, are satisfied for all values of $i>m$, where m is a number depending upon an arbitrarily chosen k, $x_i : y_i : z_i$ being the convergents to $\sqrt[3]{D^2} : \sqrt[3]{D} : 1$ obtained by Jacobi's method.

7. Mathews[6], in 1890, applying Dirichlet's method of investigation, showed in more detail than does Dirichlet, that, if an integer m can be represented by the form $F(xyz)$ at all, it can be so represented in an infinite number of ways, from which he proved, as did Dirichlet, that the equation $F_D(xyz)=1$ can be satisfied by integral values of x, y, z in an infinite number of ways, all of which may be derived from one of two reciprocally related fundamental solutions.

7a. Meissel[11] in 1891 derived the form of the reciprocal solution of $F_D(xyz)=1$, cross-product and linear recursion formulas for successive solutions, *one* solution being known, exhibited a few special forms of D for which the equation is solved by simple algebra, and obtained by tentative combinations of solutions of $F_D(x_1y_1z_1)=m_1$, $F_D(x_2y_2z_2)=m_2$, and so on, solutions for all non-cubic integral values of $D<82$. He admits the uncertainty of his method, and suggests and illustrates the use of congruences in determining whether or not a given solution is fundamental. He does not, however, give any solutions except in illustrating his method as applied to the solution of the equation for $D=29$, 71, 73, and 69. The x for $D=69$ is a 45-digit number (the greatest for any $D<101$), and Meissel remarks that the ratio x/y or y/z gives the cube root of 69 correct to 65 places of decimals.

8. Berwick[12] in 1913 developed from geometric considerations a process for deriving the expansion of a cubic irrationality, for the case of an equation having but one real root, so that periodicity ensues in every case, whence in particular from the solution of $t^3-D=0$, a solution of $F_D(xyz)=1$ is given by the last set of convergents in the period, just as in the case of the binary quadratic. In spite of the fact that Berwick's expansion is neither proper nor regular, his method leads to a unique set of convergents.

9. Lehmer[13] has shown recently that certain cubic irrationalities which are not periodic as proper continued fractions become periodic when expanded into improper continued fractions, and in particular, if the solution of the equation $F_D(xyz)=1$ is known, that a periodic expansion in an improper fraction follows at once. Berwick further shows that an ideal in a cubic corpus of negative discriminant can be linked up with an equivalent ideal by a substitution derived from the coefficients of his expansion, that every ideal is equivalent to one of a finite number of reduced ideals, which equivalence is shown in a finite number of operations. Finally he shows that the reduced ideals of each class are linked up in one closed cycle which gives rise to the primitive unit of the corpus.

10. Mathews points out in detail some of the applications of the theory of the form $F_D(xyz)=x^3+Dy^3+D^2z^3-3Dxyz$ to the solution of related problems in number theory. All numbers representable by the form $F_D(xyz)$ are cubic residues of D; the product of any number of such numbers is representable by the same form; all of the infinite set of representations of a given number may be obtained from a fundamental representation as soon as the solution of the equation $F_D(xyz)=1$ is known; the relation by which these representations are obtained is the very equation that Dirichlet had used a half a century before, and gives also an infinite set of linear transformations of the form into itself; and finally the successive sets of values of $x:y:z$ which satisfy the equation $F_D(xyz)=1$ furnish rapidly converging approximations to the ratio of $\sqrt[3]{D^2}:\sqrt[3]{D}:1$. Mathews concludes his article by giving a table of actual solutions of the equation $F_D(xyz)=1$ computed, as he says, by Jacobi's method, for $D=2$, 3, 4, 5, 7, and 11. Berwick in illustrating his method derives solutions for $D=23$, and 65.

11. In view of the importance of the equation $F_D(xyz)=1$ in the theory of cubic irrationalities, a table of fundamental solutions for successive integral values of D is as highly desirable as is such a table of solutions of the so-called Pellian equation of the second degree, and this paper aims to supply such a table to $D=100$, together with a discussion of the methods used in constructing the table. Some theorems will be proved that will establish a method of solution for an infinite set of values of D, some 60 of which lie below 100, and some 300 below 1000: other theorems will be proved that will expedite processes already known.

12. Theorem.—Among the solutions of the equation $F_D(xyz)=x^3+Dy^3+D^2z^3-3Dxyz=1$ there exists one solution, whence an infinite set of solutions, such that y is divisible by n, and z by n^2, n being any positive integer whatsoever. For if $x_1y_1z_1$ is a solution of $F_{n^3D}(xyz)=1$, then $x_2=x_1$, $y_2=ny_1$, $z_2=n^2z_1$, is one of the solutions of the equation $F_D(xyz)=1$, for $F_D(x_2y_2z_2)\equiv F_{n^3D}(x_1y_1z_1)=1$, and since there is an infinite set of solutions of $F_{n^3D}(xyz)=1$, there is a corresponding infinite set of solutions of $F_D(xyz)=1$ of the form x_1, ny_1, n^2z_1. By this theorem, if $F_D(x_1y_1z_1)=1$, y_1 being divisible by n, and z_1 by n^2, then $F_{n^3D}\left(x_1\frac{y_1}{n}\frac{z_1}{n^2}\right)=1$, and $\frac{y_1}{n}$ and $\frac{z_1}{n^2}$ are integral. Conversely, if $F_{n^3D}(x_1y_1z_1)=1$, then $F_D(x_1ny_1n^2z_1)=1$.

Again if $F_D(x_1y_1z_1)=1$, y_1 being divisible by D, then $F_{D^2}\left(x_1z_1\frac{y_1}{D}\right)=1$, and $\frac{y_1}{D}$ is integral. Conversely, if $F_{D^2}(x_1y_1z_1)=1$, then $F_D(x_1Dz_1y_1)=1$. The solution of the equation $F_{D_1}(xyz)=1$ being known, combinations of these two processes lead to a solution of the equation $F_D(xyz)=1$, where $D=m^3D_1^n$, m and n being any positive integers whatsoever, or the reciprocals of positive integers. In particular, if the solution of $F_{m^2n}(xyz)=1$ is known, then the solution $F_{mn^2}\left(x\,mz\,\frac{y}{n}\right)=1$ follows, and is integral when that solution of $F_{m^2n}(xyz)=1$ is chosen in which y is divisible by n. For example: If $m=3$ and $n=2$, the first solution for $D=m^2n=18$ in which y is divisible by 2 is $x=9073$, $y=3462$, $z=1321$, whence the solution for $D=mn^2=12$ is $x=9073$, $y=3z=3963$, $z=\frac{y}{2}=1731$.

13. Theorem.—$F_{p^3\pm\frac{3p^2}{q}}\left(p^2q^2\pm3pq+1,\ pq^2\pm2q,\ q^2\pm\frac{q}{p}\right)=1$ where p and q are positive integers, $p>1$, and q an exact divisor of $3p^2$, gives positive integral values for all of the arguments involved, except (*a*) when $p=2$, or 3, $q=1$, and $D=p^3-3\frac{p^2}{q}$; or (*b*) when z is fractional on account of q not being exactly divisible by p. Exception (*a*) is trivial, and the difficulty of exception (*b*) can be removed. For it can be shown by mathematical induction that the only term containing the fraction $\frac{q}{p}$ in the nth derived solution of $F_{p^3\pm\frac{3p^2}{q}}\left(p^2q^2\pm3pq+1,\ pq^2\pm2q,\ q^2\pm\frac{q}{p}\right)=1$ occurs in z_n, and has the numerical coefficient n : whence by taking n equal to the denominator of $\frac{q}{p}$ reduced to lowest terms, a solution is obtained, all of whose terms, including z_n are positive integers. For example: Putting $p=3$, $q=1$, $D=p^3+3\frac{p^2}{q}=54$ $x=p^2q^2+3pq+1=19$, $y=pq^2+2q=5$, $z=q^2+\frac{q}{p}=1\frac{1}{3}$, whence $x_3=61561$, $y_3=16287$, $z_3=4309$.

14. Several methods of obtaining sets of convergents, $x : y : z$, to the ratio, $t^2 : t : 1$, which are variations of Jacobi's method have been tried out in an endeavor to obtain those sets for which $F_D(xyz)=1$, but in every instance the method applied only to a limited number of cases. Jacobi's ternary fraction method applied directly to the numerical approximations to t^2 and t as obtained from Barlow's tables gave immediately solutions for every value of D below 21 except $D=17$, in which case and for $D=21$, 22, and 24, the solutions of $F_D(xyz)=1$ were obtained only indirectly as linear combinations of two sets of convergents. Jacobi's method alone does not give solutions of $F_D(xyz)=1$ directly for even such small values of D as 4 and 6, and fails to satisfy the condition imposed by Bachmann with increasing frequency as D becomes larger, though sets of values of x, y, z which do satisfy $F_D(xyz)=1$ are frequently found among simple linear combinations of neighboring sets of convergents obtained by Jacobi's process.

15. In Berwick's method alone is found a direct non-tentative process for obtaining the successive sets of convergents for which $F_D(xyz)$ has its smallest values. Even this excellent method becomes laborious if the values of x, y, z exceed 8 or 10 digit numbers, and necessitates the development of some theorems for shortening the calculation. These theorems depend upon certain relations between the arguments of the equation $F_D(xyz) \equiv x^3+Dy^3+D^2z^3-3Dxyz=m$ whereby the solution of $F_D(xyz)=1$ may be derived. A more general investigation would consider relations between the arguments of the set of equations (*a*) $F_D(xyz)=F_D(x'y'z')= \ , \ldots =m$, or (*b*) $F_D(xyz)=m$, $F_D(x'y'z')=m'$, . . . An illustration of set (*a*) is given by Mathews[6] (p. 283) where he shows that if two solutions $F_D(xyz)=F_D(x'y'z')=m$ can be found, such that $x \equiv x'$, $y \equiv y'$, $z \equiv z'$, (mod m), then a solution of $F_D(xyz)=1$ is easily derived. This is the theorem that was used by Dirichlet in his proof of the existence of a solution of $F\ (xy \ldots z)=1$ in the general case, but is occasionally useful in actually obtaining numerical solutions. A numerical illustration of set (*b*) occurs in the solution of $F_{38}(xyz)=1$. Among the sets of convergents discovered, $F_{38}(34,\ 10,\ 3)=12$, and $F_{38}(1718,\ 511,\ 152)=18$, whence $F_{38}(x'y'z')=216$ where $x'+ty'+t^2z'=(34+10t+3t^2)\ (1718+511t+152t^2)$. Moreover, $x' \equiv y' \equiv z' \equiv 0$, (mod 6), whence, as is shown in the next paragraph, $F_{38}\left(\frac{x'}{6}\frac{y'}{6}\frac{z'}{6}\right)=1$, which are the values tabulated. The investigation of relations (*a*) and (*b*) is not yet completed.

16. Theorem.—If $F_D(xyz)=m$, then $F_D(x_3y_3z_3)=m^3$, where $x_3+ty_3+t^2z_3=(x+ty+t^2z)^3$. This is only a special case of the general relation between solutions of the equation $F(xy \ldots z)=m$, proved by Dirichlet, that if $F(x_iy_i \ldots z_i)=m_i$, and if $F(x_jy_j \ldots z_j)=m_j$, then $F(x_ky_k \ldots z_k)=m_k$, where $x_k+ay_k+ \ldots +a^{n-1}z_k=(x_i+ay_i+ \ldots +a^{n-1}z_i)\ (x_j+ay_j+ \ldots +a^{n-1}z_j)$, and $m_k=m_im_j$.

17. Corollary.—If $x \equiv y \equiv z \equiv 0$. (mod m), then $\frac{x}{m}\ \frac{y}{m}\ \frac{z}{m}$ is an integral solution of $F_D(xyz)=1$, since $F_D(xyz)$ is homogeneous of the third degree in x, y, z. In particular, if $F_D(xyz)=3$, then $\frac{x_3}{3}\frac{y_3}{3}\frac{z_3}{3}$ always gives an integral solution of $F_D(xyz)=1$, since expanding $(x+ty+t^2z)^3$ it is easily shown that $x_3=F_D(xyz)+9Dxyz=9Dxyz+3$, $y_3=3D(y^2z+z^2x)+3x^2y$, and $z_3=3Dyz^2+3(zx^2+xy^2)$, whence $x \equiv y \equiv z \equiv 0$, (mod 3). The condition imposed in the general case of this corollary is one of frequent occurrence, and easily tested. When applicable, the use of this device reduces the labor of computation at least two-thirds. For example: In the case of $D=29$, $F_{29}(324876,\ 105743,\ 34418)=3$, whence expanding $(324876+105743\sqrt[3]{29}+34418\sqrt[3]{29^2})^3$, and dividing the coefficients of 1, $\sqrt[3]{29}$, and $\sqrt[3]{29^2}$ by 3, the solution as tabulated was obtained. Other numerical cases where this theorem applies are $D=4$, 6, 10, 15, 23, 25, 34, 46, 51, 74. . . Several other special cases of the general theorem are of interest and follow herewith.

18. Theorem.—If $F_D(xyz)=p$, a prime, and $D=kp$, then $x\equiv 0$, (mod p). For putting $D=kp$, $F_D(xyz)\equiv F_{kp}(xyz)=x^3+kpy+k^2p^2z^3-3kpxyz-p=0$, whence $x\equiv x^3\equiv 0$, (mod p), since p is prime.

19. Corollary.—If $F_D(xyz)=p$, a prime, and $D=kp$, then $\frac{x_3}{p}\frac{y_3}{p}\frac{z_3}{p}$ is an integral solution of $F_D(xyz)=1$ where $x_3+ty_3+t^2z_3=(x+ty+t^2z)^3$. For since $x\equiv D\equiv 0$, (mod p), $[x_3=9Dxyz+p]\equiv[y_3=3D(y^2z+z^2x)+3x^2y]\equiv[z_3=3Dyz^2+3(zx^2+xy^2)]\equiv 0$, (mod p), whence $\frac{x_3}{p}\frac{y_3}{p}\frac{z_3}{p}$ are integers. In particular, if $F\ (xyz)=D$, a prime, then $x=0$ (mod D), and y, z, $\frac{x}{D}$ is an integral solution of $F\ (xyz)=1$.

20. Theorem.—If $F_D(xyz)=p^2$, p a prime, and $D=kp\neq lp^2$, then $x\equiv y\equiv 0$, (mod p). For putting $D=kp$, $F_D(xyz)\equiv F_{kp}(xyz)=x^3+kpy+k^2p^2z^3-3kpxyz-p=0$, whence $x\equiv x^3\equiv 0$, (mod p), since p is prime. Putting $x=np$, $F_{kp}\ (npyz)=n^3p^3+kpy+k^2p^2z^3-3kp^2nyz-p=0$, whence $y\equiv y^3\equiv 0$, (mod p), since p is prime, and $k\not\equiv 0$, (mod p).

21. Corollary.—If $F_D(xyz)=p^2$, p a prime, and $D=kp\neq lp^2$, then $\frac{x_3}{p}\frac{y_3}{p}\frac{z_3}{p}$ is an integral solution of $F_D(xyz)=1$, where $x_3+ty_3+t^2z_3=(x+ty+t^2z)^3$, $t=\sqrt[3]{D}$. For since $x\equiv y\equiv 0$, (mod p), $[x_3=9Dxyz+p]\equiv[y_3=3D(y^2z+z^2x)+3x^2y]\equiv[z_3=3Dyz^2+3(zx^2+xy^2)]\equiv 0$, (mod p), whence $\frac{x_3}{p}\frac{y_3}{p}\frac{z_3}{p}$ are integers.

22. Theorem.—If $F_D(xyz)=p^2$, p a prime, and $D=kp^2$, then $x\equiv 0$, (mod p). For putting $D=kp^2$, $F_D(xyz)\equiv F_{kp^2}(xyz)=x^3+kp^2y^3+k^2p^4z^3-3kp^2xyz-p^2=0$, whence $x\equiv x^3\equiv 0$, (mod p), since p is a prime. Putting $x=np$, $F_{kp^2}(npyz)=n^3p^3+kp^2y^3+k^2p^4z^3-3kp^3nyz-p^2=0$, whence $y\equiv 0$, (mod p).

23. Corollary.—If $F_D(xyz)=p^2$, p a prime, and $D=kp^2$, then $\frac{x_3}{p^2}\frac{y_3}{p^2}\frac{z_3}{p^2}$ is a solution of $F_D(xyz)=1$, where $\frac{x_3}{p^2}\frac{y_3}{p^2}\frac{z_3}{p^2}$ are integral. For since $x\equiv 0$, (mod p), and $D\equiv 0$, (mod p^2), $[x_3=9Dxyz+p^2]\equiv[y_3=3D(y^2z+z^2x)+3x^2y]\equiv 0$, (mod p^2), and $[z=3Dyz^2+3(zx^2+xy^2)]\equiv 0$, (mod p). Putting $\frac{x_3}{p^2}\frac{y_3}{p^2}\frac{z_3}{p^2}=x'\ y'\ \frac{z'}{p}$, $F\left(x'y'\frac{z'}{p}\right)=1$, whence $x'_p\ y'_p\ z'_p$ is an integral solution of $F_D(xyz)=1$ where $x'_p+ty'_p+t^2z'_p=\left(x'+ty'+t\ \frac{z'}{p}\right)^p$ as shown in paragraph 13.

24. Two methods of solution herein described may be illustrated by the same numerical example. In the case of $D=75=3\times 5^2$, the solution for $D=45=5\times 3^2$ having been obtained as tabulated, the solution for $D=75$, by the method of paragraph 12, is $x=1477441$, $y=350340$, $z=83074\frac{4}{5}$, which raised to the fifth power as pointed out in paragraph 13, eliminates the fraction in z, giving the result as tabulated. Or otherwise: Having discovered that F_{75} (160,38,9)$=25$, cubing F_{75} (36936025,8758500,2076870)$=25^3$, by paragraph 16, whence dividing by 25, by paragraph 17, F_{75} (1477441,350340.83074$\frac{4}{5}$)$=1$, the same result as obtained by the first method.

25. To determine whether a given solution is fundamental or not, one makes use of the fact that if the solution is the nth derived solution of the fundamental solution, that is, if $x_n+ty_n+t^2z_n=(x_1+ty_1+t^2z_1)^n$, then $x_1y_1z_1$ may be determined by solving the equation $x_1+ty_1+t^2z_1=\sqrt[n]{x_n+ty_n+t^2z_n}$. In practice, n is unknown, so that different integral values are tried in order, beginning with $n=2$. If $x_2+ty_2+t^2z_2=(x_1+ty_1+t^2z_1)^2$, then $x_2=x_1^2+2Dy_1z_1\doteq 3x_1^2$, $y_2=Dz_1^2+2x_1y_1\doteq 3Dz_1^2$, $z_2=y_1^2+2z_1x_1\doteq 3y_1^2$; or $x_2\doteq 3x_1^2$, $y_2\doteq\frac{x_2}{t}$, $z_2\doteq\frac{x_2}{t_2}$, where $t=\sqrt[3]{D}$. The application of this test is further facilitated by the fact that if $F_D(x_1y_1z_1)=1$, then only values of x_1 such that $x_1^3\equiv 1$, (mod D), need be tried. Having thus discovered that the given solution is, or is not, the square of a simpler solution, either repeat the test on the simpler solution, or test the given solution to see whether or not it is the cube of a simpler solution. The relations to be used in the test as to whether a given solution is the nth power of a simpler solution arise from the equation $x_n+ty_n+t^2z_n=(x_1+ty_1+t^2z_1)^n$, and the approximate equalities, $x_1\doteq ty_1\doteq t^2z_1$. In practice the values of x soon become so small that separate tests may be made for smaller values of x, such that $x^3\equiv 1$, (mod D). Not all the tabulated solutions have been tested, owing to lack of time, but there is reason to believe that most of them are fundamental. As an illustration of this method of testing a solution to see whether it is fundamental or not, consider the solution $x=9073$, $y=3963$, $z=1731$ for $D=12$. If this is the square of a fundamental solution, $x'y'z'$, then $x'\doteq\sqrt{\frac{9073}{3}}\doteq 55$, $y'\doteq\frac{x'}{\sqrt[3]{12}}\doteq 24$, $z'\doteq\frac{x'}{\sqrt[3]{144}}\doteq 10\tfrac{1}{2}$. As a matter of fact, $F_{12}(x'y'z')=1$, but z' is fractional. The only values of $x<55$ such that $x^3\equiv 1$, (mod 12), are 1, 13, 25, 37, 49, and none of these give solutions of $F_{12}(xyz)=1$, excepting of course the solution, 1, 0, 0. For large numbers, of course, this testing is done congruentially, as Meissel[11] suggests and illustrates.

1	1	0	0
2	1	1	1
3	4	3	2
4	5	3	2
5	41	24	14
6	109	60	33
7	4	2	1
8	1	0	0
9	4	2	1
10	181	84	39
11	89	40	18
12	9073	3963	1731
13	94	40	17
14	29	12	5

15	5401	2190	888
16	16001	6350	2520
17	324	126	49
18	55	21	8
19	64	24	9
20	361	133	49
21	1705	618	224
22	793	283	101
23	2166673601	761875860	267901370
24	649	225	78
25	9375075001	3206230550	1096515424
26	9	3	1
27	1	0	0
28	9	3	1
29	102866541757601689	33481749309704842	10897883001448120
30	811	261	84
31	101209	32218	10256
32	768096001	241935080	76204775
33	15270674074129	4760876269140	1484279131362
34	334153	103146	31839
35	25776	7880	2409
36	109	33	10
37	100	30	9
38	29071	8647	2572
39	529	156	46
40	5041	1474	431
41		9311970957818975874477 29 27005174873452595403426 0 783163385334016576363 58	
42	21169	6090	1752
43	49	14	4
44	5351941	1515981	429414
45	1477441	415374	116780
46	16449049	4590798	1281255
47	562944292769	155990973316	43224852030
48	3810843073	1048593882	288531726
49	8786992	2401273	656210
50	12867751	3492845	948104
51	107846641	29081484	7841994
52	209	56	15
53	113015453598	30087022392	8009779969
54	61561	16287	4309
55	32947340560201	8663621462574	2278130361072

56	765073	199974	52269
57	1460968	379620	98641
58	929	240	62
59	2161836123797351105087 3	5553141829215933501576	1426444115533632242954
60	2161	552	141
61	3905	992	252
62	8929	2256	570
63	16	4	1
64	1	0	0
65	16	4	1
66	9505	2352	582
67	4289	1056	260
68	2449	600	147
69	404886837053487091694212951195653956127452401	98715184393700556938337454013404500951638820	24067681974543893805323831567684099602695630
70	1121	272	66
71	17883556065528164 82	431884645684316172	104299361097095425
72	1270801	303738	73011
73	99928	23910	5721
74	1025641	244297	58189
75	570211212113607886743089699946001	135212029500803659084176761891820	32062317494528092279877193074334
76	305	72	17
77	3752144035327073	881960753382348	207309411148166
78	9818137169569	2297898779424	537814730970
79	370454	86336	20121
80	28977465601	6725074038	1560751428
81	2460229201	568609218	131417204
82	6213361110928690501443 59	1430173228101452557296 01	329193076961273052907 02
83	75121376774150674740676535151 53	17221494659804811779693450111 32	3948009089467846666157610084 66
84	82988024365	18949117863	4326757632
85	130242770906883377636475068995100472 1	296219732439518970205699942411045068	67371209377351163649079349355266292
86	565	128	29
87	869659117387302940175600943029010180 01	196264899778915488534579216056495091 48	44293114526243605739960598415630292 46

88	23761	5342	1201
89	421204668989443920	94340138120314920	21130016630426161
90	58321	13014	2904
91	81	18	4
92	34447365811144441	7630624358730583	1690301325891661
93	90139756288	19895524677	4391313206
94	961302655531	211422331357	46498781564
95	3576393871601	783797980668	171776179178
96			{ 100428638580128172512127484936339201 21933124190915540772279199097914760 4790087205954739309459432490919945
97			{ 1040068758302891753134 226362669236418190113 49266029399288296684
98	3796883757	823541096	178625415
99			{ 2370504989476787061889 512423622402913328862 110768789756683516940
100	15734603821501	3389917630695	730335613992

LITERATURE CITED

[1] Dirichlet, G. L.: *Werke*, vol. 1, pp. 621-623.

[2] Euler, L.: *Corresp. Math. Phys.*, vol. 1 (1843), pp. 629-631.

[3] Lagrange, J. L.: *Mem.* Acad. Berlin, vol. 24 (1770), (*Oeuv.*, vol. 2, p. 74).

[4] Dirichlet, G. L.: *Werke*, vol. 1, pp. 627-632.

[5] Fermat, Pierre de: *Oeuv.*, vol. 2, pp. 333-335.

[6] Mathews, G. B.: *Proc.* London Math. Soc., vol. 21 (1889-90), pp. 280-287.

[7] Lagrange, J. L.: *Mem.* Akad. Berlin, vol. 24 (1770), (*Oeuv.*, vol. 2, pp. 603-615).

[8] Hermite, Ch.: *Oeuv.*, vol. 1, pp. 185-187.

[9] Jacobi, C. G. J.: *Ges. Werke*, vol. 6, pp. 385-426.

[10] Bachmann, P.: *Crelle*, vol. 75 (1873), pp. 23-24.

[11] Meissel, E.: *Beitrag zur Pell'schen Gleichung höherer Grade, Progr.*, Kiel, 1891.

[12] Berwick, W. E. H.: *Proc.* London Math. Soc., (2) vol. 12 (1913), pp. 393-429.

[13] Lehmer, D. N.: Manuscript.

For a history of the Pellian equation, the reader is referred to E. Konen, *Geschichte der Gleichung*, $t^2-Du^2=1$, Leipzig, 1901, and to L. E. Dickson, *History of the Theory of Numbers*, vol. 2, chap. 12.

For further references to the literature of the cubic form $x^3+Dy^3+D^2z^3-3Dxyz$, see L. E. Dickson, *History of the Theory of Numbers*, vol. 2, chap. 21, pp. 588-595, esp. 593-595.

UNIVERSITY OF CALIFORNIA PUBLICATIONS
IN
MATHEMATICS

Vol. 1, No. 17, pp. 371-387 November 8, 1923

A STUDY AND CLASSIFICATION OF RULED QUARTIC SURFACES BY MEANS OF A POINT-TO-LINE TRANSFORMATION

BY

BING CHIN WONG

1. The purpose of this paper is to study and classify ruled quartic surfaces by means of a point-to-line transformation. The paper will consist of four sections:

Section I. A historical discussion of the subject.

Section II. The setting up of the machinery of transformation.

Section III. The study and classification of the ruled quartics of the species from I to IX.

Section IV. A slight modification of the machinery of transformation given in Section II and the classification of the remaining three species.

SECTION I

2. Ruled quartic surfaces have been studied and classified by Cremona,[1] Sturm,[2] Cayley,[3] Salmon,[4] and others.[5] These authors classify these surfaces according to their deficiency,[6] and reclassify them according to the nature of the double curves they have on them, and once more reclassify them according to the kinds of surfaces into which they are reciprocated.

3. There are two species of ruled quartics of deficiency unity and ten species of deficiency zero. The double curve on those of deficiency unity is a pair of straight lines which, if distinct give species I, and if coincident, give species II.[7] The

[1] Bologna Accad. Sci., *Mem.*, vol. 8 (1868)

[2] *Liniengeometrie*, Bd. 1, pp. 52-61

[3] *Phil. Trans.*, vols. 154 (1864) and 159 (1869).

[4] *Analytic Geom. of Three Dim.* (ed. 5), vol. 2, chap. 16.

[5] A discussion of ruled quartic surfaces may also be found in Jessop, *The Line Complex*, Cambridge, 1903, chap. 5 and in Basset, *The Geometry of Surfaces*, Cambridge, 1910, chap. 6.

[6] The deficiency of a surface is defined as the difference between the maximum and the actual number of double points on any plane section of the surface.

[7] This discussion is based upon the works of Cremona and Sturm, but the order of the species is Sturm's. Different authors number the classes differently. Owing to the difference in the point of view, the order in this paper is yet a different one.

double curve on those of zero deficiency is either a space cubic, or a straight line and a conic, or three straight lines. Those ruled quartics with a double space cubic are of species III if their reciprocal surfaces are of the same kind, and of species IV if their reciprocal surfaces have a triple line. Those whose double curve is a straight line and a conic belong to species V if they reciprocate into surfaces of the same kind; to species VI if they reciprocate into surfaces with a triple line. Those with three double straight lines make up the remaining six species: if the three lines one of which must be a generator are distinct, the surfaces belong to species VII; if two are coincident, the distinct one being the generator, the surfaces belong to species VIII. If the three double lines are all coincident, we have species IX if the surfaces reciprocate into those with a double place cubic; species X if they reciprocate into those of the same kind; species XI if the triple line is itself a generator counted once; and finaly, species XII if the triple line is itself a generator counted twice. Species XI reciprocate into surfaces with a straight line and a conic, and species XII into those of the same kind.[8]

4. The table[9] given (p. 373) presents a view of the different orders given to the species by Sturm, Cremona, Cayley, and Salmon, and also exhibits the special features of the different classes. In this table

k^3 stands for double space cubic;
c^2 stands for double conic;
e stands for double generator;
d stands for double directrix;
$d+d$ stands for two coincident double directrices;
$d+d'$ stands for two distinct double directrices;
$d+d+d$ stands for a triple line,
$d+d+d_e$ stands for a triple line which is itself a generator counted once;
$d+d_e+d_e$ stands for a triple line which is itself a generator counted twice.

The double curves on the reciprocal surfaces are denoted by the corresponding Greek letters.

It is obvious that species IV and VI reciprocate into species IX and XI respectively, and vice versa; while the other species are their own reciprocals.

5. The classification given in the following paragraphs is from a different point of view. Here we start with certain plane curves[10] which, by a point-to-line transformation, go into quartic surfaces, and, it will be shown, all the different classes of ruled quartics can be obtained in this manner. It will also be shown that there exists a close relation between the singularities[11] on the surfaces and the position and class of the plane curves.

---

[8] Salmon divides those surfaces with a triple line into five classes, I—V according to his enumeration, making thirteen species in all, while Sturm divides them into four, IX—XII as given above. As a matter of fact, V of Salmon is only a subform of IV, that is, a subform of XII of Sturm. See § 47 below.

[9] This table, with a slight modification and an addition of Salmon's enumeration, is the combination of the two found in Jessop, *The Line Geometry*, p. 80.

[10] For our purposes, conics and plane cubics only need be employed. Plane quartics may be used but they offer nothing new. No quintic curve or curve of higher degree can give quartic surfaces. But if one were to study and classify ruled surfaces of higher degree from this point of view, one would have to resort to curves of higher order, both plane and space.

[11] Including double curves and pinch-points.

6. This method has three interesting features: (1) the one-to-one correspondence between the generators of the surfaces and the points on the plane curves; (2) the dependence of the singularities on the surfaces upon the position and class of the plane curves; and (3) the possibility of the method in studying and classifying scrolls of higher order.

Def.	Double curve	Double curve on rec. s.	Sturm	Cremona	Cayley[12]	Salmon[13]
$p=1$	$d+d'$	$\delta+\delta'$	I	11	1	11
	$d+d$	$\delta+\delta$	II	12	4	13
$p=0$	k^3	k^3	III	1	10	6
	k^3	$\delta+\delta+\delta$	IV	7	8	7
	c^2+d	$k^2+\delta$	V	2	7	8
	c^2+d	$\delta+\delta+\delta_\epsilon$	VI	4	(11)	9
	$d+d'+e$	$\delta+\delta'+\epsilon$	VII	5	2	10
	$d+d+e$	$\delta+\delta+\epsilon$	VIII	6	5	12
	$d+d+d$	k^3	IX	8	9	1
	$d+d+d$	$\delta+\delta+\delta$	X	9	3	2
	$d+d+d_e$	$k^2+\delta$	XI	3	(12)	3
	$d+d_e+d_e$	$\delta+\delta_\epsilon+\delta_\epsilon$	XII	10	6	4 (5)

SECTION II

7. Take two fixed quadrics

$$F_1 \equiv \sum_1^4 a_i x^2{}_i = 0,$$

$$F_2 \equiv \sum_1^4 b_i x^2{}_i = 0.$$

Any point $Y(y_1, y_2, y_3, y_4)$ in space is transformed into a line

$$l \equiv \begin{cases} \sum_1^4 a_i y_i x_i = 0, \\ \sum_1^4 b_i y_i x_i = 0, \end{cases}$$

[12] Cayley does not consider (11) and (12) as separate species, but as subforms of 8 and 9 respectively.

[13] Salmon begins with surfaces having the highest singularity, that is, a triple line, and divides them into five classes, and then takes up those having a proper space cubic, and then those whose cubic degenerates, etc. See note 8.

which is the intersection of the polar planes of Y with respect to F_1 and F_2. To the ∞^3 of points in space there ∞^3 of polar planes with respect to F_1 and also ∞^3 of polar planes with respect to F_2 corresponding. The space consisting of the ∞^3 of polar planes with respect to F_1 and the space consisting of the ∞^3 of polar planes with respect to F_2 are clearly collineated to each other, for to every plane of the one space there corresponds a plane of the other space, both being polar planes of the same point. Then every line 1, being a line of the intersection of the corresponding planes of the two collineated spaces, is a line of a tetrahedral complex[14] whose fundamental tetrahedron $A_1A_2A_3A_4$ is the self-polar tetrahedron common to F_1F_2. To show this algebraically [15], write the six homogeneous coördinates of l:

$$q_{12} : q_{13} : q_{14} : q_{34} : q_{42} : q_{23}$$
$$= \pi_{12}y_1y_2 : \pi_{13}y_1y_3 : \pi_{14}y_1y_4 : \pi_{34}y_3y_4 : \pi_{42}y_4y_2 : \pi_{23}y_2y_3$$

where $\pi_{ik} = a_ib_k - a_kb_i$. Then

$$\frac{q_{42}q_{13}}{q_{14}q_{23}} = \frac{\pi_{42}\pi_{13}}{\pi_{14}\pi_{23}},$$

$$\frac{q_{14}q_{23}}{q_{12}q_{34}} = \frac{\pi_{14}\pi_{23}}{\pi_{12}\pi_{34}},$$

and

$$\frac{q_{12}q_{34}}{q_{13}q_{42}} = \frac{\pi_{12}\pi_{34}}{\pi_{13}\pi_{42}},$$

Clearing fractions and adding, we have

$$Aq_{12}q_{34} + Bq_{13}q_{42} + Cq_{14}q_{23} = 0,$$

where

$$A = \pi_{13}\pi_{42} - \pi_{14}\pi_{23},$$
$$B = \pi_{14}\pi_{23} - \pi_{12}\pi_{34},$$
$$C = \pi_{12}\pi_{34} - \pi_{13}\pi_{42},$$

which is the equation of the tetrahedral complex. Since the coördinates of the point Y are not involved in this equation, *every point in space goes into a line belonging to the above complex determined by the fixed quadrics.*

8. If a point describes a straight line in space, its two polar planes with respect to F_1 and F_2 describe two projective axial pencils whose corresponding planes meet in the generators on a ruled quadric. Algebraically, let

$$(y_1 + \lambda z_1,\ y_2 + \lambda z_2,\ y_3 + \lambda z_3,\ y_4 + \lambda z_4)$$

be the coördinates of any point on a given line g through Y and Z. Then the intersections of the corresponding planes of the two projective axial pencils determined by the given line g are given by

$$\sum_1^4 a_i(y_i + \lambda z_i)x_i = 0,$$

$$\sum_1^4 b_i(y_i + \lambda z_i)x_i = 0.$$

[14] Reye, *Geometrie der Lage*, Abth. 2, erste Auflage (1868), 15. Vortrag. An exposition of this complex is also given by Sturm in his *Liniengeometrie*, Bd. 1, pp. 332-382.

[15] Jessop, *The line Complex*, Cambridge, 1903, chap. 7.

Eliminating λ, we have

$$\sum_{i=1\ k=1}^{4} \pi_{ik}p_{ik}x_ix_k \equiv \pi_{12}p_{12}x_1x_2+\pi_{13}p_{13}z_1x_3+\pi_{14}p_{14}x_1x_4+\pi_{34}p_{34}x_3x_4+\pi_{42}p_{42}x_4x_2+\pi_{23}p_{23}x_2x_3 = 0,$$

[where $p_{ik}=y_iz_k-y_kz_i$] which is the equation of a quadric surface. But the p_{ik} are the six homogeneous coördinates of the given line, and therefore, we conclude that *when the* p_{ik} *of any line are given we can at once write down the equation of its corresponding quadric in the above form.*

9. Now if we take three straight lines p'_{ik}, p''_{ik}, p'''_{ik} forming a triangle in space whose corresponding quadrics are, respectively,

$$Q' \equiv \sum_1^4 \pi_{ik}p'_{ik}x_ix_k = 0,$$

$$Q'' \equiv \sum_1^4 \pi_{ik}p''_{ik}x_ix_k = 0,$$

$$Q''' \equiv \sum_1^4 \pi_{ik}p'''_{ik}x_ix_k = 0,$$

any conic

$$A_{11}p'^2_{ik}+A_{22}p''^2_{ik}+A_{33}p'''^2_{ik}+2A_1A_2p'_{ik}p''_{ik}+2A_2A_3p''_{ik}p'''_{ik}+2A_3A_1p'''_{ik}p'_{ik}=0$$

in the plane of this triangle will go into a ruled quartic surface whose equation is

$$A_{11}Q'^2+A_{22}Q''^2+A_{33}Q'''^2+2A_{12}Q'Q''+2A_{13}Q'Q'''+2A_{23}Q''Q'''=0$$

or

$$\sum_1^4 \pi_{ik}\pi_{jl}[\Sigma A_{rs}p^{(r)}_{ik}p^{(s)}_{jl}]x_ix_kx_jx_l=0 \qquad [r,\ s=1,\ 2;\ A_{rs}=A_{sr}].$$

A plane cubic in the plane of the same triangle will go into a sextic surface whose equation is

$$\Sigma A_{rst}Q^{(r)}Q^{(s)}Q^{(t)} \equiv$$
$$\Sigma \pi_{ik}\pi_{jl}\pi_{mn}[\Sigma A_{rst}p^{(r)}_{ik}p^{(s)}_{jl}p^{(t)}_{mn}]x_ix_kx_jx_lx_mx_n=0$$
$$[A_{rst}=A_{rts}=A_{srt}=A_{str}=A_{tsr}=A_{trs};\ r,\ s,\ t=1,\ 2,\ 3].$$

Similarly, a plane curve of degree n goes into a ruled surface of degree $2n$. All the surfaces thus obtained have the four vertices of the fundamental tetrahedron in common.

10. Corresponding to the ∞^2 of points in a plane u is a congruence of lines all cutting across a cubic k^3 twice. To prove this, when u is given, we have two points Z and Z' fixed which are the poles of u with respect to F_1 and F_2. Z and Z' can be considered as two projective point systems, and the locus of the intersections of the corresponding rays which do meet is a space cubic. Every point in u will have its two polar planes, one through Z and the other through Z', intersecting in a line cutting across the cubic twice. Since every line in u has its corresponding quadric

through the four vertices A_1, A_2, A_3, A_4 of the tetrahedron of reference, this cubic k_3 must go through them also, and is therefore a cubic of the complex. Thus, *every plane in space determines a cubic of the complex uniquely. Every plane curve goes into a surface on which lies the cubic determined by the plane of the curve.*

11. A pair of intersecting lines is transformed into a pair of quadrics having in common a generator which comes from the point of intersection of the two given lines and cubic k^3 which is determined by the plane of the two given lines. To obtain the parametric equations of k^3, let (u_1, u_2, u_3, u_4) be the coördinates of a given plane u, and let p'_{ik} and p''_{ik} be any two lines lying in it. The two quadrics corresponding to these two lines

$$Q' \equiv \sum_1^4 \pi_{ik} p'_{ik} x_i x_k = 0,$$

$$Q'' \equiv \sum_1^4 \pi_{ik} p''_{ik} x_i x_k = 0,$$

have, besides the common ruling into which the intersection of p'_{ik} of p''_{ik} goes, a common cubic k^3 whose parametric equations are easily found to be, with λ as the parameter,

$$\begin{aligned}
\rho x_1 &= u_1(a_2\lambda - b_2)\ (a_3\lambda - b_3)\ (a_4\lambda - b_4),\\
\rho x_2 &= u_2(a_3\lambda - b_3)\ (a_1\lambda - b_1)\ (a_1\lambda - b_1),\\
\rho x_3 &= u_3(a_4\lambda - b_4)\ (a_1\lambda - b_1)\ (a_2\lambda - b_2),\\
\rho x_4 &= u_4(a_1\lambda - b_1)\ (a_2\lambda - b_2)\ (a_3\lambda - b_3),
\end{aligned}$$

or

$$\rho x_i = u_i \overset{4}{\underset{1}{P}} (a_j \lambda - b_j) \qquad [i = 1, 2, 3, 4;\ i \neq j].$$

These equations are independent of the coördinates of the lines but not of those of the given plane; therefore, when the coördinates of any plane are given, the equations of its corresponding cubic can be easily written.

12. Now if $u_1 = 0$, i.e., if the plane u passes through the vertex A_1 of the fundamental tetrahedron, k^3 degenerates into a conic c^2 lying in the plane x_1 and a straight line d which passes through A_1 and cuts across the conic. c^2 and d must have a point in common, for, if not, the quadric which corresponds to a line in u $(0, u_2, u_3, u_4)$ and therefore contains c^2 and d, would be intersected by a straight line lying in x_1 through P in which d pierces x_1 in three points, one being P itself and the other two being on c_2. This is impossible.

13. We can get the parametric equations of the conic c_2 by putting $u_1 = 0$ in the parametric equations of k^3 (Art. 11), or we can get them as follows:

Let (o, u_2, u_3, u_4) be the coördinates of the plane u and let any other plane $u'(u'_1, u'_2, u'_3, u'_4)$ intersect it in the line

$$q_{ik}(u'_1 u_2 : u'_2 u_3 : u'_4 : q_{34} : q_{43} : q_{23}).$$

Then corresponding to this line we have the quadric

$$\pi_{12}q_{34}x_1x_2+\pi_{13}q_{42}x_1x_3+\pi_{14}q_{23}x_1x_4 \\ -u'(\pi_{23}u_4x_2x_3+\pi_{42}u_3x_2x_4+\pi_{34}u_2x_3x_4)=0.$$

Putting $x_1=0$, we have the conic in which this quadric is intersected by the plane x_1:

$$x_1=0,\ \pi_{23}u_4x_2x_3+\pi_{42}u_3x_2x_4+\pi_{34}u_2x_3x_4=0.$$

These equations are independent of the coördinates of the plane u', and therefore represent the conic common to all the quadrics corresponding to all the lines in the plane u. This conic is none other than c^2, part of the degenerate cubic k_3.

14. Now to obtain d: Take two lines through A_1:

$$l_1\equiv\begin{cases} u(o,\ u_2,\ u_3,\ u_4),\\ u'(o,\ u'_2,\ u'_3,\ u'_4),\end{cases}$$

$$l_2\equiv\begin{cases} u(o,\ u_2,\ u_3,\ u_4),\\ u''(o,\ u''_2,\ u''_3,\ u''_4),\end{cases}$$

whose corresponding quadrics (pairs of planes)

$$Q_1\equiv x_1[\pi_{12}q'_{34}x_2+\pi_{13}q_{42}x_3+\pi_{14}q'_{23}x_4]=0,$$
$$Q_2\equiv x_2[\pi_{12}q''_{34}x_2+\pi_{13}q''_{42}x_3+\pi_{14}q''_{23}x_4]=0,$$

have, besides the plane x_1, the line

$$\pi_{12}q'_{34}x_2+\pi_{13}q'_{42}x_3+\pi_{14}q'_{23}x_4=0,$$
$$\pi_{12}q''_{34}x_2+\pi_{13}q''_{42}x_3+\pi_{14}q''_{23}x_4=0,$$

in common. This line has its homogeneous coördinates:

$$0:0:0:\pi_{13}\pi_{14}u_2:\pi_{12}\pi_{14}u_3:\pi_{12}\pi_{13}u_4$$

and has its parametric representations:

$$\rho x_1=\lambda\pi_{12}\pi_{14}u_3,$$
$$\rho x_2=\ \pi_{13}\pi_{14}u_2,$$
$$\rho x_3=\ \pi_{12}\pi_{14}u_3,$$
$$\rho x_4=\pi_{12}\pi_{13}u_4.$$

Both the coördinates and the representations of this line are independent of the coördinates of the plane u' and those of the plane u'', and therefore this line is the line d common to all the quadrics coming from all the lines in the plane u.

15. Now if $u_1=u_2=0$, i.e., if the plane u passes through the edge A_1A_2 of the fundamental tetrahedron, we have its k^3 broken up into three straight lines $d+d'+e$. d lies in the plane x_1, d' in x_2, while e being the intersection of x_1 and x_2 cuts across the other two. Let any plane

$$u'(u'_1,\ u'_2,\ u'_3,\ u'_4)$$

intersect the plane u $(0,\ 0,\ u_3,\ u_4)$ in the line whose transform is the quadric

$$\pi_{12}q_{34}x_1x_2+\pi_{13}u_4u'_2x_1x_3-\pi_{14}u_3u'_2x_1x_4-\pi_{42}u'_1u_3x_2x_4-\pi_{23}u'_1u_4x_2x_3=0.$$

When $x_1=0$, we have

$$x_2=0, \text{ and } \pi_{23}u_4x_3+\pi_{42}u_3x_4=0;$$

and when $x_2=0$, we have

$$x_1=0, \text{ and } \pi_{13}u_4x_3-\pi_{14}u_3x_4=0.$$

The equations of these three lines

$$d \begin{cases} x_1=0, \\ \pi_{23}u_4x_3+\pi_{43}u_3x_4=0; \end{cases}$$

$$d' \begin{cases} x_2=0, \\ \pi_{13}u_4x_3-\pi_{14}u_3x_4=0; \end{cases}$$

$$e \begin{cases} x_1=0, \\ x_2=0, \end{cases}$$

being independent of the coördinates of the plane u', represent the degenerate k^3 common to all the quadrics which correspond to all the lines in u. e is the common generator, while d and d' are the common directrices.

16. Every plane u, besides determining uniquely a space cubic k^3, contains a conic γ^2 of the complex, the conic to which all the complex lines in the plane are tangent. All these lines, ∞ in number, correspond to the points on k^3; or in other words, all the points on a cubic curve of the complex determined by a plane go into lines tangent to the conic of the complex in that plane.

17. To obtain the equations of this conic γ^2, write the equations of the polar planes of the points on k^3 (Art. 11) with respect to F_1 and F_2, respectively,

$$\sum_{i=1}^{4} a_iu_i \prod_{j=1}^{4} (a_j\lambda-b_j)x_i=0,$$

$$\sum_{i=1}^{4} b_iu_i \prod_{j=1}^{4} (a_j\lambda-b_j)x_i=0 \qquad [i\neq j].$$

These two planes intersect in a line which we shall designate by L. For every value of λ, there is a point on k^3 and there is corresponding to it a line given by L. *All these lines of the system L lie in the plane u*, for the matrix

$$\left\| \begin{matrix} a_1u_1B_2B_3B_4 & a_2u_2B_3B_4B_1 & a_3u_3B_4B_1B_2 & a_4u_4B_1B_2B_3 \\ b_1u_1B_2B_3B_4 & b_2u_2B_3B_4B_1 & b_3u_3B_4B_1B_2 & b_4u_4B_1B_2B_3 \\ u_1 & u_2 & u_3 & u_4 \end{matrix} \right\|$$

[where $B_i=a_i\lambda-b_i$] of the coefficients of the two equations of L and those of the equation of the plane u is of rank 2. To show this is merely algebraic work which any one may easily verify.

18. The line L pierces the plane x_1 and x_2 respectively in

$$R_1\equiv(0:\pi_{34}u_3u_4B_2:\pi_{42}u_4u_2B_3:\pi_{23}u_2u_3B_4),$$
$$R_2\equiv(\pi_{34}u_3u_4B_2:0:\pi_{41}u_4u_1B_3:\pi_{13}u_1u_3B_4).$$

The plane determined by R_1, R_2 and $A(0:0:0:1)$ is given by

$$p \equiv \pi_{41}u_1B_2B_3x_1 + \pi_{42}u_2B_3B_1x_2 + \pi_{43}u_3B_1B_2x_3 = 0$$

or

$$\sum_{i=1}^{3} \pi_{4i}u_i \prod_{j=1}^{3} (a_j\lambda - b_j)x_i = 0 \qquad [i \neq j].$$

This equation involves λ *in the second degree and is therefore a system of planes enveloping a quadric cone whose vertex is the point* $A_4(0:0:0:1)$. *Therefore, the equations*

$$p=0,\ u=0$$

represent the system of lines tangent to the conic γ^2 *of the complex in the plane* u *and these lines are lines of the complex.*

19. This system of rays is none other than the system L given by the equations of Art. 17. Thus, *when a plane is given we can easily write down the equations of the system of lines of the complex lying in it.*

20. The system of complex lines in the plane u $(0, u_2, u_3, u_4)$ is made up of two flat pencils of the first class, one of which corresponding to the points on the conic c_2 (Art. 13) in x_1 has its center at A_1 and the other corresponding to the line d (Art. 14) has its center at the point

$$A'_1 \equiv (0 : \pi_{12}\pi_{34}u_3u_4 : \pi_{13}\pi_{42}u_4u_2 : \pi_{14}\pi_{23}u_2u_3)$$

whose transform is the line d itself. To obtain the equations of the first pencil, one needs only to put $u_1=0$ in

$$p=0,\ u=0$$

given in Art. 18 or in the equations given in Art. 17. The equations of the other pencil are those of the polar planes of points of d whose coördinates are given in Art. 14 with respect to F_1 and F_2 and are given by

$$\lambda a_1u_3\pi_{12}\pi_{14}x_1 + a_1u_2\pi_{13}\pi_{14}x_2 + a_3u_3\pi_{12}\pi_{14}x_3 + a_4u_4\pi_{12}\pi_{13}x_4 = 0,$$
$$\lambda b_1u_3\pi_{12}\pi_{14}x_1 + b_2u_2\pi_{13}\pi_{14}x_2 + b_3u_3\pi_{12}\pi_{14}x_3 + b_4u_4\pi_{12}\pi_{13}x_4 = 0.$$

For every value of λ there is a ray of the pencil, and every ray of this pencil pierces the plane x_1 in the same point A_1 whose coördinates are given in Art. 20. The polar planes of this point A'_1 with respect to F_1 and F_2 are

$$a_2u_3u_4\pi_{12}\pi_{34}x_2 + a_3u_2u_4\pi_{13}\pi_{42}x_3 + a_4u_2u_3\pi_{14}\pi_{23}x_4 = 0,$$
$$b_2u_3u_4\pi_{12}\pi_{34}x_2 + b_3u_2u_4\pi_{13}\pi_{42}x_3 + b_4u_2u_3\pi_{14}\pi_{23}x_4 = 0$$

which intersect in a line which is no other than d itself.

21. Similarly, we can show without any difficulty that the system of complex lines in a plane containing an edge, A_1A_2, of the tetrahedron of reference is made up of two flat pencils, one with center at A_1 and the other with center at A_2.

22. Now we examine further the relations between the lines and points in any plane u and the points and bisecants of the space cubic k^3 which u determines. Since every point on k^3 transforms itself into a line of the complex in u and every

point in u transforms itself into a line cutting across k^3 twice, let l_1 and l_2 be two lines in u corresponding to the points L_1 and L_2 on k^3, and then the point P in which l_1 and l_2 intersect goes into the line p cutting across k^3 in L_1 and L_2. Thus, a one-to-one correspondence exists between the points in u considered as the intersection of two tangents to the conic γ_2 of the complex and the bisecants cutting across the cubic k^3 in the two points to which the tangents to γ_2 correspond. If L_1 and L_2 approach coincidence, the rays l_1 and l_2 also approach coincidence. The p becomes a tangent to k^3 while the point P in u moves up to the conic γ^2. We can easily conclude that

A point without γ^2 goes into a real bisecant of k^3; on γ^2, a tangent; within γ^2, a line with no point in common with k^3.

23. We have seen that every conic γ'^2 by our method of transformation, goes into a quartic scroll. In particular, if γ'^2 is coincident with γ^2 every point of which goes into a tangent to k^3, the surface is a developable with k^3 as the cuspidal edge. But in general, γ'^2, when not coincident with γ^2, is cut by every tangent to γ^2 in two points; therefore, every point on k^3, in general, is a double point on the corresponding quartic surface. In other words, k^3 is a double curve. We can easily see that *the ruled sextic surface coming from a plane cubic contains k^3 as triple curve*, and that, *in general, the ruled surface of degree $2n$ coming from a plane curve of degree n contains k^3 as n-fold curve.*

24. From the figure of the preceding article we see that corresponding to P_1 on the conic γ'^2, which is the intersection of the complex lines l_1 and l_2, is the ruling p_1 on the corresponding quartic, cutting across k^3 in L_1 and L_2. But the ray l_2 intersects γ'^2 again in P_2 which is on another complex line l_3; therefore, on the quartic surface, from the point L_2 on k^3 issues another ruling p_2 meeting k^3 again in the point L_3. Continuing the process, we see that *from every point on k^3 go two generators of the quartic meeting the curve again in two different points, from each of which goes another generator.* But if a ray of the complex is tangent to the given conic γ'^2, i.e., cuts it in two coincident points, then corresponding to the two coincident points P_4 and P_5 common to γ'^2 and l_5 are two coincident generators from the point L_5 on k^3 on the quartic. Then L_5 is a pinch-point on the surface. We may thus conclude that *the existence of a line of the complex tangent to a given plane curve means the existence of a pinch-point on the corresponding surface*, and that, in particular, *a ruled quartic with a double space cubic which does not degenerate can have no more than four pinch-points.* If γ'^2 intersects γ^2 in two real points, two of the pinch-points become imaginary, for the two conics can have only two real common tangents.

SECTION III

25. The most general equation of the ruled quartic from any conic is given in Art. 9. Let the lines p'_{ik}, p''_{ik}, p'''_{ik} be three lines of the complex in the plane u, i.e., form a circumscribed triangle to γ^2, and further let P', P'', P''' be the three points on k^3 to which p'_{ik}, p''_{ik}, p'''_{ik} correspond respectively. Finally, let p'_{ik} and p''_{ik} intersect in R''', p''_{ik} and p'''_{ik} in R', p'''_{ik} and p'_{ik} in R''. Without loss of generality, let a given conic in the plane pass through R'' and R''' but not R'. Then the corresponding quartic surface is given by

$$A_{11}Q'^2+2A_{12}Q'Q''+2A_{13}Q'''Q'+2A_{23}Q''Q'''=0$$

where the Q's are the left members of the equations of the quadric cones corresponding to the lines p'_{ik}, p''_{ik}, p'''_{ik} (Art. 9). This equation can be written

$$\sum_1^4 \pi_{ik}\pi_{jl}\left[\sum_1^3 A_{rs}p^{(r)}_{ik}p^{(s)}_{jl}\right]x_ix_kx_jx_l=0 \qquad [A_{22}=A_{33}=0].$$

26. Now corresponding to the points R', R'', R''' are three lines r'_{ik}, r''_{ik}, r'''_{ik} forming a triangle with its vertices P', P'', P''' on k^3. Let the plane of this triangle be denoted by u' and its complex cubic by k'^3. The above quartic surface is intersected by u' in a conic γ''^2 through P'' and P''' and two straight lines which are rulings of the surface corresponding to the points R'' and R'''. Note that R', R'', R''', being three points whose transforms are three lines in u', are three points of the space cubic k'^3 which u' determines. Then the conic γ''^2 in u' will go into a quartic surface which is intersected by u in the conic γ'^2 and the lines p''_{ik} and p'''_{ik} which are rulings of the surface. Therefore, we conclude:

27. *Any conic through two vertices of a triangle whose sides are complex lines (or through two points on any complex space cubic) goes into a quartic surface all of whose generators intersect another conic through two vertices of another triangle whose sides are complex lines (or through two points on another complex cubic). The quartic corresponding to the latter conic will have its rulings all going through the former conic.* This class of quartics we designate as class I (III of Sturm).

28. But if the given conic γ'^2 goes through R' also, we have class II (IV of Sturm). The equation of this quartic surface is

$$A_{12}Q'Q''+A_{13}Q'Q'''+A_{23}Q''Q'''=0$$

which is the same as the equation of the quartic surface given in Art. 9 after having imposed the condition that $r\neq s$. This surface is intersected by the plane u' in four straight lines three of which are r'_{ik}, r''_{ik}, r'''_{ik} and the fourth which we designate by s_{ik} is the very line whose corresponding quadric is intersected by u in the given conic γ'^2. Every plane through s_{ik} cuts the surface in three generators and this is to be expected, for there are an infinite number of triangles with vertices on γ'^2 and sides tangent to γ^2 [C. Smith, *Conic Sections*, p. 275].

29. These two classes have a space cubic for double curve. Class I has all its generators going through a conic with two points in common with the double curve and class II has all its generators going through a straight line which has no point in common with the double curve. The former reciprocates into one of the same kind, while the latter reciprocates into one with a triple line [see below, class *VIII*].

30. Now let the plane u pass through A_1, a vertex of the fundamental tetrahedron. The k^3 of this plane is made up of the conic c^2 in the plane x_1 and a straight line d, and the system of complex lines is made up of two flat pencils A_1 and A'_1 (Art. 20). Every conic γ'^2 not going through A_1 in u transforms itself into a quartic with c^2 as double conic and d as double line. From every point on d issue two generators meeting c^2 in two different points and vice versa. It can be easily seen that there are in general two pinch-points on c^2 and two on d, for there can be in general two tangents from A_1 and two from A'_1 to γ'^2.

31. Let p'_{ik} and p''_{ik} be two rays of A'_1 and $p'''_{ik}(x'''_2 : x'''_3 : x'''_4 : 0 : 0 : 0)$ be a ray of A_1, meeting p'_{ik} in R'' and p''_{ik} in R'. Without loss of generality we can let a given conic γ^2 pass through R' and R'' (but not through A_1 nor A'_1) and the equation of its corresponding quartic is the same as that of class I given at the end of Art. 25 after making the necessary changes in the p_{ik}'s. This quartic is intersected by the plane u' determined by the line d (corresponding to A'_1) and the lines r'_{ik} and r''_{ik} (corresponding to R' and R'') in a conic γ'^2 and two generators (r'_{ik} and r''_{ik}). This conic γ'^2 goes back into a quartic whose section by u is no other than the given conic γ'^2 and the two lines p'_{ik} and p''_{ik}. Therefore, *this quartic has all its generators going through a conic.* This is class III (V of Sturm). If γ'^2 is tangent to the line $A_1A'_1$, one of the pinch-points *kn* c^2 and one of the pinch-points on d fall together at the point common to c^2 and d.

32. If γ'^2 goes through A'_1 also, its transform is of class IV (VI of Sturm) whose equation can be easily obtained by putting $A_{33}=0$ in the equation of class III. The plane section by u' is made up of four straight lines r'_{ik}, r''_{ik}, the line d, and a fourth line ρ_{ik} whose corresponding quadric is cut by u in the given conic γ'^2 itself. This quartic, every one of whose generators, besides meeting c^2 and d, meets this line ρ_{ik}, reciprocates into one with a triple line [see class IX below]. Note that there is one pinch-point on d and two on c^2, for there is only one ray of the pencil A'_1 and two of the pencil A_1 tangent to γ'^2.

33. If the edge A_1A_2 of the fundamental tetrahedron lies in the plane u, any conic γ'^2 in it not going through A_1 and A_2, the centers of the two flat pencils making up the complex conic γ^2 (Art. 21), transforms itself into a quartic surface whose double cubic is $d+d'+e$. Letting p'_{ik} and p''_{ik} be two rays of pencil A_2 and p'''_{ik} be a ray of pencil A_1, we have

$$p'_{13}=p'_{14}=p'_{34}=0,\ p''_{13}=p''_{14}=p''_{34}=0,\ p'''_{23}=p'''_{42}=p'''_{34}=0;$$

and putting these in the equation of the quartic in Art. 9, we have the resulting equation representing the quartic in question. This is class V (VII of Sturm). There are in general two pinch-points on each of the lines d and d'. e, to which every point on A_1A_2 but not A_1 and A_2 is transformed, for γ'^2 has two points, real or imaginary, in common with A_1A_2. But if the given conic γ'^2 is tangent to A_1A_2 e is a cuspidal ruling and d and d' have each only one pinch-point.

34. To obtain the next class, VI (I of Sturm), of quartics which have only two double lines and no double generator, we take a cubic curve without a double point in the plane u through the points A_1 and A_2. The corresponding surface is a sextic made up of the two planes x_1 and x_2 corresponding to A_1 and A_2 respectively and a quartic.

35. Let the equation of a cubic cone with its vertex at $A_4(0:0:0:1)$ and with two elements one through A_1 and the other through A_2 be

$$\sum_1^3 A_{rst}x_rx_sx_t=0 \qquad [A_{111}=A_{222}=0].$$

The plane section of this cone by the plane $u(0:0:u_3:u_4)$ is a cubic curve through A_1 and A_2. Letting p'_{ik}, p''_{ik}, p'''_{ik} be the coördinates of the lines in which the planes x_1, x_2, x_3 intersect u, we have

$$\begin{array}{ll} p'_{ik}(0:0:0:0:p'_{42}:p'_{23}) & [\text{where } p'_{42}=u_3;\ p'_{23}=u_4], \\ p''_{ik}(0:p''_{13}:p''_{14}:0:0:0) & [\text{where } p''_{13}=u_4;\ p''_{14}=-u_3], \\ p'''_{ik}(p'''_{12}:0:0:0:0:0:) & [\text{where } p'''=u_4]. \end{array}$$

Then the quadrics corresponding to these lines are respectively

$$\begin{aligned} Q' &\equiv x_2[\pi_{42}p'_{43}x_4+\pi_{23}p'_{23}x_3]=0, \\ Q'' &\equiv x_1[\pi_{13}p''_{13}x_3+\pi_{14}p''_{14}x_4]=0, \\ Q''' &\equiv \pi_{12}p'''_{12}x_1x_2=0. \end{aligned}$$

Therefore, the sextic surface corresponding to the cubic curve is given by

$$\sum_1^3 A_{rst}Q^{(r)}Q^{(s)}Q^{(t)}=0 \qquad [A_{111}=A_{222}=0].$$

which, when expanded, is

$$x_1x_2\ [\text{a quartic factor}]=0.$$

36. The factor enclosed in the bracket equated to zero gives a quartic surface. If the cubic has a double point, this surface is of class V, for it has d and d' for double directrices and has a double generator which, however, is different from e. But if we impose upon the A_{rst}'s the condition that the cubic be without a double point, the surface will be without a double generator.

37. This surface has four pinch-points on each of the two double directrices, for, the class of the cubic curve without a double point being six, there can be drawn from a point on it four tangents exclusive of the one at the point itself.

38. Note that the equation of the degenerate sextic surface given in Art. 35 could have been obtained from the sextic equation in Art. 9 by putting A_{111}, A_{222}, p'_{12}, p'_{13}, p'_{14}, p'_{34}, p''_{12}, p''_{34}, p''_{42}, p''_{23}, p'''_{13}, p'''_{14}, p'''_{34}, p'''_{42}, p'''_{23} all equal to zero [Art. 35].

39. Now if we put in the equation of the cone in Art. 35

$$A_{111}=A_{112}=A_{113}=0,$$

the result is that of a cone with

$$x_2=0,\ x_3=0$$

as double element, thus giving a plane curve in u with a double point at A_1. With A_{111}, A_{112}, A_{113}, p'_{12}, p'_{13}, p'_{14}, p'_{34}, p''_{12}, p''_{34}, p''_{42}, p''_{23}, p'''_{13}, p'''_{14}, p'''_{34}, p'''_{42}, p'''_{23} all put equal to zero, the left member of the sextic equation in Art. 9 factors into x^2_1 [a quartic factor]. $x^2_1=0$ represents a double plane corresponding to the double point A_1. The quartic factor equated to zero represents a quartic surface which has the line d' as a triple line and d as a simple line. From every point on d' issue three rulings corresponding to the three points in which every line of the pencil A_2 intersects the given cubic; and from every point on d issues only one generator corresponding to the point in which every ray of the pencil A_1 meets the cubic besides A_1 itself. This class of quartics is class VII (X of Sturm). Since from the point A_2 which is not on the curve only four tangents can be drawn to the curve, the triple line has four pinch-points, i.e., points at which two of the three generators become coincident.

40. To get class VIII (IX of Sturm), we must return to the case where the plane u passes through the vertex A_1 of the fundamental tetrahedron. Remembering that the coördinates of A'_1 are [Art. 20].

$$(0 : \pi_{12}\pi_{34}u_3u_4 : \pi_{13}\pi_{42}u_4u_2 : \pi_{14}\pi_{23}u_2u_3)$$

and letting $A'_1A''_1$ and $A_1A''_1$ be the lines of intersection of u with x_1 and x_2 respectively, thus giving $(0 : 0 : u_4 : -u_3)$ for the coördinates of A''_1, we have

$$\begin{aligned} A'_1A''_1 &\equiv p'_{ik}(0 : 0 : 0 : u_2 : u_3 : u_4), \\ A_1A''_1 &\equiv p''_{ik}(0 : u_4 : -u_3 : 0 : 0 : 0), \\ A_1A'_1 &\equiv p'''_{ik}(\pi_{12}\pi_{34}u_3u_4 : \pi_{13}\pi_{42}u_4u_2 : \pi_{14}\pi_{23}u_2u_3 : 0 : 0 : 0). \end{aligned}$$

To these three lines correspond the following quadrics, respectively:

$$\begin{aligned} Q' &\equiv \pi_{34}u_2x_3x_4+\pi_{42}u_3x_4x_2+\pi_{23}u_4x_2x_3=0, \\ Q'' &\equiv x_1(\pi_{13}u_4u_3-\pi_{14}u_3x_4)=0, \\ Q''' &\equiv x_1(\pi^2_{12}\pi_{34}u_3u_4x_2+\pi^2_{13}\pi_{42}u_2u_4x_3+\pi^2_{14}\pi_{23}u_2u_3x_4)=0. \end{aligned}$$

Then the equation of the sextic surface corresponding to any cubic in u with a double point at A_1 is given by

$$\sum_1^3 A_{rst}Q^{(r)}Q^{(s)}Q^{(t)}=0. \qquad [A_{111}=A_{112}=A_{113}=0].$$

The result of expanding this equation which is the same as that obtained from the sextic equation in Art. 9 by putting $A_{111}=A_{112}=A_{113}=0$ and by assigning to the p_{ik}'s their respective values given above, is an equation whose left member is made up of two factors, x^2_1 and a quartic factor. $x^2_1=0$ gives a double plane corresponding to the double point A_1, and the quartic factor equated to zero represents a quartic surface which has a triple line coincident with d from every point of which issue three generators meeting the conic c^2 in x_1 in three different points. Quartics of this class have in general four pinch-points on d, and their reciprocals are of class II already considered.

41. If the cubic goes through A'_1, then $A_{222}=0$, and the corresponding quartic surface has one of its generators coincident with the triple line d. This is of class IX (XI of Sturm).

SECTION IV

42. So far we have been able to obtain nine classes of ruled quartics; to obtain the remaining classes we need to alter the machinery of transformation. Let the fundamental quadrics touch each other along a straight line, and, for the sake of simplicity, let their equations be

$$S_1 \equiv a_{11}x^2_1+a_{22}x^2_2+2a_{23}x_2x_3+2a_{14}x_1x_4=0,$$
$$S_2 \equiv b_{11}x^2_1+b_{22}x^2_2+2b_{23}x_2x_3+2b_{14}x_1x_4=0 \qquad [a_{23} : b_{23}=a_{14} : b_{14}],$$

the line of tangency being $x_1=0$, $x_2=0$. Any point Y (y_1, y_2, y_3, y_4) in space goes into the line

$$(a_{11}y_1+a_{14}y_4)x_1+(a_{22}y_2+a_{23}y_3)x_2+a_{23}y_3x_3+a_{14}y_1x_4=0,$$
$$(b_{11}y_1+b_{14}y_4)x_1+(b_{22}y_2+b_{23}y_3)x_2+b_{23}y_2x_3+b_{14}y_1x_4=0,$$

intersecting the line $x_1=0$, $x_2=0$. Therefore, all the lines thus obtained cut across the line of tangency of the two given quadrics. To the point $(0 : 0 : l : m)$ on this line of tangency correspond the common tangent planes to S_1 and S_2 at the point. Every point in x_1 goes into a ray through $A_4(0 : 0 : 0 : 1)$ and every point in x_2 goes into a ray through A_3 $(0 : 0 : 1 : 0)$. Since there is nothing special about the plane x_4 except that it is a face of the tetrahedron of reference, we shall find the remaining classes of quartics corresponding to curves in this plane.

43. The system of common tangent planes along $x_1=0$, $x_2=0$ is intersected by the plane x_4 in a flat pencil of rays with center at A_3. Since any point in x_4 is on one of these lines and hence in one of these common tangent planes, its corresponding ray must cut across the line x_1x_2 at the point at which the plane containing the given point is tangent to the two quadrics; but every point on $x_1=0$, $x_4=0$ goes into the same line $x_2=0$, $x_3=0$. Therefore, every conic γ'_2 in x_4 goes into a quartic surface with a double ruling $x_2=0$, $x_3=0$ which corresponds to the two points which γ'^2 has in common with $x_1=0$, $x_4=0$. From every point of the line of tangency of S_1 and S_2 issue two generators of the quartic surface, corresponding to the two

points in which every ray of the pencil A_3 cuts γ'^2; therefore, this line is a double line. Furthermore, it is a line along which two double lines coincide, for every ray of the pencil A_3 is a line in which two coincident planes tangent to S_1 and S_2 intersect x_4. This quartic is of class X (VIII of Sturm), and its equation is obtained as follows:

Let

$$\sum_1^3 A_{ij}x_ix_j=0$$

be the equation of any cone whose plane section in x_4 is a conic γ'^2. Any line

$$u_1x_1+u_2x_2+u_3x_3=0,\ x_4=0$$

goes into a quadric

$$(u_3\varphi_{22,\ 11}+u_2\varphi_{11,\ 23})x_1x_2+u_3\varphi_{23,\ 11}x_1x_3+u_2\varphi_{23,\ 22}x_2+u_3\varphi_{22,\ 14}x_2x_4=0$$

where

$$\varphi_{ij,\ kl}=a_{ij}b_{kl}-a_{kl}b_{ij}$$

and

$$\varphi_{ij,\ kl}=-\varphi_{kl,\ ij}.$$

Putting $u_1=1$, $u_2=u_3=0$, we have

$$Q_1\equiv\varphi_{22,\ 23}x^2{}_2=0,$$

the quadric corresponding to the line $x_1=0$, $x_4=0$. Putting $u_2=1$, $u_1=u_3=0$, we have

$$Q_2\equiv\varphi_{11,\ 23}x_1x_2=0,$$

the quadric corresponding to $x_2=0$, $x_4=0$. Finally, putting $u_3=1$, $u_1=u_2=0$, we have the quadric

$$Q_3\equiv\varphi_{11,\ 23}x_1x_2+\varphi_{11,\ 23}x_1x_3+\varphi_{14,\ 22}x_2x_4=0$$

corresponding to the line $x_3=0$, $x_4=0$. Then the equation of the quartic is given by

$$\sum_1^3 A_{ij}Q_iQ_j=0$$

or, when expanded,

$$A_{11}\varphi^2{}_{22,\ 23}x^4{}_2+2A_{13}\varphi_{11,\ 22}\varphi_{22,\ 23}x_1x^3{}_2+2A_{12}\varphi_{11,\ 23}\varphi_{22,\ 23}x_1x^3{}_3+2A_{13}\varphi_{22,\ 23}\varphi_{14,\ 22}x^3{}_2x_4$$
$$+(A_{22}\varphi^2{}_{11,\ 23}+2A_{23}\varphi_{11,\ 22}\varphi_{11,\ 23}+A_{33}\varphi^2{}_{11,\ 22})x^2{}_1x^2{}_2+A_{23}\varphi^2{}_{11,\ 23}x^2{}_1x^2{}_3+A_{33}\varphi^2{}_{14,\ 22}x^2{}_2x^2{}_4+2\varphi_{11,\ 23}(A_{33}\varphi_{11,\ 22}+A_{23}\varphi_{11,\ 23})x^2{}_1x_2x_3+2\varphi_{42}(A_{33}\varphi_{11,\ 22}+A_{23}\varphi_{11,\ 23})$$
$$x_1x^2{}_2x_4+2A_{33}\varphi_{11,\ 23}\varphi_{14,\ 22}x_1x_2x_3x_4=0.$$

44. Putting $x_2=0$, we have $x^2{}_3=0$, and vice versa, showing that the line $x_2=0$, $x_3=0$ is a double generator. Putting $x_4=0$, we have the curve of intersection of the surface in x_4, having a node at A_1 and a tacnode at A_3 with the line A_2A_3 as tangent.

45. The quartics of class XI (II of Sturm) which have two coincident double directrices but no double generator, can be obtained from a cubic in x_4, tangent to the line A_2A_3 [$x_1=0$, $x_4=0$] at A_3. This curve must not have a double point, for then the corresponding surface would have a double generator; nor must it go

through the point $A_1(1 : o : o : o)$ whose transform is the plane x_1, for then it would be impossible to obtain a quartic surface. Then to the given cubic we have corresponding a sextic surface which is made up of the plane x_2 counted twice corresponding to the point A_3 of contact between the curve and A_2A_3, and a quartic which has no double ruling.

46. Let

$$\sum_{1}^{3} A_{ijk}x_ix_jx_k=0, \quad x_4=0 \qquad [A_{333}=A_{233}=0]$$

be such a cubic curve, which transforms itself into the sextic surface

$$\sum_{1}^{3} A_{ijk}Q_iQ_jQ_k=0 \qquad [A_{333}=A_{233}=0]$$

where the Q's have the same meaning as those of Art. 43. This equation, if expanded, would be

$$x^2_2[\text{a quartic factor}]=0.$$

The bracketed factor equated to zero is the equation of the required quartic surface.

47. If, in addition, A_{133} is put equal to zero, we have a cubic curve with a node at A_3, and the corresponding surface is of class XII (XII of Sturm, which has the line $x_1=0$, $x_2=0$ for triple line, that is, a line through which three sheets of the surface pass. If the double point at A_3 becomes a cusp, only one of the sheets passes through the triple line, while the other two unite into a cuspidal sheet. This is a special case of the above but Salmon made a separate class (V according to his enumeration) of it.

48. This concludes the study and classification of ruled quartic surfaces. It may be added that classes could have been obtained from plane quartics with their singular points at one or two of the vertices of the fundamental tetrahedron, from space cubis not belonging to the complex through one vertex of the fundamental tetrahedron and also from space quartics through all the four vertices of the fundamental tetrahedron, but none can come from quintic curves or curves of higher order, plane or space.

Many thanks are due to Professor D. N. Lehmer, without whose kind and patient guidance this work would have been an impossibility.

UNIVERSITY OF CALIFORNIA PUBLICATIONS
IN
MATHEMATICS

Vol. 1, No. 18, pp. 389-400, 1 figure in text December 19, 1923

A SPECIAL QUARTIC CURVE

BY
ELSIE JEANNETTE McFARLAND

The special quartic curve whose equation is

$$F \equiv ax^4 + by^4 + cz^4 + 2fy^2z^2 + 2gx^2z^2 + 2hx^2y^2 = 0 \tag{1}$$

was studied by Hilda Fay Webb, of the University of California, and the results of her investigation are to be found in her thesis for the Master's degree, submitted in May, 1915.

Miss Webb found the quartic to be a non-singular curve with four double tangents passing through each vertex of the fundamental triangle. The six pairs so obtained form a "syzygetic sextuple." She also found that when the general quartic is represented by $UW = V^2$, the complete condition for the reducibility of the general quartic equation to the special form $F = 0$ is as follows:

Two vertices of bitangent pairs of the same set (of six pairs) must coincide; and the conics U, V, and W must cut the polar of the coincident vertices with respect to the contact conic in pairs of an involution.

It is my purpose in this paper to investigate still further the properties of this special quartic curve, and to set up other conditions for the reducibility of the general equation of the fourth degree to the special form $F = 0$.

THE BITANGENTS

The equation of the quartic being given as $ax^4 + by^4 + cz^4 + 2fy^2z^2 + 2gx^2z^2 + 2hx^2y^2 = 0$, the four double tangents through the point (0, 0, 1) will be of the form

$$\begin{aligned} x &= \pm A_1 y \\ x &= \pm A_2 y \end{aligned} \tag{2}$$

where $\pm A_1$ and $\pm A_2$ are the roots of the equation:

$$(g^2 - ac)A^4 + 2(fg - ch)A^2 + (f^2 - bc) = 0 \tag{3}$$

Those through the point (1, 0, 0) are

$$\begin{aligned} y &= \pm B_1 z \\ y &= \pm B_2 z \end{aligned} \tag{4}$$

where $\pm B_1$ and $\pm B_2$ are the roots of the equation

(5) $$(h^2-ab)B^4+2(gh-af)B^2+(g^2-ac)=0$$

Those through the point (0, 1, 0) are

(6) $$z=\pm C_1x$$ $$z=\pm C_2x$$

where $\pm C_1$ and $\pm C_2$ are the roots of the equation

(7) $$(f^2-bc)C^4+2(fh-bg)C^2+(h^2-ab)=0.$$

We see from equations (2), (4), and (6) that each set of four double tangents through a vertex contains two pairs, the lines of which are harmonically divided by the sides of the reference triangle through that vertex. For example, the lines $z=C_1x$ and $z=-C_1x$ are harmonically divided by $x=0$ and $z=0$.

The quartic $F=0$ is transformed into itself by the following collineations:

(a)	(b)	(c)	(d)
$x=x'$	$x=-x'$	$x=\ \ x'$	$x=-x'$
$y=y'$	$y=-y'$	$y=-y'$	$y=\ \ y'$
$z=z'$	$z=\ \ z'$	$z=\ \ z'$	$z=\ \ z'$

If $z=px+qy$ touches $F=0$ twice, then, since (a), (b), (c), and (d) transform F into itself, the three lines $z=-px-qy$, $z=px-qy$ and $z=-px+qy$, obtained by applying (b), (c), and (d) to $z=px+qy$, are likewise double tangents of the quartic.

Similarly, if $z=px+qy$ is an inflexion tangent, $z=-px-qy$, $z=px-qy$, and $z=-px+qy$ are likewise inflexion tangents.

Let us now take as a double tangent the line $z=px+qy$. The equation

(8) $$(a+cp^4+2gp^2)x^4+4(cp^3q+gpq)x^3y$$
$$+6\left(cp^2q^2+\frac{fp^2}{3}+\frac{gq^2}{3}+\frac{h}{3}\right)x^2y^2+4(cpq^3+fpq)xy^3+(b+cq^4+2fq^2)y^4=0,$$ obtained by substituting $px+qy$ for z in (1) must be a perfect square.

If (8) is written: $Ax^4+4Bx^3y+6Cx^2y^2+4Dxy^3+Ey^4=0$, it will be a perfect square if

(9) $$\frac{AC-B^2}{A}=\frac{CE-D^2}{E}, \text{ or } AD^2-EB^2=0$$

and $$\frac{AC-B^2}{A}=\frac{AD-BC}{2B}, \text{ or } 3ABC-2B^3-A^2D=0$$

The resulting conditions on p and q are:

(10) $(ac^2-cg^2)Q^2+(2acf-2fg^2)Q+(cf^2-bc^2)P^2+(2f^2g-2bcg)P+(af^2-bg^2)=0.$

(11) $(ac^2-cg^2)P^2Q+(ag^2-a^2c)Q+(c^2h-cfg)P^3+(3cgh-acf-2fg^2)P^2+(ach+2g^2h-3afg)P+(agh-a^2f)=0$ where $P=p^2$ and $Q=q^2$.

(12) $pq=0$. This condition merely gives us the double tangents through (1, 0, 0) and (0, 1, 0).

Eliminating Q between (10) and (11) we arrive at an equation of the sixth degree in P.

(13) $(c^5h^2-2c^4fgh+ac^4f^2-abc^5+bc^4g^2)P^6+(6ac^3f^2g-8c^3fg^2h-4ac^4fh+6c^4gh^2-2abc^4g+2bc^3g^3)P^5+(2a^2c^3f^2+13c^3g^2h^2+13ac^2f^2g^2+2ac^4h^2-20ac^3fgh-10c^2fg^3h-3abc^3g^2+bc^2g^4+2a^2bc^4)P^4+(8a^2c^2f^2g-4a^2c^3fh-32ac^2fg^2h+8ac^3gh^2+12acf^2g^3-4cfg^4h+12c^2g^3h^2+4a^2bc^3g-4abc^2g^3)P^3+(10a^2cf^2g^2+a^2c^3h^2+4cg^4h^2-10a^2c^2fgh-20acfg^3h+10ac^2g^2h^2+a^3c^2f^2+4af^2g^4+3a^2bc^2g^2-2abcg^4-a^3bc^3)P^2+(2a^3cf^2g-8a^2cfg^2h+2a^2c^2gh^2+4acg^3h^2+4a^2f^2g^3-4afg^4h-2a^3bc^2g+2a^2bcg^3)P+(a^2cg^2h^2+a^3f^2g^2-2a^2fg^3h-a^3bcg^2+a^2bg^4)=0.$

This equation, however, contains the extraneous factor $c^2P^2+2cgP+g^2$. Accordingly the equation in P reduces to

(14) $c^2(ch^2-2fgh+af^2-abc+bg^2)P^4+4c(af^2g-fg^2h-acfh+cgh^2)P^3+2(a^2cf^2+2cg^2h^2-6acfgh+a^2bc^2-abcg^2+2af^2g^2+ac^2h^2)P^2+4a(af^2g-fg^2h-acfh+cgh^2)P+a^2(ch^2-2fgh+af^2-abc+bg^2)=0.$

This equation can, of course, be solved, giving us four values of $P: P_1, P_2, P_3$, and P_4. We have then four corresponding values of $Q: Q_1, Q_2, Q_3$, and Q_4.

Sixteen double tangents are given by the equations:

(15)
$$z=\pm p_1x\pm q_1y \qquad z=\pm p_2x\pm q_2y$$
$$z=\pm p_3x\pm q_3y \qquad z=\pm p_4x\pm q_4y$$
where $p_1=\sqrt{P_1}$ and $q_1=\sqrt{Q_1}$, etc.

Let us represent the four double tangents $z=\pm p_ix\pm q_iy$ by the numbers 1, 2, 3, and 4, and the intersection of any two, say 1 and 2, by (1, 2). Then the points (1, 2) and (3, 4) lie on one side of the reference triangle, points (1, 3) and (2, 4) on another side, and points (1, 4) and (2, 3) on the third side.

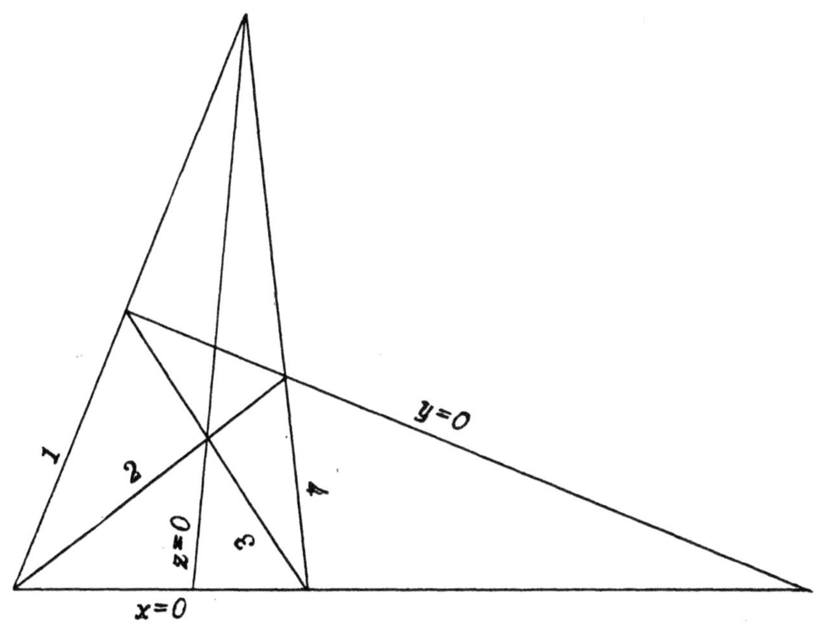

Arrangement of the Bitangents in Steiner Complexes

If we pair together the double tangents $x=A_1y$, $x=-A_1y$, and set up the identical relation $F\equiv ax^4+by^4+cz^4+2fy^2z^2+2gx^2z^2+2hx^2y^2$.

(16) $\equiv(x^2-A_1^2y^2)(lx^2+my^2+nz^2+rxy+sxz+tyz)-(l'x^2+m'y^2+n'z^2+r'xy+s'xz+t'yz)^2$, we find by equating coefficients and expressing the quantities $l, m, n, r, s, t, l', m', n', r', s', t'$ in terms of l' that we have:

$$(17)\quad \begin{cases} l=a+l'^2 \\ m=\dfrac{-b-\left\{\dfrac{f}{\sqrt{-c}}+\dfrac{A_1^2g}{\sqrt{-c}}+A_1^2l'\right\}^2}{A_1^2} \\ n=2g+2l'\sqrt{-c} \\ r=s=t=0 \end{cases} \qquad \begin{cases} l' \text{ is arbitrary.} \\ m'=-\left\{\dfrac{f}{\sqrt{-c}}+\dfrac{A_1^2g}{\sqrt{-c}}+A_1^2l'\right\} \\ n'=\sqrt{-c} \\ r'=s'=t'=0 \end{cases}$$

We then get

$$(18)\quad F\equiv[x^2-A_1^2y^2]\left[(a+l'^2)x^2-\left\{\frac{b+\left(\dfrac{f+A_1^2g}{\sqrt{-c}}+A_1^2l'\right)^2}{A_1^2}\right\}y^2+(2g+2l'\sqrt{-c})z^2\right.$$

$$\left.-\left[l'x^2-\left(\frac{f+A_1^2g}{\sqrt{-c}}+A_1^2l'\right)y^2+(\sqrt{-c})z^2\right]^2\right.$$ where l' is arbitrary and takes the place of λ as used by Weber throughout his chapter on the double tangents of a quartic.*

Rearranging according to powers of l':

$$(19)\quad \begin{aligned} F\equiv[x^2-A_1^2y^2]&\left[\left\{ax^2-\left(\frac{b+\dfrac{f^2+2fA_1^2g+A_1^4g^2}{-c}}{A_1^2}\right)y^2+2gz^2\right\}\right.\\ &\left.+2l'\left\{-\left(\frac{fA_1^2+gA_1^4}{\sqrt{-c}.A_1^2}\right)y^2+(\sqrt{-c})z^2\right\}+l'^2\{x^2-A_1^2y^2\}\right]\\ &-\left[\left\{-\left(\frac{f+A_1^2g}{\sqrt{-c}}\right).y^2+(\sqrt{-c})z^2\right\}+l'\{x^2-A_1^2y^2\}\right]^2\\ &\equiv[x^2-A_1^2y^2][V+2\lambda U+\lambda^2x_1y_1]-[U+\lambda x_1y_1]^2 \end{aligned}$$

We have expressed F in canonical form with l' taking the place of λ.

If the variable conic $V+2\lambda U+\lambda^2x_1y_1\equiv a'x^2+b'y^2+c'z^2+2f'yz+2g'xz+2h'xy=0$, it will degenerate when $\begin{vmatrix} a' & h' & g' \\ h' & b' & f' \\ g' & f' & c' \end{vmatrix}=0.$

Since the variable conic contains only x^2, y^2, and z^2 terms, the determinant becomes $\begin{vmatrix} a' & 0 & 0 \\ 0 & b' & 0 \\ 0 & 0 & c' \end{vmatrix}=a'b'c'=0.$

When $a'=0$ we get the lines $y=\pm B_1z$, $y=\pm B_2z$.

When $b'=0$ we get the lines $z=\pm C_1x$, $z=\pm C_2x$.

When $c'=0$ we get only $x=\pm A_2y$, since c' is of the first degree in l'.

We see, therefore, that the twelve double tangents through the vertices of the fundamental triangle are paired as follows to form a Steiner complex:

* Weber, H., *Lehrbuch der Algebra*, Vol. 2.

(20) (1) $x=\pm A_1y$ (2) $x=\pm A_2y$ (3) $z=\pm C_1x$
(4) $z=\pm C_2x$ (5) $y=\pm B_1z$ (6) $y=\pm B_2z$

If, now, we pair $\begin{matrix} x+A_1y=0 \\ x+A_2y=0 \end{matrix}$ and $\begin{matrix} x-A_1y=0 \\ x-A_2y-0 \end{matrix}$

we shall get a second complex, and if we pair

$$\begin{matrix} x+A_1y=0 \\ x-A_2y=0 \end{matrix} \text{ and } \begin{matrix} x-A_1y=0 \\ x+A_2y=0 \end{matrix}$$

we shall get a third, and these three complexes will contain all twenty-eight double tangents.

Let us now obtain the complex determined by the pair

(21) $\begin{matrix} x+A_1y=0. \\ x+A_2y=0. \end{matrix}$ Let $ax^4+by^4+cz^4+2fy^2z^2+2gx^2z^2+2hx^2y^2\equiv$
$[x^2+(A_1+A_2)xy+A_1A_2y^2]\ [lx^2+my^2+nz^2+rxy+sxz+tyz]$
$-[l'x^2+m'y^2+n'z^2+r'xy+s'xz+t'yz]^2$

We find that:

$$l=a+\frac{f^2+2fm'\sqrt{-c}-m'^2c}{-R^2c}+\frac{2fg}{Rc}+\frac{2gm'\sqrt{-c}}{Rc}-\frac{g^2}{c}$$

(22)

$$m=\frac{b+m'^2}{R} \qquad m' \text{ is arbitrary.}$$

$$n=\frac{2f+2m'\sqrt{-c}}{R} \qquad n'=\sqrt{-c}$$

$$r=\frac{2Sfm'}{R^2\sqrt{-c}}+\frac{2Sm'^2}{R^2}-\frac{Sb}{R^2}-\frac{Sm'^2}{R^2} \qquad r'=\frac{Sf+Sm'\sqrt{-c}}{R\sqrt{-c}}$$

$$s=t=0 \qquad s'=t'=0$$

$$l^1=\frac{f+m'\sqrt{-c}}{R\sqrt{-c}}-\frac{g}{\sqrt{-c}}$$

where $\begin{matrix} S=A_1+A_2 \\ R=A_1A_2 \end{matrix}$, if $R=\sqrt{\dfrac{f^2-bc}{g^2-ac}}$

and $S\ =\ 2\sqrt{\dfrac{f^2-bc}{g^2-ac}-\dfrac{2(gf-hc)}{g^2-ac}}$, which may be verified from equation (3).

(23)

$$F\equiv[x^2+Sxy+Ry^2]\left[\left\{\left(a-\frac{f^2}{R^2c}+\frac{2fg}{Rc}-\frac{g^2}{c}\right)x^2+\frac{b}{R}y^2+\frac{2f}{R}z^2-\frac{Sb}{R^2}\cdot xy\right\}\right.$$
$$\left.+2m'\left\{\left(\frac{g\sqrt{-c}}{Rc}-\frac{f\sqrt{-c}}{R^2c}\right)x^2+\frac{\sqrt{-c}}{R}z^2+\frac{Sf}{R^2\sqrt{-c}}xy\right\}+m'^2\left\{\frac{x^2}{R^2}+\frac{Ry^2}{R^2}+\frac{Sxy}{R^2}\right\}\right]$$
$$-\left[\left\{\left(\frac{f}{R\sqrt{-c}}-\frac{g}{\sqrt{-c}}\right)x^2+\sqrt{-c}.\ z^2+\frac{Sf}{R\sqrt{-c}}.xy\right\}+m'\left\{\frac{x^2}{R}+y^2+\frac{S}{R}.xy\right\}\right]^2$$

(24)

$$\equiv[x^2+Sxy+Ry^2]\left[\left\{\left(a-\frac{f^2}{R^2c}+\frac{2fg}{Rc}-\frac{g^2}{c}\right)x^2+\frac{b}{R}y^2+\frac{2f}{R}z^2-\frac{Sb}{R^2}\cdot xy\right\}\right.$$

$$(24) \qquad +2\lambda\left\{\left(\frac{g\sqrt{-c}}{c}-\frac{f\sqrt{-c}}{Rc}\right)x^2+\sqrt{-c}.\,z^2+\frac{Sf}{R\sqrt{-c}}.xy\right\}+\lambda^2\{x^2+Sxy+Ry^2\}\Bigg]$$

$$-\left[\left\{\left(\frac{g\sqrt{-c}}{c}-\frac{f\sqrt{-c}}{Rc}\right)x^2+\sqrt{-c}.z^2+\frac{Sf}{R\sqrt{-c}}.xy\right\}+\lambda\{x^2+Sxy+Ry^2\}\right]^2$$

which is in canonical form. $\left(\lambda=\dfrac{m'}{R}\right)$

If the variable conic is to degenerate, the following determinant must vanish:

$$\begin{vmatrix} \left(a-\frac{f^2}{R^2c}+\frac{2fg}{Rc}-\frac{g^2}{c}+\frac{2\lambda g\sqrt{-c}}{c}-\frac{2\lambda f\sqrt{-c}}{Rc}+\lambda^2\right) & \left(\frac{-Sb}{2R^2}+\frac{\lambda Sf}{R\sqrt{-c}}+\frac{\lambda^2 S}{2}\right) & 0 \\ \left(\frac{-Sb}{2R^2}+\frac{\lambda Sf}{R\sqrt{-c}}+\frac{\lambda^2 S}{2}\right) & \left(\frac{b}{R}+\lambda^2 R\right) & 0 \\ 0 & 0 & \frac{2f}{R}+2\lambda\sqrt{-c} \end{vmatrix}=0$$

$$=\left(\frac{2f}{R}+2\lambda\sqrt{-c}\right)\left\{\left(R-\frac{S^2}{4}\right)\lambda^4+\left(\frac{2g\sqrt{-c}}{c}R-\frac{2f\sqrt{-c}}{c}+\frac{S^2f\sqrt{-c}}{Rc}\right)\lambda^3\right.$$

$$(25) \qquad +\left(\frac{b}{R}+aR-\frac{f^2}{Rc}+\frac{2fg}{c}-\frac{g^2R}{c}+\frac{S^2f^2}{R^2c}+\frac{S^2b}{2R^2}\right)\lambda^2$$

$$+\left(\frac{2bg\sqrt{-c}}{Rc}-\frac{2bf\sqrt{-c}}{R^2c}-\frac{S^2bf\sqrt{-c}}{R^3c}\right)\lambda$$

$$\left.+\left(\frac{ab}{R}-\frac{bf^2}{R^3c}+\frac{2bfg}{R^2c}-\frac{bg^2}{Rc}-\frac{S^2b^2}{4R^4}\right)\right\}=0$$

If $\frac{2f}{R}+2\lambda\sqrt{-c}=0$ we find that the variable conic becomes $R^2x^2+R^3y^2-SR^2xy=0$, or $(x-A_1y)\ (x-A_2y)=0$.

If the second factor is put equal to zero, we shall get four values of λ for which the variable conic will degenerate.

If we pair $(x+A_1y)$ and $(x-A_2y)$ we replace the quantity $S=A_1+A_2$ by $D=A_1,\ -A_2$ and $R=A_1A_2$ by $-R$. Hence it will be seen at once that the remaining double tangents can be found by replacing S by D and R by $-R$ in the expression for the variable conic and in equation (25).

Since the variable conic is of the form $a'x^2+b'y^2+c'z^2+2h'xy=0$, it can degenerate into $(z\text{-}px\text{-}qy)\ (z+px+qy)=0$, or $(z-px+qy)\ (z+px-qy)=0$.

To show that the complex determined by $\begin{cases} x+A_1y=0 \\ x+A_2y=0 \end{cases}$ contains either four pairs of type $\begin{matrix} z=\ \ px+qy \\ z=-px-qy, \end{matrix}$ or four pairs of type $\begin{matrix} z=\ \ px-qy \\ z=-px+qy, \end{matrix}$ let us set up the complex determined by one of the pairs, $\begin{matrix} z=\ \ px+qy \\ z=-px-qy. \end{matrix}$ The variable conic in this case is again of form $a'x^2+b'y^2+c'z^2+2h'xy=0$, as can be determined by carrying out the process of page seven.

It will degenerate if $c'=0$, or if $a'b'+h'^2=0$.

The coefficient c' is identical with λ^2+c, so that the complex contains two pairs of lines of the form $x=Ay$.

If $a'b'+h'^2=0$, the conic degenerates into

$$(z-px-qy)\ (z+px+qy)=0,$$

where p and q must satisfy conditions (10) and (11). It will be seen then that the same complex is determined by any one of the four pairs

$$\begin{matrix} z= \ \ p_1x+q_1y \\ z=-p_1x-q_1y, \end{matrix} \quad \begin{matrix} z= \ \ p_2x+q_2y \\ z=-p_2x-q_2y, \end{matrix} \quad \begin{matrix} z= \ \ p_3x+q_3y \\ z=-p_3x-q_3y, \end{matrix} \quad \text{or} \quad \begin{matrix} z= \ \ p_4x+q_4y \\ z=-p_4x-q_4y. \end{matrix}$$

Hence one of the two complexes determined by $\begin{matrix} x+A_1y=0 \\ x+A_2y=0 \end{matrix}$ and $\begin{matrix} x+A_1y=0 \\ x-A_2y=0 \end{matrix}$ must contain four pairs of type $\begin{matrix} z= \ \ px+qy \\ z=-px-qy \end{matrix}$ and the other four pairs of type $\begin{matrix} z= \ \ px-qy \\ z=-px+qy. \end{matrix}$

If we pair $\begin{matrix} z+C_1x=0 \\ z+C_2x=0 \end{matrix}$ and set up the identity $ax^4+by^4+cz^4+2fy^2z^2+2gx^2z^2+2hx^2y^2\equiv (z^2-Sxz+Rx^2)\ (lx^2+my^2+nz^2+rxz+syz+txy) - (l'x^2+m'y^2+n'z^2+r'xz+s'yz+t'xy)^2$, we find that $t=t'=s=s'=0$. Hence the variable conic is of the form $a'x^2+b'y^2+c'z^2+2g'xz=0$. It must then degenerate into a pair of straight lines $(z+px+qy)\ (z+px-qy)=0$, or else into a pair $(z-px+qy)\ (z-px-qy)=0$. The complex then contains the pair $\begin{matrix} z-C_1x=0 \\ z-C_2x=0 \end{matrix}$, and four pairs of the form $\begin{matrix} z=-px-qy \\ z=-px+qy \end{matrix}$ or of the form $\begin{matrix} z=px-qy \\ z=px+qy. \end{matrix}$

The complex determined by $\begin{matrix} z+C_1x=0 \\ z-C_2x=0 \end{matrix}$ will contain the pair $\begin{matrix} z-C_1x=0 \\ z+C_2x=0 \end{matrix}$ and the four pairs of double tangents which do not appear in the preceding complex.

In like manner we can show that of the two complexes determined by $\begin{matrix} y+B_1z=0 \\ y+B_2z=0 \end{matrix}$ and $\begin{matrix} y+B_1z=0 \\ y-B_2z=0 \end{matrix}$, one will contain four pairs of the type $\begin{matrix} z=-px-qy \\ z= \ \ px-qy \end{matrix}$ and the other four of the type $\begin{matrix} z=-px+qy \\ z= \ \ px+qy. \end{matrix}$

Although other groupings may be made in the same manner, those already given are the most striking. They may be summarized as follows:

To the complexes determined by $\begin{matrix} x+A_1y=0 \\ x+A_2y=0 \end{matrix}$ and $\begin{matrix} x+A_1y=0 \\ x-A_2y=0 \end{matrix}$ belong the pairs $\begin{matrix} z= \ \ px+qy \\ z=-px-qy \end{matrix}$ and $\begin{matrix} z=-px+qy \\ z= \ \ px-qy. \end{matrix}$

To the two determined by $\begin{matrix} y+B_1z=0 \\ y+B_2z=0 \end{matrix}$ and $\begin{matrix} y+B_1z=0 \\ y-B_2z=0 \end{matrix}$ belong the pairs $\begin{matrix} z=-px-qy \\ z= \ \ px-qy \end{matrix}$ and $\begin{matrix} z=-px+qy \\ z= \ \ px+qy. \end{matrix}$

To the two determined by $\begin{matrix} z+C_1x=0 \\ z+C_2x=0 \end{matrix}$ and $\begin{matrix} z+C_1x=0 \\ z-C_2x=0 \end{matrix}$ belong the pairs $\begin{matrix} z=-px-qy \\ z=-px+qy \end{matrix}$ and $\begin{matrix} z=px-qy \\ z=px+qy. \end{matrix}$

REDUCTION OF THE GENERAL QUARTIC TO THE SPECIAL FORM $ax^4+by^4+cz^4+2fy^2z^2+2gx^2z^2+2hx^2y^2=0$

In his *Higher Plane Curves*, Salmon refers to this quartic, stating that its equation contains implicitly eleven independent constants. This is shown by replacing x by $lx'+my'+nz'$, y by $l'x'+m'y'+n'z'$, and z by $l''x'+m''y'+n''z'$. F can then be written

$$al^4\left(x'+\frac{m}{l}y'+\frac{n}{l}z'\right)^4+bl'^4\left(x'+\frac{m'}{l'}y'+\frac{n'}{l'}z'\right)^4+cl''^4\left(x'+\frac{m''}{l''}y'+\frac{n''}{l''}z'\right)^4$$
$$+2fl'^2l''^2\left(x'+\frac{m'}{l'}y'+\frac{n'}{l'}z'\right)^2\cdot\left(x'+\frac{m''}{l''}y'+\frac{n''}{l''}z'\right)^2+2gl^2l''^2\left(x'+\frac{m}{l}y'+\frac{n}{l}z'\right)^2\cdot$$
$$\cdot\left(x'+\frac{m''}{l''}y'+\frac{n''}{l''}z'\right)^2+2hl^2l'^2\left(x'+\frac{m}{l}y'+\frac{n}{l}z'\right)^2\cdot\left(x'+\frac{m'}{l'}y'+\frac{n'}{l'}z'\right)^2=0,$$

an expression which evidently contains eleven independent constants. Thus we see that the fourteen independent constants in the equation of the general quartic must be subject to three conditions in order that it may be reducible to the special form $F=0$.

CONDITIONS ON THE COEFFICIENTS OF THE GENERAL QUARTIC

In order that the quartic whose equation contains the full fifteen terms shall be of the type discussed in this paper, its double tangents must be such that four of them pass through some point of the plane, four more through another point, and four more through a third point.

If we represent two different double tangents of the fifteen-term quartic (which we may call Q) by $\begin{matrix} z=p_ix+q_iy \\ z=p_jx+q_jy \end{matrix}$ where $i\neq j$ and where i and j assume values from 1 to 28, the equation of degree $\frac{28\text{x}27}{2}$ giving the value of $\frac{x}{y}$ for the points of intersection of the twenty-eight double tangents, will have for coefficients the elementary symmetric functions of the various quantities $\frac{q_j-q_i}{p_i-p_j}$. But since any interchange of the subscripts of the q's requires a corresponding interchange for the p's, it can be shown that the symmetric functions of $\frac{q_j-q_i}{p_i-p_j}$ are also symmetric functions of the quantities p and q; hence that they are rational functions of the coefficients of the quartic Q.

If Q is to be of type F, the equation of degree 14x27 (which may be denoted by $V(x,y)$,) must contain a cubic factor repeated six times. For since four double tangents go through a point, six points of intersection fall together. The same thing occurs for two other sets of four double tangents.

If V has this cubic factor occurring to the sixth degree, its first derivative V' will contain the cubic factor to the fifth degree. Hence V and V' must have a common factor of degree fifteen, which is a perfect fifth power. This can be found by the rational process of division. When this factor is found its fifth root can be extracted. This fifth root is a cubic which can be solved by radicals. The roots of it give $\frac{x}{y}$ for the three points through each of which pass four double tangents. We can then determine the lines joining these three points, and taking them as sides of a new reference triangle we can find the equation of Q referred to this new triangle. If the transformed equation now contains only even powers, Q was a special quartic of the type under consideration.

If we wish to set up conditions under which Q will be reducible to form F, we must first impose the condition that V and V' shall have a common factor of degree fifteen. This means that when sufficient steps have been taken in the process of finding the greatest common divisior, the remainder of degree fourteen must vanish identically. This implies fifteen conditions, which are not, of course, independent. This common factor of degree fifteen must be a perfect fifth power and conditions for that can be set up. These conditions are not, however, sufficient—or have not been proved so—since the pairs of double tangents through a vertex of a certain triangle must also be harmonically divided by the sides of the triangle through that vertex. Also the remaining sixteen double tangents are related in a special way to the sides of this same triangle. In order to determine the additional conditions we should transform to the new reference triangle, and equate the coefficients of odd powers, if such occur, to zero.

The very large number of conditions obtained in this way must reduce ultimately to three.

GEOMETRICAL CONDITIONS ON THE CURVE

If x_1y_1, x_2y_2, and x_3y_3 are three pairs of double tangents belonging to a Steiner complex, the general quartic can be written in the form $\sqrt{x_1 \cdot y_1}+\sqrt{x_2 \cdot y_2}+\sqrt{x_3 \cdot y_3}=0$. If we take the points of intersection of x_1 with y_1, x_2 with y_2, and x_3 with y_3 for the vertices of our fundamental triangle, we can write the equation of the quartic as: $p\sqrt{(x-ay)\ (x-by)}+q\sqrt{(y-cz)\ (y-dz)}+r\sqrt{(z-fx)\ (z-gx)}=0$, or, in rational form:

$$(27)\quad (p^4-2fgp^2r^2+f^2g^2r^4)\,x^4+(q^4-2abp^2q^2+a^2b^2p^4)y^4+(r^4-2cdq^2r^2+c^2d^2q^4)z^4+(a^2p^4+b^2p^4+4abp^4-2abfgp^2r^2-2fgq^2r^2-2p^2q^2)x^2y^2+(c^2q^4+d^2q^4+4cdq^4-2abcdp^2q^2-2abp^2r^2-2q^2r^2)y^2z^2+(f^2r^4+g^2r^4+4fgr^4-2cdfgq^2r^2-2cdp^2q^2-2p^2r^2)x^2z^2+2(-ap^4-bp^4+afgp^2r^2+bfgp^2r^2)x^3y+2(-cq^4-dq^4+abcp^2q^2+abdp^2q^2)y^3z+2(-fr^4-gr^4+cdfq^2r^2+cdgq^2r^2)z^3x+2(-a^2bp^4-ab^2p^4+ap^2q^2+bp^2q^2)xy^3+2(-c^2dq^4-cd^2q^4+cq^2r^2+dq^2r^2)yz^3+2(-f^2gr^4-fg^2r^4+fp^2r^2+gp^2r^2)zx^3+2\,(-afp^2r^2-agp^2r^2-bfp^2r^2-bgp^2r^2+cfgq^2r^2+dfgq^2r^2+cp^2q^2+dp^2q^2)x^2yz+2\,(-acp^2q^2-adp^2q^2-bcp^2q^2-bdp^2q^2+abfp^2r^2+abgp^2r^2+fq^2r^2+gq^2r^2)xy^2z+2\,(-cfq^2r^2-dfq^2r^2-cgq^2r^2-dgq^2r^2+acdp^2q^2+bcdp^2q^2+ap^2r^2+bp^2r^2)xyz^2=0.$$

Let us now set up the conditions that through each vertex of the reference triangle there shall pass a third double tangent to the quartic.

A line through the point (1, 0, 0), of the form $y=mz$, will meet the quartic in four points given by:

(28) $(p^4-2fgp^2r^2+f^2g^2r^4)x^4+2\{(-ap^4-bp^4+afgp^2r^2+bfgp^2r^2)m+(-f^2gr^4-fg^2r^4+fp^2r^2+gp^2r^2)\}x^3z+\{(a^2p^4+b^2p^4+4abp^4-2abfgp^2r^2-2fgq^2r^2-2p^2q^2)m^2+2(-afp^2r^2-agp^2r^2-bfp^2r^2-bgp^2r^2+cfgq^2r^2+dfgq^2r^2+cp^2q^2+dp^2r^2)m+(f^2r^4+g^2r^4+4fgr^4-2cdfgq^2r^2-2cdp^2q^2-2p^2r^2)\}x^2z^2+2\{(-a^2bp^4-ab^2p^4+ap^2q^2+bp^2q^2)m^3+(abfp^2r^2+abgp^2r^2+fq^2r^2+gq^2r^2-acp^2q^2-adp^2q^2-bcp^2q^2-bdp^2q^2)m^2+(acdp^2q^2+bcdp^2q^2+ap^2r^2+bp^2r^2-cfq^2r^2-dfq^2r^2-cgq^2r^2-dgq^2r^2)m+(-fr^4-gr^4+cdfq^2r^2+cdgq^2r^2)\}xz^3+\{(q^4-2abp^2q^2+a^2b^2p^4)m^4+2(-cq^4-dq^4+abcp^2q^2+abdp^2q^2)m^3+(c^2q^4+4cdq^4+d^2q^4-2abp^2r^2-2q^2r^2-2abcdp^2q^2)m^2+2(-c^2dq^4-cd^2q^4+cq^2r^2+dq^2r^2)m+(r^4-2cdq^2r^2+c^2d^2q^4)\}z^4=0$, which may be written:

$$Ax^4+4Bx^3z+6Cx^2z^2+4Dxz^3+Ez^4=0.$$

If the line $y-mz$ is to be a double tangent, the above equation must be a perfect square, which will be the case if

(1) $AD^2-EB^2=0$ and

(2) $3ABC-2B^3-A^2D+0$,

or

(1) $m\{m^2-(c+d)m+cd\}\{(f+g)(a+b)(q^2-abp^2)m^2+[(f+g)^2abr^2-(f+g)(a+b)(c+d)q^2+(a+b)^2p^2]m+(f+g)(a+b)(cdq^2-r^2)\}=0.$

(2) $\{m^2-(c+d)m+cd\}\{(a+b)(-fg)m+(f+g)\}=0.$

Since m is an extraneous factor, and $m=c$ or d gives us the original double tangents $y-cz=0$ and $y-dz=0$, we may disregard the first two factors of (1) and the first of (2), and concern ourselves with the two equations:

$(f+g)(a+b)(q^2-abp^2)m^2+\{(f+g)^2abr^2-(f+g)(a+b)(c+d)q^2+(a+b)^2p^2\}m+(f+g)(a+b)(cdq^2-r^2)=0.$

and $-fg(a+b)m+(f+g)=0.$

The eliminant of these equations is:

(29) $E_2\equiv(f+g)^2(q^2-abp^2)+fg\{(f+g)^2abr^2-(f+g)(a+b)(c+d)q^2+(a+b)^2p^2\}+f^2g^2(a+b)^2(f+g)(cdq^2-r^2)=0.$

If we had chosen $x=ly$ or $z=nx$, we should have got two other eliminants:

$E_1\equiv(c+d)^2(p^2-fgr^2)+cd\{(c+d)^2fgq^2-(c+d)(f+g)(a+b)p^2+(f+g)^2r^2\}+c^2d^2(f+g)^2(c+d)(abp^2-q^2)=0.$

$E_3\equiv(a+b)^2(r^2-cdq^2)+ab\{(a+b)^2cdp^2-(a+b)(c+d(f+g)r^2+(c+d)^2q^2\}+a^2b^2(c+d)^2(a+b)(fgr^2-p^2)=0.$

It is evident from an inspection of the three eliminants that they will vanish if $a=-b$, $c=-d$, and $f=-g$. But the condition $a=-b$ is just the condition that the two double tangents through the vertex (0, 0, 1) of the reference triangle shall form a harmonic pair with the axes $x=0$, $y=0$. Similarly, if $c=-d$, the double tangents through (1, 0, 0) are harmonically divided by $y=0$, $z=0$, and finally if $f=-g$, the double tangents through (0, 1, 0) are harmonically divided by $x=0$, $z=0$.

Hence it will be possible to pass a third double tangent to the quartic through each vertex of the reference triangle (chosen as indicated on page 397), if the two original double tangents through each vertex are harmonic conjugates of the sides of the triangle meeting in that vertex.

Let us now set up the conditions that our general quartic curve as given by equation (27) shall contain only even powers of the variables. The nine conditions are as follows:

$$
(30)\quad
\begin{aligned}
&(1)\ p^2(-ap^2-bp^2+afgr^2+bfgr^2) \equiv p^2(a+b)\,(fgr^2-p^2) = 0\\
&(2)\ q^2(-cq^2-dq^2+abcp^2+abdp^2) \equiv q^2(c+d)\,(abp^2-q^2) = 0\\
&(3)\ r^2(-fr^2-gr^2+cdfq^2+cdgq^2) \equiv r^2(f+g)\,(cdq^2-r^2) = 0\\
&(4)\ p^2(-a^2bp^2-ab^2p^2+aq^2+bq^2) \equiv p^2(a+b)\,(q^2-abp^2) = 0\\
&(5)\ q^2(-c^2dq^2-cd^2q^2+cr^2+dr^2) \equiv q^2(c+d)\,(r^2-cdq^2) = 0\\
&(6)\ r^2(-f^2gr^2-fg^2r^2+fp^2+gp^2) \equiv r^2(f+g)\,(p^2-fgr^2)=0\\
&(7)\ (-afp^2r^2-agp^2r^2-bfp^2r^2-bgp^2r^2+cfgq^2r^2+dfgq^2r^2+cp^2q^2+dp^2q^2)\\
&\qquad\qquad \equiv -p^2r^2(a+b)\,(f+g)+(c+d)\,(p^2+fgr^2)q^2=0\\
&(8)\ (-acp^2q^2-adp^2q^2-bcp^2q^2-bdp^2q^2+abfp^2r^2+abgp^2r^2+fq^2r^2+fq^2r^2+gq^2r^2)\\
&\qquad\qquad \equiv -p^2q^2(a+b)\,(c+d)+(f+g)\,(q^2+abp^2)r^2=0\\
&(9)\ (-cfq^2r^2-dfq^2r^2-cgq^2r^2-dgq^2r^2+acdp^2q^2+bcdp^2q^2+ap^2r^2+bp^2r^2)\\
&\qquad\qquad \equiv -q^2r^2(c+d)\,(f+g)+(a+b)\,(r^2+cdq^2)p^2=0
\end{aligned}
$$

From inspection we see that the nine equations are satisfied if $a+b=0$, $c+d=0$, $f+g=0$, or if $p=q=r=0$, and in no other way. But we must reject the system $p=q=r=0$, since if these relations were true we should have no quartic at all. Therefore the nine conditions for the reduction of the general quartic to the special form we are studying are equivalent to three, $a+b=0$, $c+d=0$, $f+g=0$. But these conditions, as we have previously noted, are the conditions that each pair of double tangents through a vertex of the reference triangle shall be harmonically divided by the sides of the reference triangle through that vertex.

Let us first select as reference triangle one whose vertices are the intersections of x_1 with y_1, x_2 with y_2, x_3 with y_3, where x_1y_1, x_2y_2, x_3y_3 are three pairs of double tangents of the general quartic belonging to a Steiner complex. If, then, we can draw through each vertex a third double tangent to the quartic, or if—an equivalent condition—we can show that the two double tangents through each vertex are harmonic conjugates of the sides of the reference triangle through that vertex, the given quartic is reducible to the form $ax^4+by^4+cz^4+2fy^2z^2+2gx^2z^2+2hx^2y^2=0$.

Note.—We can express a, b, c, d, f, and g in terms of the moduli of the class of curves, and so obtain the conditions they must satisfy if the quartic is of this variety. Riemann, "*Zur Theorie der Abelschen Functionen für den Fall* p=3, *Gesammelte Werke*, p. 456.

SUMMARY

The quartic curve whose equation is

$$ax^4+by^4+cz^4+2fy^2z^2+2gx^2z^2+2hx^2y^2=0$$

has the following properties:

Four double tangents pass through each vertex of the reference triangle, the four consisting of two pairs which are harmonically divided by the sides of the triangle. The remaining sixteen are grouped by fours, each four consisting of the lines $z=\pm px \pm qy$. The quadrilateral formed by any one of these four-groups has for diagonals the sides of the fundamental triangle of reference.

The actual equations of the double tangents can be determined by the solution of a biquadratic equation.

The six pairs of double tangents through the vertices of the reference triangle belong to a Steiner complex. To the complexes determined by $\begin{matrix} x+A_1y=0 \\ x+A_2y=0 \end{matrix}$ and $\begin{matrix} x+A_1y=0 \\ x-A_2y=0 \end{matrix}$ belong the pairs $\begin{matrix} z+px+qy=0 \\ z-px-qy=0 \end{matrix}$ and $\begin{matrix} z+px-qy=0 \\ z-px+qy=0. \end{matrix}$ To those determined by $\begin{matrix} z-C_1x=0 \\ z-C_2x=0 \end{matrix}$ and $\begin{matrix} z-C_1x=0 \\ z+C_2x=0 \end{matrix}$ belong the pairs $\begin{matrix} z+px+qy=0 \\ z+px-qy=0 \end{matrix}$ and $\begin{matrix} z-px+qy=0 \\ z-px-qy=0. \end{matrix}$ And finally to the two determined by $\begin{matrix} y-B_1z=0 \\ y-B_2z=0 \end{matrix}$ and $\begin{matrix} y-B_1z=0 \\ y+B_2z=0 \end{matrix}$ belong the pairs $\begin{matrix} z+px+qy=0 \\ z-px+qy=0 \end{matrix}$ and $\begin{matrix} z+px-qy-0 \\ z-px-qy=0. \end{matrix}$

Under certain conditions the general quartic can be made to reduce to the special form $F=0$. Let three pairs of double tangents be selected from a Steiner complex and let their intersections be taken as the vertices of a new reference triangle. Then if it is possible to pass a third double tangent through each vertex, or if the two original double tangents are harmonically divided by the sides of the reference triangle through that vertex, the general quartic equation is reducible to the form considered in this paper.

UNIVERSITY OF CALIFORNIA PUBLICATIONS
IN
MATHEMATICS

Vol. 1, No. 19, pp. 401-423 January 24, 1924

A STUDY OF CUBIC SURFACES BY MEANS OF INVOLUTORY CUBIC SPACE TRANSFORMATIONS

BY

JOHN FREDERICK POBANZ

1. Cayley in his paper On the Rational Transformations between Two Spaces[1] was the first to call attention to birational transformations in three dimensions, as an immediate extension of Cremona's work in two dimensions. In particular he investigated a cubo-cubic transformation by means of which the coördinates of any point in one space can be expressed as rational cubic functions of the coördinates of a corresponding point in a second space, and vice versa. To the planes in either space correspond cubic surfaces in the other space. If in this transformation certain relations exist among the coefficients, the transformation is involutory.

2. The purpose of the present investigation is to obtain by involutory cubic transformations the 23 varieties of cubic surfaces as given in Cayley's memoir[2] and also the cubic cone omitted by him.

3. This paper is divided into three parts:

Part I. A discussion of the singularities and classification of cubic surfaces;

Part II. (A) A general consideration of the cubo-cubic and involutory transformations; (B) the notation used;

Part III. The special transformations necessary to produce each of the 24 varieties of cubic surfaces.

PART I

1. The order of a cubic surface without any singularities is 3 and its class is 12. The surface has upon it 27 distinct right lines lying by threes in 45 planes, which are triple tangent planes. The surface may also possess point or line singularities. Point singularities reduce the class of the surface and the number of distinct right lines. In order to account for the 27 lines, a certain multiplicity must be given to the lines through the singular point or points. The point singularities are cnicnodes (conical points), binodes and unodes.

[1] Collected Math. Papers, vol. 7, pp. 189-240. *Also* Proc. London Math. Soc., vol. 3 (1868-1873), pp. 127-180.

[2] "Memoir on Cubic Surfaces" Phil. Trans. Royal Soc., vol. 153 (1863), pp. 193-241.

2. At a cnicnode C_2 the tangent plane is replaced by a proper quadric cone. Such a point reduces the class of the surface by 2, and, therefore, a cubic surface can at most have 4 C_2. The 6 lines through a C_2 have a multiplicity of 2.

3. At a binode B the quadric cone breaks up into a plane-pair, which are called the biplanes and their line of intersection the edge. There are 4 varieties of binodes:

a) At an ordinary binode B_3 the edge does not lie on the surface. A B_3 reduces the class of a surface by 3, and, therefore, a cubic surface cannot have more than 3 B_3. The lines through a B_3 have a multiplicity of 3.

b) If the edge is a line of the surface, and if the tangent plane at each point of the edge is distinct from one of the biplanes, except at two points, the binode is a B_4. A B_4 reduces the class of a surface by 4. Such a point may be considered as resulting from the union of 2 C_2.

c) In a B_5 the tangent plane at each point of the edge coincides with one of the biplanes. A B_5 reduces the class of a surface by 5. Such a point may be considered as formed by the union of a C_2 and a B_3.

d) If one of the biplanes becomes oscular, the class of the surface is reduced by 6, and the binode is a B_6. A B_6 may be considered as resulting from the union of 3 C_2.

4. At a unode U the quadric cone becomes a coincident plane-pair, the uniplane. There are 3 varieties of unodes:

a) If the 3 lines in which the uniplane intersects the surface are distinct, the class of the surface is reduced by 6, and the U is a U_6. A U_6 may be considered as the union of 2 B_3.

b) If 2 of the 3 lines in the uniplane fall together, the class of the surface is reduced by 7 and the U is a U_7. A U_7 may be considered as resulting from the union of a B_3 and 2 C_2.

c) If all 3 lines in the uniplane fall together, the class of the surface is reduced by 8 and the U is a U_8. A U_8 results from the union of a C_2 and 2 B_3.

5. A cubic surface with a line singularity (nodal line) is necessarily a ruled surface. There are but 2 varieties:

a) the nodal line distinct from the directrix and,

b) the nodal line coinciding with the directrix.

At every point of the nodal line there are two tangent planes, which fall together at a pinch point. The nodal line of the first variety has 2 pinch points on it, while the nodal line of the 2nd variety has but one.

6. Cubic surfaces were first classified by Schläfli.[3] He finds 22 varieties depending upon the number and character of the point singularities which they possess. Each variety is further divided into species, depending on the reality of the lines on the surface.

[3] On the Distribution of Surfaces of the Third Order into Species, in reference to the absence or presence of Singular Points. Phil. Trans. Royal Soc., vol. 153 (1863), pp. 193-241.

7. Cayley's[4] classification is based upon, and is in a measure supplementary to Schläfli's memoir, but he disregards altogether the division into species depending on the reality of the lines. The surfaces are divided into 23 varieties depending upon the nature of the singularities.

PART II (A)

1. It is assumed with Cayley that the most general transformation which gives rise to a cubo-cubic transformation is given by the three bilinear equations

$$(1)\qquad \begin{cases} A\equiv\Sigma a_{ik}x_iy_k=0, \\ B\equiv\Sigma b_{ik}x_iy_k=0, \\ C\equiv\Sigma c_{ik}x_iy_k=0, \\ i,\,k=1,\,2,\,3,\,4. \end{cases}$$

where the x_i are the homogeneous coördinates of a point in the space X, and y_k those of a corresponding point in the space Y. The transformation is then expressed by

$$(2)\qquad \begin{cases} \rho x_i=\varphi_i(y_1,y_2,y_3,y_4), \\ \sigma y_k=\Psi_k(x_1,x_2,x_3,x_4), \end{cases}$$

where the φ_i and Ψ_k are rational cubic functions of (y) and (x) respectively. Cayley has shown that the φ_i have in common a twisted sextic curve $J(y)$ and the Ψ_k a similar curve $J(x)$. To a point on $J(y)$ corresponds a line in the space X meeting $J(x)$ in three points. As the point moves on $J(y)$ the corresponding line generates a surface $R(x)$ of degree 8 with $J(x)$ as a triple curve. There is a similar surface $R(y)$ in the space Y corresponding to the curve $J(x)$.

2. If in the equations (1) $a_{ik}=a_{ki}$, $b_{ik}=b_{ki}$, $c_{ik}=c_{ki}$, the transformation is involutory. Equations (1) may then be considered as the polar planes of the point (y) with respect to the three quadrics surfaces

$$\begin{aligned} A&\equiv\Sigma a_{ik}x_ix_k=0, \\ B&\equiv\Sigma b_{ik}x_ix_k=0, \\ C&\equiv\Sigma c_{ik}x_ix_k=0. \\ i,\,k&=1,\,2,\,3,\,4. \end{aligned}$$

The spaces X and Y may be thought of as superimposed upon one another in such a way that $J(x)$ coincides with $J(y)$ and $R(x)$ with $R(y)$. To any point (y) corresponds (y'), the intersection of the three polar planes of (y) with respect to the three quadrics A, B and C. To the point (y') corresponds back again the point (y). There are, however, exceptional points, namely, points for which the three polar planes meet in a line. Such points lie on J and to it corresponds the surface R of degree 8. As (y) describes a line, its corresponding point (y') describes a twisted cubic curve. As (y) describes a plane, (y') describes a cubic surface.

---

[4] *Op. cit.*, p. 1.

3. *J is the locus of vertices of cones in the general bundle*

$$\lambda_1 A+\lambda_2 B+\lambda_3 C=0.$$[5]

Those values of λ_1, λ_2, λ_3 which satisfy the equation

$$|\lambda_1 a_{ik}+\lambda_2 b_{ik}+\lambda_3 c_{ik}|=0,$$

define the cones in the bundle. If this determinant is expanded and λ_1, λ_2, λ_3 interpreted as trilinear coördinates of the points in the λ-plane, there results a plane quartic $C_4(\lambda)$ which is in one-to-one correspondence with J. Since a general bundle does not contain any composite quadrics, $C_4(\lambda)$ has no singular points. *$C_4(\lambda)$ has a deficiency of 3, hence J has the same deficiency.*[6]

PART II (B)

1. α denotes the plane which is transformed and C^3 the cubic surface corresponding to α.

2. Subscripts denote the number by which the class of a surface is diminished.

3. C_2 is a cnicnode, where the tangent plane is replaced by a proper quadric cone; $B=(B_3, B_4, B_5$ or $B_6)$ is a biplanar node, where the quadric cone is a plane-pair (the biplanes); $U=(U_6, U_7$ or $U_8)$ is a uniplanar node, where the quadric cone becomes a coincident plane-pair (the uniplane).

4. The numbers 1, 2, 3, 4, 5, 6 denote the intersections of α with J; $\overline{12}$ . . . . . . $\overline{56}$ denote the lines connecting the points 1 . . . . . . 6.

5. Roman numerals refer to conics; I, is the conic through the points 2, 3, 4, 5, 6, etc.

6. On the cubic surface the numbers denote the lines which are the transforms of the points, lines, and conics, respectively.

7. This investigation has three outstanding features:

1) the point-to-point correspondence between J and $C_4(\lambda)$;

2) the dependence of the form of J upon the form of $C_4(\lambda)$; and

3) the dependence of the singularities of the cubic surface upon the form of J.

PART III

I. General Cubic Surface.

In this case is given the general cubic surface corresponding to any arbitrary plane α. $C_4(\lambda)$ is an unrestricted plane quartic, and J a proper sextic curve. This case also gives the general plan of treatment followed throughout the remaining cases.

[5] See Snyder and Sisam, Analytic Geometry of Space, pp. 170-172. Also Salmon, Geometry of Three Dimensions, art. 238a, p. 247.

[6] The deficiency of a space curve is here defined as the number by which the number of double points, real or apparent, falls short of the maximum. Compare Salmon, Geometry of Three Dimensions, art. 353.

Let the three quadrics be

$$A \equiv \Sigma a_{ik}x_ix_k = 0,$$
$$B \equiv \Sigma b_{ik}x_ix_k = 0,$$
$$C \equiv \Sigma c_{ik}x_ix_k = 0,$$
$$i,\ k = 1,\ 2,\ 3,\ 4.$$[7]

The three polar planes of a point (y) are

$$(1) \qquad \begin{cases} \Sigma a_{ik}x_iy_k = 0, \\ \Sigma b_{ik}x_iy_k = 0, \\ \Sigma c_{ik}x_iy_k = 0. \end{cases}$$

From equations (1) we get ρx_i, which are the determinants, with the proper sign attached, of the matrix

$$\begin{Vmatrix} \Sigma a_{1k}y_k & \Sigma a_{2k}y_k & \Sigma a_{3k}y_k & \Sigma a_{4k}y_k \\ \Sigma b_{1k}y_k & \Sigma b_{2k}y_k & \Sigma b_{3k}y_k & \Sigma b_{4k}y_k \\ \Sigma c_{1k}y_k & \Sigma c_{2k}y_k & \Sigma c_{3k}y_k & \Sigma c_{4k}y_k \end{Vmatrix}$$

$C_4(\lambda) \equiv |\lambda_1 a_{ik} + \lambda_2 b_{ik} + \lambda_3 c_{ik}| = 0$ is an unrestricted quartic and, therefore, J a proper sextic curve.

The plane $\alpha \equiv \Sigma a_ix_i = 0$ transforms into a cubic surface C^3 which can be expressed in the following form

$$C^3 \equiv \Sigma\Sigma a_i(a_{lr}b_{ms}c_{nt} - a_{lt}b_{ms}c_{nr})y_ly_my_n,$$
$$i,\ l,\ m,\ n,\ r,\ s,\ t = 1\ ,2,\ 3,\ 4.\quad i \neq r,\ s,\ t.\quad r \neq s \neq t.$$

C^3 has 27 lines on it:[8]

1) 6 lines which are the transforms of the points 1, 2, 3, 4, 5, 6.

2) 6 lines corresponding to the conic sections through each 5 points I, II, III, IV, V, VI.

3) 15 lines corresponding to the lines $\overline{12}$, $\overline{13}$, ... $\overline{56}$.

II_1. Surface with 1 C_2.

If the self-polar tetrahedron of A and B is taken as the tetrahedron of reference, the equations of the three quadrics are

$$A \equiv \Sigma a_{ii}x^2{}_{ii} = 0,$$
$$B \equiv \Sigma b_{ii}x^2{}_{ii} = 0,$$
$$C \equiv \Sigma c_{ik}x_ix_k = 0.$$

The matrix is

$$\begin{Vmatrix} a_{11}y_1 & a_{22}y_2 & a_{33}y_3 & a_{44}y_4 \\ b_{11}y_1 & b_{22}y_2 & b_{33}y_3 & b_{44}y_4 \\ \Sigma c_{1k}y_k & \Sigma c_{2k}y_k & \Sigma c_{3k}y_k & \Sigma c_{4k}y_k \end{Vmatrix}$$

[7] All summations hereafter are from 1 to 4, unless otherwise stated.

[8] See Table I, pp. 38a, 38b.

from which we get

$$\begin{aligned}
\rho x_1 &= d_{34}y_3y_4\Sigma c_{2k}y_k - d_{24}y_2y_4\Sigma c_{3k}y_k + d_{23}y_2y_3\Sigma c_{4k}y_k,\\
\rho x_2 &= -d_{34}y_3y_4\Sigma c_{1k}y_k + d_{14}y_1y_4\Sigma c_{3k}y_k - d_{13}y_1y_3\Sigma c_{4k}y_k,\\
\rho x_3 &= d_{24}y_2y_4\Sigma c_{1k}y_k - d_{14}y_1y_4\Sigma c_{2k}y_k + d_{12}y_1y_2\Sigma c_{4k}y_k,\\
\rho x_4 &= -d_{23}y_2y_3\Sigma c_{1k}y_k + d_{13}y_1y_3\Sigma c_{2k}y_k - d_{12}y_1y_2\Sigma c_{3k}y_k,
\end{aligned}$$

where $d_{ij}=a_{ii}b_{jj}-a_{jj}b_{ii}$. $[i, j=1, 2, 3, 4]$.

$C_4(\lambda)\equiv |\lambda_1 a_{ik}+\lambda_2 b_{ik}+\lambda_3 c_{ik}|=0$, $[a_{ik}=b_{ik}=0$, if $k\neq i]$ which is a proper quartic curve. J, therefore, is a proper sextic going through the four vertices of the tetrahedron of reference.

To get a C^3 with one conical point now take for α the plane $x_4=0$. This plane contains the ruling of R corresponding to the point $P_1(0:0:0:1)$. α goes into the following C^3:

$$\begin{aligned}
C^3\equiv d_{32}[c_{11}y_1y_2y_3+c_{12}y^2{}_2y_3+c_{13}y_2y^2{}_3+c_{14}y_2y_3y_4]+d_{13}[c_{12}y^2{}_1y_3+c_{22}y_1y_2y_3+c_{23}y_1y^2{}_3\\
+c_{24}y_1y_3y_4]+d_{21}[c_{13}y^2{}_1y_2+c_{23}y_1y^2{}_2+c_{33}y_1y_2y_3+c_{34}y_1y_2y_4]=0.
\end{aligned}$$

This C^3 has a C_2 at $P_1(0:0:0:1)$. The tangent plane at this point is replaced by a proper quadric cone, the equation of which may be obtained by writing $y_4=1$ in the terms containing y_4 in the above equation.

On C^3 we have the lines:

A. Lines through the $C_2=(0:0:0:1)$.

a) Lines corresponding to the 3 points lying on the ruling of R. 1, 2, 3.

b) Lines $\overline{56}$, $\overline{46}$, $\overline{45}$.

B. Free lines.[9]

a) 4, 5, 6.

b) I, II, III.

c) $\overline{14}$, $\overline{15}$ . . . . . . $\overline{36}$.

II_2. A C^3 with 1 C_2 may also be obtained by letting one of the quadrics be unrestricted and the other two be pairs of planes with their double lines intersecting. $C_4(\lambda)$ is then a line and a cubic. J is made up of three concurrent lines and a plane cubic. As any plane α will cut the plane cubic of J in three points lying on a right line, the 6 points will have the same relative positions as given in II_1. The C_2 on C^3 is the triple point of J. To it corresponds in the λ-plane any point on the line of $C_4(\lambda)$ except the 3 intersections of this line and the cubic of $C_4(\lambda)$. The lines on the surface are the same as in II_1.

[9] A free line is a line on C^3 which does not go through the singular point or points.

II_3. A C^3 with 1 C_2 may also be obtained by taking α tangent at any point of J. The lines on C^3 are as follows:

A. Lines through the C_2.

a) Line corresponding to (11).

b) $\overline{12}$, $\overline{13}$, $\overline{14}$, $\overline{15}$.

c) Line corresponding to conic section I.

B. Free lines.

a) $\overline{24}$, $\overline{25}$, $\overline{23}$, $\overline{34}$, $\overline{35}$, $\overline{45}$, $\overline{11}$.

b) II, III, IV, V.

c) 2, 3, 4, 5.

II_4. For a fourth way to obtain a surface with 1 C_2, let A be unrestricted and B and C quadrics tangent along a ruling. $C_4(\lambda)$ is a quartic with a cusp. J is made up of a double line and a quartic. Any plane α will transform into a C^3 having a C_2 on the double line. The distribution of the six points in α and the corresponding lines on C^3 are as given in II_3.

III_1. Surface with 1 B_3.

Let the three quadrics A, B and C be taken as in II, with $C_{34}=0$. Then the plane $\alpha=x_4=0$ goes into the surface $C^3=d_{21}c_{13}y^2{}_1y_2+d_{21}c_{23}y_1y^2{}_2+(d_{21}c_{33}+d_{13}c_{22}+d_{32}c_{11})$ $y_1y_2y_3+d_{13}c_{12}y^2{}_1y_3+d_{13}c_{23}y_1y^2{}_3+d_{32}c_{12}y^2{}_2y_3+d_{32}c_{13}y_1y^2{}_3+d_{13}c_{24}y_1y_2y_4+d_{32}c_{14}y_1y_3y_4=0$.[10]

$P_1(0:0:0:1)$ is a singular point on C^3. Since the line of intersection of the biplanes $y_1=0$, $d_{13}c_{24}y_2+d_{32}c_{14}y_3=0$ is not on C^3 the point is a B_3. On C^3 are the lines:

A. Lines through B_3.

(1) 2, 3, $\overline{14}$, $\overline{15}$, $\overline{45}$.

B. Free lines.

4, 5, $\overline{24}$, $\overline{25}$, $\overline{34}$, $\overline{35}$ ($\overline{11}$) and II, III.

III_2. A surface with a B_3 may also be obtained in the following manner. Letting A, B and C be unrestricted, and taking α osculating J at any point, the corresponding C^3 will have a B_3. The B_3 may be found by getting the point of intersection of the line from the point of osculation and the line from the tangent line at the point of osculation.[11] The lines on C^3 are

A. Lines through B_3.

a) Line corresponding to (1).

b) $\overline{12}$, $\overline{13}$, $\overline{14}$, ($\overline{11}$).

c) Line corresponding to conic (11) 234.

---

[10] d_{ij} has the same meaning as in II_1.

[11] In general, if a curve goes through n-1 or n consecutive points of J, its transform will go through the singular point on C^3.

B. Free lines.

a) 2, 3, 4.

b) $\overline{23}$, $\overline{24}$, $\overline{34}$.

c) Lines corresponding to conics (111) 34; (111) 24; (111) 23.

IV_1. Surface with 2 C_2.

Let the three quadrics be

$$A \equiv a_{11}x^2_1 + a_{22}x^2_2 + a_{33}x^2_3 + a_{44}x^2_4 = 0,$$
$$B \equiv b_{11}x^2_1 + b_{22}x^2_2 + b_{33}x^2_3 + b_{44}x^2_4 = 0,$$
$$C \equiv c_{11}x^2_1 + c_{22}x^2_2 + c_{33}x^2_3 + c_{44}x^2_4 + 2c_{12}x_1x_2 = 0.$$

The matrix is

$$\begin{Vmatrix} a_{11}y_1 & a_{22}y_2 & a_{33}y_3 & a_{44}y_4 \\ b_{11}y_1 & b_{22}y_2 & b_{33}y_3 & b_{44}y_4 \\ c_{11}y_1 + c_{12}y_2 & c_{12}y_1 + c_{22}y_2 & c_{33}y_3 & c_{44}y_4 \end{Vmatrix}$$

$$\rho x_1 = (c_{22}d_{34} + c_{33}d_{42} + c_{44}d_{23})y_2y_3y_4 + c_{12}d_{34}y_1y_3y_4,$$
$$\rho x_2 = (c_{11}d_{43} + c_{33}d_{14} + c_{44}d_{31})y_1y_3y_4 + c_{12}d_{43}y_2y_3y_4,$$
$$\rho x_3 = (c_{11}d_{24} + c_{22}d_{41} + c_{44}d_{12})y_1y_2y_4 + c_{12}d_{41}y^2_1y_4 + c_{12}d_{24}y^2_2y_4,$$
$$\rho x_4 = (c_{11}d_{32} + c_{22}d_{13} + c_{33}d_{21})y_1y_2y_3 + c_{12}d_{31}y^2_1y_3 + c_{12}d_{23}y^2_2y_3,$$

$C_4(\lambda)$ in this case is made up of the two lines $a_{44}\lambda_1 + b_{44}\lambda_2 + c_{44}\lambda_3 = 0$, $a_{33}\lambda_1 + b_{33}\lambda_2 + c_{33}\lambda_3 = 0$ and the conic $(a_{11}\lambda_1 + b_{11}\lambda_2 + c_{11}\lambda_3)$ $(a_{22}\lambda_1 + b_{22}\lambda_2 + c_{22}\lambda_3) - c^2_{12}\lambda^2_3 = 0$. The two lines each intersect the conic in two distinct points. J is made up of 6 lines. One has two triple points and the other skew to it has four double points.

Any plane $\alpha \equiv \Sigma\alpha_ix_i = 0$ goes into the surface $C^3 \equiv [\alpha_1(c_{22}d_{34} + c_{33}d_{42} + c_{44}d_{23}) + \alpha_2c_{12}d_{43}]\ y_2y_3y_4 + [\alpha_2(c_{11}d_{43} + c_{33}d_{14} + c_{44}d_{31}) + \alpha_1c_{12}d_{34}]\ y_1y_3y_4 + \alpha_3(c_{11}d_{24} + c_{22}d_{41} + c_{44}d_{12})\ y_1y_2y_4 + \alpha_3c_{12}d_{41}y^2_1y_4 + \alpha_3c_{12}d_{24}y^2_2y_4 + \alpha_4(c_{11}d_{32} + c_{22}d_{13} + c_{33}d_{21})y_1y_2y_3 + \alpha_4c_{12}d_{31}y^2_1y_3 + \alpha_4c_{12}d_{23}y^2_2y_3 = 0$.

It is easily seen that the C^3 has a C_2 at $P_1(0:0:0:1)$ and one at $P_2(0:0:1:0)$. On C^3 we have the lines:

Through P_1 the lines: 1, 2, 3, $\overline{45}$, $\overline{56}$.
Through P_2 the lines: 1, 5, 6, $\overline{34}$, $\overline{24}$.
Free lines: 4, $\overline{14}$, $\overline{25}$, $\overline{26}$, $\overline{35}$, $\overline{36}$, I.

IV_2. A C^3 with 2 C_2 may be obtained by taking A, B and C as in I and α through 2 rulings of R. The positions of the 6 points in α and the corresponding lines on C^3 are the same as in IV_1.

IV_3. As a third way to obtain the above C^3, let A be unrestricted, B and C be two pairs of planes with the plane of one of the two pairs containing the double line of the other. $C_4(\lambda)$ will then be made up of a cubic and a line tangent to the cubic. J consists of a single line, a double line and a plane cubic. Any arbitrary plane α

will transform into a C^3 having a fixed double point and another one on the double line. As the plane α changes its position, one of the C_2 remains fixed while the other moves on the double line.

The lines on C^3 are as follows:

Through the fixed C_2: 1, 2, 3, $\overline{45}$, $\overline{55}$.
Through the moving C_2: (5), $\overline{15}$, $\overline{25}$, $\overline{35}$, $\overline{45}$.
Free lines: 4, $\overline{14}$, $\overline{24}$, $\overline{34}$, I, II, III.

V. Surface with a B_4.

Let the quadrics A, B and C be taken as in II_1, but let $c_{34}=c_{24}=0$. The plane $\alpha\equiv x_4=0$ then osculates J, and corresponding to it we have the surface $C^3\equiv d_{21}c_{13}y^2{}_1y_2 + d_{21}c_{23}y_1y^2{}_2 + (d_{21}c_{33}+d_{13}c_{22}+d_{32}c_{11})y_1y_2y_3 + d_{13}c_{12}y^2{}_1y_3 + d_{13}c_{23}y_1y^2{}_3 + d_{32}c_{12}y^2{}_2y_3+d_{32}c_{13}y_1y^2{}_3+d_{32}c_{14}y_1y_2y_4=0$, where $d_{ij}=a_{ii}b_{jj}-a_{jj}b_{ii}$. $P_1(0:0:0:1)$ is a B_4, since the line of intersection of the biplanes $y_1=y_2=0$ lies on the surface.[12]

On C^3 are the lines:

A. Through B_4

a) (1), 2, 3,

b) $\overline{14}$, I.

B. Free lines, 4, $\overline{24}$, $\overline{34}$, III, II.

VI. Surface with B_3+C_2.

To get this surface, let A be an unrestricted quadric, B a pair of planes, and C a pair of planes with the double line a ruling on A. The equations are

$$A\equiv\Sigma a_{ik}x_ix_k=0,$$
$$B\equiv 2x_3x_4=0,$$
$$C\equiv 2x_1x_2=0.$$
$$i,\ k=1,\ 2,\ 3,\ 4.\quad a_{33}=a_{44}=a_{34}=0.$$

The matrix is

$$\left\|\begin{matrix}\Sigma a_{1k}y_k & \Sigma a_{2k}y_k & \Sigma a_{3k}y_k & \Sigma a_{4k}y_k\\ 0 & 0 & y_4 & y_3\\ y_2 & y_1 & 0 & 0\end{matrix}\right\|$$

from which we get

$$\rho x_1=y_1[y_3\Sigma a_{3k}y_k-y_4\Sigma a_{4k}y_k],$$
$$\rho x_2=y_2[y_4\Sigma a_{4k}y_k-y_3\Sigma a_{3k}y_k],$$
$$\rho x_3=y_3[y_2\Sigma a_{2k}y_k-y_1\Sigma a_{1k}y_k],$$
$$\rho x_4=y_4[y_1\Sigma a_{1k}y_k-y_2\Sigma a_{2k}y_k].$$

$C_4(\lambda)\equiv a_1\lambda^4{}_1+a_2\lambda^3{}_1\lambda_2+a_3\lambda^2{}_1\lambda_2\lambda_3+a_4\lambda^2{}_1\lambda^2{}_2+a_5\lambda_1\lambda^2{}_2\lambda_3+\lambda^2{}_2\lambda^2{}_3=0$, where a_j ($j=1, 2, 3, 4, 5$) are functions of the coefficients of A, B and C. This quartic has a node and a cusp. J, therefore, is made up of a double line, a single line skew to the double line, and a twisted cubic.

[12] It is easily shown that the edge is torsal.

Any plane $\alpha \equiv \Sigma \alpha_i x_i = 0$ transforms into a C^3 having a double point on the double line. The equation written out in full is
$C^3 \equiv (\alpha_1 a_{13} - \alpha_3 a_{11}) y^2_1 y_3 - (\alpha_1 a_{14} - \alpha_4 a_{11}) y^2_1 y_4 - (\alpha_2 a_{23} - \alpha_3 a_{22}) y^2_1 y_3 + (\alpha_2 a_{24} - \alpha_4 a_{22}) y^2_2 y_4 - \alpha_3 a_{13} y_1 y^2_3 + \alpha_3 a_{23} y_1 y^2_3 + \alpha_4 a_{14} y_1 y^2_4 - \alpha_4 a_{24} y_2 y^2_4 + (\alpha_1 a_{23} - \alpha_2 a_{13}) y_1 y_2 y_3 - (\alpha_1 a_{24} - \alpha_2 a_{14}) y_1 y_2 y_4 - (\alpha_3 a_{14} - \alpha_4 a_{13}) y_1 y_3 y_4 + (\alpha_3 a_{24} - \alpha_4 a_{23}) y_2 y_3 y_4 = 0$. If the α_i have such values that α is tangent to the twisted cubic, C^3 has 2 C_2. If α osculates the twisted cubic at any point, C^3 has a B_3 and a C_2.

On C^3 the lines are arranged as follows:

Through B_3: 1, $\overline{12}$, $\overline{11}$, I.
Through C_2: $\overline{12}$, (2), $\overline{23}$, II.
Free lines: 3, $\overline{22}$, III, making a total of 11 distinct lines.

VII. Surface with a B_5.

Let A be unrestricted, B and C pairs of planes, one plane of C containing the double line of B. The equations are

$$A \equiv \Sigma a_{ik} x_i x_k = 0,$$
$$B \equiv x^2_1 - b x^2_2 = 0,$$
$$C \equiv 2 x_1 x_3 = 0.$$
$$a_{ik} = 1,\ 2,\ 3,\ 4.\quad a_{44} = 1,\ a_{14} = a_{24} = a_{34} = 0.$$

The matrix is

$$\left\|\begin{matrix} \Sigma a_{1k} y_k & \Sigma a_{2k} y_k & \Sigma a_{3k} y_k & y_4 \\ y_1 & -b y_2 & 0 & 0 \\ y_3 & 0 & y_1 & 0 \end{matrix}\right\|$$

from which we get

$$\rho x_1 = -b y_1 y_2 y_4,$$
$$\rho x_2 = -y^2_1 y_4,$$
$$\rho x_3 = b y_2 y_3 y_4,$$
$$\rho x_4 = (a_{11} b + a_{22}) y^2_1 y_2 + a_{12} b y_1 y^2_2 + a_{12} y^3_1 - a_{33} y_2 y^2_3 - a_{23} b y^2_2 + a_{23} y^2_1 y_3.$$

$C_4(\lambda) \equiv \lambda_1[a_1 \lambda^3_1 + a_2 \lambda^2_1 \lambda_2 + a_3 \lambda_1 \lambda^2_2 + a_4 \lambda^2_1 \lambda_3 + a_5 \lambda_1 \lambda^2_3 + a_6 \lambda_2 \lambda^2_3 + a_7 \lambda_1 \lambda_2 \lambda_3] = 0$, where the a_i ($i = 1 \ldots\ldots 7$) are functions of the a_{ik}. The line intersects the cubic in the point $0:0:1$ and is tangent at $0:1:0$. J, therefore, is made up of a single line, a double line, and a plane cubic.

Corresponding to the plane $\alpha \equiv \alpha_1 x_1 + \alpha_4 x_4 = 0$, we have the surface
$C^3 \equiv \alpha_1 b y_1 y_2 y_4 - \alpha_4 (a_{11} b + a_{22}) y^2_1 y_2 - \alpha_4 a_{12} y_1 y^2_2 - \alpha_4 y^3_1 + \alpha_4 a_{33} y_2 y^2_3 + \alpha_4 a_{23} y^2_2 y_3 - \alpha_4 a_{23} y^2_1 y_3 = 0.$

Putting $y_4 = 1$, we have $y_1 = y_2 = 0$ as the biplanes and $P_1(0:0:0:1)$ is a B_5. It is also easily shown that the tangent plane coincides with one of the biplanes. This surface has the lines:

Through $P_1(0:0:0:1)$, (1), (2), 3.
Free lines: $\overline{11}$, $\overline{22}$, III.

VIII. Surface with 3 C_2.

Let

$$A \equiv 2a_{12}x_1x_2 + 2a_{13}x_1x_3 + 2a_{23}x_2x_3 = 0,$$
$$B \equiv 2b_{12}x_1x_2 + 2b_{14}x_1x_4 + 2b_{24}x_2x_4 = 0,$$
$$C \equiv 2c_{23}x_2x_3 + 2c_{24}x_2x_4 + 2c_{34}x_3x_4 = 0,$$

from which we get the matrix

$$\left\| \begin{matrix} a_{12}y_2+a_{13}y_3 & a_{12}y_1+a_{23}y_3 & a_{13}y_1+a_{23}y_3 & 0 \\ b_{12}y_2+b_{14}y_4 & b_{12}y_1+b_{24}y_2 & 0 & b_{14}y_1+b_{24}y_2 \\ 0 & c_{23}y_3+c_{24}y_4 & c_{23}y_2+c_{34}y_4 & c_{24}y_2+c_{34}y_3 \end{matrix} \right\|$$

and

$$\rho x_1 = (a_{13}y_1+a_{23}y_2)\ (b_{14}y_1+b_{24}y_2)\ (c_{23}y_2+c_{24}y_4) - (a_{13}y_1+a_{23}y_2)\ (b_{12}y_1+b_{24}y_4)\ (c_{24}y_2+c_{34}y_3) - (a_{12}y_1+a_{23}y_3)\ (b_{12}y_1+b_{24}y_2)\ (c_{23}y_2+c_{34}y_4),$$
$$\rho x_2 = (a_{13}y_1+a_{23}y_2)\ (b_{12}y_2+b_{14}y_4)\ (c_{24}y_2+c_{34}y_3) + (a_{12}y_2+a_{13}y_3)\ (b_{14}y_1+b_{24}y_2)\ (c_{23}y_2+c_{34}y_4),$$
$$\rho x_3 = (a_{12}y_2+a_{13}y_3)\ (b_{12}y_1+b_{24}y_4)\ (c_{24}y_2+c_{34}y_3) - (a_{12}y_1+a_{23}y_3)\ (b_{12}y_2+b_{14}y_4)\ (c_{24}y_2+c_{34}y_3) - (a_{12}y_2+a_{13}y_3)\ (c_{23}y_2+c_{24}y_4)\ (b_{14}y_1+b_{24}y_2),$$
$$\rho x_4 = (a_{12}y_1+a_{23}y_3)\ (b_{12}y_2+b_{14}y_4)\ (c_{23}y_2+c_{34}y_4) - (a_{12}y_2+a_{13}y_3)\ (b_{12}y_1+b_{24}y_4)\ (c_{23}y_2+c_{34}y_4) - (a_{13}y_1+a_{23}y_3)\ (b_{12}y_2+b_{14}y_4)\ (c_{23}y_2+c_{24}y_4).$$

$C_4(\lambda) \equiv |a_{ik}\lambda_1+b_{ik}\lambda_2+c_{ik}\lambda_3| = 0$, $[i,\ k=1,\ 2,\ 3,\ 4.\ \ i \neq k,\ a_{4k}=b_{3k}=c_{1k}=0]$, is a quartic with three double points. J, therefore, consists of three skew lines and a twisted cubic. The skew lines are bisecants of the cubic.

Corresponding to $\alpha \equiv x_2 = 0$, we have the cubic surface

$$C^3 \equiv (a_{13}b_{12}c_{24} + a_{12}b_{14}c_{23})\ y_1y_2^2 + (a_{23}b_{12}c_{24} + a_{12}b_{24}c_{23})\ y_2^3 + (a_{13}b_{14}c_{24} + a_{12}b_{14}c_{34})\ y_1y_2y_4 + (a_{23}b_{14}c_{24} + a_{12}b_{24}c_{34})\ y_2^2y_4 + (a_{13}b_{12}c_{34} + a_{13}b_{14}c_{23})\ y_1y_2y_3 + (a_{23}b_{12}c_{34} + a_{13}b_{24}c_{23})\ y_2^2y_3 + 2a_{13}b_{14}c_{34}y_1y_3y_4 + (a_{23}b_{14}c_{34} + a_{13}b_{24}c_{34})y_2y_3y_4 = 0.$$

On C^3 are the lines:

A. a) (1), (2), (3),
   b) $\overline{12}$, $\overline{13}$, $\overline{23}$,
   c) I, II, III.

B. Free lines:

   a) $\overline{11}$, $\overline{22}$, $\overline{33}$.

IX. Surface with 2 B_3.

To get this surface let the quadrics be taken as follows:

$$A \equiv a_{11}x_1^2 + a_{22}x_2^2 + x_4^2 + 2a_{12}x_1x_2 + 2a_{13}x_1x_3 = 0,$$
$$B \equiv x_1^2 - bx_2^2 = 0,$$
$$C \equiv 2x_1x_3 = 0.^{13}$$

---

[13] The above equation is of the same form as Cayley's except that he has replaced the singular tangent planes which touch the surface along the axes by the planes $y_1 = 0$, $y_3 = 0$, $y_4 = 0$ giving him the form $y_1^3 + (y_2+y_3+y_4)y_1^2 + 4ay_1y_3y_4 = 0$. a must not be equal to -1, or the surface will have 4 C_2.

The matrix is

$$\left\|\begin{matrix} a_{11}y_1+a_{12}y_2+a_{13}y_3 & a_{12}y_1+a_{22}y_2 & a_{13}y_1 & y_4 \\ y_1 & -by_2 & 0 & 0 \\ y_3 & 0 & y_1 & 0 \end{matrix}\right\|$$

$$\begin{aligned} \rho x_1 &= -by_1y_2y_4, \\ \rho x_2 &= -y^2{}_1y_4, \\ \rho x_3 &= by_2y_3y_4, \\ \rho x_4 &= (a_{11}b+a_{22})y^2{}_1y_2+a_{12}by_1y^2{}_2+a_{12}y^3{}_1. \end{aligned}$$

$C_4(\lambda)\equiv\lambda_1[a_{22}\lambda_1-b\lambda_2]\ [a_{13}\lambda_1+\lambda_3]^2=0$, which is a double line and two single lines. To the point $0:0:1$ corresponds the line $x_1=x_3=0$, a single line. To the point $0:1:0$ corresponds the double line $x_1=x_2=0$. To the line $a_{22}\lambda_1-b\lambda_2=0$ corresponds $x_1=x_4=0$, and to the double line $a_{13}\lambda_1+\lambda_3=0$ the conic in the $x_4=0$ plane. The conic is tangent to the line $x_1=x_4=0$ at the point $0:0:1:0$.

The plane $\alpha\equiv\alpha_3x_3+\alpha_4x_4=0$ goes into the surface

$$C^3\equiv\alpha_3by_2y_3y_4+\alpha_4[(a_{11}b+a_{22})y^2{}_1y_2+a_{12}by_1y^2{}_2+a_{12}y^3{}_1]=0.$$

$P_1(0:0:0:1)$ and $P_2(0:0:1:0)$ are biplanar points since the lines of intersections of their biplanes are not on the surface. C^3 has 7 lines on it; viz: (1), (2), 3, 4, $\overline{13}$, $\overline{14}$, $\overline{22}$.

X. Surface with B_4+C_2.

Let all three quadrics be pairs of planes, and let the double lines of A and B each intersect the double line of C in a distinct point. The equations of the quadrics are

$$\begin{aligned} A&\equiv 2x_1x_2=0, \\ B&\equiv 2x_3x_4=0, \\ C&\equiv 2(a_1x_1+a_2x_2)\ (a_3x_3+a_4x_4)=0. \end{aligned}$$

$$\left\|\begin{matrix} y_2 & y_1 & 0 & 0 \\ 0 & 0 & y_4 & y_3 \\ a_1a_3y_3+a_1a_4y_4 & a_2a_3y_3+a_2a_4y_4 & a_1a_3y_1+a_2a_3y_2 & a_1a_4y_1+a_2a_4y_2 \end{matrix}\right\|$$

from which we have

$$\begin{aligned} \rho x_1 &= y_1y_4(a_1a_4y_1+a_2a_4y_2)-y_1y_3(a_1a_3y_1+a_2a_3y_2), \\ \rho x_2 &= -y_2y_4(a_1a_4y_1+a_2a_4y_2)+y_2y_3(a_1a_3y_1+a_2a_3y_2), \\ \rho x_3 &= y_1y_3(a_1a_3y_3+a_1a_4y_4)-y_2y_3(a_2a_3y_3+a_2a_4y_4), \\ \rho x_4 &= -y_1y_4(a_1a_3y_3+a_1a_4y_4)+y_2y_4(a_2a_3y_3+a_2a_4y_4). \end{aligned}$$

$C_4(\lambda)\equiv\lambda_1\lambda_2(\lambda_1\lambda_2-4a_1a_2a_3a_4\lambda^2{}_3)=0$, which is a conic and two lines tangent to it. J, therefore, is made up of two skew double lines and two skew single lines each meeting each double line.

$\alpha\equiv\alpha_1x_1+\alpha_2x_2+\alpha_3x_3+\alpha_4x_4=0$ goes into the surface

$$\begin{aligned} C^3\equiv{}&\alpha_1a_1a_4y^2{}_1y_4+(\alpha_1a_2a_4-\alpha_2a_1a_4)y_1y_2y_4-\alpha_1a_1a_3y^2{}_1y_3-(\alpha_1a_2a_3-\alpha_2a_1a_3)y_1y_2y_3-\alpha_2a_2a_4y^2{}_2y_4 \\ &+\alpha_2a_2a_3y^2{}_2y_3+a_1a^2{}_3y_1y^2{}_3-a_2a^2{}_3y_2y^3{}_3-a_1a^2{}_4y_1y^2{}_4+a_2a^2{}_4y_2y^2{}_4=0. \end{aligned}$$

It is easily shown that $P_1(0:0:a_4:a_3)$ is a B_4 and $P_2(a_2:a_1:0:0)$ a C_2. On C^3 are the lines: (1), 2, 3, $\overline{12}$, $\overline{23}$, $\overline{22}$, III.

X_1. A second way to obtain the above surface is to let A, B and C have a twisted cubic in common. $C_4(\lambda)$ is then a double conic and J the twisted cubic of A, B and C counted twice. Any plane through a tangent line of the twisted cubic will transform into a cubic surface with a B_4+C_2. The distribution of the lines on the surface is:

Through B_4: the lines (1), $\overline{12}$, $\overline{11}$, I.
Through C_2: the lines $\overline{12}$, (2) II,
and the free line $\overline{22}$.

XI. Surface with a B_6.

Let the equations of the three quadrics be

$$A\equiv x^2_1+x^2_2+2a_3x_1x_2+2a_4x_1x_3+2a_6x_2x_3+2a_7x_2x_4=0,$$
$$B\equiv 2x_3x_4=0,$$
$$C\equiv 2x_1x_2=0,$$

from which we have the matrix

$$\left\|\begin{array}{cccc} y_1+a_3y_2+a_4y_3 & a_3y_1+y_2+a_6y_3+a_7y_4 & a_4y_1+a_6y_2 & a_7y_2 \\ 0 & 0 & y_4 & y_3 \\ y_2 & y_1 & 0 & 0 \end{array}\right\|$$

and

$$\rho x_1=y_1[a_4y_1y_3+a_6y_2y_3-a_7y_2y_4],$$
$$\rho x_2=-y_2[a_4y_1y_3+a_6y_2y_3-a_7y_2y_4],$$
$$\rho x_3=y_3[y^3_2+a_6y_2y_3+a_7y_2y_4-y^3_1-a_4y_1y_3],$$
$$px_4=-y_4[y^3_2+a_6y_2y_3+a_7y_2y_4-y^3_1-a_4y_1y_3].$$

In this case $C_4(\lambda)\equiv a^2_4a_7\lambda^4_1+(2a_6a_7-2a_3a_4a_7)\lambda^3_1\lambda_2-2a_4a_7\lambda^2_1\lambda_2\lambda_3+a^2_3\lambda^2_1\lambda^2_2+2a_3\lambda_1\lambda^2_2\lambda_3-\lambda_1\lambda^3_2+\lambda^2_2\lambda^2_3=0$, is a quartic with a double point at $0:1:0$ and a cusp at $0:0:1$. J is made up of a double line, a single line skew to the double line and a twisted cubic.

Taking $\alpha\equiv x_1-x_2-a_6x_3=0$ tangent at P_1 where the single line cuts the cubic and through P_2 where the double line cuts the cubic, we have the cubic surface $C^3\equiv(a_4+a_6)y^2_1y_3-a_7y^2_2y_4+a_4a_6y_1y^2_3-a^2_6y_2y^2_3+(a_4+a_6)y_1y_2y_3-a_7y_1y_2y_4-a_6a_7y_2y_3y_4=0$. $P_1(0:0:0:1)$ is a B_6, the biplanes are $y_2=y_1+y_2+a_6y_3=0$. On C^3 we have the lines through P_1, (1), (2), $\overline{12}$. There are no free lines.

XII. Surface with a U_6.

To get this surface, let us take the quadrics as in VII. Then $\alpha\equiv a_2x_2+a_4x_4=0$ transforms into the surface

$$C^3\equiv a_2y^2_1y_4-a_4(a_{11}b+a_{22})y^2_1y_2-a_4a_{12}by_1y^2_2+a_4a_{33}by_2y^2_3+a_4a_{23}by^2_2y_3-a_4a_{12}y^3_1-a_4a_{23}y^2_1y_3=0.$$

Putting $y_4=1$, we have the $y_1=0$ plane counted twice and $P_1(0:0:0:1)$ is a U_6. On C^3 are the lines:

A. Through U_6: (1), 2, 3.
B. Free lines: 4, $\overline{24}$, $\overline{34}$.

XIII. Surface with $B_3+2\,C_2$.

If in VIII (Surface with 3 C_2) the following relations exist among the coefficients of the quadrics, viz.:

$$(a_{23}b_{14}-a_{13}b_{24})(a_{12}c_{34}-a_{13}c_{24})=0,$$

$P_1(0:0:0:1)$ will go over into a B_3. In VIII $C_4(\lambda)$ was a quartic with three double points and the corresponding sextic J consisted of three skew lines and a twisted cubic. But if the relations as given above are satisfied, the quartic has two double points and a cusp. J is a double line, two single lines and a conic. The equation of the surface is

$$C^3\equiv(a_{13}b_{12}c_{24}+a_{12}b_{14}c_{23})y_1y^2_2+(a_{23}b_{12}c_{24}+a_{12}b_{24}c_{23})y^3_2+(a_{13}b_{14}c_{24}+a_{12}b_{14}c_{34})y_1y_2y_4+ (a_{23}b_{14}c_{24}+a_{12}b_{24}c_{34})y^2_2y_4+(a_{13}b_{12}c_{34}+a_{13}b_{14}c_{23})y_1y_2y_3+(a_{23}b_{12}c_{34}+a_{13}b_{24}c_{23})y^2_2y_3 +2a_{13}b_{14}c_{34}y_1y_3y_4+(a_{23}b_{14}c_{34}+a_{13}b_{24}c_{34})y_2y_3y_4=0.$$

On the C^3 are the lines: (1), (2), (3), $\overline{13}$, $\overline{23}$, $\overline{22}$, $\overline{33}$, I.

XIV. Surface with B_5+C_2.

Let two of the quadrics be cones and the third a pair of planes with the double line through the vertex of one of the cones. Their equations are

$$A\equiv a_{11}x^2_2+a_{33}x^2_3+2a_{24}x_2x_4=0,$$
$$B\equiv 2x_1x_2=0,$$
$$C\equiv c_{22}x^2_2-2c_{13}x_1x_3=0.$$

The matrix is

$$\left\|\begin{matrix} 0 & a_{11}y_2+a_{24}y_4 & a_{33}y_3 & a_{24}y_2 \\ y_2 & y_1 & 0 & 0 \\ -c_{13}y_3 & c_{22}y_2 & -c_{13}y_1 & 0 \end{matrix}\right\|$$

from which we get

$$\rho x_1=-a_{24}c_{13}y^2_1y_2,$$
$$\rho x_2=a_{24}c_{13}y_1y^2_2,$$
$$\rho x_3=a_{24}c_{22}y^3_2+a_{24}c_{13}y_1y_2y_3,$$
$$\rho x_4=-a_{33}c_{22}y^2_2y_3-a_{33}c_{13}y_1y^2_3-a_{11}c_{13}y_1y^2_2-a_{24}c_{13}y_1y_2y_4.$$

$C_4(\lambda)\equiv y^2_1\lambda^2_3=0$, namely two intersecting double lines. To the point of intersection corresponds the 4-fold line of J $x_1=x_2=0$; to $\lambda_1=0$, the double line $x_2=x_3=0$; and to $\lambda_3=0$ a family of cones with vertex at $0:0:0:1$.

$\alpha\equiv x_4=0$ goes into the surface

$C^3\equiv a_{33}c_{22}y^2_2y_3+a_{33}c_{13}y_1y^2_3+a_{11}c_{13}y_1y^2_2+a_{24}c_{13}y_1y_2y_4=0$. This cubic surface has a B_5 at $0:0:0:1$ and a C_2 at $1:0:0:0$ as can easily be shown. On the surface are the lines: (1), (2), $\overline{12}$, I.

XV. Surface with a U_7.

Let the quadrics be taken as follows:

$$A \equiv a_{33}x^2_3 + x^2_4 + 2a_{12}x_1x_2 = 0,$$
$$B \equiv x^2_1 - bx^2_2 = 0,$$
$$C \equiv 2x_1x_3 = 0.$$

The matrix in this case is

$$\left\|\begin{matrix} a_{12}y_2 & a_{12}y_1 & a_{33}y_3 & y_4 \\ y_1 & -by_2 & 0 & 0 \\ y_3 & 0 & y_1 & 0 \end{matrix}\right\|$$

from which we get

$$\rho x_1 = -by_1y_2y_4,$$
$$\rho x_2 = -y^2_1y_4,$$
$$px_3 = by_2y_3y_4,$$
$$\rho x_4 = a_{12}by_1y^2_2 - a_{33}by_2y^2_3 + a_{12}y^3_1.$$

$C_4(\lambda) \equiv \lambda_1[a_{33}b\lambda_1\lambda^2_2 - b\lambda_2\lambda^2_3 + a^2_{12}a_{33}\lambda^3_1] = 0$; the line intersects the cubic in the point $0:0:1$, to which corresponds the single line of J, $x_1 = x_3 = 0$, and is tangent at $0:1:0$. To the point of tangency corresponds the double line $x_1 = x_2 = 0$. The other points of the cubic are in point-to-point correspondence with a cuspidal cubic in the $x_4 = 0$ plane.

$\alpha \equiv \alpha_2x_2 + \alpha_4x_4 = 0$, through the cusp, goes into the surface

$$C^3 \equiv \alpha_2y^2_1y_4 - \alpha_4a_{12}by_1y^2_2 + \alpha_4a_{33}by_2y^2_3 - \alpha_4a_{12}y^3_1 = 0.$$

This surface has a U_7 at $0:0:0:1$. On C^3 we have the lines:

Through U_7, (1), $\overline{12}$,
and the free line 2.

XVI. Surface with 4 C_2.

The simplest way to obtain this surface is to let the three quadrics be pairs of planes with their double lines lying in the same plane. The equations are

$$A \equiv x^2_1 - ax^2_4 = 0,$$
$$B \equiv x^2_2 - bx^2_4 = 0,$$
$$C \equiv x^2_3 - cx^2_4 = 0.$$

The matrix is

$$\left\|\begin{matrix} y_1 & 0 & 0 & -ay_4 \\ 0 & y_2 & 0 & -by_4 \\ 0 & 0 & y_3 & -cy_4 \end{matrix}\right\|$$

from which we have

$$\rho x_1 = -ay_2y_3y_4,$$
$$\rho x_2 = -by_1y_3y_4,$$
$$\rho x_3 = -cy_1y_2y_4,$$
$$\rho x_4 = -y_1y_2y_3.$$

$C_4(\lambda)\equiv\lambda_1\lambda_2\lambda_3[a\lambda_1+b\lambda_2+c\lambda_3]=0$ is composed of the four sides of a quadrilateral. To the six points of intersection correspond the six edges of a tetrahedron. To the other points on the four lines correspond the $4C_2$ respectively.

Any plane $\alpha\equiv\alpha_i x_i=0$ goes into a surface having a C_2 at each vertex of the tetrahedron. The equation of the surface is

$$C^3\equiv\alpha_1 a y_2 y_3 y_4+\alpha_2 b y_1 y_3 y_4+\alpha_3 c y_1 y_2 y_4+\alpha_4 y_1 y_2 y_3=0.$$

On C^3 are the six edges of the tetrahedron and the lines $\overline{15}$, $\overline{24}$, $\overline{36}$.

XVII. Surface with $2\ B_3+C_2$.

If in VIII (Surface with $3\ C_2$) relations a) and b) given below, or what amounts to the same thing, in XIII (Surface with $B_3+2\ C_2$) relation b) exist among the coefficients of the three quadrics, namely:

$$\text{a)}\quad (a_{23}b_{14}-a_{13}b_{24})(a_{12}c_{34}-a_{13}c_{24})=0,$$
$$b)\quad (b_{12}c_{34}-b_{14}c_{23})(a_{13}b_{24}-a_{23}b_{14})=0,$$

VIII or XIII will go over into a C^3 with $2\ B_3+C_2$. $C_4(\lambda)$, which is a quartic with a cusp and two double points in XIII, goes into a quartic with two cusps and one double point. J, therefore, is made up of two double lines, corresponding to the two cusps, a single line corresponding to the double point and a single line corresponding to the other points on the quartic.

The equation of the surface is

$$C^3\equiv(a_{13}b_{12}c_{24}+a_{12}b_{14}c_{23})y_1y^2_2+(a_{23}b_{12}c_{24}+a_{12}b_{24}c_{23})y^3_2+(a_{13}b_{14}c_{24}+a_{12}b_{14}c_{34})y_1y_2y_4+(a_{23}b_{14}C_{24}+a_{12}b_{24}c_{34})y^2_2y_4+(a_{13}b_{12}c_{34}+a_{13}b_{14}c_{23})y_1y_2y_3+(a_{23}b_{12}c_{34}+a_{13}b_{24}c_{23})y^2_2y_3+2a_{13}b_{14}c_{34}y_1y_3y_4+(a_{23}b_{14}c_{34}+a_{13}b_{24}c_{34})y_2y_3y_4=0.$$

On C^3 $P_1(0:0:0:1)$ and $P_2(0:0:1:0)$ are B_3 if conditions a) and b) are satisfied. $P_3(1:0:0:0)$ is a C_2. The surface has the following lines: (1), (2), (3), $\overline{13}$, $\overline{33}$.

XVIII. Surface with $B_4+2\ C_2$.

Let A be a cone, B and C each a pair of planes with their double lines intersecting. The equations are

$$A\equiv a_{22}x^2_2-2a_{13}x_3x_4=0,$$
$$B\equiv x^2_1-bx^2_2=0,$$
$$C\equiv 2x_2x_3=0.$$

The matrix is in this case

$$\left\|\begin{matrix} 0 & a_{22}y_2 & -a_{13}y_4 & -a_{13}y_3 \\ y_1 & -by_2 & 0 & 0 \\ 0 & y_3 & y_2 & 0 \end{matrix}\right\|$$

from which we get

$$\rho x_1 = a_{13} b y^2_2 y_3,$$
$$\rho x_2 = a_{13} y_1 y_2 y_3,$$
$$\rho x_3 = -a_{13} y_1 y^2_3,$$
$$\rho x_4 = a_{13} y_1 y_3 y_4 + a_{22} y_1 y^2_2.$$

$C_4(\lambda) \equiv \lambda^2_1 \lambda_2 [a_{22}\lambda_1 - b\lambda_2] = 0$, a double line and two single lines intersecting in the point $0:0:1$. To this point corresponds the triple line of J, $x_2 = x_3 = 0$. To the point $0:1:0$ corresponds the double line of J, $x_1 = x_2 = 0$, and to the line $a_{22}\lambda_1 - b\lambda_2 = 0$ the line $x_1 = x_3 = 0$.

$\alpha \equiv \alpha_1 x_1 + \alpha_4 x_4 = 0$ transforms into the surface
$C^3 \equiv \alpha_1 a_{13} b y^2_2 y_3 + \alpha_4 a_{13} y_1 y_3 y_4 + \alpha_4 a_{22} y_1 y^2_2 = 0.$

This equation is the same, except for the coefficients, as that which Cayley gives for a C^3 with a $B_4 + 2\,C_2$. The singular points are

$$P_1(0:0:0:1) \text{ a } B_4,$$
$$P_2(0:0:1:0) \text{ a } C_2,$$
$$P_3(1:0:0:0) \text{ a } C_2.$$

On C^3 are the lines (1), (2), (3), $\overline{13}$, and the free line II.

XIX. Surface with $B_6 + C_2$.

Let two of the quadrics be cones and the third a pair of planes with the double line through the vertices of the cones. Then

$$A \equiv a_{11} x^2_1 + a_{33} x^2_3 + 2a_{12} x_1 x_2 = 0,$$
$$B \equiv 2x_1 x_2 = 0,$$
$$C \equiv c_{22} x^2_2 - 2c x_1 x_4 = 0.$$

The matrix is

$$\left\| \begin{matrix} a_{11}y_1 + a_{12}y_2 & a_{12}y_1 & a_{33}y_3 & 0 \\ y_2 & y_1 & 0 & 0 \\ -c_{14}y_4 & c_{22}y_2 & 0 & -c_{14}y_1 \end{matrix} \right\|$$

$$\rho x_1 = a_{33} c_{14} y^2_1 y_3,$$
$$\rho x_2 = -a_{33} c_{14} y_1 y_2 y_3,$$
$$\rho x_3 = -a_{11} c_{14} y^3_1,$$
$$\rho x_4 = -a_{33} c_{22} y^2_2 y_3 - a_{33} c_{14} y_1 y_3 y_4.$$

$C_4(\lambda) \equiv \lambda_1 \lambda^3_3 = 0$. J is made up of a 5-fold line and a single line cutting it. To the plane $\alpha \equiv \alpha_3 x_3 + \alpha_4 x_4 = 0$ corresponds the surface

$$C^3 \equiv \alpha_3 a_{11} a_{14} y^3_1 + \alpha_4 a_{33} c_{22} y^2_2 y_3 + \alpha_4 a_{33} c_{14} y_1 y_3 y_4 = 0.$$

This equation is equivalent to Cayley's. C^3 has a B_6 at $0:0:0:1$ and a C_2 at $0:0:1:0$. The two lines are (1) and (2).

XX. Surface with a U_8.

Let the quadrics be

$$A \equiv a_{11}x^2_1 + a_{44}x^2_4 + 2a_{23}x_2x_3 = 0,$$
$$B \equiv x^2_2 - cx^2_3 = 0,$$
$$C \equiv 2x_1x_3 = 0,$$

from which follows the matrix

$$\left\| \begin{matrix} a_{11}y_1 & a_{23}y_3 & a_{23}y_2 & a_{44}y_4 \\ 0 & y_2 & -cy_3 & 0 \\ y_3 & 0 & y_1 & 0 \end{matrix} \right\|$$

and

$$\rho x_1 = a_{44}y_1y_2y_4,$$
$$\rho x_2 = -a_{44}cy^2_3y_4,$$
$$\rho x_3 = -a_{44}y_2y_3y_4,$$
$$\rho x_4 = -a_{11}y^2_1y_2 + a_{23}cy^3_3 + a_{23}y^2_2y_3.$$

$\alpha \equiv \alpha_2x_2 + \alpha_4x_4 = 0$ transforms into the surface
$C^3 \equiv \alpha_2a_{44}cy^2_3y_4 + \alpha_4a_{11}y^2_1y_2 - \alpha_4a_{23}cy^2_3 - \alpha_4a_{23}y^2_2y_3 = 0.$

This surface has a U_8 at $0 : 0 : 0 : 1$ and but one line on it, $y_2 = y_3 = 0$, corresponding to the point (1).

XXI. Surface with 3 B_3.

If in VIII (Surface with 3 C_2) the following relations exist among the coefficients of the three quadrics, viz.:

a) $(a_{23}b_{14} - a_{13}b_{24})(a_{12}c_{34} - a_{13}c_{24}) = 0,$
b) $(b_{12}c_{34} - b_{14}c_{23})(a_{13}b_{24} - a_{23}b_{14}) = 0,$
c) $(b_{14}c_{23} - b_{12}c_{34})(a_{13}c_{24} - a_{12}c_{34}) = 0,$

each of the 3 C_2 goes over into a B_3. $C_4(\lambda)$, which in VIII has 3 nodes, now has 3 cusps. J, therefore, is made up of three double lines intersecting in a point. The equation of the surface is

$$C^3 \equiv (a_{13}b_{12}c_{24} + a_{12}b_{14}c_{23})y_1y^2_2 + (a_{23}b_{12}c_{24} + a_{12}b_{24}c_{23})y^3_1 + (a_{13}b_{14}c_{24} + a_{12}b_{14}c_{34})y_1y_2y_4 + (a_{23}b_{14}c_{24} + a_{12}b_{24}c_{34})y^2_2y_4 + (a_{13}b_{12}c_{34} + a_{13}b_{14}c_{23})y_1y_2y_3 + (a_{23}b_{12}c_{34} + a_{13}b_{24}c_{23})y^2_2y_3 + 2a_{13}b_{14}c_{34}y_1y_3y_4 + (a_{23}b_{14}c_{34} + a_{13}b_{24}c_{34})y_1y_3y_4 = 0.$$

The three lines on C^3 are (1), (2), (3).

XXII. Ruled Surface. Nodal line of the first kind.

To obtain this surface in the simplest manner, let

$$A \equiv 2x_1x_3 = 0,$$
$$B \equiv 2x_2x_4 = 0,$$
$$C \equiv x^2_1 - bx^2_2 = 0.$$

The matrix is

$$\begin{Vmatrix} y_3 & 0 & y_1 & 0 \\ 0 & y_4 & 0 & y_2 \\ y_1 & -by_2 & 0 & 0 \end{Vmatrix}$$

from which we get

$$\rho x_1 = -by_1y^2{}_2,$$
$$\rho x_2 = -y^2{}_1y_2,$$
$$\rho x_3 = by^2{}_2y_3,$$
$$\rho x_4 = y^2{}_1y_4.$$

$C_4(\lambda)$ reduces to $\lambda_1=0$, $\lambda_2=0$, each counted twice. They intersect in the point $0:0:1$; to this corresponds the 4-fold line $x_1=x_2=0$ of J. To the point $0:1:0$ corresponds the line $x_2=x_4=0$ and to $1:0:0$ the line $x_1=x_3=0$ of J.

$\alpha\equiv\alpha_3x_3+\alpha_4x_4=0$, a plane through the directrix, transforms into the ruled surface

$$C^3\equiv\alpha_3by^2{}_2y_3+\alpha_4y^2{}_1y_4=0,$$

which is Cayley's equation except for the coefficients. $x_1=x_2=0$ is a nodal line of the first kind. $0:0:0:1$ and $0:0:1:0$ are the two pinch points, i.e., the two tangent planes fall together at each of these points.

XXIII. Ruled Surface. Nodal line of the second kind.

Let the three quadrics have a twisted cubic in common. For the sake of simplicity they may be taken as follows:

$$A\equiv 2x_1x_4-2x_2x_3=0,$$
$$B\equiv 2x^2{}_2-2x_2x_3-2x_1x_3+2x_1x_4=0,$$
$$C\equiv 2x^2{}_3-2x_2x_3+2x_1x_4-2x_2x_4=0.$$

The matrix is

$$\begin{Vmatrix} y_4 & -y_3 & -y_2 & y_1 \\ -y_3+y_4 & y_2-y_3 & -y_1-y_2 & y_1 \\ y_4 & -y_3 & -y_2+y_3 & y_1-y_2 \end{Vmatrix}$$

from which we get

$$\rho x_1 = 3y_1y_2y_3-y^2{}_1y_4-2y^3{}_2,$$
$$\rho x_2 = -y_1y_2y_4-y^2{}_2y_3+2y_1y^2{}_3,$$
$$\rho x_3 = y_1y_3y_4+y_1y^2{}_3-2y^2{}_2y_4,$$
$$\rho x_4 = 2y^3{}_2+y_1y^2{}_4-3y_2y_3y_4.$$

$C_4(\lambda)\equiv[\lambda^2{}_1+\lambda^2{}_2+\lambda^2{}_3+2\lambda_1\lambda_2+2\lambda_1\lambda_3+\lambda_2\lambda_3]^2=0$, which is a double conic. This conic is in one-to-one correspondence with the twisted cubic

$$x_1:x_2:x_3:x_4=\lambda^3:\lambda^2:\lambda:1.$$

This cubic is the common curve of intersection of A, B and C.

$\alpha\equiv x_1=0$ osculates the cubic at $0:0:0:1$ and to it corresponds the surface

$$C^3\equiv 3y_1y_2y_3-y^2_1y_3-2y^3_2=0.$$

$y_1=y_2=0$ is a nodal line of the second kind having a pinch point at $0:0:0:1$.

XXIV. Cubic Cone.

To obtain the cubic cone, let one of the quadrics be unrestricted and the other two be pairs of planes with their double lines intersecting. The equations are

$$A\equiv a_{11}x^2_1+a_{22}x^2_2+a_{33}x^2_3+x^2_4+2a_{12}x_1x_2+2a_{13}x_1x_3+2a_{23}x_2x_3=0,$$
$$B\equiv x^2_1-bx^2_2=0,$$
$$C\equiv x^2_1-cx^2_3=0.$$

The matrix is

$$\left\| \begin{array}{cccc} a_{11}y_1+a_{12}y_2+a_{13}y_3 & a_{12}y_1+a_{22}y_2+a_{23}y_3 & a_{13}y_1+a_{23}y_2+a_{33}y_3 & y_4 \\ y_1 & -by_2 & 0 & 0 \\ y_1 & 0 & -cy_3 & 0 \end{array} \right\|$$

from which we get

$$\rho x_1=bcy_2y_3y_4,$$
$$\rho x_2=cy_1y_3y_4,$$
$$\rho x_3=by_1y_2y_4,$$
$$\rho x_4=(a_{11}bc+a_{33}b+a_{22}c)y_1y_2y_3+a_{12}bcy^2_2y_3+a_{13}bcy_1y^2_3+a_{13}by^2_1y_2+a_{23}by_1y^2_2+a_{12}cy^2_1y_3+a_{23}cy_1y^2_3.$$

$C_4(\lambda)$ is made up of a cubic and a line. J, therefore, consists of three concurrent lines and a plane cubic. $\alpha\equiv x_4=0$, the plane of the plane cubic transforms into the surface.

$$C^3\equiv(a_{11}bc+a_{33}b+a_{22}c)y_1y_2y_3+a_{12}bcy^2_2y_3+a_{13}bcy_1y^2_3+a_{13}by^2_1y_2+a_{23}by_1y^2_2+a_{12}cy^2_1y_3+a_{23}cy_1y^2_3=0.$$

C^3 is a cubic cone with vertex at $0:0:0:1$. The points of the plane cubic go into the rulings through $0:0:0:1$. All other points in the plane go into this vertex.

This concludes the study of the 24 varieties of cubic surfaces. The preceding results are summarized in the following table in which the surfaces are given in the same order as they appear in this paper. Column 2 gives the plane quartic $C_4(\lambda)$ and column 3 the space sextic J in point-to-point correspondence with $C_4(\lambda)$. In column 4 the selection of the plane α is given; column 5 gives the six intersections of J with α; in column 6 the singularity or singularities on the corresponding cubic surface are given and column 7 gives the number of distinct right lines on the surface.

TABLE I

Surface	$C_4(\lambda)$	J	α-plane	Position of the 6 points	Sing.	Dist. Lines
I	Unrestricted quartic	Proper sextic curve	Arbitrary	.1 .2 .3 .5 .6 .4	None	27
II_1	Unrestricted quartic	Proper sextic curve	Through a ruling of R	.3 .4 .2 .1 .6 .5	C_2	21
II_2	Unrestricted quartic	Proper sextic curve	Tangent to J	. . (11) .2 .5 .3 .4	C_2	21
III_1	Unrestricted quartic	Proper sextic curve	Through a ruling of R and tangent to J	. . (11) .2 .5 .3 .6	B_3	15
III_2	Unrestricted quartic	Proper sextic curve	Osculating J	∴ (111) .4 .2 .3	B_3	15
IV_1	2 lines and a conic	J made up of 6 lines. One has 4 double points and the other 2 triple points	Arbitrary	.3 .2 .6 .5 .1 .4	2 C_2	16
IV_2	Unrestricted quartic	Proper sextic curve	Through 2 rulings of R	Same as IV_1		
IV_3	Cubic and a line tangent to it	J made up of a single line, a double line and a plane cubic	Arbitrary	.3 .2 .1 . . (44) .5	2 C_2	16

TABLE I—(*Continued*)

Surface	$C_4(\lambda)$	J	α-plane	Position of the six points	Sing.	Dist. Lines
V	Unrestricted quartic	Proper sextic curve	Through a ruling of R and osculating J	. · · (111) .2 .1 .4	B_4	10
VI	Quartic with one double and one cuspidal point	J made up of a double line, a single line skew to the double line and a twisted cubic	Osculating the twisted cubic	. · · (111) .3 . · (22)	B_3+C_2	11
VII	Cubic and a line tangent to it	Single line, double line and a plane cubic	Through 0 : 1 : 0 : 0 and 0 : 0 : 1 : 0	. ·· (111) .3 . · (22)	B_5	6
VIII	Quartic with 3 double points	3 skew lines and a twisted cubic	Through vertices of cones	. · (11) . . (22) · . (33)	3 C_2	12
IX	A double line, and 2 single lines	A double line, a single line, a conic, and a line tangent to the conic	Through 1 : 0 : 0 : 0 and 0 : 1 : 0 : 0	. . (22) .3 .4 . · (11)	2 B_3	7
X_1	A conic and 2 lines each tangent to the conic	Two skew double lines and two skew lines each meeting each double line	Through the point where a single line meets the double line	. ·· (111) .3 . · (22)	B_4+C_2	7
X_2	Double conic	Double twisted cubic	Tangent to cubic	:: (1111) . · (22)	B_4+C_2	7
XI	Quartic with a cusp and a node	Double line, a single line skew to the double line and a twisted cubic	Plane tangent where single line cuts the cubic and through the point where the double line cuts the cubic	·.·(111) · · . (222)	B_6	3
XII	Same as VII		Through 1 : 0 : 0 : 0 and 0 : 0 : 1 : 0	.·.(111) 2 3 . . .4	U_6	6
XIII	Quartic with 2 nodes and a cusp	Double line, single line and a conic	Same as VII	. . (11) : (22) · . (33)	$B_3+2\ C_2$	8

TABLE I—(*Concluded*)

Surface	$C_4(\lambda)$	J	α-plane	Position of the 6 points	Sing.	Dist. Lines
XIV	Two double lines	4-fold line and a double line cutting it	$x_4=0$	. . (1111) . . (22)	B_5+C_2	4
XV	Cubic and a line tangent to it	Single line, double line and a cuspidal cubic	Through cusp	. . . . . (11111) .2	U_7	3
XVI	4 sides of a quadrilateral	J made up of the 6 edges of a tetrahedron	Arbitrary	.3 .4 .2 .6 .5 .1	4 C_2	9
XVII	Quartic with one node and two cusps	Two double lines and two single lines	Same as VIII	. . (33) . . (22) . . (11)	2 B_3+C_2	5
XVIII	Double line and 2 single lines intersecting in one point	A triple line, a double line and a single line intersecting in one point	Through 1 : 0 : 0 : 0 and 0 : 1 : 0 : 0	. . . (111) ·3 . . (22)	$B_4+2\ C_2$	5
XIX	Triple line and a single line	5-fold line and a single line cutting it	Through 0 : 1 : 0 : 0 and 0 : 0 : 1 : 0	. . (11111) . . . ·2	B_6+C_2	2
XX	Cuspidal cubic and inflexional tangent	Triple line and a cuspidal cubic	Through cusp	. . . . . . (111111)	U_8	1
XXI	Quartic with 3 cusps	Three double lines	Same as VIII	. · (33) . . (22) . . (11)	3 B_3	3
XXII	Two double lines	4-fold line and 2 single lines cutting it	Through directrix	. (1) .3 .2	Nodal line and directrix distinct	
XXIII	Double conic	Double twisted cubic	Osculating cubic	.(1)	Nodal line and directrix coincide	
XXIV	Cubic and a line	Three concurrent lines and a plane cubic	3 concurrent lines and a plane cubic	∞ number of points of plane cubic	Cone with vertex at 0 : 0 : 0 : 1	

UNIVERSITY OF CALIFORNIA PUBLICATIONS
IN
MATHEMATICS

Vol. 1, No. 20, 425-443, 1 figure in text May 17, 1924

THE HYPERSPACE GENERALIZATION OF THE LINES ON THE CUBIC SURFACE.

BY

DANIEL VICTOR STEED

I. INTRODUCTION

Of all general algebraic surfaces in space of three dimensions the cubic surface is unique in one respect, namely, there exists upon it a finite number of straight lines. The discovery of this property of the cubic surface was made by Cayley[1] in 1849, and the number of the lines was determined by Salmon in the same year. Since that time the remarkable configuration which these lines form has been the subject of much important research not only in pure and analytical geometry but also in the theory of groups.

Considering the very extensive mathematical literature which has grown up around this configuration, the existence of which can from a certain point of view be looked upon as a property of three dimensional space, it would seem that geometers interested in hyperspace would very naturally inquire: Is there an analogous property for space of four dimensions? Or to put the question more fully: Is there in space of four dimensions an algebraic hypersurface which possesses the unique property of containing upon it a finite number of straight lines? If so, what is the order of this hypersurface, and how many lines are there on it? Finally, what are the positions of these lines relative to each other?

The first two of these questions were answered by Schubert, who proved that there are 2875 lines on the general hypersurface of order five in space of four dimensions. Schubert's result seems, however, to have been obtained not as a result of a direct attempt to answer the questions asked above, but rather as incidental to another investigation in which that mathematician was engaged, namely, the determination of the number of lines having six point contact with the general hypersurface of order m in space of four dimensions.[2] Now a line having six points in

[1] *The Cambridge and Dublin Mathematical Journal*, vol. 4, p. 132.

[2] Schubert, "Die n-dimensional Verallgemeinerungen der Anzahlen für die vielpunktig berührenden Tangenten einer punktallgemeinen Fläche n-ten Grades," *Mathematische Annalen*, vol. 26, pp. 26-52.

common with a hypersurface of order five lies entirely on it. Hence for $m=5$ the formula for the number of six point tangents gives the number of lines on the quintic hypersurface.

In 1919, however, Professor D. N. Lehmer, being then unacquainted with Schubert's result, made a more direct inquiry into the question of what in space of four dimensions corresponds to the existence of lines on the cubic surface, and came to the conclusion that, as far as the existence of lines upon it is concerned, the hypersurface of order five was the analogue of the cubic surface. The method by which Lehmer arrived at this result was as follows: Let (x_1, x_2, x_3, x) be the cartesian co-ordinates of a point in space of four dimensions. The equations of any line can be put into the form

$$x_1=a_1x+a_2,\ x_2=b_1x+b_2,\ x_3=c_1x+c_2. \tag{1}$$

If these values of x_1, x_2, and x_3 be substituted in the general equation of the algebraic hypersurface of order m, there results an equation in x,

$$A_1x^m+A_2x^{m-1}+\dots\dots+A_{m+1}=0 \tag{2}$$

whose roots are the x values corresponding to the m points of intersection of the line (1) with the hypersurface, and whose coefficients are polynomials of degree m in the quantities a_1, a_2, b_1, b_2, c_1, c_2. If now the line (1) is to lie entirely on the hypersurface, these six quantities must satisfy the $m+1$ equations

$$A_i=0\ (i=1, 2, 3, \dots\ m+1). \tag{3}$$

This system of equations, containing as it does six unknowns, will in general have a finite number of solutions when and only when there are six equations, i.e., when $m+1=6$ or $m=5$. Thus the only general hypersurface containing a finite number of lines on it is that of order five. For $m=5$ (3) consists of six equations each of degree five. The number of lines will equal the degree of the resultant of this system of equations.

Professor Lehmer proposed to me the problem of enumerating these lines, his discovery and suggestion thus leading to the present investigation. The results given in this paper were all obtained before I discovered that Schubert had already arrived at the result in the manner briefly described above. They constitute an independent verification of Schubert's work, the method differing from Schubert's in that no use is made of the notion of tangency.

It is conceivable that there may be values of n and r for which the general hypersurface of order r in space of n dimensions contains on it a finite number of linear spaces of dimension d. The question whether such values of n and r actually exist is considered in Part II, where a relation between r, n, and d is obtained, the existence of which is a necessary condition for the existence of a finite number of linear spaces of dimension d on the general hypersurface of order r in space of n dimensions. The determination of the values of n and r for which this condition is also sufficient is a problem requiring further investigation, which might lead to interesting results.

In Part V recursion formulae are developed, successive applications of which make possible the determination of the number of lines on the general hypersurface of order $2n-3$ in space of n dimensions. These formulae are then applied in Part VI in actually determining the number of lines on the corresponding hypersurfaces for spaces of dimensions four to seven, inclusive.

II. GENERALIZATION

The problems of the existence of a finite number of lines on the general cubic surface in S_3, and of lines on the general quintic hypersurface in S_4, are special cases of a more general problem for space of n dimensions which may be formulated as follows: To determine whether or not there exists on the general hypersurface of order r in space of n dimensions a finite number of linear spaces of dimension d.

In this section we shall derive an equation connecting r, n, and d, the existence of which is a necessary condition for the existence of spaces of dimension d satisfying the condition stated in the problem as formulated above.

Let the equation of the general hypersurface of order r in space of n dimensions be

$$f(x_1, x_2, x_3, \dots x_n)=0. \tag{1}$$

The equations of any space[3] S_d of dimension d consist of $n-d$ linear equations in the n cartesian coördinates $x_1, x_2, x_3, \dots x_n$ of n dimensional space. These can be solved for $n-d$ of the variables in terms of the remaining d, and the equations can thus be put in the form

$$x_i=a_{i1}x_{n-d+1}+a_{i2}x_{n-d+2}+a_{i3}x_{n-d+3}+ \dots +a_{id}x_n+a_{id+1} (i=1, 2, \dots n-d). \tag{2}$$

The result of substituting these values of x_i into (1) is an equation of degree r in the d variables x_{n-d+k} $(k=1, 2, 3, \dots d)$. If the space defined by equations (2) lies entirely on the hypersurface, then this equation in the d variables must vanish identically, and hence each of its coefficients must vanish. The vanishing of each of these coefficients imposes a condition upon the coefficients in (2) (which may be called the coördinates of S_d). If these coördinates be thought of as unknowns, then the number of unknowns is evidently $(n-d)(d+1)$. And since the general equation of degree r in d variables contains $\binom{r+d}{d}$ or

$$\frac{(r+d)(r+d-1)\dots(r+2)(r+1)}{d!}$$

terms the number of equations of condition is $\binom{r+d}{d}$. In order that the number

[3] For the sake of brevity the word 'space' will be used hereafter to denote a 'linear space.'

of solutions of these equations be finite, it is necessary that the number of equations equal the number of unknowns. Hence the

Theorem. A necessary condition for the existence of a finite number of linear spaces of dimension d on the general hypersurface of order r in space of n dimensions is that r, n, and d satisfy the equation $\binom{r+d}{d}=(n-d)\ (d+1)$.

It follows immediately from the theorem that the general hypersurface of order r in space of n dimensions does not contain upon it a finite number of lines unless $r=2n-3$. By taking $d=2$ it will be seen that a necessary condition for the existence of planes on the hypersurface is that r satisfy the equation

$$r^2+3r+14=6n,$$

and hence that either $r\equiv 1 \pmod 3$ or $r\equiv 2 \pmod 3$. When $r=1$, $n=3$, which is the case of the plane in space of three dimensions, and the number of planes on the surface is obviously finite, being one. One is naturally lead to inquire whether the condition of the theorem is not also a sufficient one. That it is not can be seen by taking $r=2$, $n=4$, values which satisfy the equation above. That the general quadric hypersurface Q in space of four dimensions does not contain upon it a finite number of planes can be easily shown as follows: Suppose that Q contains upon it the plane π. Then any hyperplane through π meets Q in a quadric surface, which contains π and therefore consists of π and a second plane π' which is also on Q. Thus for every hyperplane through π there is a plane on Q distinct from π. Hence if there is one plane on Q there must be an infinity of them.

Of course the assumption that there is one plane on the hyperquadric involves assuming a special hyperquadric.[4] For $d=1$, $n=4$, $r=2$, then, we can conclude that the six equations are in general inconsistent, their resultant taking the form

$$0\cdot x_4+c=0 \tag{3}$$

where c is a function of the coefficients of the equation of the hypersurface. Now if these coefficients are so selected that c vanishes, then (3) will be satisfied by any finite value of x_4. Hence (3) expresses the condition that the hyperquadric contain an infinity of planes on it. This property of the hyperquadric might be described as a "poristic property of space of four dimensions." For although the number of equations necessary to determine the coördinates of a plane on the hyperquadric equals the number of these coördinates, there exists in general no plane on it, and, secondly, if we take a hyperquadric containing one plane on it, the hyperquadric will then necessarily contain an infinity of planes.

[4] This special hyperquadric need not be degenerate, i.e., consist of a pair of hyperplanes. Such a hypersurface would be a still more special case than the one here considered.

III. THE ENUMERATIVE METHOD OF SCHUBERT FOR SPACE OF n DIMENSIONS

It has been pointed out in Part I that the enumeration of the lines on the quintic hypersurface is equivalent to the determination of the degree of the resultant of a system of six equations, each of degree five. The equations of this system, however, are special, and hence we cannot say that the degree of the resultant is 5^6. All that we can conclude without further investigation is that 5^6 is an upper limit for the degree of the resultant. The writer has been unable to find in the theory of elimination a method for the determination of this degree. It seems that its determination by purely algebraic methods would be a more difficult task than the solving of the original problem by geometric methods.

The solution of the three dimensional analogue of the problem now under consideration, i.e., of the problem of determining the number of lines on the cubic surface, is easily obtained by determining the number of double tangent planes that can be drawn to the cubic surface from an arbitrary fixed point P not on the cubic surface. For any such plane meets the surface in a cubic curve having two double points, and therefore consisting of a line and a conic. Conversely every plane through P and a line on the surface is a double tangent plane, and hence the number of lines on the surface is equal to the number of double tangent planes from P.

Now in space of four dimensions, any plane through a point P and a line on the quintic hypersurface F meets the hypersurface in a quintic curve having four double points, namely, the four points of intersection of the line with the residual quartic. Hence such a plane is a fourfold tangent plane. But the converse is not true. A fourfold tangent plane does not necessarily contain a line on F. For a quintic curve can have four double points without degenerating. Hence, although from a point not on F a finite number of fourfold tangent planes can be drawn to the hypersurface, the number of such planes exceeds the number of lines on the quintic hypersurface.

Having come to the conclusion that a more powerful tool would be necessary for the solution of the problem for four dimensions than was needed for the corresponding problem in space of three dimensions, I began the study of Schubert's *Kalkül der Abzählenden Geometrie.*[5] For an understanding of the following development, a knowledge of the fundamental concepts of enumerative geometry for space of three dimensions made use of by Schubert and explained in the textbook just mentioned, is essential and will be presupposed in this paper. The remainder of this discussion will be devoted to an extension of this method to $n-$ dimensional space and its application to the particular problem of the present investigation. The method is that of pure rather than analytic geometry, and what follows will there-

[5] Leipzig, 1879.

fore require and presuppose in addition to what has been mentioned above a knowledge of $n-$ dimensional geometry equivalent to what will be found in Schoute, *Mehrdimensional Geometrie*, vol. 1, part 1.

In what follows, the word 'space' will always refer to a 'linear space,' i.e., a space which, if it contains two points of a line, contains the entire line. The word 'hyperplane' will denote a space of dimension of $n-1$ where n is the dimension of the space of operation. A hypersurface of order r is a manifold of dimension $n-1$ which in general has r points of intersection with an arbitrary line. Analytically it is the locus of a single equation of degree r in the coördinates of the space of operation. The word 'variety' will be used for a point locus of dimension greater than one and less than $n-1$. Analytically it is defined by means of a set of not more than $n-2$ equations, the dimension of the variety being $n-d$ where d is the number of defining equations.

A Notation for Line Conditions

The adoption of a suitable notation is of fundamental importance in the extension we are about to make. A notation which will be found to have several advantages and which will be used throughout this paper is given in the following:

Definition 1. *The symbol* $g\ [i\ k_1k_2k_3\ \ldots\ k_j(n)]$ *where* $n>i\geqq o$ *and* $i-2<k_\lambda<n$ $(\lambda=1, 2, 3, \ldots j)$ *denotes the condition that the line* g *lie in a space* S_{n-i} *of dimension* $n-i$ *and meet* j *linear spaces of dimension* $n-k_\lambda-1$, *these spaces all being contained in the space* S_{n-i}; *and the same symbol will denote also the number of lines satisfying this condition. The letter enclosed in parenthesis and indicating the dimension of the space of operation may be omitted when its value is n or is known from the context. The letters* i *and* k_λ *will be called arguments.*

When the number of arguments is small, they will sometimes be written as subscripts and the brackets omitted. Thus the condition $g\ [1\ 2\ (3)]$ will be written $g_{1\,2\,(3)}$.

The following rule will be seen to be applicable to the above notation. *If for any* λ, $k_\lambda=i$ *or* $i-1$, *then the argument* k_λ *can be omitted; and conversely, an argument* k_λ *where* $k_\lambda=i$ *or* $i-1$ *can be supplied without the numerical value of the condition being thereby altered.*

For if $k_\lambda=i$ the presence of the argument k_λ expresses the condition that g meet a space of dimension $n-i-1$ contained in the space S_{n-i} which contains g. If the space S_{n-i} be thought of for the moment as the space of operation, then the argument expresses the condition that g meet a given hyperplane. This condition, however, being satisfied by any line in S_{n-i}, does not impose any restriction upon g, and the number of lines satisfying the original condition is not affected by the omission or presence of the argument. In similar manner the rule for the case when $k_\lambda=i-1$ can be verified.

Fundamental Theorems on Line Conditions

Theorem 1. *When* $i<k_1+k_2+1-n$, *the condition* $g\ [i\ k_1k_2]$ *is equivalent to the condition* $g[j\ k_1k_2]$, *where* $j=k_1+k_2+1-n$.

For let M and N be the two linear spaces of dimension $n-k_1-1$ and $n-k_2-1$ respectively which the line g meets. Then g has two points in common with the space R determined by M and N, and therefore lies entirely in that space. Now if M and N intersect the dimension x of the intersection[6] is given by

$$2n-2-k_1-k_2-x=n-i$$

and hence

$$x=n+i-(k_1+k_2+2).$$

By the hypothesis of the theorem, x is consequently negative, that is M and N do not intersect. Hence the dimension of R is $n-j$,
where

$$n-j=n-k_1-1+n-k_2-1+1,$$

whence

$$j=k_1+k_2+1-n.$$

Since then any line that satisfies the condition of meeting both M and N must lie in the space R of dimension $n-j$, the argument i can be replaced by the greater number j.

Theorem 2. *The product of two line conditions* $g\ [i\ k_1k_2k_3\ \ldots\ k_a]$ *and* $g[j\ l_1l_2l_3\ \ldots\ l_\beta]$ *is* $g[i+j\ \ i+l_1\ \ i+l_2\ \ldots\ i+l_\beta\ \ j+k_1\ \ j+k_2\ \ldots\ j+k_a]$.

A line satisfying both of the factor conditions lies in one space of dimension $n-i$ and in another of dimension $n-j$ and hence lies in their intersection, which will be denoted by R. If the dimension of R is $n-x$, then

$$n-i+n-j-(n-x)=n$$

and hence $x=i+j$, and the first argument of the product is accounted for.

To any argument k_λ corresponds the condition that g meet a space M of dimension $n-k_\lambda-1$ contained in the space S_{n-i} of the first factor condition. But since, as has just been shown, g lies in a space R of dimension $n-i-j$, g must meet M in a point which is in the intersection of M and R. M and R are however both contained in S_{n-i} and hence the dimension of their space of intersection is $n-y-1$, where

$$n-i-j+n-k_\lambda-1-n+y+1=n-i$$

and consequently

$$y=j+k_\lambda.$$

The presence of the subscripts $i+l_\lambda$ in the product condition is accounted for in like manner.

[6] *Cf.* Schoute, *Mehrdimensionale Geometrie*, vol. 1, p. 14, et seq.

If definition 1 only is assumed, then the application of theorem 2 will sometimes lead to a symbol which is undefined, since we may have either $i+j>n-i$, $j+k_\lambda \geq n$ or $i+l_\lambda \geq n$. In the second and third cases the spaces M and R have no point in common, as is evident from the fact that $n-y-i$ is negative. Hence, since there can be no line in R which meets M, there is no line satisfying both factor conditions, i.e., their product is zero. In the first case it is evident that the dimension of R does not exceed zero, and hence no line can satisfy the condition of being in both S_{n-i} and S_{n-j}. Thus in this case also the product of the given conditions is zero, and the following definition is justified.

Definition 2. *The symbol* $g[i\ k_1 k_2 k_3\ \ldots\ k_j\ (n)]=0$ *when any one of the arguments* i *or* k *exceeds* $n-1$.

In addition to the above theorems we shall make use of the relation

$$g[i\ k_1 k_2 k_3\ \ldots\ k_a(n)]=g[0\ k_1-i\ k_2-i\ k_3-i\ \ldots\ k_a-i(n-i)]$$

which follows immediately from the definition of line conditions.

The Dimension of a Line Condition

The order of infinity of lines in space of n dimensions is $2n-2$. If c is the order of infinity of lines satisfying a given line condition, then the dimension[7] of the condition is defined as being $2n-2-c$. It will be seen that the dimension of the line condition $g[i\ k_1 k_2 k_3\ \ldots\ k_a]$ can be expressed in terms of the arguments of the condition. On account of the rule stated in a preceding paragraph we can and will assume that no k is less than $i+1$. By theorem 2

$$g[ik_1]\ g[0\ k_2-i]\ g[0\ k_3-i]\ \ldots\ g[0\ k_a-i]=g[ik_1 k_2 k_3\ \ldots\ k_a].$$

Let M be the space of dimension $n-k_1-1$ of the condition $g[ik_1]$ contained in the space S_{n-i}. Then through each point of M an $n-i-1$ fold infinity of lines can be drawn, each of which satisfies the condition $g[ik_1]$. The order of infinity of lines satisfying the condition $g[ik_1]$ is therefore $n-k_1-1+n-i-1=2n-k_1-i-2$, and hence the dimension of the condition is

$$2n-2-(2n-k_1-i-2)=i+k_1.$$

Similarly the dimension of $g[0\ k_\lambda-i]$ is $k_\lambda-i$. Hence, since the dimension of a product is equal to the sum of the dimensions of the factor conditions,[8] the dimension δ of the line condition $g[ik_1 k_2 k_3\ \ldots\ k_a]$ is $i+k_1+k_2-i+k_3-i+\ \ldots\ +k_a-i$ or

$$\delta=\Sigma\ k_\lambda-(a-2)i\ \ (\lambda=1, 2, 3, \ \ldots\ a).$$

When $a=2$ the dimension is simply k_1+k_2, and when $i=0$ the dimension is $\Sigma\ k_\lambda$. Since in an equation between conditions the terms must all be of the same dimension, this simple means of determining the dimension of a line condition will be found useful.

---

[7] Schubert, *loc. cit.*, p. 7. [8] Schubert, *loc. cit.*, p. 9.

As an example consider the condition $g[1222(4)]$. The condition is that in space of operation of four dimensions a line g lie in a given hyperplane and intersect three given lines in that hyperplane. By the formula for δ given above, the dimension of the condition is 5. Since there is a sixfold infinity of lines in space of four dimensions, there is a single infinity of lines satisfying the given condition, a fact otherwise obvious since the lines satisfying the condition belong to a regulus.

A Formula for the Evaluation of Line Conditions

Consider the line condition $g^2[0\ k(n)]$ where $k<n/2$. Let the two spaces of dimension $n-k-1$ of the condition be denoted by M and N. If the positions of the spaces M and N are specialized by taking them in the same hyperplane, then the dimension d of their intersection R is given by $2n-2k-2-d=n-1$ and hence $d=n-2k-1$. Consequently either R is a point or d exceeds zero, and the lines that satisfy the condition of meeting both M and N are (1) any line which intersects R, and (2) any line which is in the hyperplane and meets both M and N. The first of these conditions is expressed by the symbol $g[0\ 2k]$ and the second by $g[1\ kk]$. Hence

I $$g^2[0\ k]=g[1\ kk]+g[0\ 2k],\quad (k<n/2).$$

By Schubert's *Princip von der Erhaltung der Anzahl* (hereafter referred to as the *principle of the permanence of number*), this formula which is true for the special position of M and N is true in general, provided that the order of infinity of lines satisfying the given condition has not been increased by virtue of the assumption of a special position for M and N. But the formula for finding the dimension of a condition shows that each condition of the formula is of dimension $2k$. Hence if an algebraically defined system of lines of dimension $2k$ is given, a finite number of lines will satisfy each of the conditions of the formula. Thus by the principle just mentioned the formula is true for an arbitrary choice of the spaces of the condition in the left-hand member of the equation.

It may be worth while to note here a case in which by specializing the positions of the spaces of a condition the order of infinity of the lines satisfying the condition is increased. Consider the condition that in space of four dimensions a line intersect two given lines M and N, a condition expressed by $g^2[0\ 2(4)]$. Clearly there is a twofold infinity of lines satisfying this condition. Let the position of M and N be specialized by supposing them to intersect in P. Then the lines meeting M and N are 1° any line in the plane MN and 2° any line through P. The lines of 1° and 2° are the lines satisfying $g[2]$ and $g[0\ 3]$ respectively. But a threefold infinity of lines satisfies $g[0\ 3]$. Hence we cannot say that $g^2[0\ 2]$ is equal to the sum of $g[2]$ and $g[0\ 3]$. That such a relation does not exist follows immediately from an application of the formula for the dimension of the line condition which shows that the dimensions of the three conditions are 4, 4, and 3, respectively.

It is to be noticed that we have taken a case where k does not satisfy the provision $k<n/2$ of formula I. In such a case, however, theorem 1 is applicable. For suppose $2k=n+j$, $(j\geqq 0)$. Then by theorem 1,

$$\begin{aligned} g^2[0\ k] &= g[2k+1-n\ kk] \\ &= g[j+1\ kk(n)] \\ &= g[0\ k-j-1\ k-j-1\ (n-j-1)] \\ &= g[0\ k-j-1\ (n-j-1)]. \end{aligned}$$

Formula I is now applicable to this last condition, since

$$k-j-1=\frac{n}{2}-\frac{j}{2}-1<\frac{n-j-1}{2}$$

The Incidence Formulae

It is necessary for our purposes to develop a formula for the enumeration of the geometric forms which consist of a line and a point on the line. In this paragraph the word 'form' will be used to designate this figure. The point will be called 'the point of the form' and denoted by g. p will also express the condition that the point p lie in a given hyperplane, and denote the number of forms whose points satisfy this condition.

In accordance with these definitions and the notation for a line condition, $pg[0\ k]$ denotes the number of forms each of which has its point in a given hyperplane H, and its line intersecting a given space M of dimension $n-k-1$. Let the space M be taken in the hyperplane H. Then the lines satisfying the condition $pg[0\ k]$ consist of two groups. First, there are those forms whose points lie in M. These are the forms which satisfy the condition p^{k+1}, since in general $k+1$ hyperplanes intersect in a space of dimension $n-k-1$, and hence the condition that a point lie in $k+1$ hyperplanes is equivalent to the condition that it lie in a space of dimension $n-k-1$. The second group of forms consists of those whose lines lie in H and intersect M, and which consequently satisfy the condition $g[1\ k]$. Thus for this special position of the spaces M and H

II $$pg[0\ k]=p^{k+1}+g[1\ k]$$

and consequently by the principle of the permanence of number this relation holds for an arbitrary choice of the spaces of the condition.

It will be noticed that if this formula is applied to space of three dimensions it becomes for $k=1$, and $k=2$, the *incidenz-formeln I and II* respectively of Schubert.[9] Thus the notation here introduced makes it possible to express these two formulae in one. For space of n dimensions the formula above includes the $n-1$ formulae for which $k=1, 2, 3, \ldots\ldots n-1$.

---

[9] *Loc. cit.*, pp. 25 and 26.

As an application of the above theorems and formulae consider the $n-$dimensional analogue of the problem in space of three dimensions of determining the number of lines which intersect four given lines. In the present notation the number for space of three dimensions is denoted by $g^4{}_{01}$, and its value is well known to be two. Now in space of n dimensions the condition $g^{2n-2}{}_{01}$ is of dimension $2n-2$, as shown by the formula for the dimension of a line condition. Hence a finite number of lines satisfy this condition. Consider the case $n=4$.

$$g^6{}_{01}=(g^2{}_{01})^3=(g_{111}+g_{02})^3=(g_1+g_{02})^3 \quad \text{Formula I and the rule preceding theorem 1.}$$
$$=g^3{}_1+3g^2{}_1g_{02}+3g_1g^2{}_{02}+g^3{}_{02}$$
$$=g_3+3g_2g_{02}+3g_1g_{022}+g^3{}_{02} \quad \text{Theorem 2.}$$

But $g_2g_{02}=g_{24}$ by theorem 2 and $g_{24}=0$ by definition 2.

By theorem 2, $g_1g_{022}=g_{133}=g_{022(3)}=1$.

$$g^3{}_{02}=g_{02}(g_{122}+g_{04})=g_{1322}+g_{06}$$
$$=1 \text{ by definitions 1 and 2.}$$

Hence $g^6{}_{01}=1+3\cdot 0+3\cdot 1+1=5$, or the number of lines which in space of four dimensions intersect six planes is five. For space of five dimensions

$$g^8{}_{01}=(g_1+g_{02})^4=g^4{}_1+4g^3{}_1g_{02}+6g^2{}_1g^2{}_{02}+4g^3{}_1g_{02}+g^4{}_{02}.$$

Now $g^4{}_1=g_4=1$, $g^3{}_1g_{02}=g_3g_{02}=g_{35}=0$ by definition 2, and

$$g^2{}_1g^2{}_{02}=g_2g_{022}=g^2{}_{02(3)}=1$$
$$g_1g^3{}_{02}=g_{1333}=g^3{}_{02(4)}=1 \quad \text{(See above.)}$$
$$g^4{}_{02}=(g_{122}+g_{04})^2=g_{23333}+2g_{1522}+g_{044}$$
$$=g^4{}_{01(3)}+0+g_4 \text{ (by Theorem 1).}$$
$$=2+1.$$

Hence $g_{801}=1+0+6+4+3=14$, or the number of lines which in space of five dimensions intersect eight given linear spaces of dimension three is fourteen. The same method is applicable to the analogous enumerative problems for space of any dimension.

IV. THE LINES ON THE HYPERSURFACE OF ORDER $2n-3$

In space of n dimensions, the order of infinity of forms consisting of a line and a point on the line is $2n-1$. For any line contains a single infinity of these forms, namely, any point on the line can be taken as the point of the form and the line itself is the line of the form. Thus, since there is a $2n-2$ fold infinity of lines in space, the order of infinity of the forms under consideration is $2n-1$.

Denoting by p the condition that the point p lie in a given hyperplane, then if p is the point of the form mentioned above, this condition is a condition imposed upon the form. It is required to determine the dimension of the condition p. Let P be any point of the hyperplane H of the condition p. Then the form which consists of the point P and any line through P satisfies the condition. Hence there is an $n-1$ fold infinity of forms whose point is P. The order of infinity of the points P being $n-1$, it follows that the order of infinity of forms satisfying the condition is $2n-2$, and consequently the dimension of the condition is one.

Consider the form which consists of a line and the r points of intersection of the line with a general hypersurface of order r. The word 'form' will in the remainder of this article refer exclusively to this figure, excepting that later r will assume the particular value $2n-3$. The line of the form will be denoted by g, and the r points by $p_1, p_2, p_3, \ldots . p_r$. In any line in space the number of forms is $r!$, there being one form for each of the r ways that the letters can be assigned to the r points of intersection of the given line with the hypersurface. Hence the order of infinity of forms in space is the same as that of lines, namely, $2n-2$.

The product $p_1p_2p_3 \ldots . p_r$ expresses the condition that the point p_1 lie in a given hyperplane H_1, that p_2 lie in a given hyperplane H_2, etc., that p_r lie in a given hyperplane H_r, and that the r points be distinct. Consider in particular the case in which the given hypersurface, which will be denoted by F, is of order $2n-3$. Then the dimension of the condition above being r, there is a single infinity of forms satisfying the condition $p_1p_2p_3 \ldots . p_{2n-3}$. The lines of the forms satisfying this condition are therefore the rulings of a surface[10] which will be denoted by V. The order of V is the number of points of intersection of V with a space of dimension $n-2$, and is equal to the number of rulings of V which meet a space of that dimension. The order of V is therefore the number of forms which satisfy the condition $g_{01}p_1p_2p_3 \ldots . p_{2n-3}$ and is itself expressed by this symbol. Let H be any hyperplane distinct from the hyperplanes $H_1, H_2, H_3, \ldots . H_{2n-3}$. Then H intersects the hypersurface F in a variety f of order $2n-3$, and intersects V in a curve C of order $g_{01}p_1p_2p_3 \ldots . p_{2n-3}$. By Bezout's theorem the curve C has $(2n-3)g_{01}p_1p_2p_3 \ldots . p_{2n-3}$ points in common with f. Each of these points is a point on the given hypersurface F, and through each of them there is a ruling of V. Let P be any one of these points and let l be the ruling that goes through it. Then l meets F in the $2n-3$ distinct points $p_1, p_2, p_3, \ldots . p_{2n-3}$ and also in P. Thus the line has $2n-2$ distinct points in common with the hypersurface of order $2n-3$ and will therefore be entirely on it, provided that P is distinct from each of the distinct points $p_\lambda (\lambda = 1, 2, 3 \ldots . 2n-3)$. Now if P coincides with p_1, then p_1 satisfies the condition (expressed by p^2_1) of lying in both H_1 and H. And the form whose line is l must therefore satisfy the condition $p^2_1p_2p_3 \ldots . p_{2n-3}$. Similarly if P concides with p_2 the form must satisfy the condition $p_1p^2_2p_3 \ldots . p_{2n-3}$, and so on. Hence if λ_n is the number of lines on F, then

$$\lambda_n = (2n-3)g_{01}p_1p_2p_3 \ldots . p_{2n-3} - p^2_1p_3p_3 \ldots . p_{2n-3} - p_1p^2_2p_3 \ldots . p_{2n-3} - p_1p_2p_3 \ldots . p^2_{2n-3}.$$

[10] The word 'surface' will be used for a variety of dimension two, whether or not the variety is in a hyperplane.

Since in this equation the terms following the first are all equal to each other,

$$\lambda_n=(2n-3)g_{01}p_1p_2p_3 \ldots\ldots p_{2n-3}-(2n-3)p^2{}_1p_2p_3 \ldots\ldots p_{2n-3}.$$

Now for $k=1$ the incidence formula II becomes

$$g_{01}p_1=p^2{}_1+g_1.$$

And when both members of this equation are multiplied symbolically by $p_2p_3 \ldots\ldots p_{2n-3}$, we have

$$g_{01}p_1p_2p_3 \ldots\ldots p_{2n-3}=p^2{}_1p_2p_3 \ldots\ldots p_{2n-3}+g_1p_2p_3p_4 \ldots\ldots p_{2n-3}.$$

On account of this relation the expression for λ_n becomes

$$\lambda_n=(2n-3)p^2{}_1p_2p_3 \ldots\ldots p_{2n-3}+(2n-3)g_1p_2p_3p_4 \ldots\ldots p_{2n-3}$$
$$-(2n-3)p^2{}_1p_2p_3 \ldots\ldots p_{2n-3}$$

or

$$\lambda_n=(2n-3)g_1p_1p_2p_3 \ldots\ldots p_{2n-4}.$$

Denote $g_1p_1p_2p_3 \ldots\ldots p_{2n-4}$ by ρ_n and the hyperplanes of the conditions g_1 and $p_\kappa(\kappa=1, 2, 3, \ldots\ldots 2n-4)$ by S_{n-1} and H_κ respectively. Then the hyperplanes H_κ intersect S_{n-1} in spaces of dimension $n-2$ which are hyperplanes in the space S_{n-1} considered as a space of operation. On account of the general nature of the investigation, the intersection of S_{n-1} with F is a general hypersurface, f, of order $2n-3$ in space of $n-1$ dimensions. And ρ_n is the number of forms (each consisting of a line and the $2n-3$ points in which it intersects f) which in S_{n-1} have $2n-4$ points p_κ in $2n-4$ given hyperplanes h_κ respectively. Hence from the last equation above follows the

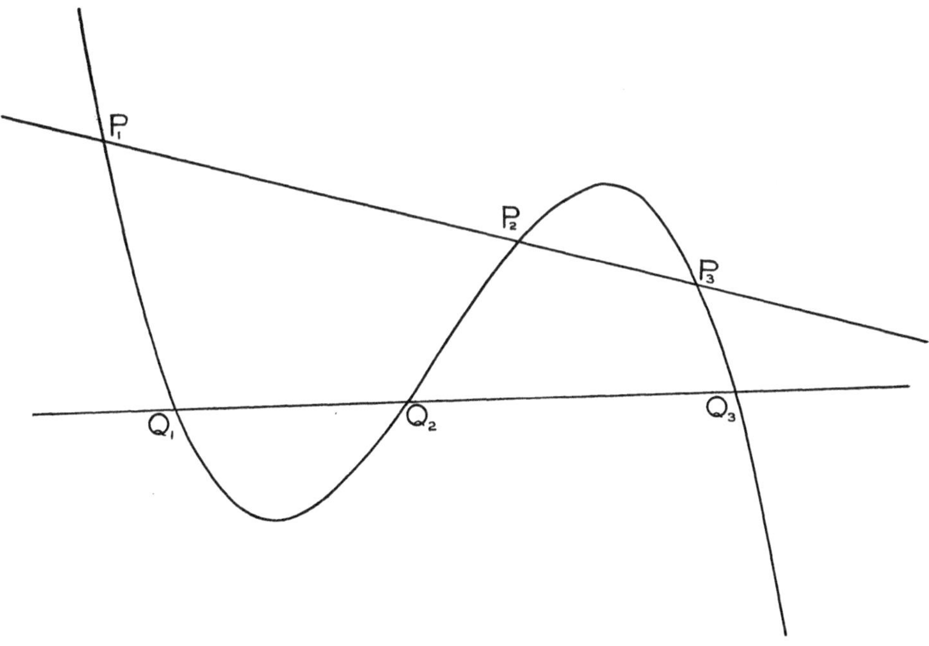

Theorem. *On the general hypersurface of order $2n-3$ in space of n dimensions there exist a finite number of lines, the number of these lines being $(2n-3)\ \rho_n$ where ρ_n is defined as follows: In a space of operation of dimension $n-1$ take a general hypersurface of order $2n-3$, then ρ_n is the number of lines which meet $2n-4$ given hyperplanes in $2n-4$ distinct points, each of which is also a point on the hypersurface.*

By means of this theorem the problem of determining the number of lines on the cubic surface is reduced to a simple problem in plane geometry. Let the line p meet the non-singular plane cubic curve C in the points P_1, P_2 and P_3, and let the line q meet C in Q_1, Q_2 and Q_3. Then ρ_3 is the number of lines which can be drawn to meet p and q in points which are also on C. But the 9 lines P_iQ_j where $i, j=1, 2, 3$, and only these lines satisfy this condition. Hence $\rho_3=9$ and the number of lines on the cubic surface is

$$\lambda_3=(2n-3)\rho_3=3\times 9=27.$$

V. RECURSION FORMULAE FOR THE FUNCTION $L_n\ (r)$

The number of lines which meet $2n-2$ hyperplanes in points which are also on the hypersurface F of order r is finite and is a function of r. This function will be denoted by $L_n(r)$, or more briefly by L_n. On account of the theorem of the preceding section,

$$(2n-1)L_n(2n-1)=\lambda_{n+1}$$

and

$$(2n-3)L_{n-1}(2n-3)=\lambda_n,$$

where λ_n is the number of lines on the general hypersurface of order $2n-3$ in space of n dimensions. The remainder of this paper will be devoted to the determination of recursion formulae by means of which the function L_n can be determined for successive values of n, and to the actual determination of these functions for particular values of n.

The word 'form' from now on will denote the figure consisting of a line g and $2n-3$ of its intersections $p_1, p_2, p_3 \ldots\ldots p_r$ with a general hypersurface F of order r, where $r>2n-4$. In any given line there are

$$P_{r,\ 2n-3} \text{ or } r(r-1)\ (r-2)\ (r-3)\ \ldots\ldots\ (r-2n+4)$$

such forms, namely, one for each of the ways that the letters p_i $(i=1, 2, 3, \ldots\ldots 2n-3)$ can be assigned to the r points of intersection of the given line with F.

The following abbreviations will be used:

$$g_i p_1^k p_2 p_3 \ldots\ldots p_{2n-2i-1-k}=q_{n-i}^k \qquad (2i+k>1)$$

$$g_{ik} p_1 p_2 p_3 \ldots\ldots p_{2n-i-k-2}=q_{n-i\ k-i} \qquad (i+k>0).$$

The reasoning of the preceding section carried out for the hypersurface of order r instead of $2n-3$ gives

$$L_n = rq_{n-1} + (r-2n+3)q_n^2 \qquad \text{or}$$

$$(1) \qquad L_n = r(r-2n+4)L_{n-1} + (r-2n+3)q_n^2.$$

We shall now derive an expression for q_n^k $(k>1)$. In any line g which contains a form satisfying the condition q_n^k, there are $P_{r-2n+k+1,\ k-1}$ forms satisfying the same condition, namely, one for each of the ways that the $k-1$ points, which in addition to the points $p_1, p_2, p_3 \ldots . p_{2n-k-1}$ constitute the points of the form, can be selected from the $r-2n+k+1$ remaining points of intersection of g with F. Hence the number Q_n^k of distinct lines which contain forms satisfying q_n^k is

$$Q_n^k = \frac{q_n^k}{P_{r-2n+k+1,\ k-1}}$$

Similarly the number of lines containing forms satisfying the condition q_{nk} is

$$Q_{nk} = \frac{q_{nk}}{P_{r-2n+k+2,\ k}}$$

Let the linear space of intersection of the k hyperplanes of the condition p_1^k be denoted by S_{n-k}, and let H_i denote the hyperplanes of the conditions p_i $(i=1, 2, 3, \ldots . 2n-k-1)$ respectively. The lines satisfying the condition $p_2p_3p_4 \ldots . p_{2n-k-1}$ are the k fold infinity rulings of a variety V, whose order is the number of rulings which meet a space of dimension $n-k-1$, and is therefore equal to Q_{nk}. S_{n-k} intersects V in a curve C. For the rulings of V which meet S_{n-k} satisfy the condition $g_{0\ k-1}\ p_2p_3p_4 \ldots .p_{2n-k-1}$, and since the dimension of this condition is $2n-1$, a single infinity of forms satisfies it. The points in which these lines meet S_{n-k} are the points of C. The order of C, being the same as that of V, is Q_{nk}. Hence by Bezout's theorem, C intersects F in rQ_{nk} points. Let p_1 be any one of these points. Through p_1 there goes a ruling of V which meets H_i in p_i $(i=1, 2, 3, \ldots . 2n-k-1)$, and which is the line of a form satisfying the condition q_n^k, provided that p_1 does not coincide with p_2, or with p_3, etc. If p_1 coincides with p_2 then g is the line of a form satisfying the condition $p_1^{k+1}p_3p_4 \ldots . p_{2n-k-2}$; if p_1 coincides with p_3 then g is the line of a form satisfying the condition $p_1^{k+1}p_2p_4 \ldots . p_{2n-k-2}$, etc. The number of lines containing forms satisfying q_{nk} is therefore equal to rQ_{nk} decreased by $2n-k-2$ times the number of lines containing forms satisfying q_n^{k+1}. Hence the equation

$$Q_n^k = rQ_{nk} - (2n-k-2)Q_n^{k+1}.$$

If both members of this equation be multiplied by $P_{r-2n+k+1,\ k}$ the result is

$$(r-2n+k+2)q_n^k = rq_{nk} - (2n-k-2)q_n^{k+1}.$$

On account of the incidence formula

$$g_{0k}p_1 = g_{1k}p_1 + p_1^{k+1},$$

we can replace q_{nk} by $q_{(n-1)(k-1)} + q_n^{k+1}$ and obtain

$$(r-2n+k+2)q_n^k = rq_{(n-1)(k-1)} + (r-2n-k-2)q_n^{k+1}.$$

Successive applications of the formula

$$q_{(n-i)(k-i)} = q_{(n-i-1)(k-i-1)} + q_{n-i-1}^{k-i+1}$$

lead to the result

(3) $$(r-2n+k+2)q_n^k = rq_{(n-d)(k-d)} + (r-2n+k+2)q_n^{k+1} + r\Sigma q_{n-i}^{k-i+1},$$
$$(i = d-1,\ d-2,\ \ldots\ 3,\ 2,\ 1)$$

For $d = n-2$ this becomes

(4) $$(r-2n+k+2)q_n^k = rq_{(2)(k-n+2)} + (r-2n+k+2)q_n^{k+1} + r\Sigma q_{n-i}^{k-i+1},$$
$$(i = n-3,\ n-4,\ \ldots\ 3,\ 2,\ 1).$$

In particular placing $k = n-1$, we have

(5) $$(r-n+1)q_n^{n-1} = rq_{21} + (r-n+1)q_n^n + r\sum_{n-3}^{1} q_{n-1}^{n-1}.$$

Now if P is the point of intersection of the n hyperplanes of the condition p_1^n, then P is not in general a point of F, and hence $q_n^n = 0$, and similarly $q_{n-i}^{n-i} = 0$. Referring to the definition of the abbreviated notation (p. 438) it will be seen that q_{21} is the condition that the form lie in a given plane, have its line passing through a given point in that plane, and have one of its points in a given hyperplane H, and therefore in the line of intersection of H with the plane. Denote the plane by α, the given point by P, and the line $H\alpha$ by a. α meets F in a curve f of order r. If A is any one of the points of intersection of a with F, then the line PA is a line containing $P_{r-1,\ 2n-4}$ forms satisfying the condition. There being r such lines, it follows that

$$q_{21} = rP_{r-1,\ 2n-4} = P_{r,\ 2n-3}.$$ Hence from (5)

(6) $$(r-n+1)q_n^{n-1} = rP_{r,\ 2n-3}.$$

In similar manner by means of equation (4) the value of q_n^{n-2} can be found in terms of r and n. For $k = n-2$, (4) is

(7) $$(r-n)q_n^{n-2} = rq_2 + (r-n)q_n^{n-1} + r\sum_{n-3}^{1} q_{n-i}^{n-i-1}.$$

Now q_2 is the number of forms which lie in a given plane and have one point in a given line in that plane and a second point in a second given line in that plane. Hence $q_2 = r^2P_{r-2,\ 2n-5}$. If the symbol q_{n-i}^{n-i-1} were defined for the form consisting of a line and $2n-2i-3$ of its intersections with F, (5) would be applicable, and the value of the symbol would be $rP_{r,\ 2n-2i-3} \div (r-n-i+1)$. But for every form satisfying the condition defined in this way there are $P_{r-2n+2i+3,\ 2i}$ forms satisfying the condition expressed by the same symbol but defined for the form consisting of a line and $2n-3$ points of intersection with F, namely one for each of the ways that the $2i$ additional points can be selected. Hence

$$(r-n+i+1)q_{n-1}^{n-i-1}=rP_{r,\ 2n-2i-3}\ P_{r-2n+2i+3,\ 2i}$$
$$=rP_{r,\ 2n-3}.$$

Applying this result and the value for q_2 to equation (7) we obtain

$$(8)\qquad (r-n)q_n^{n-2}=r^3P_{r-2,\ 2n-5}+\frac{r(r-n)}{r-n+1}P_{r,\ 2n-3}+r^2\sum_{n-3}^{1}\frac{P_{r,\ 2n-3}}{r-n+i+1}$$
$$=r^3P_{r-2,\ 2n-5}+\frac{r\ (r-n)}{r-n+1}P_{r,\ 2n-3}+r^2P_{r,\ 2n-3}\sum_{n-3}^{1}\frac{1}{r-n+i+1}.$$

Proceeding in a similar manner it would be possible to obtain formulae for q_n^{n-3}, for q_n^{n-4} and indeed for q_n^{n-k} for any particular value of k. It is evident, however, that the expressions become quite complicated as k increases. For the determination of the functions L_n for particular values of n it will perhaps be simpler to revert to the recursion formula (3).

In applying the above formulae it will be found necessary to distinguish between the symbol q_d^k when defined for space of n dimensions and the same symbol when defined for space of any other dimension say of d dimensions. We adopt the notation $q_{d[d}^k$ for the condition q_d^k defined for a space of operation of dimension d, the letter or number following the bracket indicating the dimension of the space of operation. When no bracket occurs in the symbol it will be understood that the symbol is defined for space of n dimensions. It is to be noted that the symbol q_d^k refers to the form consisting of a line and $2n-3$ of its intersections with a general hypersurface of order r, whereas the symbol $q_{d[d}^k$ has reference to the form which consists of a line and $2d-3$ of its intersections with the general hypersurface of order r.

Referring to the definitions of the abbreviated notation (p. 438) it will be seen that

$$q_d^k=g_{n-d}p_1^kp_2p_3\ .\ .\ .\ .\ p_{2d-k-1}\qquad (k>1).$$

The forms satisfying this condition therefore lie in a space of dimension d. Now for each form satisfying the condition $q_{d[d}^k$ there are $\gamma=P_{r-2d+3,\ 2n-2d}$ forms satisfying the condition q_d^k, namely one for each of the γ ways that the $2n-3-(2d-3)$ $=2n-2d$ additional points can be selected from the $r-2d+3$ points other than $p_1, p_2, p_3\ .\ .\ .\ .\ p_{2d-3}$ in which the line of the form intersects F. Hence

$$(9)\qquad q_d^k=P_{r-2d+3,\ 2n-2d}\ q_{d[d}^k\qquad (k>1).$$

In similar manner it can be shown that

$$(10)\qquad q_d=P_{r-2d+2,\ 2n-2d-1}\ L_d(r).$$

VI. THE DETERMINATION OF $L_n(r)$ FOR $n=3, 4, 5,$ AND 6

The Function $L_3(r)$

On account of (1) and (6) respectively,

$$L_3(r)=r(r-2)r^2+(r-3)q^2_3$$

and

$$(r-2)q^2_3=r^2(r-1)(r-2),$$

and hence[11]

$$L_3(r)=2r^4-6r^3+3r^2$$

The Function $L_4(r)$

For $n=4$ (1) becomes

$$L_4(r)=r(r-4)(2r^4-6r^3+3r^2)+(r-5)q^2_4$$

For the determination of q^2_4, (8) can be used with the result

$$q^2_4=3r^5-16r^4+23r^3-8r^2.$$

Consequently[12]

$$L_4(r)=5r^6-45r^5+130r^4-135r^3+40r^2.$$

The Function $L_5(r)$

$$L_5(r)=r(r-6)L_4(r)+(r-7)q^2_5.$$

For $d=k=2$, $n=5$, equation (3) becomes

$$(r-6)q^2_5=rq_3+(r-6)q^3_5+rq^2_4$$

where each of the q's is to be interpreted for space of operation of five dimensions. The value of q^3_5 obtained by means of (8) is

$$q^3_5=4r^7-59r^6+310r^5-705r^4+660r^3-180r^2$$

By means of (9) and (10) q^2_4 and q_3 can be found from the values of $q^2_{4[4}$ and $L_3(r)$ respectively, which are given above. The results are

$$rq_3=(2r^7-24r^6+101r^5-183r^4+140r^3)(r-6)$$

and

$$rq^2_4=(3r^7-31r^6+103r^5-123r^4+40r^3)(r-6).$$

Substituting these values, together with the value of L_4 in the equation above for L_5, we obtain

$$L_5(r)=14r^8-252r^7+1712r^6-5524r^5+8767r^4-6300r^3+1260r^2.$$

---

[11] *Cf.* Schubert, *Abzählenden Geometrie*, p. 236.

[12] The expression here denoted by L_4 was given by Schubert without proof in a footnote to p. 72 of *Math. Annalen.*, vol. 26.

The Function $L_6(r)$

For the determination of L_6 equation (1) gives

$$L_6(r)=r(r-8)L_5(r)+(r-9)q^2{}_6.$$

Equation (3) becomes for $n=6$, $k=d=2$,

$$(r-8)q^2{}_6=rq_4+(r-8)q^3{}_6+rq^2{}_5$$

and for $n=6$, $k=d=3$,

$$(r-7)q^3{}_6=rq_3+(r-7)q^4{}_6+rq^2{}_4+rq^3{}_5$$

and from these it follows that

$$(r-7)(r-8)q^2{}_6=r(r-7)q_4+r(r-8)q_3+(r-7)(r-8)q^4{}_6+r(r-7)q^2{}_5+r(r-8)q^2{}_4 \\ +r(r-8)q^3{}_5$$

where each of the q's is to be interpreted for space of six dimensions. Applying formulae (9) and (10) the equation for $q^2{}_6$ becomes

$$q^2{}_6=r(r-6)(r-7)L_4+r(r-4)(r-5)(r-6)(r-8)L_3+q^4{}_6+r(r-7)q^2{}_{5[5} \\ +r(r-5)(r-6)(r-8)q^2{}_{4[4}+r(r-8)q^3{}_{5[5}$$

The values of all the q's in the right member of this equation have been determined above, with the exception of $q^4{}_6$ the value of which by equation (8) is

$$q^4{}_6=5r^9-141r^8+1585r^7-9081r^6+28114r^5-45714r^4+34304r^3-8064r^2.$$

Hence

$$q^2{}_6=28r^9-644r^8+5814r^7-26300r^6+63078r^5-77952r^4+44492r^3-8064r^2$$

and consequently

$$L_6(r)=42r^{10}-1260r^9+15338r^8-97846r^7+352737r^6-722090r^5+797720r^4 \\ -418572r^3+72576r^2.$$

On account of the theorem of Part IV,

$$\lambda_4=\ 5L_3(5)\ =23\cdot 5^3 \qquad =2875$$
$$\lambda_5=\ 7L_4(7)\ =2035\cdot 7^3 \qquad =698005$$
$$\lambda_6=\ 9L_5(9)\ =419949\cdot 9^3 \qquad =306142821$$
$$\lambda_7=11L_6(11)=158557045\cdot 11^3=211039426895.$$

The results may be stated as follows: *On the general hypersurface of order five in space of four dimensions there are 2875 lines; on the general hypersurface of order seven in space of five dimensions there are 698,005 lines; on the general hypersurface of order nine in space of six dimensions there are 306,142,821 lines; and on the general hypersurface of order eleven in space of seven dimensions there are 211,039,426,895 lines.*

UNIVERSITY OF CALIFORNIA PUBLICATIONS
IN
MATHEMATICS

Vol. 1, No. 20, pp. 425-443, 1 figure in text May 17, 1924

THE HYPERSPACE GENERALIZATION OF THE LINES ON THE CUBIC SURFACE

BY

DANIEL VICTOR STEED

UNIVERSITY OF CALIFORNIA PRESS
BERKELEY, CALIFORNIA

Note.—The University of California Publications are offered in exchange for the publications of learned societies and institutions, universities and libraries. Complete lists of all the publications of the University will be sent upon request. For sample copies, lists of publications or other information, address the Manager of the University of California Press, Berkeley, California, U. S. A. All matter sent in exchange should be addressed to The Exchange Department, University Library, Berkeley, California, U. S. A.

MATHEMATICS.—Derrick N. Lehmer, Editor. Price per volume, $5.00.
Cited as Univ. Calif. Publ. Math.

Vol. 1.
1. On Numbers which Contain no Factors of the Form p(kp+1), by Henry W. Stager. Pp. 1–26. May, 1912 .... $0.50
2. Constructive Theory of the Unicursal Plane Quartic by Synthetic Methods, by Annie Dale Biddle. Pp. 27–54, 31 text-figures. September, 1912 .... .50
3. A Discussion by Synthetic Methods of Two Projective Pencils of Conics, by Baldwin Munger Woods. Pp. 55–85. February, 1913 .... .50
4. A Complete Set of Postulates for the Logic of Classes Expressed in Terms of the Operation ''Exception,'' and a Proof of the Independence of a Set of Postulates due to Del Rè, by B. A. Bernstein. Pp. 87–96. May, 1914 .... .10
5. On a Tabulation of Reduced Binary Quadratic Forms of a Negative Determinant, by Harry N. Wright. Pp. 97–114. June, 1914 .... .20
6. The Abelian Equations of the Tenth Degree Irreducible in a Given Domain of Rationality, by Charles G. P. Kuschke. Pp. 115–162. June, 1914 .... .50
7. Abridged Tables of Hyperbolic Functions, by F. E. Pernot. Pp. 163–169. February, 1915 .... .10
8. A List of Oughtred's Mathematical Symbols, with Historical Notes, by Florian Cajori. Pp. 171–186. February, 1920 .... .25
9. On the History of Gunter's Scale and the Slide Rule during the Seventeenth Century, by Florian Cajori. Pp. 187–209. February, 1920 .... .35
10. On a Birational Transformation Connected with a Pencil of Cubics, by Arthur Robinson Williams. Pp. 211–222. February, 1920 .... .15
11. Classification of Involutory Cubic Space Transformations, by Frank Ray Morris. Pp. 223–240. February, 1920 .... .25
12. A Set of Five Postulates for Boolean Algebras in Terms of the Operation ''Exception,'' by J. S. Taylor. Pp. 241–248. April, 1920 .... .15
13. Flow of Electricity in a Magnetic Field. Four Lectures, by Vito Volterra. Pp. 249–320, 40 figures in text. May, 1920 .... 1.25
14. The Homogeneous Vector Function and Determinants of the P-th Class, by John D. Barter. Pp. 321–343. November, 1920 .... .35
15. Involutory Quartic Transformations in Space of Four Dimensions, by Nina Alderton. Pp. 354–358. November, 1923 .... .25
16. On the Indeterminate Cubic Equation $x^3 + Dy^3 + D^2z^3 - 3Dxyz = 1$, by Clyde Wolfe. Pp. 359–369. December, 1923 .... .25
17. A Study and Classification of Ruled Quartic Surfaces by means of a point-to-line Transformation, by Bing Chin Wong. Pp. 371–387. November, 1923 .... .25
18. A Special Quartic Curve, by Elsie Jeannette McFarland. Pp. 389–400, 1 figure in text. December, 1923 .... .25
19. A Study of Cubic Surfaces by Means of Involutory Cubic Space Transformations, by John Frederick Pobanz. Pp. 401–423. January, 1924 .25
20. The Hyperspace Generalization of the Lines on the Cubic Surface, by Daniel Victor Steed. Pp. 425–443, 1 figure in text. May, 1924 .... .25

PHILOSOPHY.—George P. Adams and J. Loewenberg, Editors. Volumes 1, 2, 3, and 4 are completed. Volume 5 is in progress.

Vol. 1. Studies in Philosophy prepared in Commemoration of the seventieth birthday of Professor George Holmes Howison, November 29, 1904. 262 pages.

1. McGilvary, The Summum Bonum $0.25
2. Mezes, The Essentials of Human Faculty .... .25
3. Stratton, Some Scientific Apologies for Evil .... .15
4. Rieber, Pragmatism and the *a priori* .... .20
5. Bakewell, Latter-Day Flowing-Philosophy .... .20
6. Henderson, Some Problems in Evolution and Education .... .10
7. Burks, Philosophy and Science in the Study of Education .... .15
8. Lovejoy, The Dialectic of Bruno and Spinoza .... .35
9. Stuart, The Logic of Self-Realization .... .30
10. de Laguna, Utility and the Accepted Type .... .20
11. Dunlap, A Theory of the Syllogism .... .10
12. Overstreet, The Basal Principle of Truth-Evaluation .... .25

Vol. 2. 1909–1921.

1. Overstreet, The Dialectic of Plotinus .25
2. Hocking, Two Extensions of the Use of Graphs in Formal Logic. 20 figures in text .15
3. Hocking, On the Law of History .20
4. Adams, The Mystical Element in Hegel's Early Theological Writings .35
5. Parker, The Metaphysics of Historical Knowledge .85
6. Boas, An Analysis of Certain Theories of Truth 1.00

Vol. 3. 1918–1921.

1. Rieber, Footnotes to Formal Logic. Pp. 1–177. Cloth, $2.00; paper, $1.50. Carriage extra. Weight, cloth, 2 lbs.; paper, 1½ lbs.
2. Prall, A Study in the Theory of Value. Pp. 197–290 1.00

Vol. 4. 1923.

Issues and Tendencies in Contemporary Philosophy. Lectures delivered before the Philosophical Union, University of California, 1922–1923. 224 pp. 3.75

INDEPENDENT VOLUMES*

Bacon, Leonard, and Rose, Robert S.	The Lay of the Cid. xiv + 130 pages. Cloth, $1.35. Carriage extra. Weight, 1½ lbs.
Daniel, J. Frank	The Elasmobranch Fishes. xi + 334 pages, 260 illustrations. Cloth, $5.50. Carriage extra. Weight, 2 lbs., 10 oz.
Glover, T. R.	Herodotus. xv + 301 pages. Cloth, $3.25. Weight, 3 lbs.
Grinnell, Joseph Bryant, Harold C. Storer, Tracy I.	The Game Birds of California. x + 642 pages, 16 colored plates, 94 line drawings. Cloth, $6.00. Carriage extra. Weight, 4½ lbs.
Hart, Walter M.	Kipling the Story Writer. 225 pages. Cloth, $2.25. Carriage extra. Weight, 1½ lbs.
Lewis, C. I.	Survey of Symbolic Logic. vi + 407 pages. Cloth, $3.00. Carriage extra. Weight, 2¾ lbs.
McCormac, Eugene I.	James K. Polk: A Political Biography. x + 746 pages, 2 plates. Cloth, $6.00 Carriage extra. Weight, 3 lbs., 10 oz.
McEwen, George F.	Ocean Temperatures: Their Relation to Solar Radiation and Oceanic Circulation. 130 pages. Paper, $1.50.
Morley, S. Griswold	Anthero de Quental: Sonnets and Poems. xxiv + 133 pages. Quarter-vellum, $2.25.
Moses, Bernard	Spain's Declining Power in South America. xx + 440 pages. Cloth, $4.00. Carriage extra. Weight, 2 lbs., 4 oz.
Petersson, Torsten	Cicero: A Biography. 699 pages. Cloth, $5.00. Carriage extra. Weight, 4 lbs.
Scott, John A.	The Unity of Homer. 275 pages. Cloth, $3.25. Carriage extra. Weight, 1 lb.
Slate, Frederick	The Fundamental Equations of Dynamics. xii + 225 pages. Cloth, $2.00. Carriage extra. Weight, 1½ lbs.
Tolman, Richard C.	Theory of the Relativity of Motion. ix + 242 pages. Cloth, $1.75. Carriage extra. Weight, 1¾ lbs.

* The works listed as independent volumes are also on sale at The Baker and Taylor Co., 354 Fourth Avenue, New York City.

www.ingramcontent.com/pod-product-compliance
Lightning Source LLC
LaVergne TN
LVHW020551110826
845149LV00002B/234

* 9 7 8 1 4 1 8 1 8 4 4 4 5 *